6795

Continued

VALUE	PROBABILITY	VALUE	PROBABILITY
.49	.688	−.49	.312
.50	.691	−.50	.309
.51	.695	−.51	.305
.52	.698	−.52	.302
.53	.702	−.53	.298
.54	.705	−.54	.295
.55	.709	−.55	.291
.56	.712	−.56	.288
.57	.716	−.57	.284
.58	.719	−.58	.281
.59	.722	−.59	.278
.60	.726	−.60	.274
.61	.729	−.61	.271
.62	.732	−.62	.268
.63	.736	−.63	.264
.64	.739	−.64	.261
.65	.742	−.65	.258
.66	.745	−.66	.255
.67	.749	−.67	.251
.674	.750**	−.674	.250**
.68	.752	−.68	.248
.69	.755	−.69	.245
.70	.758	−.70	.242
.71	.761	−.71	.239
.72	.764	−.72	.236
.73	.767	−.73	.233
.74	.770	−.74	.230
.75	.773	−.75	.227
.76	.776	−.76	.224
.77	.779	−.77	.221
.78	.782	−.78	.218
.79	.785	−.79	.215
.80	.788	−.80	.212
.81	.791	−.81	.209
.82	.794	−.82	.206
.83	.797	−.83	.203
.84	.800	−.84	.200
.85	.802	−.85	.198
.86	.805	−.86	.195
.87	.808	−.87	.192
.88	.811	−.88	.189
.89	.813	−.89	.187
.90	.816	−.90	.184
.91	.819	−.91	.181
.92	.821	−.92	.179
.93	.824	−.93	.176
.94	.826	−.94	.174
.95	.829	−.95	.171
.96	.831	−.96	.169
.97	.834	−.97	.166
.98	.836	−.98	.164
.99	.839	−.99	.161
1.00	.841	−1.00	.159
1.01	.844	−1.01	.156

(continued on back cover)

Statistics and Data Analysis

AN INTRODUCTION

SECOND EDITION

Statistics and Data Analysis

AN INTRODUCTION

Andrew F. Siegel
Charles J. Morgan

JOHN WILEY & SONS, INC.
New York • Chichester • Brisbane • Toronto • Singapore

ACQUISITIONS EDITOR Brad Wiley II
MARKETING MANAGER Debra Riegert
SENIOR PRODUCTION EDITOR Tony VenGraitis
DESIGNER Kevin Murphy
MANUFACTURING MANAGER Mark Cirillo
ILLUSTRATION COORDINATOR Jaime Perea
PRODUCTION SERVICES Susan L. Reiland
COVER PHOTO Alex S. MacLean/Landslides

This book was set in Gill Sans Light by Ruttle, Shaw & Wetherill and printed and bound by R. R. Donnelley and Sons. The cover was printed by Lehigh Press.

Recognizing the importance of preserving what has been written, it is a policy of John Wiley & Sons, Inc. to have books of enduring value published in the United States printed on acid-free paper, and we exert our best efforts to that end.

The paper in this book was manufactured by a mill whose forest management programs include sustained yield harvesting of its timberlands. Sustained yield harvesting principles ensure that the number of trees cut each year does not exceed the amount of new growth.

Library of Congress Cataloging-in-Publication Data

Siegel, Andrew F.
Statistics and data analysis : an introduction / by Andrew F. Siegel and Charles J. Morgan. — 2nd ed.
p. cm.
Includes bibliographical references.
ISBN 0-471-57424-4 (cloth : alk. paper)
1. Statistics. I. Morgan, Charles J. (Charles John), 1945– . II. Title.
QA276.12.S553 1996
519.5—dc20 95-12110

Printed in the United States of America

10 9 8 7 6 5 4 3 2 1

Preface

Our objective in writing this book is to provide an introductory undergraduate statistics textbook for the nontechnical student that integrates the traditional foundations of statistical inference with the more modern ideas of data analysis. In this way, exploratory data analysis takes its proper place in the introductory curriculum: not as a special chapter on its own, but instead as a way of giving new life and meaning to the traditional subjects.

An Intuitive and Direct Approach

Throughout the book, we have tried to make it straightforward for the reader to learn and understand statistics. The use of examples helps motivate and focus attention. The student should always know where we are heading and why. "How-to" explanations are presented in addition to discussions of the deeper issues and theory. We make sensible use of what students know coming into the course, and of what they have already learned as the course progresses. Wherever possible, we try to build directly on material that has already been covered. One example of this is in hypothesis testing. Since confidence intervals were previously covered in Chapter 10, we perform hypothesis testing (Chapter 11) initially by simply looking to see whether the hypothesized value is in the interval or not. This always gives the same answer as the traditional approach (compare the test statistic to the critical value in the table), but gives it immediately and within an already meaningful context. Only after this has been learned (and after the reader has a basic understanding of hypothesis testing) is the more traditional approach to hypothesis testing covered in detail.

Formulas and Step-by-Step Instructions

Some students taking introductory statistics may not feel entirely comfortable with mathematics and may be unsure of their ability to interpret formulas. Please be open-minded and don't worry. We always provide step-by-step instructions for solving problems. We explain what the formula says, what it means, and how to work with it. Formulas themselves are presented both in an English-language shorthand and in standard mathematical notation, with the shorthand providing a bridge between the step-by-step procedures and the standard notation.

Exploratory Data Analysis and Transforming Data

Let's face it: Just looking at numbers isn't going to bring enlightenment and understanding. It's the role of pictures and graphs to bring to life the message in the data. Besides telling us "what's in the data," exploratory techniques help us understand what is going on when other statistical methods are applied. For example, it is more meaningful to test for a significant difference in the mean when this concept can be visualized than when it is just represented by numbers.

We have included an entire chapter on transformation of data using logarithms and square roots. This material is important because many real data sets are strongly skewed and the ability to transform them allows the use of powerful techniques that would otherwise be inappropriate.

The Computer

The results of computer analysis are provided throughout the text. The purpose is to show how computer output is interpreted and to recognize the important role of automatic computation in the efficient analysis of data. We've chosen Minitab statistical software because it is so widely used in the university environment; however, any other statistical package can be used instead for working out the problems. In fact, the book does not require the use of a computer, except for some problems that can be skipped if there is no computer available.

Exercises and Problems

There are lots of exercises and problems for the student; in addition to the end-of-chapter problems, we have also included exercises at the end of many sections. These end-of-section exercises are typically straightforward applications of the specific topic just covered and students are expected to quickly find the correct answer. The objective is to provide immediate practice on the basic procedures before moving on to more detailed discussion or more complex problems.

You will find a generous supply of varied end-of-chapter problems. Some, like the end-of-section exercises, are fairly simple and straightforward, whereas others are more challenging. Our intent is to meet instructor and student needs for a variety of levels of difficulty and a variety of topics.

Models and Assumptions

Many statistical methods are based on one or another model of the situation; they make certain assumptions about the nature of the data and how they were obtained (e.g., random sampling from a normal distribution). If the model is inappropriate to the real situation because the assumptions are not met, the results of the statistical calculation can be very misleading. We have tried to bring out the nature of the assumptions required for the different procedures, to show ways to check whether they are satisfied, to explain how they can fail in actual applications of statistics, and to show what can be done to help the data satisfy these assumptions.

Structure of the Book

Following the introductory chapter, the book is divided into three parts. Part One is concerned with data in general and with describing groups of numbers in particular. Chapter 2 discusses data structures, categorical data, and numerical data (distribution shapes displayed using stem-and-leaf plots and histograms). The next two chapters involve using sets of statistical summaries. Chapter 3 shows how to describe any distribution using the median, the interquartile range, the five-number summary, and the box plot. Chapter 4 uses the average and the standard deviation to summarize a normal distribution, shows how to use them to find percentiles, and interprets the average for binary data. Transformations are presented in Chapter 5 as a technique that can reduce skewness by working with logarithms or square roots of data values. Chapter 6 discusses the problem of choosing a description (e.g., average or median).

Part Two develops the ideas of randomness, probability, and statistical inference. Chapter 7 introduces the basic terminology of probability: random experiment, outcomes, events, probabilities, combinations of events, conditional probabilities that reflect new information, independence, and helpful solution techniques such as probability trees, Venn diagrams, and joint probability tables. Random variables (numerical outcomes), including discrete distributions, the binomial distribution, normal distributions, and the central limit theorem are covered in Chapter 8. Random sampling, sampling variability, and standard errors are presented in Chapter 9, completing the basis for statistical inference in the following two chapters: confidence intervals in Chapter 10 and hypothesis testing in Chapter 11.

Part Three moves forward, applying these ideas to more complex data structures and the analysis of relationships. The comparison of two groups of numbers is covered in Chapter 12. For several groups of numbers, Chapter 13 presents data exploration and statistical inference using the one-way analysis of variance (ANOVA). Chi-squared analysis of categorical data is presented in Chapter 14. The analysis of relationships in bivariate data (Chapter 15) demonstrates data exploration, correlation analysis, regression/prediction analysis, and residual analysis.

What's New in the Second Edition?

You will now find many more problems at the end of each chapter, as well as exercises at the end of selected sections. There are also more examples, and many of the previous examples have been updated. Computer output is included systematically throughout the book.

Some of the material has been rearranged; here are some highlights. The median and interquartile range are now presented together in Chapter 3. The average and standard deviation are presented together in Chapter 4 along with *z*-scores and percentiles for a normal distribution. Material on transformations (Chapter 5) has been shortened but still includes the essentials. The topic of categorical data is now covered more extensively (e.g., Chapter 2) along with a discussion of levels of measurement. The binomial distribution is now presented in more detail (Chapter 8). The regression material (Chapter 13) is now treated more thoroughly, particularly regarding conventional summaries and statistical inference.

Statistical techniques for understanding the vast and varied collection of information out there have become even more important in recent years. We have enjoyed putting this book together for you. Students will find it a useful reference for the future.

Acknowledgments

We are grateful to the many people who helped us with this project in one way or another. We would like to acknowledge all who read and offered constructive comments on material leading up to this edition; these include (but are not necessarily limited to) John W. Tukey, Peter Bloomfield, June Morita, Gary Biddle, Larry Schall, Robert W. Stephenson, Alan Fask, Douglas G. Kelly, Josephine Gervase, Homer W. Austin, Christina M. Yuengling, Louise J. Clark, Robert Hayden, Gerry Hobbs, Jim Albert, Jim Cotts, Robert McCuiston, James A. Condor, Pam Jackson, Trania J. Aquino, James C. Curl, Margaret L. Weaver, Barbara Wainwright, Syed Kirmani, Chris Fields, and Katherine Bowen. Special thanks are due to Armand Siegel for extensive and detailed comments on several drafts; Andy found it a special satisfaction to be able to work so closely with his father on a project like this. For their help in turning this project from a manuscript into a real book, we gratefully acknowledge the assistance of Ann Meader, Elizabeth Meder, Toni Reineke, and Susan Reiland. Finally, for his constant encouragement during all phases of this work, it is a special pleasure to acknowledge the contribution of our editor, Brad Wiley II.

Andrew F. Siegel
Charles J. Morgan

Contents

PART TWO Probability, Sampling, and Tests of Statistical Significance 195

chapter

Introduction

Welcome to the world of statistics, the art of drawing conclusions from imperfect data. We will begin this introductory chapter with examples demonstrating the important role that statistical methods play in many fields of study, followed by a definition of *statistics.* Then, we will explain the basic tasks involved in a statistical study, from the initial planning stages through the end results. Finally, we will present an overview of the main concepts of statistics, the fundamental ideas that give us the basis to apply the statistical approach in solving a wide variety of problems.

1.1 Why Statistics?

We need methods for extracting information from observed or collected data to obtain a deeper understanding about the situations they represent. Some techniques of statistics and data analysis are surprisingly simple to learn and use, even when the mathematical theory that explains them is very complicated and involved. In this book, we will concentrate on the ideas, concepts, and methods of statistics, probability, and data analysis.

Even professional statisticians have trouble understanding a data set (a collection of numerical information) merely by looking at it (see Figure 1.1), so do not be upset if a data set looks to you like just a list of numbers. There are many statistical methods whose sole purpose is to help you understand data. One of the main reasons for the existence of statistical methods is to make clear and apparent all the important and interesting features of a data set.

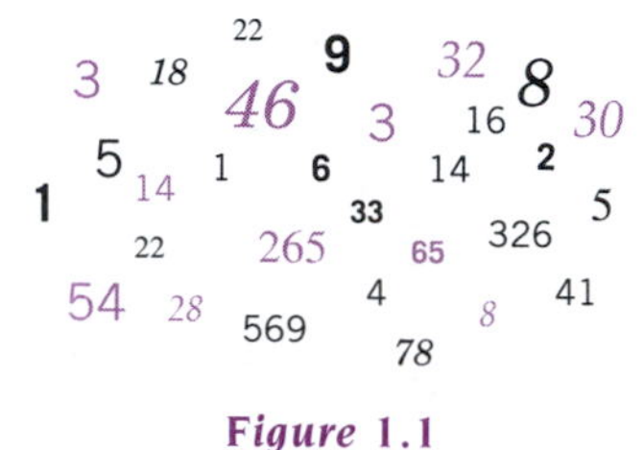

Figure 1.1

Statistics and data analysis provide methods that can help in the understanding of nearly every field of human experience. One well-known example is the collection and analysis of data for the U.S. census every 10 years, with the many political, economic, and policy implications that arise from population changes. The U.S. government collects and studies many other kinds of statistics relating to the economy, the environment, the legal system, and other areas. Another clear example of the use of statistics is in the natural sciences, where data are collected either in the laboratory from carefully designed experiments or "in the field" from careful observation.

These examples show that statistics is used in several fields. But to justify our claim that data analysis permeates nearly *every* field, some surprising examples are now in order.

EXAMPLE 1.1 Shakespeare's Words

Sometimes we have the chance to observe and study a natural situation for a while. We might spend the time observing butterflies in a field, for example. At the end of the study period, we know what we have seen, but we do not know what was there that we have not seen. We know which species of butterflies we saw, but we do not know which were present that we failed to see. Techniques have been developed that use information about what we did see to estimate how many species were present but remained unseen.

In an interesting use of tools developed in one area to answer questions in another, techniques developed in field biology to estimate the number of unseen species were used to estimate the number of words that Shakespeare knew.[1] The study's authors made use of the detailed structure of word counts from all the known works of the playwright. As explained in their report,

> Shakespeare wrote 31,534 different words, of which 14,376 appear only once, 4,343 twice, etc. The question considered is how many words he knew but did not use.... We conclude that... Shakespeare knew at least 35,000 more words.

One interesting feature of statistical methods is that seemingly unrelated problems will often have a common underlying structure, so that the same basic techniques can be used on each problem. In this example of word counts, the techniques used apply equally well to the problem of estimating the number of unobserved species of butterflies as to the number of unused words known to Shakespeare.

Much of statistical analysis operates on general questions concerning data, setting aside the obvious differences such as between words and butterflies, and focusing on the common features of the processes generating the observations. This ensures that, for example, methods developed for government analysis of comparative energy use will also work for comparing fertilizers.

EXAMPLE 1.2 *The Federalist Papers*

Here is another surprising example of the use of statistics in an unexpected situation.

We often want to decide whether one group of observations is *really* different from another, or whether their apparent differences are just due to chance variation—the "luck of the draw." The techniques used to answer this question have been used to decide the authorship of the anonymous Federalist papers published in 1787–1788, a problem belonging to the realm of the historian.

These Federalist papers are important to the study of the ratification of the U.S. Constitution. Although historians agreed on the authorship of most of these papers, it seems that about 15 could have been written by either Alexander Hamilton or James Madison or both. Statistical analysis has helped to resolve the authorship problem, while detailed political and historical analysis has been ambiguous. The statistical analysis used word counts of such common, non-political words as *while, by, from,* and *no.* The frequency with which these simple words are used can vary markedly from one author to another while remaining roughly constant for a given author. The results of Mosteller and Wallace (1972)[2] indicated that Madison was probably the author of these disputed papers.

> Our data independently supplement the evidence of the historians. Madison is extremely likely, in the sense of degree of belief, to have written the disputed Federalist papers, with the possible exception of paper number 55, and there our evidence yields odds of 80 to 1 for Madison—strong, but not overwhelming. Paper 56, next weakest, is a very strong 800 to 1 for Madison. The data are overwhelming for all the rest, including the two papers historians feel weakest about, papers 62 and 63.

EXAMPLE 1.3 *Behavior of an Extinct Animal*

The fields of psychology and sociology use statistics to develop and test theories of behavior and social organization. For example, *Archaeopteryx* is an extinct animal that had some features of birds and some of dinosaurs. Some scientists think it was a feathered dinosaur that lived on the ground; others think it was a bird that lived in trees. How can statistics help resolve a debate like this?

Alan Feduccia approached the issue by looking at the shapes of the claws of *Archaeopteryx* (see Figure 1.2 on page 4) and comparing them to the claws of birds with known habits.[3] There are three fossil specimens of *Archaeopteryx* that are complete enough to allow the arches of the claws to be measured. Three foot (*pes*) claws have arcs of 125°, 120°, and 115°. (An arc of 180° means the claw curves over a half circle; an arc of 90° means the claw curves over a quarter circle.) *Archaeopteryx* also had claws on the wings. Measurement of these front hand

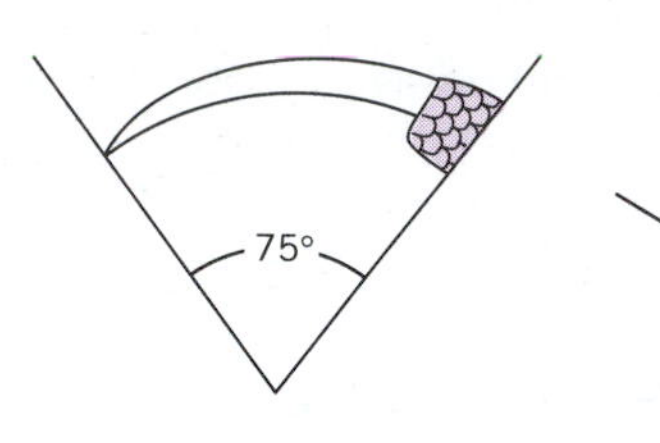

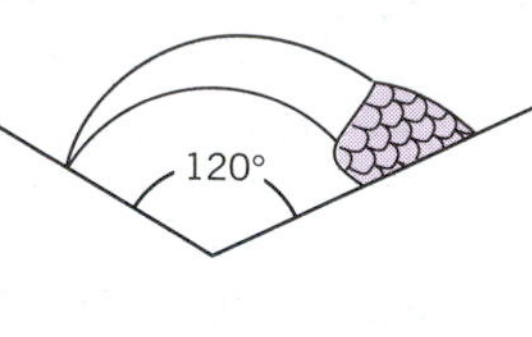

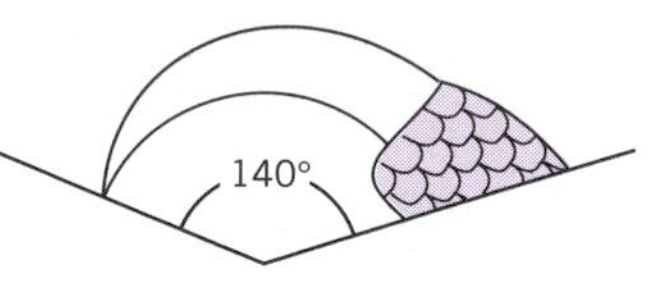

Figure 1.2 (a) Cursorial claw (b) Perching claw (c) Climbing claw

(*manus*) claws showed arcs of 155°, 142°, and 145° for an average of 147.3°. Feduccia then measured the claw arc of a group of living birds. He found that the birds that perched in trees had an average claw arc of 116.3°, birds that climbed tree trunks had an average of 148.7°, and ground-dwelling birds had an average of 64.3°. There is more to the analysis, but the point here is that by comparing the claw arcs of an extinct animal with the claws of living animals, Feduccia was able to draw some conclusions about how the extinct animal lived:

> Examination of claw geometry . . . shows that (i) modern ground- and tree-dwelling birds can be distinguished on the basis of claw curvature, in that greater arcs characterize tree dwellers and trunk-climbers, and (ii) the claws of the pes (hind foot) and manus (front hand) of *Archaeopteryx* exhibit degrees of curvature typical of perching and trunk-climbing birds, respectively. On this basis, *Archaeopteryx* appears to have been a perching bird, not a cursorial predator.

1.2 What Is Statistics?

Statistics is the art of making inferences and drawing conclusions from imperfect data. Data values are often imperfect, in that they convey useful information but do not tell the whole story. Statistical methods can be (and should be) used in all parts of a study from beginning to end. For example, data may be obtained in various ways—from experiments in a laboratory or from public opinion surveys, among others. In each case, the study design should be developed statistically, before any data have been collected, so that the results will have the greatest chance of increasing our knowledge. After the data are collected, statistical concepts guide the analysis so that as much useful information as possible can be extracted from the data. Finally, a report of the data may be prepared. Statistical graphics, plots, and charts are extremely useful in the final report when the main results must be communicated to others.

Imperfections in data arise for many reasons. Measurement error is a common cause. Imagine the difficulty of precisely determining the arc of a fossil claw, as Feduccia did. Indeed, looking at the reported figures, it appears they have been rounded to the nearest 5°. An effort to reproduce the measurements might well produce slightly different results. There also may be limitations in the amount of data available. Only three *Archaeopteryx* fossils are preserved well enough to measure claw arcs. More commonly, limitations of time or funding for a study restrict the amount of available data. Perhaps only 100 people will be interviewed in a survey when we would like to draw conclusions about a community of 100,000. Often there is a basic randomness in the problem itself, as in psychological studies of the behavior of people, who can be quite unpredictable even in a controlled situation. In the case of the behavior of *Archaeopteryx*, there are differences from fossil to fossil. Not only are the fossils different, but the modern birds used for comparison also differ from each other. We can think of data as containing a core reality—a main trend or a basic relationship—mixed with **random variation**, and we would like to infer the underlying core of the situation.

DATA	=	CORE REALITY	+	RANDOMNESS
What we have		What we want to know		Imperfections, errors, and natural variation, which get in the way

EXAMPLE 1.4 *Study Design*

The many different places where statistical considerations enter a study can be seen by considering the problem of testing a new medication.

Suppose a new brand of cold medicine is to be tested against an old brand of cold medicine to see if it is any better at relieving symptoms. The first problem is to design the experiment so that any real differences between the two drugs can be found. The classic danger in drug trials is that the new drug might be given just to the sickest people, "since they need it the most." But suppose it does not seem to work. Does this mean that the drug was ineffective, or that the people receiving it were too sick to benefit? The general solution is to use a random-assignment procedure to give the experiment a fair chance by helping to eliminate other possible explanations.

For this hypothetical experiment, suppose we have 20 patients. Each will take both drugs and decide which they prefer. Half will take the new drug on day 1, and the old drug on day 2; the other half will take the old drug first, followed by the new drug. Who gets which drug is determined by random assignment. By doing it this way, we eliminate possible explanations such as "colds always get worse the second day." Even if that is true, it applies equally to both groups.

A second problem is controlling the perceptions of the people observing and recording the results of the experiment. Patients who know they are receiving the experimental drug may report feeling better just because they believe they are getting the best available treatment. Similarly, the doctors writing progress reports may be particularly sensitive to signs of improvement if they know the patient is receiving the experimental drug. We can help control the effects of perception by administering the drugs in a "double-blind" manner, which means that neither the patient nor the physician knows which of the two drugs is being administered. Who does know? The statistician—and others who will analyze the data later (but who cannot influence the outcome). These careful procedures help to guarantee the validity of the results.

Suppose the recorded preferences of the 20 patients were as shown in the following chart.

	NUMBER OF PATIENTS
New drug better	12
Standard drug better	8

Can we conclude that the new drug is better because more people preferred it? This is an important statistical question because we need to know if "12 to 8" here represents a real superiority of the new drug over the old drug, or if it could be explained as the random variation we would expect when the two treatments are equally effective. If the two treatments were in fact equally effective, we might expect to see the results "10 to 10" with equal numbers of people preferring each drug each time the test is performed. Unfortunately, such a neat and perfect result does not happen very often; random variation would lead instead to numbers close to "10 to 10," such as the "12 to 8" obtained here. The problem (which will be addressed in detail in a later chapter) is to decide how far from exact equality we have to be before declaring the new drug better than the old one.

Notice how much this problem resembles a coin-flipping problem. If we flip a coin 20 times, what are the chances of getting 12 heads and 8 tails? Indeed, games and pastimes like flipping coins or rolling dice provide simple models for the development of many statistical ideas. We return to this in the problems at the end of the chapter.

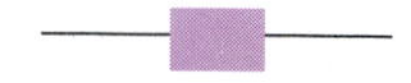

1.3 Statistical Tasks

To better understand the use of statistical methods, it is helpful to consider the various kinds of activities that are considered statistical. The tasks that statistical methods help with can be divided into four categories:

1. **Designing** a good study, be it an experiment, a survey, or a review of existing data
2. **Exploring** data
3. **Estimating** a quantity
4. **Testing** a hypothesis

We will consider each of these tasks in turn.

Experimental Design

Experimental design is the important activity of deciding which data should be collected, with the goal of ensuring that the data will be as useful as possible. Data collection often requires great effort; consequently, special care in the planning stage will be well worth the extra trouble. A well-designed experiment will be a pleasure to analyze. A poorly designed experiment will be difficult to analyze at best and may well result in data that are incapable of answering the questions that originally motivated the study.

Of course, not every statistical endeavor involves designing an experiment. When studying historical data, for example, we are restricted to the available recorded information. But even when an experiment is not explicitly performed, the principles of good experimental design may nonetheless be useful in selecting a reasonable set of data from all of the possible data sets available that relate to the question under study.

In the study of *Archaeopteryx,* there was no experiment in the classical sense, and the data available on *Archaeopteryx* were limited to the three well-preserved specimens. What was available was an abundance of data on living birds. The selection of modern birds for comparison data called for a knowledge of both ornithology and statistics. The knowledge of ornithology was required to classify birds by niche—as ground dwellers, perchers, or tree climbers—and the statistical techniques were required to select a sample of individuals whose claw arcs would be measured and compared to those of *Archaeopteryx.* Finally, the exact way the comparison would be done and the demonstration that the claws of the three groups were in fact different relied on statistical techniques.

The most important first step in designing and planning an experiment is to identify its purpose. However, this may not be as easy as it sounds. Especially in the beginning stages of a study, the purpose may be somewhat vague such as, for example, to find out the health implications of a policy change. Before an experiment can be performed, such a vague purpose would have to be narrowed down to a reasonable scope by the identification of the diseases that are in question, the population in question, the time frame, and so on. A good time to think about the purpose of an experiment is *before* it is performed, especially if it would be costly to do it over again.

The most important principle of good experimental design might be summarized as "good common sense." Although statistical methods are very important, an experiment should not be planned on the basis of statistical principles alone. The other realities of the situation should be taken into account. Costs, legalities, ethics, and relevance of each possible course of action should be considered as well as the current knowledge and beliefs in the field. If at all possible, a small experiment, often called a pilot study, should be performed first to identify and solve problems before the final version of the experiment is run.

The most important statistical principle of good experimental design is **randomization.** It would be very difficult to evaluate the effectiveness of a new medical procedure if the standard treatment were given only to the patients with mild cases of a disease while patients with severe cases received the new remedy, because the two groups are so different. If possible, patients should be assigned one of the treatments without regard to how serious their condition is. By assigning treatments randomly (for example, by tossing a coin for each patient with heads indicating the standard treatment and tails the new remedy), the patients can be divided into two groups in a fair way. In the ideal experiment, things are randomly assigned to eliminate the possibility that some unexpected factor will distort the results.

Exploratory Data Analysis

Exploratory data analysis gives us a close look at the data using summaries and graphics. Such data exploration should always be the first step once data are available because:

1. It may be that our basic question can be answered with a simple exploratory analysis of the data.
2. Many of the classic statistical techniques assume that the data have certain characteristics. Exploration of the data tells us whether it meets these assumptions.
3. We may use the exploration to develop ideas that we will later test.

The methods of exploratory data analysis help present the data to us so that its structure is clear. There are many clever and interesting ways of arranging and analyzing data to produce graphic displays (pictures) that will aid us in seeing the trends as well as the unexpected features of the data. Another way to explore data is to examine various summaries of it, an approach crucial to the understnading of a large data set that cannot easily be examined in detail all at once.

One important purpose of exploratory data analysis is to identify structure in the presence of randomness. If we think of data as consisting of some interesting core structural aspects together with some uninteresting irrelevant random components (such as measurement error), the object of exploratory data analysis is to separate and identify all of the important, interesting components of data, leaving only randomness behind.

Estimation

Statistical **estimation** is educated guessing of an unknown quantity when we have some observed information about that quantity. For example, test scores estimate the ability of students, poll results estimate political attitudes (see Figure 1.3 on page 8), the Consumer Price Index estimates the rate of inflation, and the fossil record estimates conditions on the earth many years ago. When Feduccia used the average claw arcs of three fossil *Archaeopteryx* to stand in for all *Archaeopteryx,* he was doing statistical estimation.

Figure 1.3

The quality of estimates varies widely, some being very controversial while others are quite exact. In fact, anything can be thought of as an estimate of anything else if the error is defined as their difference! Of course, we want to choose an estimate that makes this error as small as possible, but we are constrained by the realities of the situation, and perfection is usually unattainable. We accept the fact that statistical estimates are (nearly) always wrong, in the sense that they are not perfectly exact. But they are nonetheless useful and usually represent the best possible assessment under the circumstances. An important contribution of statistics is an indication of the approximate error involved in making an estimate.

Hypothesis Testing

Hypothesis testing uses data to decide between competing theories. Hypothesis testing is used routinely in many fields. The Food and Drug Administration certifies medical treatments as effective based on a test of the hypothesis that the treatment is indeed effective. (Note that this is a strong test, in the sense that the treatment is assumed ineffective until shown to be worthwhile.) An advertising company decides whether a new television commercial will have a strong impact on consumer buying habits by testing the hypothesis that more consumers are positively affected by the new commercial than the older ones. A manufacturing plant will use hypothesis testing to decide whether a new quality control process is worth the extra cost of implementing it. Scientists use these methods routinely to test new ideas in their fields.

Care must be taken in hypothesis testing because biases and preconceptions, whether hidden or not, can affect the results. Even if no active deception has taken place, subtle aspects of the testing solution can affect the results one way or the other. Consequently, in interpreting test results we should consider (1) whether the group that did the testing has any preference that it come out one way instead of the other and (2) whether the details of the test seem to favor some outcomes over others.

1.4 Fundamental Concepts of Statistics

Although there are many different kinds of approaches to statistical problems in different situations, certain concepts arise so often they are nearly universal. Although their use may change form from one problem to the next, the underlying concepts are fundamental. These include

1. **Models** and **assumptions**
2. **Robust methods** for dealing with unusual observations in the data
3. **Samples** from **populations**

4. **Probability**
5. **Central value summary**
6. **Variability**

We will consider each of these concepts in turn.

Models and Assumptions

A **model** is a simplified description of reality. Because the actual situation under study is usually very complex and it is usually not necessary (or even possible) to understand all of the subtleties involved, working with a simplified model of the situation can be a help. Statistician George Box has said, "All models are wrong; some models are useful." The object of choosing a good model is to find one that captures the important basics while remaining simple enough to work with. Following are some examples.

1. "On the average, each cigarette smoked each day reduces the life span by a certain amount of time, which will be estimated based on a large set of data."

   A model like this ignores many factors, such as a person's age, gender, lifestyle, and so on. Nonetheless, it may provide a good simple summary of the effects of smoking on health.
2. "Taking a speed-reading course will increase your reading speed by an average of 200 words per minute, according to a recent study."

   This model also ignores many individual items, such as person's former reading speed, age, intelligence, and so on. To understand the meaning of the 200 extra words per minute, it would be necessary to know what kind of people were involved in the study. It may well be that this is a bad model for people in general, but possibly a good one for students who have some trouble with large reading assignments.
3. "Tomorrow's weather depends on today's weather if we take into account the barometric pressure, humidity, cloud formations, and wind speeds."

   This may not be a bad model for weather prediction. Nonetheless, it *is* a model, rather than reality itself, and we all know that such a model is not always perfect in its predictions.

Statistical procedures are themselves based on models. These models make assumptions about the situation, such as mathematical properties representing how the data were collected. Generally, if more assumptions are made (and if these assumptions are correct), then the conclusions of the analysis will be stronger because more information (data with assumptions) will have entered the reasoning process. When fewer assumptions are made, conclusions tend to be weaker (for the same data set). Making appropriate assumptions is an essential part of the modeling process; the strongest correct assumptions will lead to the strongest valid conclusions.

If more assumptions lead to stronger conclusions, then why not make as many assumptions as we possibly can? We do not do so because if some assumptions are wrong, then the conclusions may well end up wrong also. A conclusion can be strong and wrong at the same time; it can be very specific and definite, but incorrect. One challenge of the analyst is to find and use as many *correct* assumptions as possible while avoiding incorrect assumptions. This is often difficult to do in practice because with imperfect or incomplete data, it may be hard to tell whether a given assumption is correct or not.

The assumptions should naturally be taken into account when interpreting the results of the analysis. This is one of the first areas to look at when criticizing a study or when trying to understand how it is that two different studies came up with such different conclusions. Even standard statistical assumptions, such as those that we will use later in this book, are open to question and should not be automatically applied without some efforts to be sure that they are appropriate to the situation at hand.

Even when the same data set is analyzed, different assumptions may produce different results. This is why it is sometimes said that if you give two different statisticians the same problem, you will end up with three answers. Naturally, the need for assumptions is stronger when data are scarce, because when data are plentiful, this extra information can help the numbers to "speak for themselves." Assumptions and models can be thought of as a way of placing the data into a framework so that the data's message comes across clearly.

Robust Methods

The concept of **robustness** of a statistical method describes its ability to withstand mistakes in the data or violations of assumptions (that is, to be robust against them). Mistakes do happen, at least occasionally, even with careful handling, and it would be reassuring to know that one or two mistakes in a data set of hundreds of numbers would not affect the conclusions very much. Unfortunately, some of the standard statistical methods may be affected very strongly by even one lone error in the data, and, to make matters worse, some methods tend to cover up the existence of any problem. Other statistical methods are robust and therefore will continue to do the right thing even when a few strange data values are present or assumptions are violated some other way.

A strange data value that stands out because it is not like the rest of the data in some sense is called an **outlier.** Often, a mistake in the data will result in an outlier being present. However, not all outliers are mistakes, for it could be that one or a few data values are in a class by themselves (see Figure 1.4) and do not behave like the rest of the data because of this. Robust methods also work well in this case by making it easy to identify the outliers, which should probably be treated separately. Unfortunately, nonrobust methods can make it difficult to detect these outliers.

If robust methods are so good at dealing with outliers in data, then would we ever consider using a nonrobust statistical procedure? The answer is a definite "yes" because robustness often can be achieved only by giving up some sensitivity. That is, it is difficult to resist the effects of outliers without also losing some sensitivity to some of the good, regular parts of the data set. Robust methods tend to be somewhat less sensitive than nonrobust methods—they are less likely to find *small* differences between cold remedies, for instance—but the protection against mistakes in the data or violations of assumptions can be well worth some loss of sensitivity.

Samples from Populations

Most statistical situations involve the need to make general statements about a large *population* based on specific data from a smaller *sample* chosen from the population. **Population** refers simply to the collection of all things under study. It might be a population of people, but it might

Figure 1.4

also be one of animals, manufacturing plants, volcanoes, and so on. Because populations can be large and possibly difficult to pin down, often a smaller group called the **sample** will be studied instead. The ability of statistical reasoning to reach conclusions about a larger population based on a smaller sample is one of the most powerful and useful features of statistics.

For example, consider a study to be done about people in general concerning economics, politics, psychology, or sociology. Because it is practically impossible to study the population of all people, a manageable sample of people will be studied in its place. Any conclusions drawn from this data set might seem to apply only to the sample actually studied. However, if the sample was chosen so as to represent the population, it may be possible to draw more general conclusions about the population, within certain limits. It may even be possible to say what those limits are.

Feduccia's study of *Archaeopteryx* used several samples. There was, first, the sample of three *Archaeopteryx* that were well preserved enough to get claw measurements from. They were being used to represent the larger population of all *Archaeopteryx*. The assumption was being made that the sample was representative of the larger population. There were also a series of samples of modern birds, used to represent different adaptations and claw shapes.

In a given statistical situation in which there is a population and a sample, each interesting quantity will appear in two forms: as a *population value*, called a **parameter,** and a *sample value*, called a **statistic.** It is important to keep in mind the distinction between these two related quantities to avoid conceptual difficulties. For example, to the population quantity "the percent of people who will vote for me in the election" corresponds the sample quantity "the percent of the 241 people we interviewed who said they would vote for me." Note that these are very distinct quantities. In many situations, the population value is thought of as exact, whereas the sample value is viewed as a useful estimate of the population value.

The selection of a good sample from a population is an example of a problem in study design discussed earlier. The principle of randomization is very important here. A **random sample** is a sample chosen from the population so that every possible sample of the given size has an equal chance of being selected. This is very different from a haphazard sample, gathered as convenient. Use of a random sample ensures that the sample is, on the average, representative of the population and helps us to assess the degree of accuracy of our conclusions.

Probability

Methods of probability help us understand the behavior of random systems. The **probability** of an event is an indication of how likely it is that the event will happen and gives a measure of the event's predictability. Probability and statistics complement each other. Probability may be thought of as the study of the kinds of data that a known random system will tend to generate, whereas statistics goes the other way around and tries to reconstruct the unknown system based on a set of data that it generated.

For example, consider the problem of finding the percent of U.S. voters who approve of the current president's performance. The basic solution is to ask a sample of voters whether they approve and then use their answers to estimate the answers for all voters. The results will almost certainly be slightly different from the true percent of all voters who approve, just because of accidents of random sampling. They can be very far off if there were mistakes in the survey design. If the sample of voters was systematically different from the whole population of voters—say, they were systematically better educated—their answers might be systematically different.

The task of probability theory in such a situation is to start with a model of the situation and tell what types of survey results we should see, and how far off our estimates are likely to be. The job of the statistician is, first, to design a survey that meets the assumptions of the probability model and, second, to take the survey results and work backward to a description of the population.

Central Value Summary

The most basic kind of statistical summary is the **central value summary,** which provides the answer to the first natural question that arises when one is surrounded by data: "How big are these numbers, anyway?" or, "What single number would best describe this entire data set?" The answer should represent a typical value, one that we might expect to see in the situation under study. Because some values in the data set will be larger and some will be smaller than the summary value, the term *central value* expresses the fact that this number should be near the middle of the data in some sense. (See Figure 1.5.)

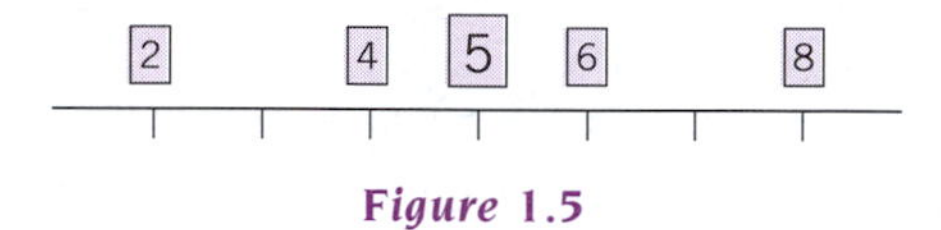

Figure 1.5

For example, in reviewing Feduccia's study of *Archaeopteryx,* we reported the average claw arcs of their feet and hands and the average claw arcs of modern birds with different adaptations. These averages are central value summaries, used to indicate values that would be typical for their respective populations.

To complicate matters, it is not always clear which central value should be chosen to represent an entire group. Thus there are several different methods for calculating a central value. For example, the *average,* the *median,* and the *mode* are examples of central value summaries. As we will see, each is a valid but different way of summarizing a list of numbers. It will be useful to understand the properties of each so that the most appropriate summary can be chosen for each problem. Perhaps it should be expected in a situation with randomness that there will be some uncertainty in the selection of a good summary value and that different valid methods can disagree with one another.

Variability

The second kind of statistical summary to consider is the amount of **variability** in the data, a measure of how different the numbers are from one another. If only the central value is known, we have no idea of how the numbers are randomly spread out or distributed about that average. At one extreme, they could all equal the central value and show no variation at all. At the other extreme, the data could consist entirely of some very large and some very small numbers. The central value alone does not reveal which is the case. For example, if we read that prices rose at an average rate of 10% last year, this average increase could have been due to all prices rising by 10% (which is unlikely but possible), or to some prices rising very little and others (such as energy perhaps) rising very much.

Let us look, briefly, at Feduccia's data for two ground-dwelling birds. Ten claw-arc measurements for *Gallus varius* are shown here:

42 56 56 59 59 64 64 67 68 70

To start with the completely obvious, notice that not all the numbers are the same. That is basically what we mean by variability. Somewhat more technically, it happens that the average of these ten numbers is 60.5. We will want to develop a summary of how much they vary around this average value of 60.5.

There is nearly always some variation in data. In later chapters, we will see some ways of measuring it such as the *standard deviation,* the *variance,* and the *interquartile range.* As is common in statistics, there is more than one method to choose from. Familiarity with the properties of each method will help guide the choice in any particular application.

Summary

Statistics and *data analysis* are techniques that are useful in nearly every field for the extraction of information from data. Because data are often imperfect, due to randomness or measurement error or other sources, we can think of statistics as the art of making inferences and drawing conclusions from imperfect data by searching for the message hidden in the noise.

The first task of statistics is to design an experiment or study, if this is possible, even before data are collected. Common sense helps ensure that the data will be appropriate to the problem being examined, and randomization helps ensure that statistical reasoning will be valid and that the data set represents the situation under study. Once data have been collected, they should first be *explored* with the help of graphics and summaries to check their validity and to look for structure. Statistical *estimation* is used to compute an educated guess of the value of an unknown quantity. *Hypothesis testing* is used to decide between competing theories to see which one best explains the data.

Here are some of the fundamental underlying concepts of statistics.

- Statistical procedures are based on *models,* which are simplified pictures of reality that are very useful in the analysis of complicated situations. These are powerful methods, but they are often based on *assumptions* (for example, that the data were obtained as a random sample) that must be taken into account when interpreting the results.
- *Robust methods* give up some sensitivity to the data in order to resist the effects of mistakes in the data or violations of assumptions. A strange data value, whether a mistake or not, is called an *outlier.*
- An important concept in statistics is that of a *sample* chosen from a *population.* The word *population* refers generally to the collection of items under study, which may or may not be humans. Because populations can only rarely be studied in their entirety, small samples are chosen from the population for study. If the sample is representative of the population (for example, if it is a *random sample*), then conclusions may legitimately be drawn about the general population.
- *Probability* is the study of random systems and the kind of outcomes and data that they are likely to produce. With probability, we suppose that the system is known and that the data are randomly produced. In contrast, statistics tries to reconstruct the unknown system based on an observed set of data that the system has generated.
- The most basic statistical summary is the *central value* summary, which gives a single number that represents a data set. There are several different ways to measure a central value, each with advantages and disadvantages, depending on the situation.
- The second most important statistical summary is *variability,* which expresses how random or spread out the numbers are. Nearly always some randomness exists in data, and there are several approaches to the problem of quantifying it.

Problems

There are three basic ideas in this chapter that we start using right away: randomness, variability, and sampling. We have designed one problem to illustrate each, and more attention will be given to each as the book develops.

Since these are real-world problems, we do not know what will happen or what answers you will get. Thus, your instructors will never be sure which of you students really do the exercises. But as statisticians, they will be able to say something about how probable the class's results are if everyone really does the assignments.

These problems illustrate important points that are difficult to appreciate without some experience. Because the problems give you some of that experience, doing them will make the rest of the course easier. Save your answers; you will need them later. (By the way, you might want to work these with classmates as teams. One person can flip a coin and another record the results, then you can switch roles. But each team member should have a personal set of answers.)

Finally, since we like to think of experimental design as applied common sense, we include a question asking you to critique a study.

1. **Randomness.** The essential idea is that in some situations several outcomes are possible, and which one will actually occur is unpredictable. People usually try to find more order in their affairs than this; the psychological impulse is to look for reasons that something happened even when it was just a chance occurrence. This problem will give you a little systematic experience with randomness.
   - **a.** Take a coin, flip it 20 times, and write down "H" for each head and "T" for each tail.
   - **b.** Count the number of H's in your list. How many are there? What percent of the list are they?
   - **c.** Go through the list and find the longest string of H's. How long is it?
   - **d.** Go through the list and find the longest string of T's. How long is it?
   - **e.** What is the longest string of either H's or T's?
2. **Variability.** Obtain two dice somewhere.
   - **a.** Roll the two dice 50 times. For each roll, add up the face values and write down the sum.
   - **b.** Go through the list of summed face values, counting the number of times you got 2, 3, 4, and so on. (The easiest way to do this is to write the numbers 2 through 12 on a sheet of paper, then put a mark beside the appropriate number as you go down the list. You could use the technique of marking them off in blocks of five, with four vertical marks and then a slash through them for the fifth mark.)
   - **c.** What number (or numbers) did you get most often? How many times did you get it (or them)?
   - **d.** What number (or numbers) occurred least often? How many times did you get it (or them)?
   - **e.** How many times did the sum equal 7?
   - **f.** How many times did the sum equal 2?
   - **g.** How many times did the sum equal 12?
3. **Sampling.** Find a group of advertisements (for example, from the Sunday newspaper classifieds) for something that interests you. Write down all the prices of the item advertised. (Save this list.) The objective is to find a few advertised prices that let us talk about the whole set of advertised prices.
   - **a.** Now pick a few (at least 10, but no more than 30) prices that you think represent the whole group of advertised prices. The representative group is your *sample*; the whole group is your *population.*
   - **b.** What is a typical value for your sample? (Name a specific value; we will not argue about how you decide. You will add a margin of error in the next couple questions.)
   - **c.** What is a typical middle-low value? What is a typical middle-high value?

d. Looking at the typical value you nominated in part (b) and the middle-low and middle-high values you nominated in part (c), about how far off would you reasonably expect to be if you used the overall typical value to guess the price of the item if you had no other information?

4. Study design. Minitab, a software company that graciously allowed us to use many of the data sets included in this text, distributes a data set called VOICE.MTW. Here is Minitab's description of the study that generated the data:

> A professor in a theater department was interested in determining if a particular voice training class improved a performer's voice quality. The professor studied 20 students, 10 of whom took the class and 10 of whom did not. Six judges rated the subjects' voice quality on a scale of 1–6 (6 = best) before and after the class. The data set includes scores for the ten subjects who took the class. (Included as a SAMPLE DATA SET help file with Minitab for Windows.)

Notice that Minitab does not distribute the data for the subjects who did not take the class. Write a brief statement of what additional information you would need beyond that included in the data set and its description to determine whether the class actually helped students.

PART ONE

Describing Groups of Numbers

chapter 2

The Shape of a Group of Numbers

There are many different structures that a data set might have. We will begin by considering data sets with a simple structure, such as a single group of numbers or categories, reserving more complex data sets for later. Our first goal is to introduce a vocabulary that lets us talk about data values. Then we go on to develop ways to describe and visualize the important features of a data set.

If the data are recorded as categories, then we can consider how often each category occurs (and its percentage) and draw bar charts to visualize the data set.

If we have a group of measurement numbers, then we can draw a quick picture to easily identify its overall features. We will see that our pictures of groups of numbers have different shapes. There are noticeable general features, and overlaid on these general features there is randomness. We will describe and name the more important general shapes, and then we will show examples of how randomness can be expected to distort these idealized shapes.

2.1 Variables and Cases: The Structure of a Data Set

Because data sets can come from a very wide variety of problems, it is natural that they can have many different types of structures. Although some data sets are simple (for example, the ages of your ten closest friends), others are quite complex (for example, all the information gathered by the U.S. census). Somewhere between in complexity might be the closing stock prices on the New York Stock Exchange for the past year, the results of a medical experiment to determine the safety and effectiveness of an experimental treatment, and the information needed for the preparation of an environmental impact statement.

To discuss the structure of a data set, we need to introduce two ideas:

A **case** is the entity—a person, place, or thing—of interest.
A **variable** is any observable characteristic of a case.

EXAMPLE 2.1 A Complex Data Set: Population, Life Expectancy, and Health Expenditures

Notice that the complexity of a data set is not measured by its size. Consider the data representing Western European nations in the late 1970s found in Table 2.1.[1]

Although this table contains only 21 numbers, it is nonetheless rather complex. There are seven cases (Austria, Belgium, and so on), and each case has three measured variables: population, life expectancy, and health expenditures. Notice that it would make no sense to treat all 21 numbers as if they measured the same thing. They do not. Thus they need additional classifying information to be properly interpreted, studied, and analyzed.

Table 2.1

COUNTRY	POPULATION (MILLIONS)	LIFE EXPECTANCY AT BIRTH (YEARS)	HEALTH EXPENDITURES (DOLLARS PER CAPITA)
Austria	7.5	71	376
Belgium	9.8	71	399
France	53.1	73	546
Germany (West)	62.1	71	113
Luxembourg	.4	71	102
Netherlands	13.8	74	560
Switzerland	6.5	73	486

EXAMPLE 2.2 A Simple Data Set: Voting in New Jersey

Now, consider a simpler set of data. The list of numbers in Table 2.2 represents the percent of the vote received by presidential candidate Bill Clinton in 1992 in each of the 21 counties of New Jersey.[2]

Table 2.2

COUNTY	PERCENT VOTING FOR CLINTON
Atlantic	44
Bergen	42
Burlington	43
Camden	50
Cape May	36
Cumberland	43
Essex	57
Gloucester	41
Hudson	55
Hunterdon	29
Mercer	49
Middlesex	45
Monmouth	37
Morris	32
Ocean	35
Passaic	43
Salem	36
Somerset	36
Sussex	26
Union	46
Warren	31

Although there are just as many numbers here (21) as in the previous example, this data set has a much simpler structure. There is only one measured variable: the percent voting for Clinton. These percentages can be viewed as a single group of numbers. Thus, these number can all be grouped together for analysis.

A Single Group of Numbers

Much of this book concerns how to analyze a single group of numbers. By a **single group of numbers** we mean a list of numbers that are all measured in the same units on the same variable. The units might be dollars, grams, people, or miles. This will be our only restriction for now. In particular, at this point we are not concerned with how the observations were obtained, be it by a careful survey, a laboratory experiment, or an informal catch-as-catch-can operation. This definition clearly rules out the data table of Western European nations, whereas the voting data for New Jersey counties certainly qualifies under this definition as a single group.

Exercise Set 2.1

Examine Table 2.3 on page 22.

1. **a.** What are the three variables being measured?

   b. Are the heights (66″, 70″, 75″) a single group of numbers by our definition?

2. The following numbers are all Tom's:

   66 inches, 120 pounds, 18 years

Table 2.3

	HEIGHT (inches)	WEIGHT (pounds)	AGE (years)
Tom	66	120	18
Dick	70	150	19
Harry	75	180	20

a. Are they a single group of numbers by our definition?

b. Do they all pertain to the case of Tom, but measure different variables?

3. How many single groups of numbers are there in Table 2.3?

Levels of Measurement

The types of analysis we will be able to perform depend, in part, on certain characteristics of the variable we are concerned with, especially how it was measured. There are, for our purposes, three levels of measurement: categorical, ordered categorical, and numerical.[3]

Many problems we are interested in are basically **categorical**; the issue is how many items are in each of several categories. The categories can be male and female; Republican and Democrat; Blue, Red, Green, and Yellow; and so on. The point is that the categories are essentially just labels; they have no *mathematical* information beyond membership, albeit they may have a great deal of social information. There is no way to do arithmetic on the labels; we cannot add blue and green.

What we can do is *count* the number of members in each category. We can then do some mathematics with the counts. For instance, we can *compare* the counts to each other, and we can convert them to percentages to simplify the comparison.

In **ordered categories** the categories have a natural order. For example, the quality measurements "poor, fair, good, excellent" are ordered categories representing progressively better quality. We can add the counts in different categories, and because they are ordered we can ask questions such as, "How many ratings were less than 'good'?"

In **numerical measurement** we measure something and assign a number of units as its value. For example, a house costs so many dollars, or a sprinter runs 100 meters in so many seconds. We can do arithmetic on the measurements themselves. We can find the average value of houses or the average speed of a group of runners, and the answers make sense.

Exercise Set 2.2

4. Suppose a wildlife biologist does an inventory of the different birds in a city, counting the number of individuals and classifying them by species (robin, sparrow, and so on). Is the variable "species"' a categorical, ordered categorical, or numerical variable?

5. A psychologist asks students to rate their feelings about mathematics on a 3-point scale, with A = very bad, B = neutral, and C = great. Are these ratings categorical, ordered categorical, or measured numerical?

6. A medical investigator gets height, weight, and blood pressure readings from a group of patients. Are these categorical, ordered categorical, or measured numerical variables?

7. Write a (very) brief description of yourself, then name the variables you used in the description and their level of measurement.

8. Suppose an investigator asks people to rate how happy they are, using a 5-point scale like this:

VERY UNHAPPY		PRETTY HAPPY		VERY HAPPY
1[ ]	2[ ]	3[ ]	4[ ]	5[ ]

What is the variable, and at what level is it being measured?

9. Suppose an investigator asks a college senior to rate his or her alcohol consumption on a typical Saturday night using this scale:

NONE	SOME		AVERAGE			A LOT
0[ ]	1[ ]	2[ ]	3[ ]	4[ ]	5[ ]	6[ ]

What is the variable, and at what level is it being measured?

10. Suppose an investigator, after first obtaining informed consent, accompanies a college senior on a typical Saturday night and carefully records the actual number of alcoholic drinks this student consumes. What is the variable and at what level is it being measured?

Discussion

We have now developed a basic vocabulary that we can use to discuss data analysis and statistics. For now, the only restriction we will put on our data, before we start analyzing it, is that it form a single group of numbers. Later on, when we wish to generalize from our data, we may require the assumption that they were collected by random sampling, so that the data accurately represent a larger population (for example, so that the results of a carefully designed survey of 1,500 people can be said to represent the thoughts of all Americans). Initially, we will see that many questions can be studied without these extra assumptions.

Unfortunately, data sets usually do not come with labels on them that tell us how many variables there are or what their level of measurement is. In fact, it is often difficult to assemble the data into a reasonable form at all. Sometimes, the same data set can be looked at in several different ways; this is actually an advantage because different analyses will tend to shed light on different types of problems. Your job as analyst is to carefully consider the types of questions that need to be answered and to select the type or types of analysis that will be the most useful in answering them. For now, be alert to single groups of numbers either by themselves or embedded within larger and more complex data sets.

2.2 Describing Categorical Data

"Sex," "religion," and "race" are all categorical variables that might be used to describe a person; "form of ownership" is a categorical variable that might be used to describe a business; "flavor" is a categorical variable that might describe a dessert. Categorical data are often developed in opinion surveys. In this section we develop the basic ways of presenting these data—as tables, percents, bar charts, and pie charts.

EXAMPLE 2.3 *Counting Answers*

Suppose a teacher conducts a survey of the members of a class and records each student's preferred ice cream flavor. Here are the first eight answers:

chocolate
vanilla
chocolate
mint chocolate chip
double fudge chocolate chunk
vanilla
super fudge chunk chocolate
vanilla

Accepting these answers as they are given, the teacher counts the number of times each flavor is mentioned. The result is a little work sheet, as shown in Display 2.1. The teacher went down the list and made a tally mark each time a flavor was mentioned. Then, the teacher went back and counted the number of tallies for each flavor. This provides the data for a table like Table 2.4.

DISPLAY 2.1

FLAVOR	TALLY
chocolate	//
vanilla	///
mint chocolate chip	/
double fudge chocolate chunk	/
super fudge chunk chocolate	/

Table 2.4

FLAVOR	COUNT
chocolate	2
vanilla	3
mint chocolate chip	1
double fudge chocolate chunk	1
super fudge chunk chocolate	1

We can use the table to answer basic questions such as, "What is the favorite ice cream flavor among these students?" It seems to be vanilla, which got three votes compared to the runner up, chocolate, which got two votes. But, are double fudge chocolate chunk and super fudge chunk chocolate really different flavors? At least, are they different at the same level that chocolate is different from vanilla? At a minimum, it seems they should both be considered types of chunky fudge chocolate, in which case they tie with chocolate for second place. Or perhaps they should be counted as types of chocolate ice cream. If they are types of chocolate, then our table will look like Table 2.5, where chocolate is the winner.

We do not want to get into a substantive discussion of whether or not fudge flavors should be counted as types of chocolate. The point we want to make is that the results are sensitive

Table 2.5

FLAVOR	COUNT
chocolate	4
vanilla	3
mint chocolate chip	1

to how we define our categories. In general, we pick categories such that each observation can be fit into one and only one category.

But let's take category definitions as given. How do we analyze and present data on a categorical variable? The most basic answer has already been given, at least by example: We count the number in each category and present the results. Some methods for presenting the results will be discussed next.

EXAMPLE 2.4 Charts For Categorical Data: Religions

Table 2.6 gives the numbers of people, in millions, belonging to different religions in Africa. Reading a one-variable table like this is completely straightforward. It says that in Africa there are 317 million Christians, 269 million Muslims, and so on. (Religions with fewer than a million members are not shown.) To simplify comparisons, we also have a column giving the percent of the population belonging to each religious group.

Table 2.6 *Millions of members in various religious groups on the African continent*

GROUP	NUMBER (IN MILLIONS)	PERCENT
Christians	317	48%
Muslims	269	41%
Tribal religionists	68	10%
Other	8	1%

Bar Charts. A graphic display called a **bar chart** can be used to show categorical data. We draw bars with lengths proportional to the number of cases in each category. By tradition, the bars do not touch each other. Figure 2.1 (page 26) shows the bar chart for African religious membership, for religions with more than a million members.

The great virtue of a display like this is that it gives us a clear sense of the relative frequencies of the different religions. We see right away that Christians and Muslims are far and away the most common, with Christians slightly more common than Muslims. Tribal religions are third most common.

Pie Charts. Newspapers and news magazines often present categorical data in **pie charts,** where the "slices" of a circular "pie" are drawn so that they are in the same proportions as the category counts. The information contained in a pie chart is the same as in a bar chart. A pie chart of African religious membership is shown in Figure 2.2.

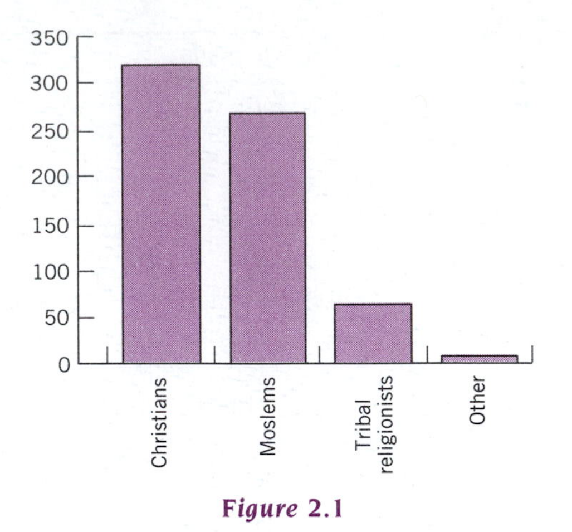

Figure 2.1

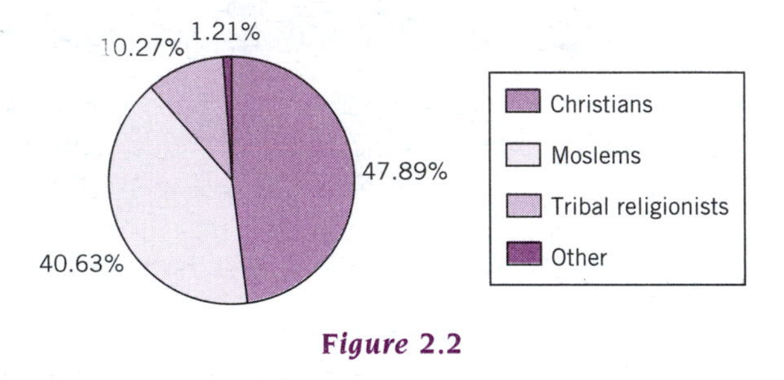

Figure 2.2

2.3 The Shape of a Data Distribution: Stem-and-Leaf Plots and Histograms of Numerical Data

Now we will look at ways to analyze numerical measurements. In this section, we will learn how to look at the way a single group of measured numerical values is distributed. Our goal is to be able to answer questions like these about the data:

- What is a typical value for this group of numbers?
- Are the data values very much the same?
- Are they very different from one another?
- How different are they?
- In what way are they different?
- How can we describe any patterns or trends?
- Is there one cohesive group?
- Are there several clusters of numbers?
- Do a few numbers differ greatly from the others?

These questions cannot be answered easily by looking at a group of numbers—even experienced statisticians see only a collection of numbers when they look at a data table. Instead, we will use pictures to represent the group of numbers in such a way that important features of the data can be seen easily. Such pictures can give an observer an impression of the overall structure and allow us to see far beyond the details of the individual numbers. This ability is really very special: Even now, computers cannot be programmed to duplicate or even come close to the mind's ability to interpret a scene.

The first method we will study is called the **histogram,** which is designed to show us how the data values are distributed. Histograms come in two general styles: the stem-and-leaf plot, which we sometimes call the quick histogram, and the classic histogram. We explain them below, by example.

Stem-and-Leaf Plots

Table 2.7 shows a complex set of data on major earthquakes from 1983 to 1991 (inclusive).[4] Bear in mind that at this point we are engaged in data exploration. We do not have a specific hypothesis in mind, we just want to see what we can find out about major earthquakes (those that are widely reported) by looking at some data.

For now we want to focus on one single group of numbers, the magnitudes of the earthquakes as measured on the Richter scale:

5.5 7.7 7.1 7.8 8.1 7.3 6.5 7.3 6.8 6.9 6.3 6.5 7.7 7.7 6.8

To create a stem-and-leaf plot of these 15 earthquake magnitudes, we will start with a vertical line with a scale indicated beside it, as shown in Display 2.2. We call the numbers to the left of the line the *stems.* We got the stem numbers by scanning our list and noticing that the smallest value starts with a 5 and the largest with an 8. These stem numbers represent the "ones" place of each number (the 5 of 5.5, the first 7 of 7.7, and so on). The "tenths" place digit, just to the right of the decimal point, will be placed to the right of the line. The first number, 5.5, will be recorded by placing a second 5 beside the scale number 5 as shown in Display 2.3 (page 28).

DISPLAY 2.2

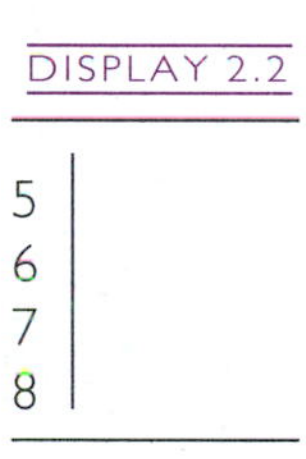

Table 2.7 *Major earthquakes*

COUNTRY	YEAR	SIZE	DEATHS
Colombia	1983	5.5	250
Japan	1983	7.7	81
Turkey	1983	7.1	1,300
Chile	1985	7.8	146
Mexico	1985	8.1	4,200
Ecuador	1987	7.3	4,000
India/Nepal	1988	6.5	1,000
China/Burma	1988	7.3	1,000
Armenia	1988	6.8	55,000
U.S.A.	1989	6.9	62
Peru	1990	6.3	114
Romania	1990	6.5	8
Iran	1990	7.7	40,000
Philippines	1990	7.7	1,621
Pakistan/Afghanistan	1991	6.8	1,200

DISPLAY 2.3

5	5
6	
7	
8	

Display 2.4 shows the next two numbers, 7.7 and 7.1, recorded.

DISPLAY 2.4

5	5
6	
7	7 1
8	

Continuing with the remaining numbers, we obtain the stem-and-leaf plot seen in Display 2.5. The name derives from the idea that the scale numbers on the left are "stems" from which the "leaves" on the right are hung.

DISPLAY 2.5

5	5
6	5 8 9 3 5 8
7	7 1 8 3 3 7 7
8	1

With a little practice, these can be drawn very quickly using only pencil and paper. Thus, there is no excuse for not drawing one! This will be the *first step* with any new data set to develop an overall impression of it.

In the spirit of exploration, we can tell a lot about our data from this picture. We see immediately that all magnitudes range from 5.5 to 8.1; the extreme smallest and largest data values will always be found in the top and bottom rows. The length of each line indicates how frequently major earthquakes were that approximate size. In this case, we see that most major earthquakes were of magnitude 6 or 7. We also see that roughly half the earthquakes were less than 7.0 and roughly half were 7.0 or greater.

These facts could have been determined by examining the numbers themselves and doing computations, but it would have involved much more work. For a first examination of the data, a good method should be quick and easy to do (this improves the chances that it will be done, even if you are in a hurry) and also very broad and general in its ability to find and show structure. It is generally easier to look at a display of data and identify structure than it is to complete detailed calculations and comparisons. Moreover, any unexpected features of the data will be immediately apparent when looking at a stem-and-leaf plot, whereas they might be overlooked in computations designed for more specific purposes. Computations are important also, as we will see later on, but a broad first look at the data is essential.

Refinements. When constructing a stem-and-leaf plot from an unorganized list of numbers, we need not worry about arranging the leaf values in order. However, we may want to go back later and put them in order. Computer programs typically arrange the leaves in order.

When constructing a stem-and-leaf plot by hand, people often truncate the number to two digits, rather than rounding. Thus 199 is shown as 1 | 9, rather than rounding to 200 and showing it as 2 | 0. Truncating is faster and less error prone, and makes it easier to go back to the list and find the original value. Computer programs typically round, rather than truncate.

If there are too many leaf values to fit comfortably on a single line, the stem value can be repeated. Then the leaf values can be sorted between the higher and lower stems. This is called *scaling* or *stretching* a display using split stems.

Exercise Set 2.3

11. Construct a stem-and-leaf plot of this list;

55 98 66 73 87 53 75 83 66

12. Make a stem-and-leaf plot of these quiz scores:

25 29 30 31 36 38 41 43 48

13. Make a stem-and-left plot of this list:

012, 085, 135, 191, 233, 252, 291, 312, 222

Hint: Use the hundreds digits as stems.

14. Construct a stem-and-leaf plot of this list:

−25, −10, −12, −5, −3, 4, 6, 10, 12, 25

Hint: Use a −0 stem for −3 and −5.

15. Suppose that the students in an introductory statistics class are surveyed, and that this is a stem-and-leaf plot of their ages:

```
1 | 6 7 7 7
1 | 8 8 8 8 9 9 9 9 9
2 | 0 0 0 0 0 1 1 1 1
2 | 2 2 2 3 3 3
2 | 4 4 5
2 |
2 | 9
3 | 0
```

a. How many students are in the class?

b. How old is the youngest?

c. How old is the oldest?

d. How many are 24 or 25 years old?

e. How many are 26 or 27 years old?

The Classic Histogram

The classic histogram for numerical variables has a horizontal axis with values marked along it. Rectangles indicate the number (or percent) of cases in each interval between values. A stem-and-leaf plot and a classic histogram of the earthquake magnitude data from Table 2.7 are shown in Display 2.6 and Figure 2.3 (page 30), respectively.

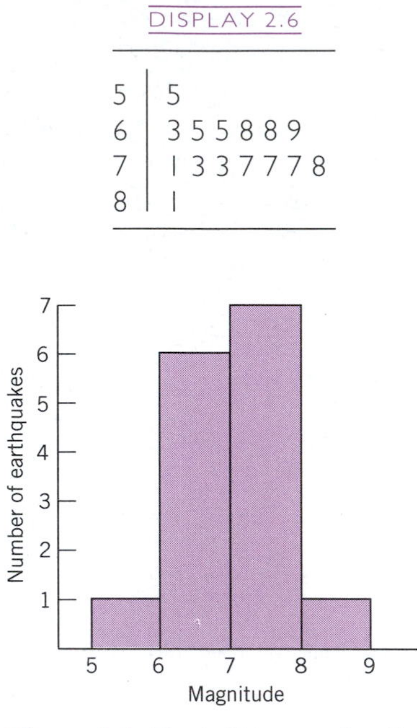

DISPLAY 2.6

5	5
6	3 5 5 8 8 9
7	1 3 3 7 7 7 8
8	1

Figure 2.3 *Classic histogram of earthquake data*

These two displays are basically very similar. (Turn the stem-and-leaf plot sideways to compare them.) Either method immediately conveys the same essential visual information, telling us at what magnitudes the data values are concentrated. Thus, we can tell which values are common, which are rare, approximately where the middle value is, approximately how spread out the numbers are, and so on.

Notice that the rectangles can touch each other at the sides. Traditionally, this indicates that we are dealing with a numerical variable. (This is different from a bar chart for categorical data, where the bars do not touch and can be arranged in any order.)

PROCEDURE: DRAWING A HISTOGRAM FROM THE DATA IN A STEM-AND-LEAF PLOT

1. Draw a horizontal axis and label it with the stem values. Put the stem values at the cutpoints of the categories, rather than in the middle of the category.
2. Count the number of entries in each row.
3. Draw a vertical axis and label it, allowing room for the greatest number of entries.
4. Take the number of entries in each row of the stem-and-leaf plot and draw a bar that height over the corresponding interval of the horizontal axis.

Here is why we put the stem values at the cutpoints of the rectangles. Notice in our example above that the bars run from one stem value to the next—from 5.0 to 5.99, then from 6.0 to 6.99, and so on. It might seem more natural to enter the bar over the stem value, but this would not be technically correct. If we *centered* the first bar, with height 1, over the stem value 5, we would be implying that there was one value between 4.50 and 5.49. There is not. The value we are plotting is 5.5, and it belongs *between* 5.00 and 5.99.

Exercise Set 2.4

16. Draw a classic histogram from this stem-and-leaf plot:

5	3 5
6	6 6 7
7	3 5 6 8
8	7 3
9	8

17. Draw a classic histogram of the data presented in Exercise Set 2.3, exercise 12.

18. Draw a classic histogram of the data presented in Exercise Set 2.3, exercise 14.

Discussion

A histogram can be thought of as an "archive" of the data in the sense that the approximate data values can be reconstructed from the display. For this purpose, the stem-and-leaf plot is preferable to the classic histogram because the data values can be read more accurately. From the stem-and-left plot, we can reconstruct (in order from smallest to largest) the exact values 5.5, 6.3, 6.5, 6.5, 6.8, 6.8, 6.9, 7.1, 7.3, 7.3, 7.7, 7.7, 7.7, 7.8, and 8.1. Only the approximate magnitudes are available, however, from the classic histogram. The best we can discern from it is that there was one magnitude between 5 and 6, six from 6 to (but not including) 7, seven from 7 to 8, and one from 8 to 9.

The stem-and-leaf plot is easier and faster to draw by hand than the classic histogram. Indeed, doing a stem-and-leaf plot is an efficient way of organizing the information we need to do a classic histogram. When then should we go on to the step of preparing a classic histogram? One instance may be for presenting an analysis to an audience that may not be familiar with stem-and-leaf plots. Unless the point of the presentation is to teach a new technique, we would want to avoid taking attention away from the data by explaining the technique. It may also be that the stem-and-leaf plot preserves too much detail for presentation. If we want our audience to focus on the general features of the group of numbers rather than the individual numbers, we may be better off showing them a classic histogram. Finally, stem-and-leaf plots work best with relatively small groups of numbers. If we are dealing with, say, several hundred numbers, we probably should prepare a classic histogram.

2.4 Computer Computations: Graphic Displays

The great virtue of computers is that they allow us to do many analyses quickly—once we have figured out how to turn the machine on and constructed a data file. In this section we will interpret computer output and generally become familiar with the types of analysis that can be done by computer.

We have used Minitab. In addition to showing you the output, we have also included the commands as a guide in case you wish to repeat the analysis. One advantage of a menu system, available on some computers, is that you do not have to remember the commands because the computer proposes a choice of actions at each stage.

EXAMPLE 2.5 *Computer Analysis of Earthquake Magnitudes*

The earthquake data from Table 2.7 have been entered into a file and saved. The magnitudes are stored as a variable called "size." Display 2.7 (page 32) shows the stem-and-leaf plot produced by the computer.

DISPLAY 2.7

```
MTB > Retrieve 'EQUAKES.MTW'.
MTB > Stem-and-Leaf 'size';
SUBC>  Increment 1.

Stem-and-leaf of size                N = 15
Leaf Unit = 0.10

  1     5 5
  7     6 355889
 (7)    7 1337778
  1     8 1
```

The first three lines are the commands to retrieve the stored file and to produce a stem-and-leaf plot of "size." The next line is produced by the computer and tells us this is a stem-and-leaf of "size" and that there are 15 data values (N = 15). The next line tells us that the leaf numbers represent tenths. Finally, we have the stem-and-leaf plot itself.

There are two differences between this stem-and-leaf plot and the one we drew by hand. There is no vertical line separating the stems and the leaves, and an extra column of numbers appears on the left. The extra column is a running count of the numbers of data points that have appeared in the stem-and-leaf plot, counting in from each end. The count for the row containing the middle observation is enclosed in parentheses.

A simple histogram can be produced using printer characters. Display 2.8 shows an example, where the first three lines are commands to produce the histogram.

DISPLAY 2.8

```
MTB > Histogram 'size';
SUBC>  Start 5.5;
SUBC>  Increment 1.

Histogram of size                N = 15

     Midpoint        Count
         5.50            1  *
         6.50            6  ******
         7.50            7  *******
         8.50            1  *
```

There are three major differences between this histogram and a classic histogram. First, of course, it is sideways. Second, an extra column containing the count for each row is included. Third, the categories are defined by their midpoints rather than their boundaries.

Producing a classic histogram by computer requires graphic capabilities. The histogram shown in Figure 2.4 was produced using Minitab 9.0.

A third computer graphic display, the dot plot, draws and numbers a horizontal line to represent the variable, and then puts a dot at the appropriate place for each observation. Display 2.9 shows a dot plot of the earthquake magnitudes.

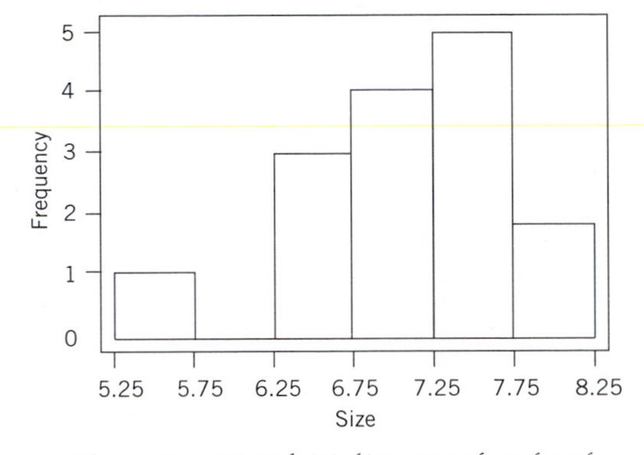

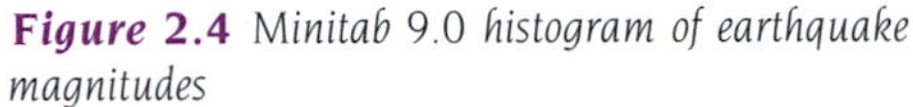

Figure 2.4 *Minitab 9.0 histogram of earthquake magnitudes*

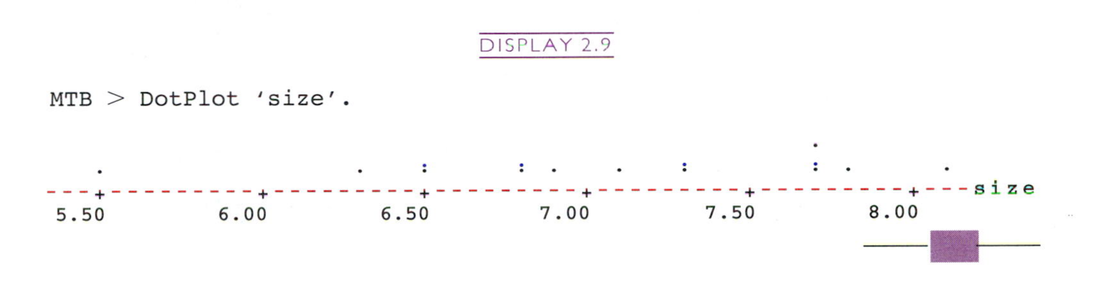

2.5 Recognizing the Shape of a Distribution

The pattern we see in a histogram, stem-and-leaf plot, or dot plot is a reflection of the distribution of numbers. This distribution shape is a fundamental property that is important in describing, summarizing, and understanding the data. In fact, identifying the correct shape can be a crucial step in determining which statistical methods to apply to a data set. Although there are many different possible distribution shapes, many are fairly easily recognized as one of the basic types presented in this section. We will begin with some general principles.

Symmetric Distributions

A **symmetric distribution** is one that behaves the same way on its left side as it does on its right. It has the symmetry of a mirror reflection about its middle value. There are many different types of symmetry. Some examples of perfectly symmetric distributions are shown in Display 2.10 and Figure 2.5 (page 34).

These distributions are actually *too* symmetric to be realistic and rarely occur. We would be very surprised to see such perfection in the more random data that are likely to occur in everyday life. Thus when the distribution is approximately symmetric, we will consider that to be close enough. This is an example of what we mean by "seeing structure" (symmetry, in this case) in a situation that contains randomness.

DISPLAY 2.10

THREE SYMMETRIC DATA SETS

Stem	Leaves
5	9
6	2 5 6
7	1 3 4 6 7 9
8	4 5 8
9	1
2	3
3	5
4	2 4 7 8 8
5	2 2 3 6 8
6	5
7	7
1	1 1 3 8 9
2	2 5 5
3	5 5 8
4	1 2 7 9 9

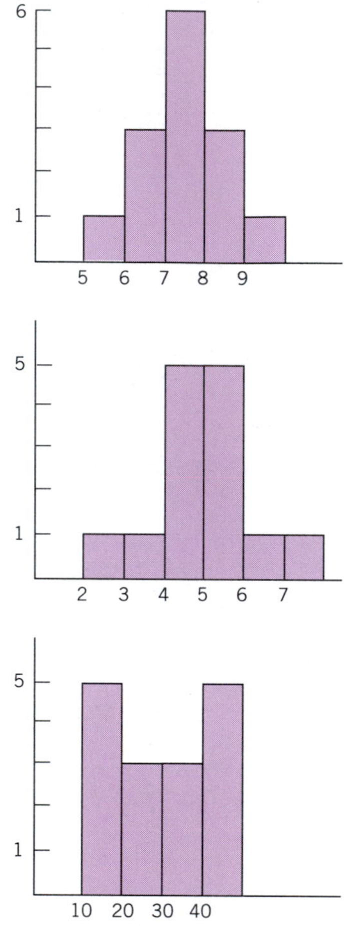

Figure 2.5

EXAMPLE 2.6 *Symmetric Distributions: Rolling a Ball, Iris Flowers, Dice Tossing*

Some realistic data sets that we would consider, except for randomness, to be basically symmetric are shown in Display 2.11 and Figure 2.6.

The Normal or Gaussian Distribution

As you can see, there are many different types of symmetric distributions: Some are flat on top, some go down at the left and right edges, and a few even go up at the edges. Of these, one very special kind of symmetric shape is called a **normal distribution** (or a **Gaussian distribution,** after K. F. Gauss).[6] It plays a central role in the field of statistics. The word *normal* is used here in a technical sense; there is nothing wrong with data that follow a different distribution.

DISPLAY 2.11

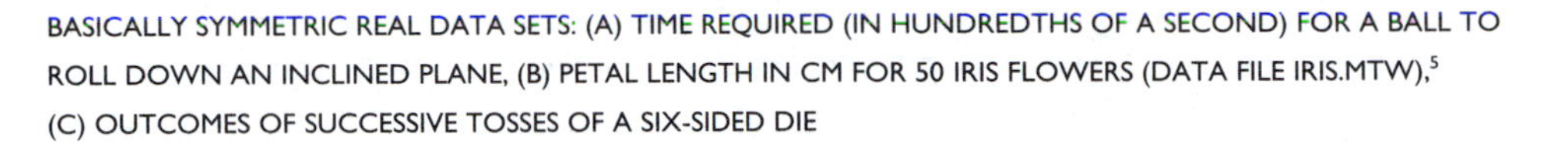
BASICALLY SYMMETRIC REAL DATA SETS: (A) TIME REQUIRED (IN HUNDREDTHS OF A SECOND) FOR A BALL TO ROLL DOWN AN INCLINED PLANE, (B) PETAL LENGTH IN CM FOR 50 IRIS FLOWERS (DATA FILE IRIS.MTW),[5] (C) OUTCOMES OF SUCCESSIVE TOSSES OF A SIX-SIDED DIE

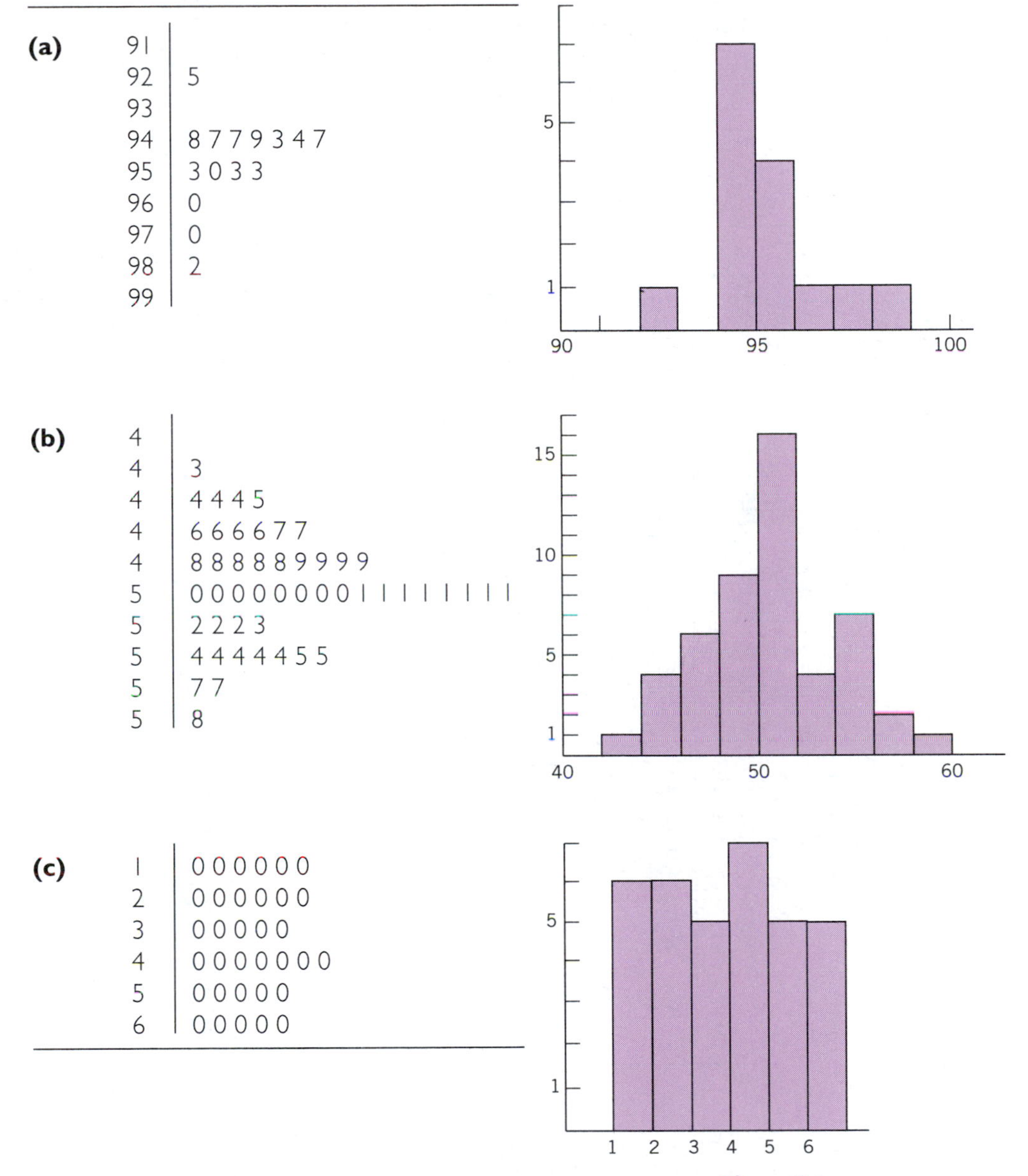

Figure 2.6

The Normal Curve

The normal distribution is the well-known bell-shaped curve that you may already have heard of. In its idealized state, with no randomness, the curve is smooth. Normal curves follow a precise formula, but can look different from each other because they are drawn to different scales.

A realistic data set may approach this smooth curve more or less closely, depending in part on how finely things are measured and in part on "the luck of the draw." Display 2.12 and Figure 2.7 illustrate a normally distributed data set with no randomness.

DISPLAY 2.12

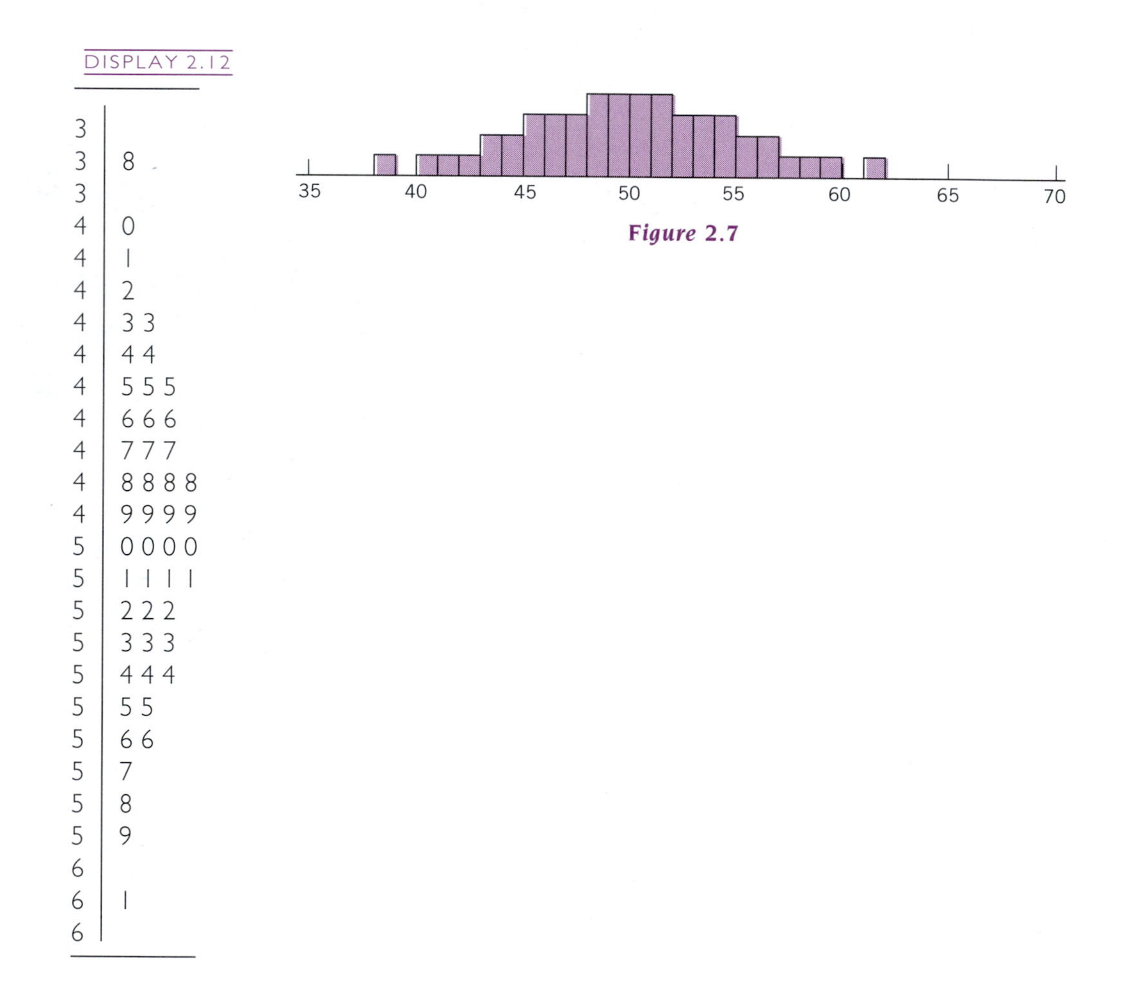

Stem	Leaves
3	
3	8
3	
4	0
4	1
4	2
4	3 3
4	4 4
4	5 5 5
4	6 6 6
4	7 7 7
4	8 8 8 8
4	9 9 9 9
5	0 0 0 0
5	1 1 1 1
5	2 2 2
5	3 3 3
5	4 4 4
5	5 5
5	6 6
5	7
5	8
5	9
6	
6	1
6	

Figure 2.7

The normal distribution is considered especially well behaved partly because the numbers tend to cluster in a symmetric way around a central value. Although the numbers can be somewhat spread out, there are successively fewer and fewer numbers as we go outward from the middle value. Also, there are no numbers that are very far away from the rest of the data.

This idealized nonrandom normal distribution is shown on a slightly smaller scale in Display 2.13 and Figure 2.8.

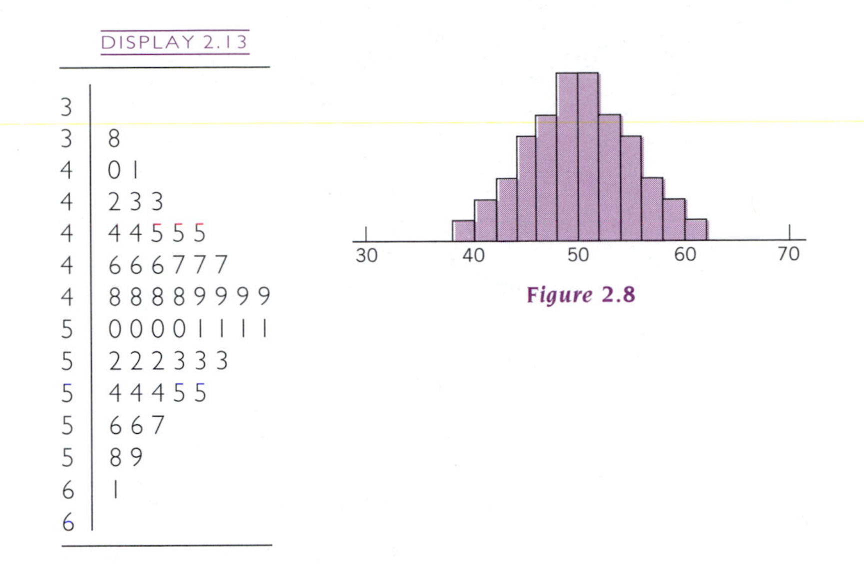

DISPLAY 2.13

Stem	Leaves
3	
3	8
4	0 1
4	2 3 3
4	4 4 5 5 5
4	6 6 6 7 7 7
4	8 8 8 8 9 9 9 9
5	0 0 0 0 1 1 1 1
5	2 2 2 3 3 3
5	4 4 4 5 5
5	6 6 7
5	8 9
6	1
6	

Figure 2.8

On an even smaller scale, the idealized normal distribution might look like what we see in Display 2.14 and Figure 2.9.

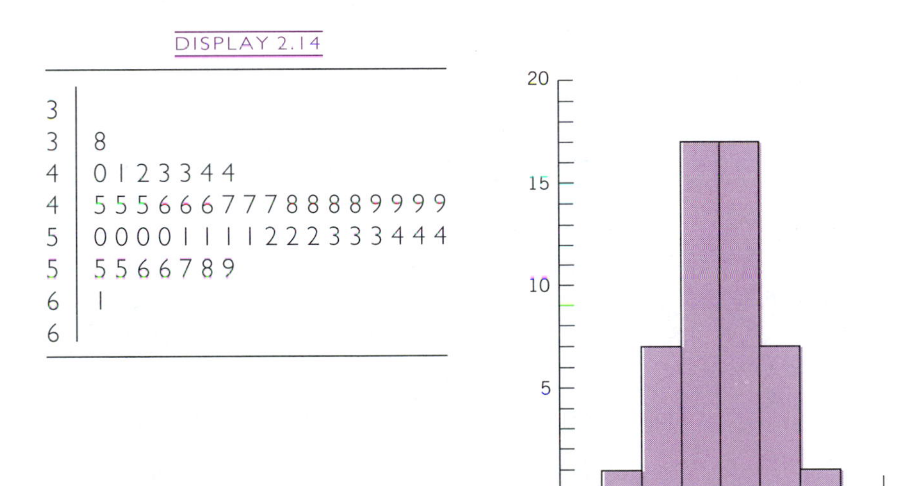

DISPLAY 2.14

Stem	Leaves
3	
3	8
4	0 1 2 3 3 4 4
4	5 5 5 6 6 6 7 7 7 8 8 8 8 9 9 9 9
5	0 0 0 0 1 1 1 1 2 2 2 3 3 3 4 4 4
5	5 5 6 6 7 8 9
6	1
6	

Figure 2.9

These examples are themselves only approximations to the ideal mathematical normal distribution: Figure 2.10 shows several shapes under different scale factors.

Skewed (Nonsymmetric) Distributions

One way a distribution can fail to be normal is by not being symmetric. If a distribution falls or trails off more slowly on one side than the other, we say it is a **skewed distribution.** There are two basic kinds of skewed distributions: *skewed toward high values* (also called skewed to the right) and *skewed toward low values* (also called skewed to the left). We say the distribution

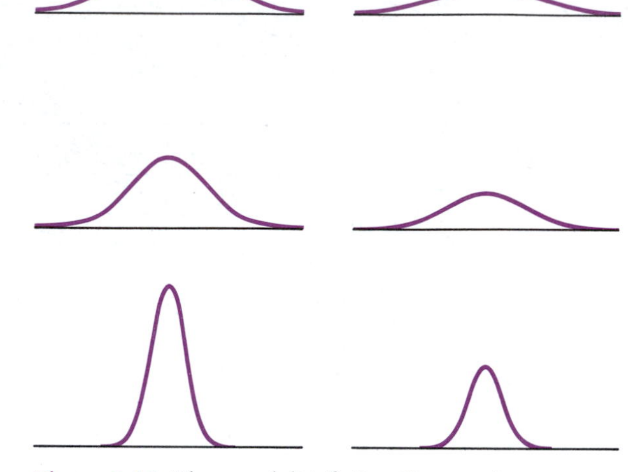

Figure 2.10 *The normal distribution: Six examples*

is skewed in the direction of the long "tail," which points *away* from the bulk of the cases. Displays 2.15 and 2.16, along with Figures 2.11 and 2.12, show two examples.

DISPLAY 2.15

A DISTRIBUTION THAT IS SKEWED TOWARD HIGH VALUES:
1991 REVENUES OF STATES IN THE UNITED STATES,
IN TENS OF BILLIONS OF DOLLARS (U.S. CENSUS)

Stem	Leaves
0	1 1 1 1 2 2 2 2 3 3 3 3 4 4 4 4 5 5 5 6 7 7 7 8 9 9 9 9 9 9
1	0 0 2 2 3 3 4 4 5 6 8
2	4 4 4 5 7
3	1 3
4	
5	
6	5
7	
8	
9	0

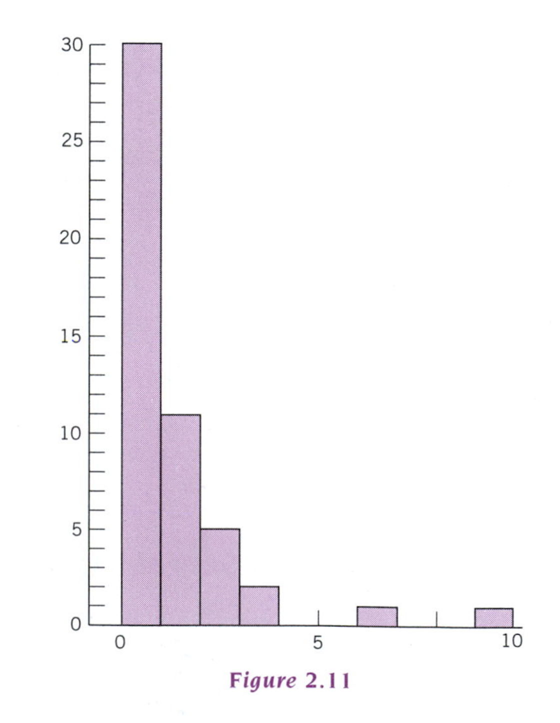

Figure 2.11

DISPLAY 2.16

MIDTERM EXAMINATION SCORES

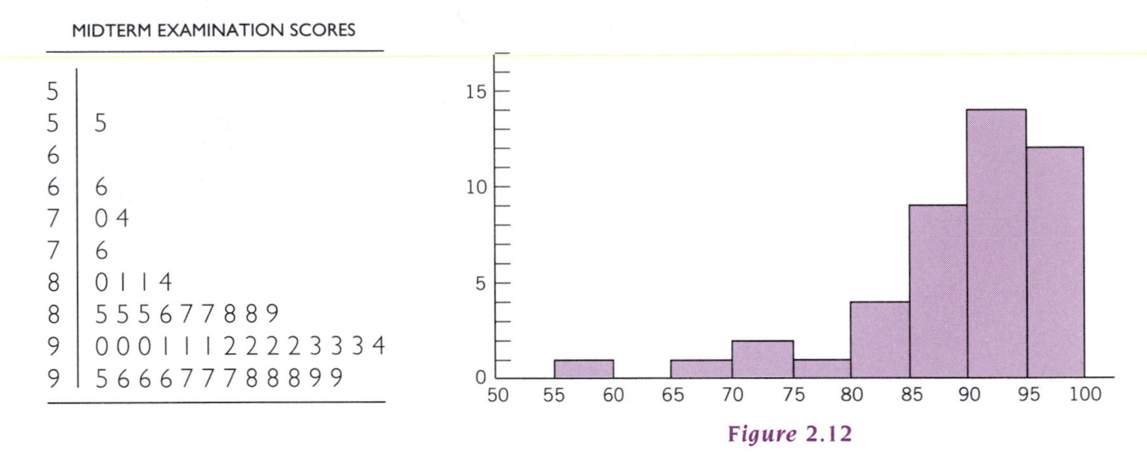

Figure 2.12

The first example, state revenues, is skewed toward high values, as you can see by the more gentle decline on the right than on the left. This indicates that there are a few states with large revenues but many more states with small revenues.

The second example, midterm examination scores, is skewed toward low values, as you can see by the more gentle decline on the left than on the right. This indicates that there are a few students with low scores but many more students with high scores (that is, most students do their work and do it well, but there are always a few who have trouble).

There are many kinds of data that are skewed toward high values, and this is more common than skewness toward low values. One reason is that skewness toward high values often happens for counts or amounts, which are not permitted to take on negative values. These distributions cannot spread out on the left past the zero point, but they can (and sometimes do) spread out toward the right. Some examples that may behave this way include incomes, population sizes, net worth of companies, areas of regions, and attendance at sporting events.

Long-Tailed Distributions

A distribution can be symmetric but still not normal. This can happen if the left and right sides of the distribution fall off in the same way (ensuring symmetry), but at a different rate than the graceful bell-shaped curve of the normal distribution. Perhaps because the sides of the distribution look like an animal's tail (a dinosaur?), they are called the left and right *tail* of the distribution.

Display 2.17 and Figure 2.13 (page 40) illustrate a histogram of hypothetical data to show a nearly ideal case of a symmetric but long-tailed distribution. Note how the data values form tails at both sides, extending far away from most of the data.

EXAMPLE 2.7 *Long-Tailed Distribution: Marine Corps Officers*

Consider the number of U.S. Marine Corps officers each year, from 1916 to 1935, as a single group of numbers.[7] This is a very long-tailed distribution, with most of the data concentrated very near the middle and with long tails extending on both sides. (See Display 2.18 and Figure 2.14 on page 40.)

DISPLAY 2.17

A FAIRLY SYMMETRIC LONG-TAILED DISTRIBUTION

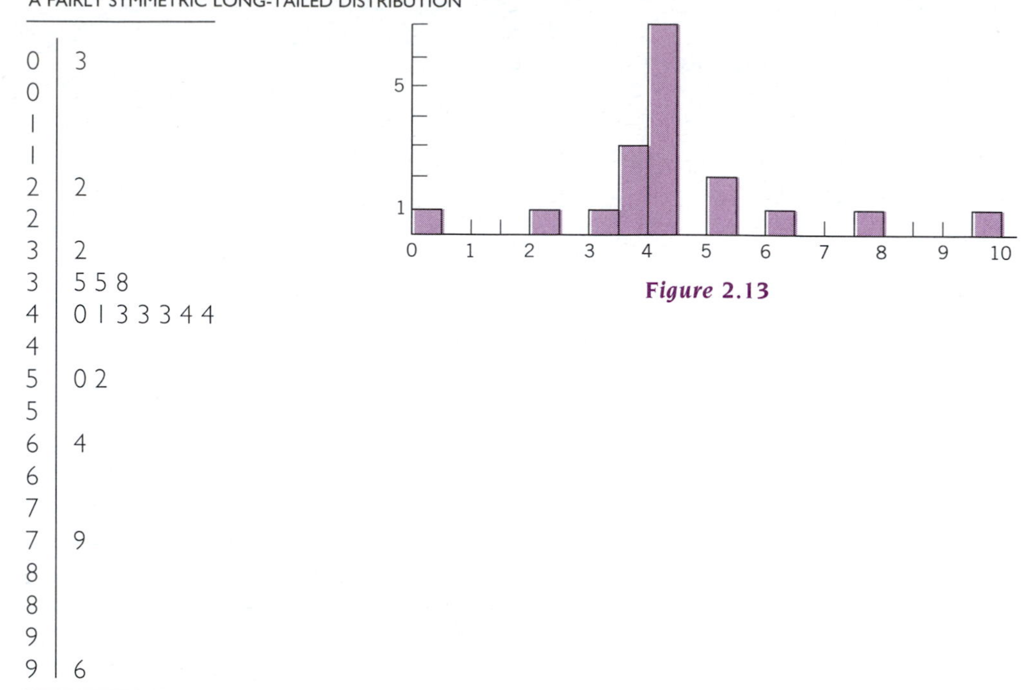

Figure 2.13

DISPLAY 2.18

A SYMMETRIC LONG-TAILED DISTRIBUTION: THE NUMBER OF U.S. MARINE CORPS OFFICERS EACH YEAR FROM 1916 TO 1935

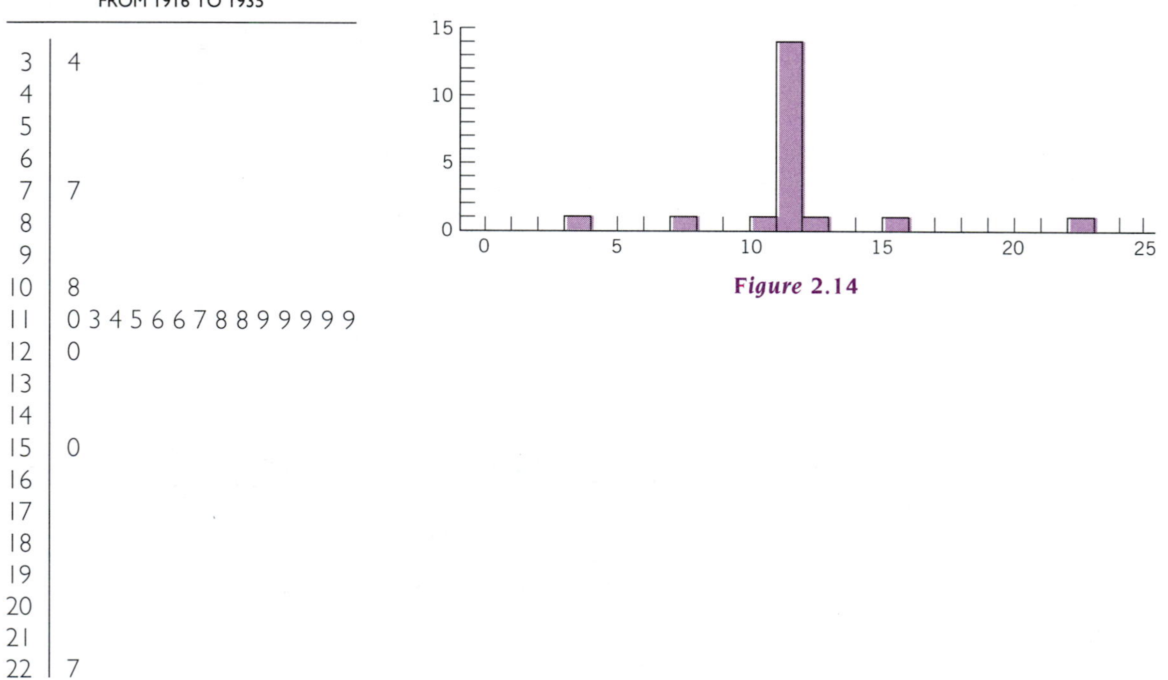

Figure 2.14

This histogram merely displays the data as a group, without indicating any time trends. However, a close look at the original data set would show that the long tail at the low end represents the period prior to World War I, whereas the long tail at the high end represents World War I itself.

Rectangular Distributions

A distribution can also be too *short-tailed* to be normal, but still be relatively symmetric. The tails of a distribution can be so short that they are hardly discernable at all. Such a histogram is relatively flat on top (except for randomness) and then drops quite suddenly at the sides.

EXAMPLE 2.8 Rectangular Distribution: Dice Tossing

Consider the successive tosses of an ordinary six-sided die. This should give us a data set consisting of the numbers 1 through 6, with each number expected to occur just as often as any other number. Thus, in the ideal case, this histogram should be perfectly flat on top with sudden abrupt corners at the sides. In reality, due to randomness (because each time you toss the die, it does not "know" enough to even out its past behavior), the results would be more like what is shown in Display 2.19 and Figure 2.15.

DISPLAY 2.19

A SHORT-TAILED SYMMETRIC DISTRIBUTION:
100 TOSSES OF A SIX-SIDED DIE

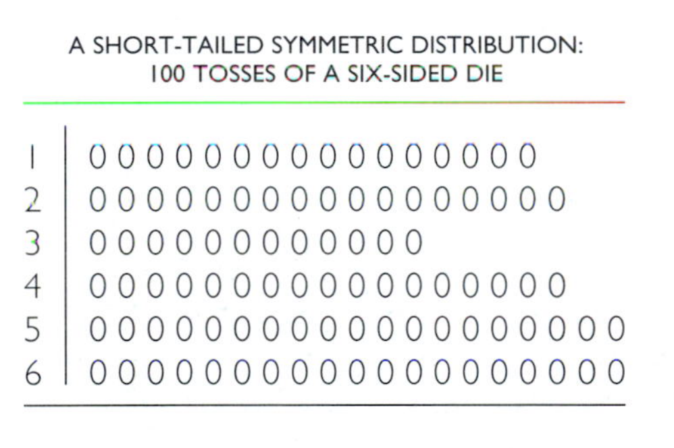

1	0000000000000000
2	00000000000000000
3	000000000000
4	00000000000000000
5	0000000000000000000
6	0000000000000000000

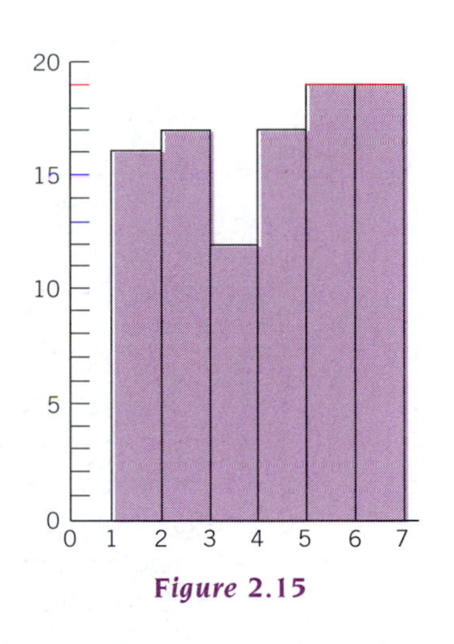

Figure 2.15

Bimodal Distributions: Two Groups

Sometimes, a data set will look as if it is made up of more than one group and will have more than one clear peak or *mode* in the histogram. Normally distributed data will have only one mode, which will be just about in the center of the data. Only when there are separate, distinct, and fairly high peaks (ignoring fluctuations of one or two data values due to randomness) will we declare that there is more than one mode. A distribution with exactly two modes is called **bimodal.**

EXAMPLE 2.9 *Bimodal Distribution: Golf Prize Money*

Consider Display 2.20, a histogram of the prize money involved in professional golf tournaments in 1991.[8] The two modes here, one at about $60,000 and the other at about $180,000, are distinct and separate enough that we can be fairly sure that these data have two modes, are not normally distributed, and are bimodal. (Remember that the modes do not have to be exactly as frequent.) Often in a bimodal distribution, the data represented can be divided into two subgroups on some criterion. If the two groups are different enough from each other, we will see more than one cluster of points in a histogram or stem-and-leaf plot.

DISPLAY 2.20

A DATA SET WITH TWO MODES: PRIZE MONEY FOR PROFESSIONAL GOLF TOURNAMENTS IN 1991. LEAF UNIT = $10,000.

Stem	Leaves
0	4 5 5 5
0	6 6 6 6 6 6 6 6 6 6 6 7 7 7 7
0	8 9
1	0 1 1
1	2 2
1	4 4 5
1	6 6 6
1	8 8 8 8 8 8 8 8 8 8 8 8 8 8 8 8 8 9 9 9 9 9
2	1 1
2	2 2 3 3
2	5
2	
2	8

When we see a data set with more than one mode, we should immediately consider the possibility that more than one group is being represented. We need to try to identify the groups, which may require creativity, imagination, and detective work. An analysis might then be more effective if each group were analyzed separately, rather than all together. In any case, we are better off if we are aware of peculiarities like this in the data.

In the golf prize money example, for instance, is it possible that there are two separate groups? Let us look at separate displays for men and for women to see whether this can explain the bimodality. (See Displays 2.21 and 2.22.)

Now we see where the two modes came from: There is one for men at about $180,000 and one for women at about $60,000. Clearly, the prizes are larger for men. Nonetheless, there is some overlap. The largest prize for a woman ($160,000) is larger than the prizes for several men (all prizes from $120,000 through $150,000).

The groups in this example are fairly well separated from each other compared to how spread out each group is. Naturally, the more overlap there is, the harder it will be to tell the two (or more) component groups apart from one another, as they may seem to have just a single mode or peak.

To make this point even more clearly, let's look at some dot plots of these data. The first (Display 2.23) shows all of the prizes as a single group of numbers. The next two dot plots (Display 2.24 on page 44) show the prizes for men (sex 1) separately from the prizes for women (sex 2).

DISPLAY 2.21

STEM-AND-LEAF PLOT OF PRIZE MONEY
FOR MEN'S PROFESSIONAL
GOLF TOURNAMENTS IN 1991.
LEAF UNIT = $10,000.

Stem	Leaves
1	2 2
1	4 4
1	6 6
1	8 8 8 8 8 8 8 8 8 8 8 8 8 8 8 8 8 8 8 9 9 9 9 9
2	1 1
2	2 2 3 3
2	5
2	
2	8

DISPLAY 2.22

STEM-AND-LEAF PLOT OF PRIZE MONEY
FOR WOMEN'S PROFESSIONAL GOLF TOURNAMENTS
IN 1991. LEAF UNIT = $10,000.

Stem	Leaves
0	4 5 5 5
0	6 6 6 6 6 6 6 6 6 6 6 6 7 7 7 7
0	8 9
1	0 1 1
1	
1	5
1	6

DISPLAY 2.23

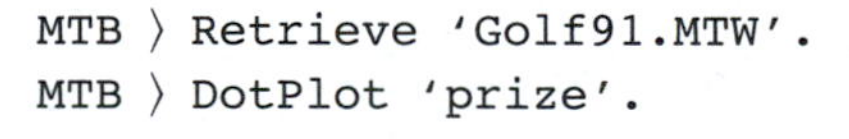

```
MTB > Retrieve 'Golf91.MTW'.
MTB > DotPlot 'prize'.
```

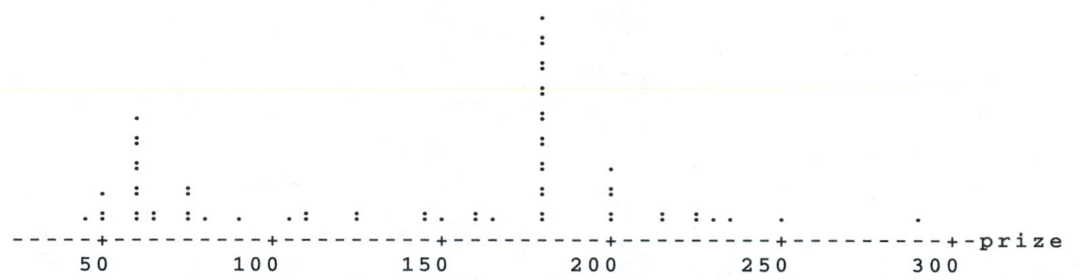

DISPLAY 2.24

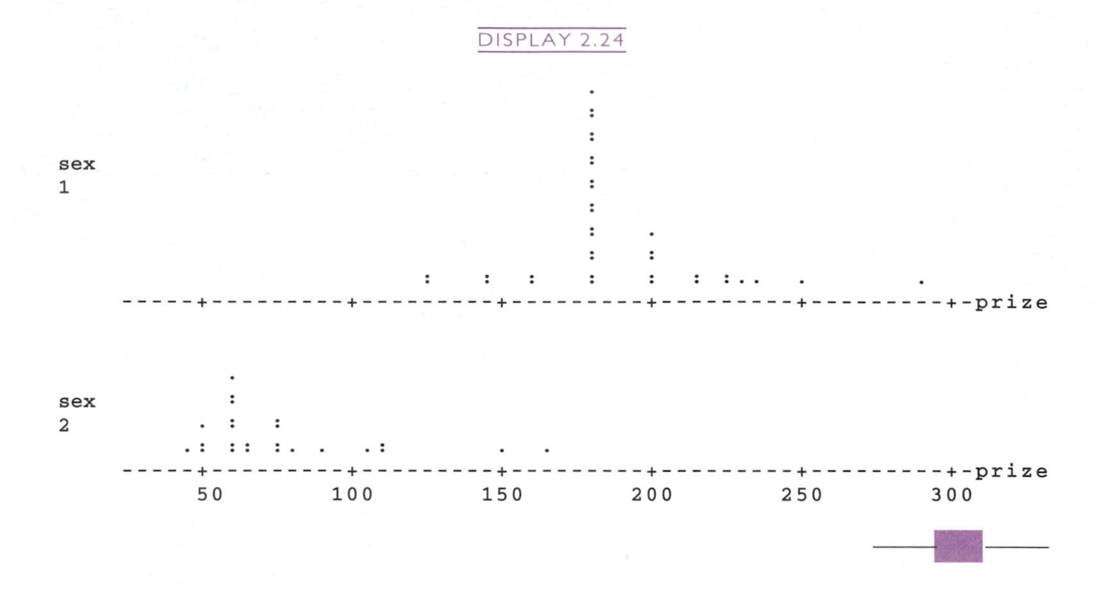

Exercise Set 2.5

19. Six distributions are shown in Figure 2.16.

a. Which is (1) bimodal? (2) very short-tailed? (3) very long-tailed? (4) apparently normal? (5) skewed to the right? (6) skewed to the left?

b. Which ones are symmetrical?

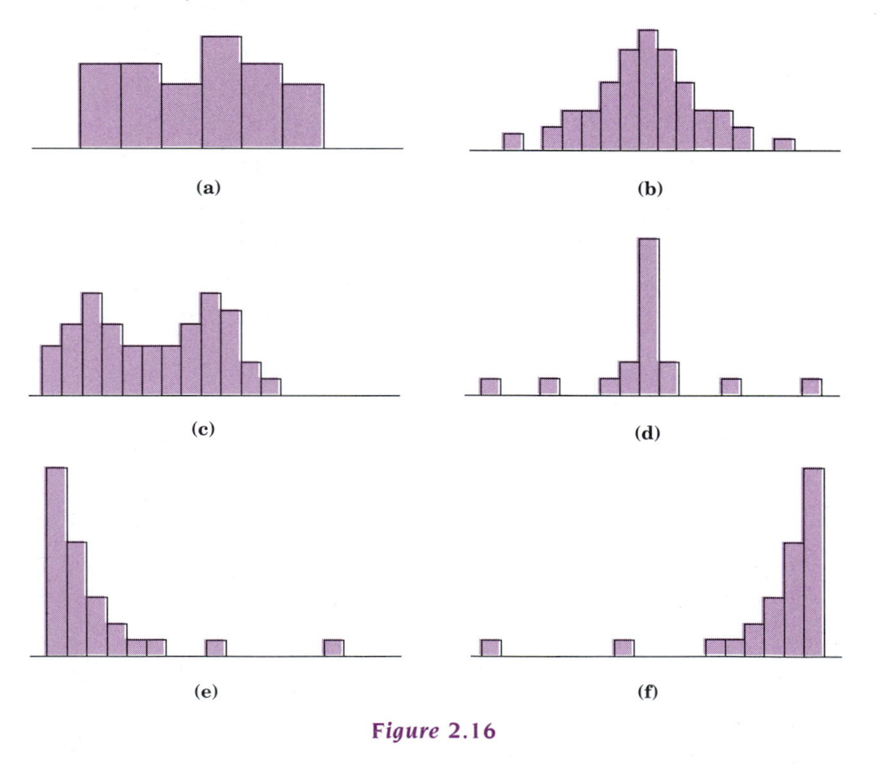

Figure 2.16

20. Is the distribution in Display 2.25 (p. 46) symmetrical, skewed toward high values, or skewed toward low values?

21. Here are some numbers: 1, 1, 1, 1, 2, 2, 2, 2, 2, 3, 3, 6, 6, 10, 20, 146. Is their distribution symmetrical, skewed toward high values, or skewed toward low values?

22. Here are some numbers: 10, 50, 75, 90, 92, 93, 93, 95, 96, 97, 98. Is their distribution symmetrical, skewed toward high values, or skewed toward low values?

23. Women's professional golf tournaments in 1979 paid the following prizes:

THOUSANDS OF DOLLARS	NUMBER OF TOURNAMENTS
10–14	3
15–19	20
20–24	6
25–29	0
30–34	7

a. Construct a histogram of this data distribution.

b. Describe the shape of this distribution.

24. Men's professional golf tournaments in 1979 paid the following prizes:

THOUSANDS OF DOLLARS	NUMBER OF TOURNAMENTS
30–34	1
35–39	2
40–44	0
45–49	11
50–54	15
55–59	0
60–64	4
65–69	0
70–74	2

a. Construct a histogram of this data distribution.

b. Describe the shape of this distribution.

25. Suppose the data from Exercises 23 and 24 were combined in a single data set, then plotted. What would the distribution look like?

Outliers

An **outlier** is a data value that is very different from most other values in a data set. Sometimes outliers arise because of important underlying differences among cases; sometimes they reflect a population distribution that is skewed; and sometimes they are just typing mistakes. In any case they require special attention.

EXAMPLE 2.10 *An Outlier: Growth in Food Production*

Consider Display 2.25 (page 46), which shows the growth in food production, in percent increase per year, for developed countries from 1969 to 1971.[9] The country with the highest growth rate, 7.7%, stands out from the rest. A closer look at the table that this data came from might show this to be a mistake. (Perhaps it should have been 2.7% but was copied incorrectly.) However, in this case it is not an error; it is the correct value for the country of Israel. With further research one might find out how and why Israel achieved such an impressive growth rate.

DISPLAY 2.25

GROWTH IN FOOD PRODUCTION, IN PERCENT PER YEAR, FOR DEVELOPED COUNTRIES, 1969–1971

0	9 6 6 7 7 7 8 8 9 9 9
1	3 1 2 4 5 5 7 6 7
2	0 0 0 0 2 2 4
3	0 3 3 5
4	0
5	
6	
7	7

2.6 Effects of Randomness on Apparent Distribution

A particular data set can often be looked at as a sample from some larger data set, the population. Sometimes this larger data set is potentially observable ("all New Hampshire voters"), and sometimes it is hypothetical ("all potential earthquakes"), but we can think and talk about either kind. How well the sample data set represents this larger set is an important question. There can be substantial apparent differences just on the basis of chance.

Let's consider three examples: a normal population distribution, a population distribution that is skewed to the right, and a rectangular population distribution.

The histogram (drawn to a fine scale) of a normal distribution is shown in Figure 2.17. First we had a computer draw three random samples from this distribution. Each sample had 1,000 numbers in it, as shown in Figure 2.18. Notice that these sample distributions closely resemble the population distribution.

Next we draw three random samples of 100, as shown in Figure 2.19. Notice that the resemblance to the parent distribution is becoming erratic. This is an effect of the random sampling.

Continuing, we drew samples of 50, 20, and 10, which are illustrated in Figures 2.20, 2.21, and 2.22, respectively. Notice that as the samples get smaller it is harder to see the original shape.

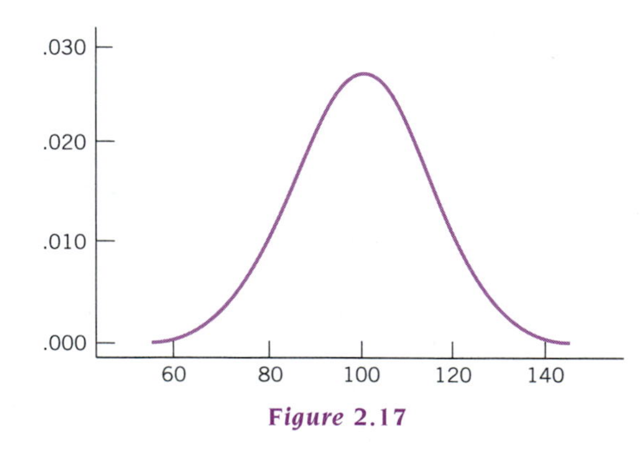

Figure 2.17

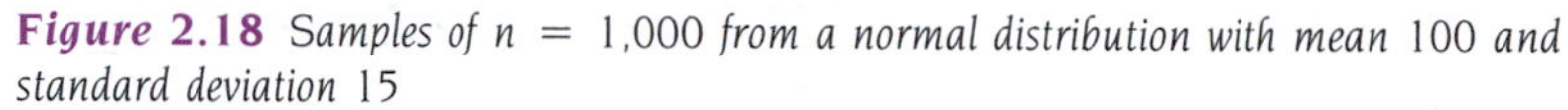

Figure 2.18 *Samples of n = 1,000 from a normal distribution with mean 100 and standard deviation 15*

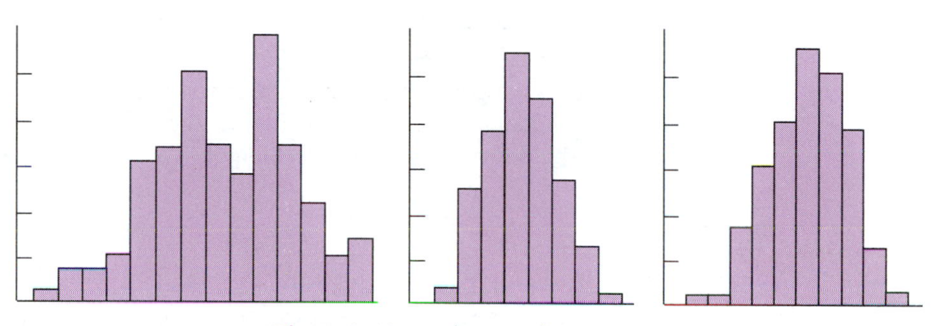

Figure 2.19 *Random samples, n = 100*

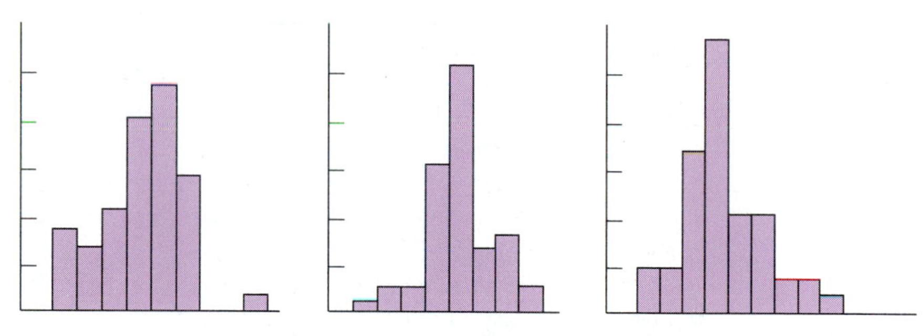

Figure 2.20 *Random samples, n = 50*

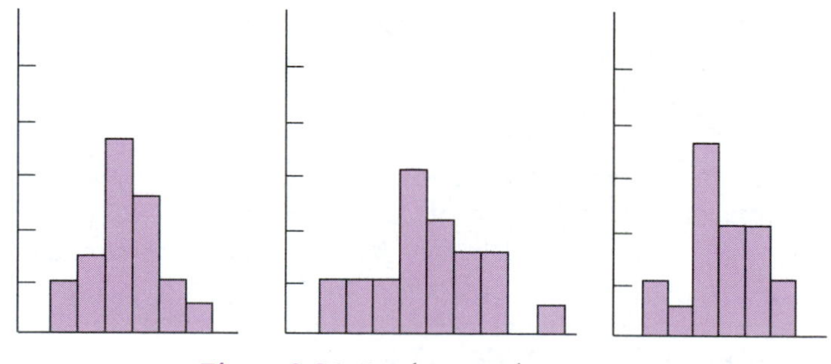

Figure 2.21 *Random samples, n = 20*

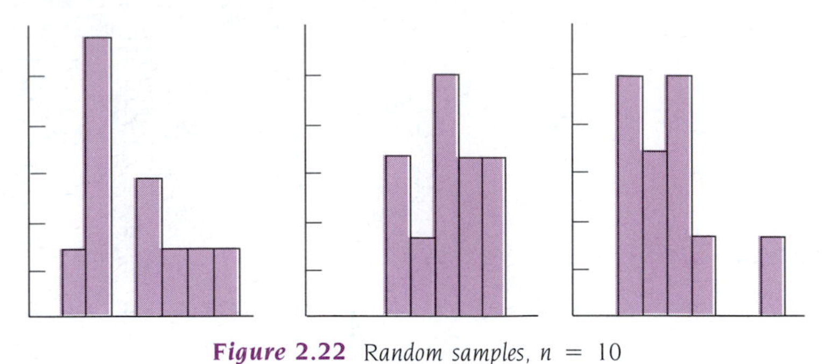

Figure 2.22 *Random samples, n = 10*

Now we repeat this experiment for a distribution that is skewed to the right, and present the results in Figure 2.23.

Finally, we repeat the experiment for a rectangular distribution and show the results in Figure 2.24 (page 50). Again, notice that as the samples get smaller, it is harder to detect the shape of the original population distribution. For the smallest samples ($n = 10$), sometimes it is not at all clear whether the original population is normal, skewed, or rectangular.

Summary

A *case* is an entity—a person, place, or thing—of interest, whereas a *variable* is any observable characteristic of a case. A data set typically consists of one or more variables observed for each of a collection of cases. Each variable has a level of measurement: *categorical* (the variable tells which of several categories a case belongs to), *ordered categorical* (the categories have a natural order), or *numerical* (the variable specifies a number).

Categorical data may be described by counting the number (and percentage) of times each category occurs. A bar chart gives a picture of the situation.

Numerical measurements have a pattern in which the numbers are spread out, called the *distribution* of the data. The best way to analyze the shape of the distribution is to examine a display such as a *stem-and-leaf plot,* a *classic histogram,* or a *dot plot,* all of which show you how frequently various ranges of data values occur.

The sides or edges of a distribution are sometimes called its *tails.* These tails may be long or short on either side. *Symmetric distributions* have left and right tails that are mirror images of each other, falling off in the same way on each side. A *normal* or *Gaussian distribution* has a particular symmetric bell-shape curve for which data values near the middle of the distribution are more frequent than those that are high or low. *Skewed distributions* are not symmetric, but have a longer tail on one side than the other. They are said to be *skewed toward high values* (or to the right) if the right-hand tail is longer and *skewed toward low values* (or to the left) if the left-hand tail is longer.

A distribution may have one or more peaks, called *modes.* Multiple modes often indicate the presence of more than one identifiable group within the data. A data value that is very much bigger or smaller than the others is called an *outlier.* Outliers should always be identified and checked to make sure that they are not simply mistakes in the data.

It is important to assess the distribution of the data as an early step in a more complex analysis. Many of the more detailed procedures that we will learn about later on will require that the distribution have a certain shape (for example, that it be normally distributed). The randomness that we expect to see in real data can make it somewhat difficult to tell whether the underlying distribution is, say, normal or skewed, especially with smaller data sets.

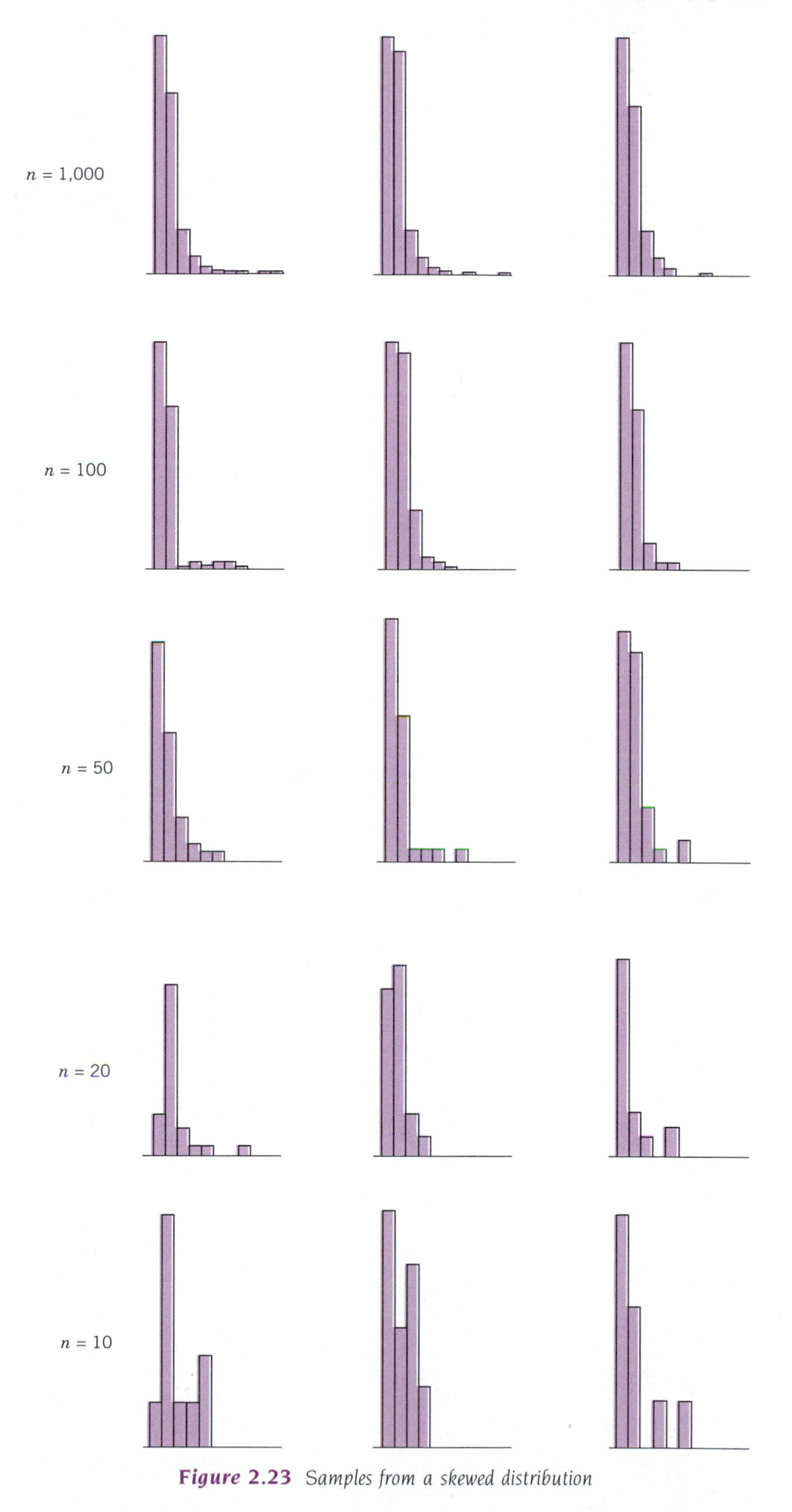

Figure 2.23 *Samples from a skewed distribution*

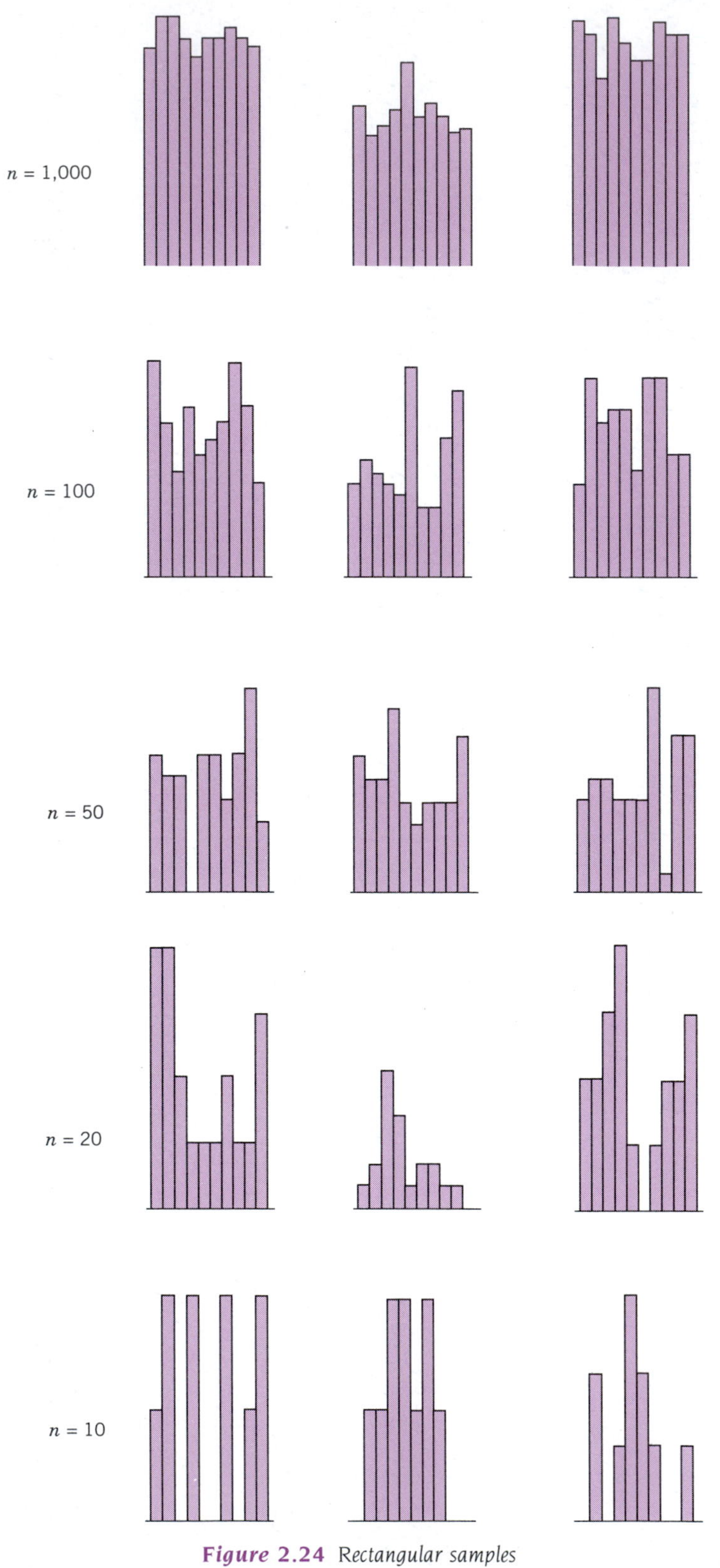

Figure 2.24 Rectangular samples

Problems

Basic Reviews

1. What is the smallest value in the distribution shown in Display 2.26?

DISPLAY 2.26

Stem	Leaves
0	2 3 8
1	2 3 5 5 8
2	2 5 8 8 9
3	0 1
4	0 1
5	5

Leaf unit = 1

2. What is the largest value in the distribution shown in Display 2.26?
3. How many times does the data value 8 appear in the distribution shown in Display 2.26?
4. How many times does the value 28 appear in the distribution shown in Display 2.26?
5. Figure 2.25 shows a histogram of a numerical distribution, with the numbers grouped into four ranges: 0–7, 8–15, 16–23, and 24–31. How many of the numbers in this distribution are 24 or higher?

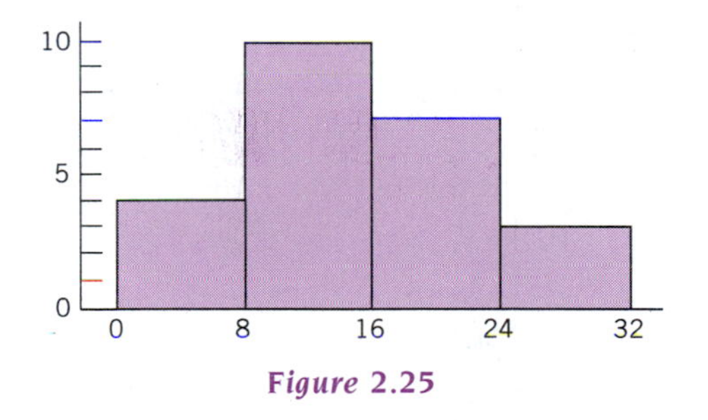

Figure 2.25

6. There are 24 numbers in the distribution shown in Figure 2.25. What percent of them are 24 or higher?
7. What percent of the values in the distribution shown in Figure 2.25 are in the interval 16–23?

Analysis

8. Table 2.8 (page 52) contains some data on the Whistler Mountain ski area, outside of Vancouver, British Columbia.
   - **a.** Construct a stem-and-leaf plot of the number of seats.
   - **b.** Does one lift appear to be an outlier? If so, which one?

Table 2.8 *Whistler Mountain ski area data: Speed is "Hi" if high, blank otherwise.*

LIFT	LENGTH (meters)	VERTICAL RISE (meters)	NUMBER SEATS	SPEED
Peak Chair	1,060	401	3	
Blue Chair	1,048	267	2	
Green Chair	1,833	424	4	Hi
Black Chair	1,752	540	3	
Olympic Chair	729	142	3	
Express Gondola	5,012	1,157	10	Hi
Quicksilver Quad	2,050	647	4	Hi
Alpine T-Bar Two	841	200	2	
Alpine T-Bar One	943	224	2	
Red Line Quad	2,495	553	4	Hi
Orange Chair	1,238	386	2	
Platter Lift	171	19	1	
Scampland Handle				
Tow 1	117	22	1	
Tow 2	150	35	1	

c. Construct a stem-and-leaf plot of the vertical rise in meters. (You are free to choose your own scale, of course, but we suggest using stem values of 100 m and leaf values of 10 m.)

d. Does one lift appear to be an outlier? If so, which one?

9. We have data on 1,301 sailboats that were for sale in the Puget Sound area in the summer of 1992. Using Minitab, we constructed the stem-and-leaf plot of their lengths given in Display 2.27. Notice that the longest rows end in plus signs, meaning that not all of the cases have been displayed in those rows.

a. How long is the longest boat in the data set?

b. How long is the shortest boat in the data set?

c. How many boats in the data set are 35 to 39 feet long, inclusive? (*Hint:* Recall that Minitab includes an extra column of numbers on the left, giving the total number of cases, counting in from each end. For the row containing the middle case, the number of cases is shown in parentheses.)

10. Figure 2.26 shows a classic histogram of the boat lengths, using the same data set as Display 2.27. Each interval is 5 feet wide. Each interval includes the boats whose length put them on the lower boundary. (Looking at it another way, a boat that falls on a boundary line is counted in the higher category.)

a. If a boat is exactly 25 feet long, is it in the interval 20–25 feet or the interval 25–30 feet?

b. Looking at Figure 2.26, which interval has the most boats in it?

c. What is the modal interval?

d. About how many boats are in the interval 25–30 feet? (Do this by eye; you do not have to be exactly right.)

e. About how many boats are in the interval 20–25 feet?

DISPLAY 2.27

STEM-AND-LEAF PLOTS OF SAILBOAT LENGTHS, PROBLEM 9

```
MTB > Stem-and-Leaf 'Length';
SUBC> Increment 5.

Stem-and-Leaf of Length     N  = 1301
Leaf Unit = 1.0             N* =    1


    2    0  88
    6    1  2444
   16    1  6677888999
  110    2  00000000000111122222222222222222222222222222233333333333333333333+
  375    2  55555555555555555555555555555555555555555555556666666666666666666+
  636    3  00000000000000000000000000000000000000000000000000000000000000000+
(245)    3  55555555555555555555555555555555555555555555555555555666666666666+
  420    4  00000000000000000000000000000000000000000000000000000000000000000+
  179    4  55555555555555555555555555555555666666666667777777777888888888888+
  100    5  0000000000000000000000000000000001111111111122223333444444444
   39    5  55566666777777777889999
   16    6  000444
   10    6  568
    7    7  4
    6    7
    6    8  04
    4    8  5
    3    9  01
    1    9
    1   10
    1   10
    1   11  2
```

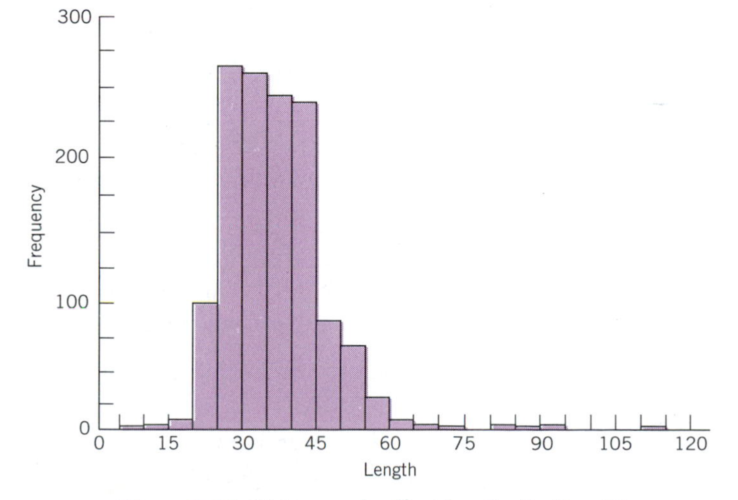

Figure 2.26 *Histogram of sailboat lengths, Problem 10*

f. About how long is the longest boat (or boats)? (*Hint:* Be careful; it barely shows.)

g. Describe the shape of the distribution.

11. What are the relative merits of the stem-and-leaf plot, compared with the classic histogram, for the boat length data set? Which did a better job of showing all the data? Which did a better job of showing the most extreme values?

12. Consider the seating capacities of baseball stadiums in the National League, as shown in Table 2.9.[10]

Table 2.9

Atlanta Braves	52,003
Chicago Cubs	38,170
Cincinnati Reds	52,952
Colorado Rockies	approx. 50,000
Florida Marlins	47,226
Houston Astros	54,816
Los Angeles Dodgers	56,000
Montreal Expos	43,739
New York Mets	55,601
Philadelphia Phillies	62,382
Pittsburgh Pirates	58,727
St. Louis Cardinals	56,227
San Diego Padres	59,022
San Francisco Giants	58,000

a. Draw a stem-and-leaf plot, using only the first two digits of each number.

b. Draw the corresponding classic histogram.

c. Use split stems to expand the scale: Write each stem twice, and hang smaller leaves from one stem and larger leaves from the other.

d. Now repeat each stem five times. Write leaves of 0 and 1 after the first stem, leaves of 2 and 3 after the second, and so on.

e. Which scale do you prefer? Why? (This is a matter of opinion; there are no wrong answers, but do give a sensible reason for your choice.)

f. Discuss a few things you have learned about baseball stadium capacities by looking at the distribution through these graphic techniques.

13. Get out the list of 10 to 30 advertised prices you compiled for Problem 3, Chapter 1.

a. Construct a stem-and-leaf plot of the data by hand.

b. Create a classic histogram of the data, by hand or, if possible, by computer.

c. Describe the shape of the distribution: Is it skewed or symmetric? If skewed, in what direction? What is the modal interval of values? What is the lowest value? What is the highest? Are there any outliers?

14. Table 2.10 contains the list prices of a group of "exotic" skis.[11]

a. Construct a stem-and-leaf plot of the prices. (We suggest using 100-dollar stems and 10-dollar leaves, but you are free to try other scales.)

b. Is the distribution symmetric or skewed? If skewed, in what direction?

Table 2.10

LENGTH (cm)		SUGGESTED
MINIMUM	MAXIMUM	RETAIL PRICE
175	205	400
180	203	600
180	210	475
175	207	440
185	207.5	550
180	210	465
190	205	650
185	210	550
—	—	595
185	205	450
170	203	420
180	207	475
170	204	660
170	205	995[a]
175	207	430
175	204	425
180	204	465
180	205	495
180	205	400
175	195	450
175	200	550

[a] Nishizawa Premier LTD-X: "The Premier especially thrives when you step quickly on and off an edge—yielding the sort of transcendent performance that begins to put the ski's price tag into a somewhat saner light."

c. Does one value appear to be an outlier? If so, which one?

d. Given the information provided in the footnote to the table, would you say that (a) the outlier value is a typing or transcription error, or (b) a real observation?

15. Here are some measurements of the wing length, in millimeters, of "killer bees" from French Guiana. These are bees from Africa that look very much like ordinary honeybees, but are much more dangerous. Only by using a detailed statistical analysis is it possible to classify a dead bee as an ordinary honeybee or a killer bee.[12]

8.56 8.51 8.51 8.56 8.96 8.82 8.39 8.54 8.62

a. Make a graphic display (stem-and-leaf plot, dot plot, or classic histogram) of the data.

b. Is the distribution (approximately) symmetrical?

16. A. A. Michelson is famous for being the first to accurately measure the speed of light. Good experimental scientists know that it is crucial to obtain many measurements, and Michelson did repeat his experiments many times. Table 2.11 (page 56) shows the results, in kilometers per second, of some of his measurements from a series of experiments performed in 1882 (part of Stigler [1977]).[13]

a. Create a graphic display of the data.

b. Is the distribution approximately symmetrical?

Table 2.11 *Determinations of the speed of light by Michelson, in kilometers per second*

299,960	299,940
299,960	299,940
299,880	299,800
299,850	299,880
299,900	299,840

17. Bell Telephone was divided into independent companies as the result of an agreement between the company and the U.S. government, which changed the way the industry was regulated. At about this time, the different divisions of the Bell System owed various amounts of money, a factor that could influence the ability of these divisions to survive as independent companies. These debt positions are shown in Table 2.12.[14]

Table 2.12 *Debts of the Bell System's local operating companies in billions of dollars at the end of 1980*

Bell Telephone of Pennsylvania	$1.5
Chesapeake and Potomac Telephone (DC)	.2
Chesapeake and Potomac Telephone (MD)	.7
Chesapeake and Potomac Telephone (VA)	.7
Chesapeake and Potomac Telephone (WV)	.2
Cincinnati Bell	.2
Diamond State Telephone	.1
Illinois Bell Telephone	1.7
Indiana Bell Telephone	.5
Michigan Bell Telephone	1.2
Mountain States Telephone	2.0
New England Telephone	1.7
New Jersey Bell Telephone	1.0
New York Telephone	3.4
Northwestern Bell Telephone	1.3
Ohio Bell Telephone	.9
Pacific Northwest Bell Telephone	1.1
Pacific Telephone	5.8
South Central Bell Telephone	2.7
Southern Bell Telephone	3.2
Southern New England Telephone	.6
Southwestern Bell Telephone	4.6
Wisconsin Telephone	.5

a. Construct a graphic display of the data.

b. Describe the shape of the distribution.

18. The region of Appalachia, defined by the Appalachian Regional Development Act of 1965, represents low-income portions of 13 states. The average yearly income per person for this portion of each state was as shown in Table 2.13 when President Johnson declared war on poverty.[15]

Table 2.13 *Per capita income, Appalachian portion of each state, in thousands of dollars (1965 dollars)*

Alabama	4.1	Ohio	3.8
Georgia	3.8	Pennsylvania	4.4
Kentucky	3.2	South Carolina	4.1
Maryland	4.1	Tennessee	3.8
Mississippi	3.1	Virginia	3.7
New York	4.1	West Virginia	4.0
North Carolina	4.0		

a. Do a graphic display of the data. (Expand the scale until you can see some skewness.)

b. Describe the shape of the distribution. In what direction is it skewed?

c. Can you think of a plausible reason for this kind of skewness, given that these regions were chosen, in part, because their income fell below poverty level? (*Hint:* Picture this as part of a more normal distribution where we see only values smaller than a certain low threshold.)

19. Table 2.14 shows data on heights of boys who participated in a study of the effect of diet on their growth. This group consisted of those who ate a macrobiotic diet (a special kind of vegetarian diet).

Table 2.14 *Heights of boys on macrobiotic diets, in centimeters*

718	883	853
630	768	840
1,002	830	820
560	810	862
715	813	785
1,010	1,105	1,132
1,085	860	1,084
750	1,090	1,152
510	615	490
1,112	912	781
1,093	684	1,000

a. Construct a graphic display of the data.

b. Is the distribution approximately symmetric?

20. Consider the data shown in Table 2.15 (page 58) on the total deposits of the ten largest U.S. commercial banks, about 1979.[16]

a. Create a graphic display of the data.

b. Describe the shape of the distribution.

c. Would you usually expect to see skewness toward high values in a data set like this that represents positive amounts of money? Why or why not?

d. Count the number of these banks in each city. Which city seems to be the financial center of the United States?

Table 2.15 *Deposits, in dollars, of the ten largest commercial banks in the United States, about 1979*

Bank	Deposits
Bank of America, San Francisco	$85,069,944,000
Citibank, New York	72,429,528,000
Chase Manhattan Bank, New York	56,173,381,000
Manufacturers Hanover Trust Co., New York	39,000,618,000
Morgan Guaranty Trust Co., New York	35,485,994,000
Chemical Bank, New York	30,521,757,000
Continental Illinois National Bank, Chicago	25,566,850,000
Bankers Trust Co., New York	22,121,919,000
First National Bank, Chicago	20,784,969,000
Security Pacific National Bank, Los Angeles	19,590,576,000

21. Consider the data in Table 2.16 on assets (primarily deposits) of the ten largest U.S. commercial banks in 1991. Amounts are rounded to the nearest billion.

Table 2.16 *Deposits, in billions of dollars, of the ten largest commercial banks in the United States, 1991*

Bank	Deposits
Citicorp, New York	217
Chemical Bank, New York	139
BankAmerica, San Francisco	116
NationsBank Corp., Charlotte, North Carolina	110
J. P. Morgan & Co., New York	103
Chase Manhattan, New York	98
Security Pacific Corp., Los Angeles	76
Bankers Trust, New York	69
Wells Fargo & Co., San Francisco	54
First Chicago Corp., Chicago	49

a. Construct a graphic display of the data.

b. Is the distribution approximately symmetric?

c. How does this distribution compare with the distribution of Problem 20? Are the shapes the same? How about the amounts of money involved?

22. Consider the seating capacities of baseball stadiums in the National League, given in Table 2.9.[17]

a. Make a stem-and-leaf plot of the data, if you have not already.

b. Is the distribution approximately symmetric?

23. Table 2.17 shows the per capita gross national products of ten countries in middle south Asia, in about 1975, prior to some recent political and military actions.

a. Construct a graphic display of the data.

b. Does there appear to be an outlier? If so, which country (by name) is it?

c. The outlier might be due to an error in copying, in which the number 106.0 could have been copied as 1,060 by leaving out the decimal point. Does this seem like a reasonable possibility, in the sense that 106.0 would be like most of the other data?

Table 2.17 *Per capita gross national products in middle south Asia, about 1975*

COUNTRY	AMOUNT
Afghanistan	100
Bangladesh	100
Bhutan	70
India	130
Iran	1,060
Maldive Islands	90
Nepal	110
Pakistan	130
Sikkim	90
Sri Lanka	130

d. Suppose the figures have been checked and it is determined that there are no copying errors. Is there some other explanation for the outlier? (*Hint:* Is the country a major exporter of some important natural resource?)

24. In Table 2.18 is an extract of a moderately larger data set, "Alfalfa," distributed with Minitab. The variables in the columns are grain yields (in tons/acre) per variety of alfalfa. (There are six varieties, numbered 1 to 6. These numbers just serve as labels and have no mathematical content.) Test plots were planted in four different fields, labeled 1 to 4. The yield data are shown as a dot plot in Display 2.28 and as a stem-and-leaf plot in Display 2.29 (page 60).

Table 2.18 *Tons per acre of alfalfa*

FIELD	VARIETY	YIELD	FIELD	VARIETY	YIELD
1	1	3.22	3	1	3.26
1	2	3.04	3	2	3.27
1	3	3.06	3	3	2.93
1	4	2.64	3	4	2.59
1	5	3.19	3	5	3.11
1	6	2.49	3	6	2.38
2	1	3.31	4	1	3.25
2	2	2.99	4	2	3.20
2	3	3.17	4	3	3.09
2	4	2.75	4	4	2.62
2	5	3.40	4	5	3.23
2	6	2.37	4	6	2.37

a. Describe the distribution.

b. Does it appear that this distribution may be composed of different groups?

c. Use the information in Table 2.18 to make an argument supporting your answer to part (b). Construct at least one graphic display supporting your argument.

DISPLAY 2.28

```
MTB > Retrieve 'C:\MTBWIN\DATA\ALFALFA.MTW'.

MTB > DotPlot 'YIELD'.

 :.        .    . ..      .         .   . ....   .:..:. .   .
---+---------+---------+---------+---------+---------+---YIELD
 2.40      2.60      2.80      3.00      3.20      3.40
```

DISPLAY 2.29

```
MTB > Stem-and-Leaf 'YIELD'.

Stem-and-leaf of YIELD     N = 24
Leaf Unit = 0.010

       3    23 | 7 7 8
       4    24 | 9
       5    25 | 9
       7    26 | 2 4
       8    27 | 5
       8    28 |
      10    29 | 3 9
     (3)    30 | 4 6 9
      11    31 | 1 7 9
       8    32 | 0 2 3 5 6 7
       2    33 | 1
       1    34 | 0
MTB >
```

25. Table 2.19 shows a section of the sailboat data file used in Problem 9.

a. What variables are included, and what are their levels of measurement?

b. What is a case in this file?

c. How are the data arranged? (That is, how are the cases ordered?)

d. Speculate on how this particular set of cases was chosen out of the larger data set.

26. Hambrecht et al. (1993)[18] conducted a study in which heart patients were randomly assigned to a special exercise program, which included exercise classes and the loan of an exercise bike for home. Patients filled out questionnaires about the amount of exercise they did. These data were used to estimate the number of calories (kilocalories) burned in leisure-time exercise per week.

The authors also measured the "ventilatory threshold" (VT) of each participant. The ventilatory threshold is determined by measuring the ratio of carbon dioxide to oxygen in exhaled breath and corresponds (approximately) to the point at which people exercising experience themselves as being "out of breath." (More technically, it is a noninvasive way of finding the point at which exercise exceeds the aerobic capacity of the exerciser so that the muscles start using anaerobically produced energy.) The authors are particularly interested in changes in the ventilatory threshold produced by exercise. If you participate in their exercise program, how much more can you do before you are out of breath? In math,

Table 2.19 *Portion of sailboat data file*

LENGTH	TYPE	YEAR	POWER	PRICE	BROKER
25	Hunter25.5	1986	OB	16,500	ABC Yacht
26	Clark	1975	D	7,900	ABC Yacht
26	CheoyLee	1968	D	13,950	ABC Yacht
33	Sounder	1977	D	42,500	ABC Yacht
36	Catalina	1990	D	72,000	ABC Yacht
29	C&C	1983	D	35,500	Active Yt
30	G&SCustom	1987	D	48,000	Active Yt
30	Andrews	1990	D	92,000	Active Yt
30	Wavelength	1982	D	35,000	Active Yt
31	KauluaCat	1981	OB	45,000	Active Yt
32	C&C	1984	D	49,750	Active Yt
33	C&C	1977	G	29,900	Active Yt
33	Abbott	1984	D	30,000	Active Yt
33	J/33	1988	D	69,900	Active Yt
34	Hunter	1983	D	37,800	Active Yt
35	ThomasT_35	1989	D	79,900	Active Yt
36	S_2	1979	D	59,900	Active Yt
37	C&C	1984	D	89,900	Active Yt
38	Mull	1981	D	59,000	Active Yt
38	Baltic	1983	D	145,000	Active Yt
38	Catalina	1980	D	52,000	Active Yt
38	Hood	1984	D	115,000	Active Yt
38	Catalina	1983	D	68,000	Active Yt
39	Beneteau	1987	D	79,900	Active Yt
40	C&C	1980	D	80,000	Active Yt
40	N/M 1_Ton	1988	D	85,000	Active Yt
40	Tripp	1991	D	200,000	Active Yt
41	C&C	1984	D	99,500	Active Yt
41	Bianca414	1980	D	77,900	Active Yt
42	Allied	1972	D	79,000	Active Yt
43	J/NCustom	1986	D	135,000	Active Yt
44	C&C	1987	D	149,000	Active Yt
44	J/Boat	1990	D	295,000	Active Yt
45	Frers	1990	D	240,000	Active Yt
45	Morgan	1983	D	147,500	Active Yt
48	C&C	1975	D	165,000	Active Yt

Note: "Length" is in feet; "Type" is maker and, sometimes, model; "Year" is year of manufacture; "Power" is type of auxiliary power, where OB = outboard, D = diesel, and G = gas; "Price" is in dollars.

a change in value is often indicated by the Greek letter *delta,* a small triangle. (In English, this is simply read as "delta.")

Data approximating the results of Hambrecht et al. are given in Table 2.20 (page 62). (These values were obtained by reading values off a graph in the original article, which introduced some additional error.)

a. What constitutes a case in this study? What variables have the authors tried to measure? What levels of measurement are possible with these variables?

b. Examine the distribution of calories per week. Describe its shape and, in particular, discuss whether there is clear evidence (from the shape alone) that there are two different groups represented in this distribution.

Table 2.20 Changes in ventilatory threshold (VT) with different amounts of exercise, over one year (data file Cal_VT.MTW)

CAL/WK	DELTA VT	CAL/WK	DELTA VT
3,600	.375	1,300	.01
3,400	.05	1,300	.045
3,400	.175	1,250	−.2
2,975	.04	1,250	−.025
2,875	.05	1,200	.02
2,850	.2	1,200	−.5
2,850	.24	1,200	−.11
2,750	.12	1,200	.025
2,350	.1	1,150	−.1
2,250	.1	1,100	−.125
2,200	.375	1,050	−.05
2,100	.13	1,000	−.1
2,000	.02	1,000	.075
1,950	0	900	−.125
1,900	−.35	900	.12
1,900	.08	900	.45
1,900	.4	850	−.125
1,875	.12	800	−.24
1,750	.075	750	.03
1,750	.3	700	.175
1,750	.4	675	−.075
1,600	−.3	650	−.7
1,600	.03	650	−.125
1,575	−.05	400	−.2
1,500	.8	300	−.225
1,450	0	200	−.3
1,400	.125	100	−.123
1,300	−.3	100	−.03

27. Display 2.30 shows the age at which a sample of runaway and homeless girls reported first having had sex. Exactly what "having sex" means was left to the individual respondent to decide. The data are part of a database assembled for a study of adolescent pregnancy, and four individuals who reported never having had sex were excluded.[19] (These data are also available as the variable "agesex" in the file RUNAWAY.MTW.) Describe the shape of this distribution. (*Note:* Some individuals reported having first had sex as toddlers. In some cases these instances of childhood abuse had been previously reported.)

28. The RUNAWAY.MTW data file contains information on the number of different places each of 125 runaway girls have lived. These data are shown in Display 2.31. Describe the shape of this distribution.

29. Who do kids with problems live with? The family compositions of 30 adolescents (ages 13 to 18) with emotional or behavioral disabilities are given in Figure 2.27. [Data are from Smith (1993), p. 368.[20]] Write a very brief paragraph putting the information in the figure into words.

DISPLAY 2.30

AGE OF FIRST SEXUAL EXPERIENCE FOR SELECTED INDIVIDUALS, PROBLEM 27

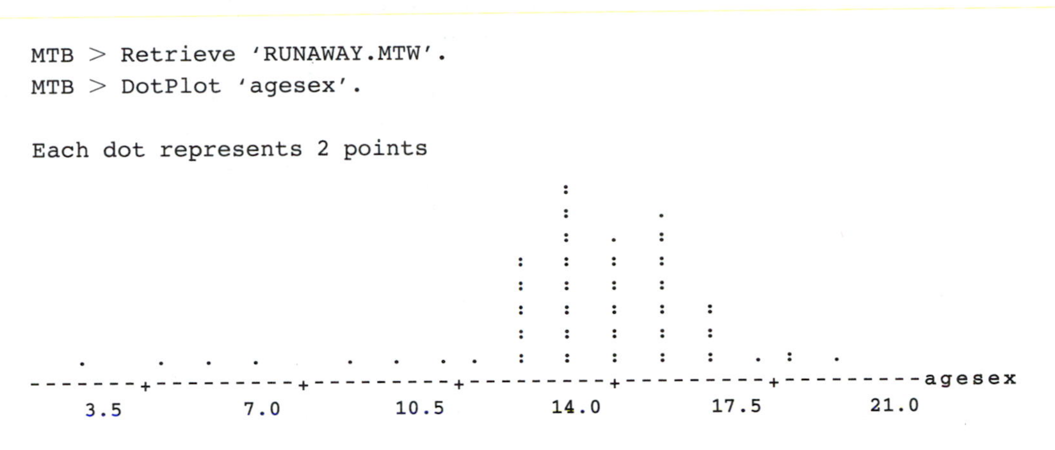

DISPLAY 2.31

NUMBER OF DWELLING PLACES OF SELECTED RUNAWAYS, PROBLEM 28

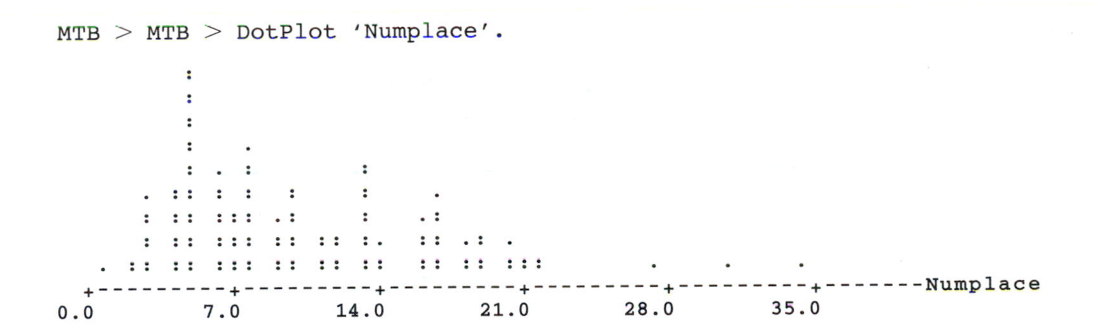

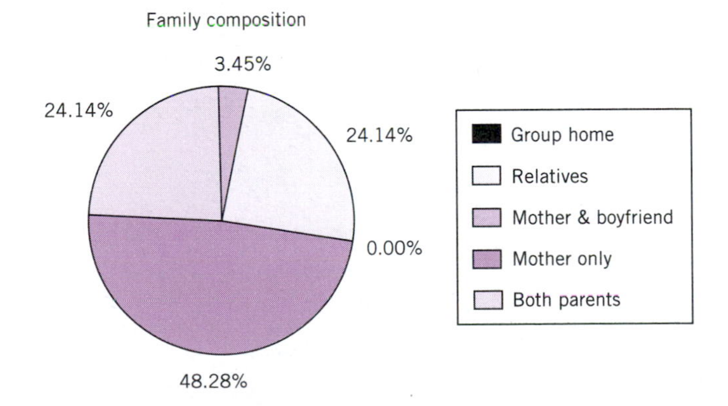

Figure 2.27 *Family composition of adolescents with emotional or behavioral disabilities, Problem 29*

30. Data file WHALES.MTW (also shown in Table 2.21[21]) contains information on the body measurements of male bowhead whales, collected by the National Marine Fisheries Service. The two measurements are total body length, in centimeters, and the length of the jawbone or mandible (mouth), also in centimeters.

Table 2.21 *Total body length and mandible (mouth) length of male bowhead whales, in centimeters*

TOTAL	MOUTH	TOTAL	MOUTH
610	250	854	271
670	287	869	283
671	217	980	328
691	218	1,021	350
750	315	1,100	380
762	272	1,120	384
762	250	1,144	419
813	254	1,235	437
813	264	1,402	464
823	274	1,468	430

a. Prepare a graphic display of the body-length data and describe the shape of the distribution.

b. Prepare a graphic display of the data on mandible length (mouth). Describe the shape of the distribution.

31. Levy et al. (1993) studied the effect of exercise on heart filling rate in older (age 60 to 82) and younger (age 24 to 32) healthy males. (Exercise improved heart fill rates for both groups.) As part of the study, they looked at the relationship between maximum oxygen consumption—the ability to exercise—and heart fill rate at the start of the study. Data approximately matching theirs are provided in Table 2.22.[22] (These data are also in data file HEART.MTW.)

Table 2.22 *Heart fill rate and maximum oxygen use in older (60–82) and younger (24–32) males*

FILL RATE (ml/s)	O_2 USE (ml/kg/min)	AGEGRP[a]	FILL RATE (ml/s)	O_2 USE (ml/kg/min)	AGEGRP[a]
52.0	21	1	125.0	37	2
49.5	25	1	160.0	37	2
49.0	27	1	120.0	40	2
50.0	33	1	155.0	39	2
75.0	27	1	105.0	43	2
80.0	28	1	115.0	45	2
78.0	26	1	120.0	47	2
80.0	28	1	115.0	52	2
95.0	27	1	145.0	45	2
100.0	28	1	175.0	43	2
140.0	33	1	212.0	42	2
115.0	34	1	200.0	50	2
116.0	38	1	205.0	49	2
81.0	27	1	205.0	48	2
90.0	28	1	240.0	46	2
140.0	34	2			

[a] 1 = older, 2 = younger.

a. Describe the shape of the distribution of heart filling rate. In particular, does it appear to be bimodal?

b. Describe the shape of the distribution of maximum oxygen consumption. In particular, does the distribution appear to be bimodal?

c. Describe the shape of the distribution of heart fill rate for younger men (age group 2).

d. Describe the shape of the distribution of maximum oxygen consumption for younger men (age group 2).

chapter 3

Describing Distributions

Summarization is the process of going from extensive and complicated details to simple informative generalizations while preserving some of the essentials of a situation. Although a summary is not a complete description of the underlying situation, summaries are extremely useful in understanding and communicating the most important characteristics of a data set.

When we talk about a summary description of a distribution, we are primarily concerned with three questions:

1. What is the shape of the distribution?
2. Where is the center of the distribution?
3. How spread out is the distribution? (that is, How much variability does it have?)

In this chapter we introduce the *mode* as a summary of center that can be used with any data distribution. We also introduce the *median,* the *range,* and the *interquartile range,* a set of summaries that can be used with distributions of ordered categorical or numerical data, regardless of the shape of the distribution. Finally, we use these summaries to construct a very informative graphic display called a *box plot*.

Other summary measures will be introduced in Chapter 4.

3.1 Where Is the Center?

We start with the problem of finding a representative *central value* for the distribution. Can we just look at the distribution shape in a histogram and identify the middle? This technique will often get us fairly close to what looks like the center of a symmetric data set, but the ambiguity due to the randomness of the data makes it difficult to say exactly where the middle is. It is even harder to say where the middle is when the distribution is skewed because it looks different on the two sides (see Figure 3.1).

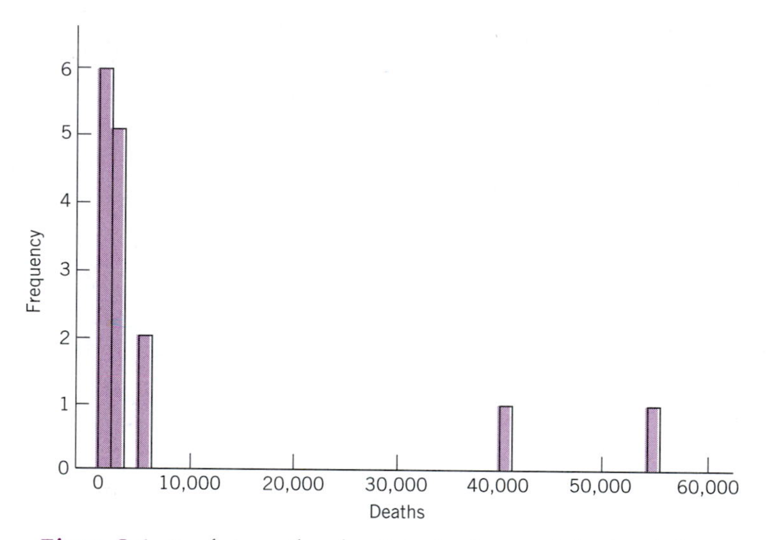

Figure 3.1 *Deaths in earthquakes. Note that because of the skewed shape, there is no clear "center" to the distribution.*

Because of this uncertainty, there is no unique answer to the problem of identifying the best central value. Instead, there are several answers to this simple problem. For a symmetric data set, different measures of center should produce (at least approximately) the same value. However, due to randomness, different methods will often produce slightly different summaries for the same data set even when there appears to be a well-defined center seen clearly in the histogram.

When the distribution is skewed, there is no clear center. In this case, the different summarization methods focus on different concepts of "center" and result in different summary values. In addition, different methods respond to outliers in different ways.

There are three commonly used measures of the center of a distribution:

1. The **arithmetic average** or **mean** is calculated by adding up all the numbers and dividing this sum by the number of numbers. This is probably the most widely used

measure of the center, and we assume that you already know how to calculate it. We discuss it at length in Chapters 4 and 6.

2. The **mode** is the most common value or category in a distribution. It is most useful with categorical data.
3. The **median** is a value that splits the distribution in half, so that half the values are above it and half are below it. It is easy to calculate and can be used with most distributions.

3.2 Summarizing a Categorical Variable: Percentages and the Mode

One central value summary, the mode, can be used with variables measured at any level. The data value, or small range of data values, that occurs most often in the data set is called the **mode**. With a categorical variable, it is the most common category. With a measured numerical variable, it is at a high point or peak on a histogram. The mode is an especially useful summary for data sets that consist not of numbers but of names or categories. With categorical data we sometimes call the most common category the *modal category*. It is sometimes useful to consider the possibility that the data have more than one mode, in the sense that there is more than one prominent set of values. In such instances, we do not require the modes to be equally frequent.

To summarize a categorical variable, we can simply report the total number of observations and/or the percentage of the total represented by each category. The mode is then the category with the largest number of data values.

For example, consider the birthplaces of a class of 29 students, shown in Table 3.1. These percentages have a simple and direct interpretation: 51.7% of the students are from California, 10.3% are from Massachusetts, and so on. The mode in this case is California because it is the category with the largest number of students.

Table 3.1

BIRTHPLACE	NUMBER OF STUDENTS	PERCENT OF TOTAL
California	15	51.7
Massachusetts	3	10.3
New York	8	27.6
Oregon	1	3.4
Wisconsin	2	6.9

Watch Out for Small Samples. Be careful when interpreting percentages based on small amounts of data. For example, 42.9% is more meaningful (that is, more exact) when it represents 252 out of 588 than when it represents 3 out of 7. Consider also the following apocryphal story:

> Our study has found that 33% of high-school students prefer new-wave music and that 33% prefer rock and roll. The third teenager refused to comment.

EXAMPLE 3.1 *Voting and the Mode*

Consider the results of an election to choose a single representative from four candidates whose names are Davidson, Efron, Solomon, and Fisher. The data set consists of ballots, each of which names one of these people. The entire data set consists of 36,628 ballots, forming a very long list of names starting like the following:

> Davidson, Efron, Solomon, Davidson, Fisher, Efron, Solomon, Solomon, Davidson, Efron, Fisher, Solomon, Davidson, Davidson, Efron, Solomon, Davidson, Fisher, Solomon, Efron, Fisher, Fisher...

Because it is simply the name that occurs most often, the mode is the name of the person who wins the election. In a case like this, not only is the mode computable, it is probably the most important single summary value that there is!

Exercise Set 3.1

1. What is the modal first initial in the following list of names?

Bernie, Barry, Beatrice, Betty, Jack, Jim, Jared, Jeremy, Josh, Jessica, Mary, Mike, Sam, Sally

2. What is the modal color in the following list?

blue, blue, green, purple, red, red, red

3. What is the modal value in the following list?

10, 10, 12, 15, 15, 15, 17, 20

4. Study Table 3.2.[1]

- **a.** According to this table, what percent of the bachelor's degrees conferred were in computer sciences and engineering?
- **b.** What was the most common degree area?

Table 3.2 *Bachelor's degrees conferred, by field of study, 1988*

DEGREE AREA	PERCENT
Humanities and social/behavioral sciences	28.7
Natural sciences	7.1
Computer sciences and engineering	12.4
Education	9.2
Business	24.5
Other technical/professional	17.9

5. Construct a bar chart of the percentage of degrees by area of study using the data in Table 3.2.

3.3 Summarizing Ordered and Numerical Data

The Median

With ordered categories and measured numerical data we can still compute the mode, but now we can do more. Because we can arrange the data in order, we can talk about the position of any given observation; we can ask where in the distribution a particular observation falls. To start, we will ask which observation is in the middle of the distribution.

The **median** is the middle value of a group of numbers in the sense that when the numbers are arranged in order from smallest to largest, the median will be the middle number in the ordered list. Generally, half of the data will be larger than the median and half will be smaller than the median.

We will develop the procedure first with very small data sets, then apply it to the earthquake data from the previous chapter.

PROCEDURE: FINDING THE MEDIAN OF A SINGLE GROUP OF NUMBERS

1. First, put the numbers in order from smallest to largest.
2. Then, find the median as follows:
   a. If the number of data values is an *odd* number (one whose rightmost digit is 1, 3, 5, 7, 9), then the median is simply the middle number.
      In the following example there is an odd number, 5, of data values.

$$\begin{aligned}&\text{Median}(1, 8, 4, 6, 8)\\&\quad = \text{Median}(1, 4, 6, 8, 8)\\&\quad = 6\end{aligned}$$

   The median value is 6, since half the values are greater than 6 and half smaller.

   b. If the number of data values is an *even* number (one whose rightmost digit is 0, 2, 4, 6, or 8), then the median must be found as the average of the two middle numbers.
      In the following example, there is an even number, 6, of data values.

$$\begin{aligned}&\text{Median}(1, 9, 4, 6, 8, 13)\\&\quad = \text{Median}(1, 4, 6, 8, 9, 13)\\&\quad = \text{Average}(6 + 8) = 7\end{aligned}$$

   We use parentheses for "of," so "Average(6 + 8)" stands for "Average of 6 and 8." Our use of English-based notation here is consistent with computer programming traditions. You will see similar notation if you use Excel or any other computer spreadsheet program.

Ranks of the Data

With very small data sets it is easy enough to find the middle by eye. With larger data sets we have to formalize the procedure and do a little arithmetic. To do this, we use the idea of *rank*. As we arrange data in order, their **rank** is their position in that order: first, second, third, and so on. There are two kinds of ranks: ranks from the top (from largest to smallest) and ranks from the bottom (from smallest to largest).

The median can then be defined as the data value whose rank is halfway down the list of values. The rank of the median is about half the number of data values. A precise formula that can be used for computing the median rank follows.

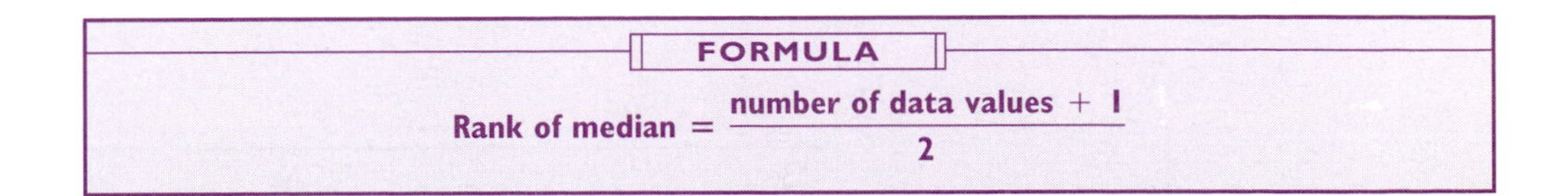

FORMULA

$$\textbf{Rank of median} = \frac{\textbf{number of data values} + 1}{2}$$

We have to add 1 to the number of data values to find the middle value. If we only have three values, for example, we know that the second is in the middle. But 2 is not half of 3; it is half of 4, which is 3 + 1.

With the median as a measure of the center of a distribution, we are now able to obtain more precise information from the data.

EXAMPLE 3.2 *Finding the Median: Earthquake Magnitudes*

Following are the earthquake magnitudes from Table 2.7.[2]

5.5	7.7	7.1	7.8	8.1
7.3	6.5	7.3	6.8	6.9
6.3	6.5	7.7	7.7	6.8

To determine just how big a typical major earthquake is, we will use the median value. We want to find the value that divides the list in half, with bigger earthquakes on one side and smaller ones on the other. We first arrange these in order from smallest to largest:

5.5	6.3	6.5	6.5	6.8
6.8	6.9	7.1	7.3	7.3
7.7	7.7	7.7	7.8	8.1

Now, just to make this discussion a little easier, we arrange them in a vertical column with the biggest value on top (see Table 3.3).

With respect to rank from the bottom, we see from this table that the data value of 6.5 has a rank of 3, telling us that it ranks as the third smallest value. Similarly, the value 7.7 has the rank of 3 from the top. (It also happens to have the ranks of 4 and 5 from the top.)

Using the formula for the median rank, we find that the rank of the median for the earthquake

Table 3.3

DATA VALUES	RANK (FROM THE TOP)	RANK (FROM THE BOTTOM)
8.1	1	15
7.8	2	14
7.7	3	13
7.7	4	12
7.7	5	11
7.3	6	10
7.3	7	9
7.1	8	8
6.9	9	7
6.8	10	6
6.8	11	5
6.5	12	4
6.5	13	3
6.3	14	2
5.5	15	1

data is (15 + 1)/2 = 16/2 = 8. Counting in eight from either the top or the bottom, we find the median value is 7.1. This is our answer. So we can say that a typical major earthquake has a magnitude of 7.1 on the Richter scale.

Even-numbered Data Sets

If we have an even number of data values to start with, say, 10, we get a median rank that is not a whole number, such as

$$\frac{10 + 1}{2} = \frac{11}{2} = 5.5$$

But how do we deal with a rank that is not a whole number? This is a signal that we should take the average of the two data values on either side. Because $5\frac{1}{2}$ is halfway between 5 and 6, we will take as the median the average of the data values of rank 5 and 6. From the definition of the median given earlier, we know that with an even number of data values we have to average the two middle numbers. The use of ranks helps us find these middle numbers.

Note that the same answer will always be found whether we use ranks from the top or ranks from the bottom because we will average the same two numbers (sometimes in the reverse order), giving the same final answer in either case.

EXAMPLE 3.3 *Ranking Data and Finding the Median: Ski Prices*

What is a typical price for new all-terrain skis? The November 1991 issue of *Ski Magazine* has test reports on 22 all-terrain skis. Their retail prices are shown in Table 3.4 on page 74.[3]

We can use the median price as our answer. Here are the list prices of the all-terrain skis arranged in order.

400	425	425	425	445	450	450	460	465	475	475
475	475	475	475	490	495	498	500	540	550	825

The corresponding stem-and-leaf plot is shown in Display 3.1. Counting, we find that there are 22 prices, so

$$\text{Median rank} = \frac{22 + 1}{2} = 11.5$$

The median rank is 11.5, so the median value is going to be the average of the values at rank 11 and 12:

$$\begin{aligned}\text{Median value} &= \frac{\text{value(rank 11)} + \text{value(rank 12)}}{2} \\ &= \frac{475 + 475}{2} \\ &= 475\end{aligned}$$

Thus, the median price for these all-terrain skis is $475.

Table 3.4 *Lengths and prices of all-terrain skis*

MODEL	LENGTH (cm)	PRICE ($)
Atomic Arc 837 Exir	180–207	450
Authier Zurbriggen GS	180–210	500
Blizzard V20 ATV R30	180–205	475
Dynamic VR27 Equipc SLT	180–204	475
Dynastar Mondial 5.0	180–203	475
Dynastar Pantara 5.0	160–195	475
Elan MBX 16	185–207	550
Fisher Vacuum Technic SLS	175–205	425
Hart Extreme Ultimate SL	185–205	498
Head CE6	180–205	450
Kastle Extreme	193–213	425
Kneissl White Star Steep	175–207	465
K2 CdS	180–204	475
La Croix Reference	175–207	825
Olin OTSL	180–207	445
PRE 860	180–204	495
RD Locomotion	180–204	400
Rossignol DV6S	168–205	475
Salomon Force 9 35	—	490
Tyrolia Comp 30X	180–205	460
Volant FX-1	175–200	425
Volkl Renntiger R	175–210	540

DISPLAY 3.1

4	0 2 2 2 4
4	5 5 6 6 7 7 7 7 7 7 9 9 9
5	0 4
5	5
6	
6	
7	
7	
8	2
8	

Exercise Set 3.2

6. Here are 5 numbers:

12 14 16 18 20

a. What is the median *rank*?

b. What is the median *value*?

7. Here are 6 numbers:

12	14	16	18	20	22

a. What is the median *rank*?

b. What is the median *value*?

8. Here are the EPA city gas mileage figures for two-seat car models available in the United States in 1992:

8	9	10	13	13	15	15	15	16	17
17	17	18	21	22	24	25	28	32	

a. What is a "typical" mileage estimate for these car models?

b. Describe how you found your answer to part (a).

c. In what sense is this a "typical" value?

3.4 Variability

The median is one way to describe the center of a distribution. It describes a "typical" value for the numbers in the data set, but does not tell us how close together or spread out the numbers are. This characteristic of the distribution is called **variability.** Measures of variability provide us with ways to estimate how accurate—or inaccurate—statistical estimates or survey results are likely to be.

In this section we introduce two measures of variability: the *range* and the *interquartile range*.

The Range

The simplest possible measure of variability is the **range,** which is the width of the distribution as measured from the smallest to the largest data value. The range measures the span of the group of numbers and is found by subtracting the smallest data value from the largest data value. Note that because it is computed in this way, the range can never be a negative number.

PROCEDURE: COMPUTING THE RANGE OF A SINGLE GROUP OF NUMBERS

1. Find the largest and smallest number in the data set.
2. Subtract the smallest from the largest.

$$\text{Range} = \text{largest} - \text{smallest}$$

EXAMPLE 3.4 *Range of Ski Prices*

How different are the ski prices shown in Table 3.4? The lowest price for any of the all-terrain skis was \$400 and the highest was \$825. The *range* is their difference:

$$\text{Range} = \$825 - \$400 = \$425$$

So the amount we spend on a pair of skis can vary by as much as \$425.

EXAMPLE 3.5 Range of Fuel Efficiency

How much difference *can* it make in our estimated gas mileage, if we buy one two-seat car model instead of another? Here are the EPA city gas mileage figures for two-seat car models available in the United States in 1992:

8	9	10	13	13	15	15	15	16	17
17	17	18	21	22	24	25	28	32	

The lowest value is 8 mpg, the highest is 32 mpg. So

$$\text{Range} = 32 - 8 = 24 \text{ mpg}$$

There is a 24 mpg range in the fuel efficiency of these cars. In other words, the gas mileage difference from one model to another can be as much as 24 miles per gallon.

Discussion

The range—or, rather, the extreme values—are often used in the popular press to give a quick idea of possible options. The newspaper often reports high and low prices for a stock, high and low gas mileage estimates are posted on new cars, and so on. For small data sets and nontechnical audiences, the range is useful in giving a general idea of the variability.

But the range has some undesirable features. Because it is the difference between the two most extreme values in the data set, and outliers (if there are any) will show up in the extremes, it follows that the range is extremely sensitive to measurement or recording errors. If the distribution is skewed, it is sensitive to which high values happen to be observed. The range ignores where most of the data are and concentrates entirely on the extreme values. Thus, the range is neither robust against outliers nor sensitive to most of the data, representing the worst of both worlds.

There is one job for which the range and the extreme values are entirely suited: editing and checking a data set to see if there are outliers or related problems. Although it would be better to actually look at the distribution if time were available, when time is short and many data sets must be examined, the range can provide a useful shortcut to checking the validity and reasonableness of the data.

Exercise Set 3.3

9. What is the range of the following numbers?

50	60	70

10. On May 13, 1992, the highest U.S. temperature was 105°F in Borrego Springs, CA, and the lowest was 22°F in Sun Valley, ID. What was the temperature range?

11. Here are the elevations (in feet) of the first 15 mountains mentioned in the Appalachian Mountain Club's *White Mountain Guide*:

6,288	5,532	5,715	5,798	5,363	2,404	5,585	5,470
3,450	5,385	5,004	4,761	4,310	4,052	3,919	

a. Compute a measure of the variation in these elevations.

b. Write a sentence describing the variation of these elevations.

12. Here are the high elevations (in feet) of the first 15 hikes listed in *Best Hikes with Children in Western Washington*:

6,000	5,699	5,200	6,250	5,200	4,550	4,300	4,800
900	5,537	5,392	5,475	6,200	6,254	7,000	

a. Compute a measure of the variation in these maximum elevations.

b. Write a sentence describing the variation of these elevations.

The Interquartile Range: A Robust Variability Measure

The interquartile range is a more sophisticated way of describing the variation in data. Like the range, it is easily computed and easily interpreted. Unlike the range, it is robust against outliers. By *robust,* we mean that one or two outliers do not change the results very much.

To proceed, we must first define *quartiles.*

DEFINITION

Quartiles **are data values that divide an ordered data set into four quarters, in the same sense that the median is the data value that divides it in two. (The median is sometimes called the middle quartile.)**

For example, we will look again at the earthquake magnitude data and, essentially, divide it into four groups: a lower quarter, a low middle quarter, an upper middle quarter, and an upper quarter, as shown in Display 3.2.

DISPLAY 3.2

5.5	6.3	6.5	6.5	6.8	6.8	6.9	7.1	7.3	7.3	7.7	7.7	7.7	7.8	8.1
lower				low middle				upper middle				upper		

There are different computational procedures used to find the value of the quartile. Some computer programs use a slightly different procedure from the one we do. The procedure we use always gives an answer that equals one of the observed data values, or that is halfway between two observed data values. Quartiles computed in this way are sometimes referred to as the "hinges" of the data set.

PROCEDURE: FINDING THE QUARTILES OF A SINGLE GROUP OF NUMBERS

1. Arrange the data in order.
2. Find the median value.
3. Divide the data set into a lower half and an upper half. If the median equals an observed value (rather than being an average of two values), put it in both groups.
4. Find the median of the lower half. This is the *lower quartile.*
5. Find the median of the upper half. This is the *upper quartile.*

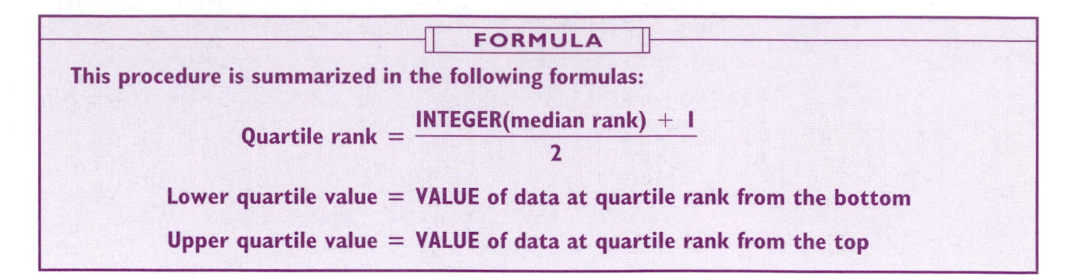

FORMULA

This procedure is summarized in the following formulas:

$$\textbf{Quartile rank} = \frac{\textbf{INTEGER(median rank)} + 1}{2}$$

Lower quartile value = VALUE of data at quartile rank from the bottom

Upper quartile value = VALUE of data at quartile rank from the top

The notation "INTEGER(median rank)" tells us to discard any fractional part (that is, to drop any .5 from the median rank) before proceeding.

For example, for 10 data values, we have:

$$\text{Median rank} = \frac{10 + 1}{2} = 5.5$$

and

$$\text{Quartile rank} = \frac{\text{INTEGER}(5.5) + 1}{2} = \frac{5 + 1}{2} = \frac{6}{2} = 3$$

Finding the Interquartile Range (IQR)

Once the quartiles have been found, the interquartile range (IQR) is defined simply as the difference (or the distance) between them.

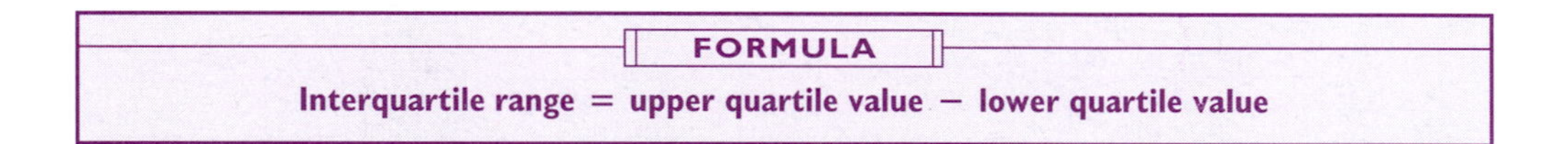

FORMULA

Interquartile range = upper quartile value − lower quartile value

The lower quartile is sometimes called the "first quartile," the upper quartile the "third quartile," and the median the "second quartile." Following this tradition, Minitab labels the lower and upper quartiles "Q1" and "Q3." If we agree that the bottom quarter of a group of values is "low," and the top quarter is "high," then the quartiles mark the boundaries between "low," "medium," and "high" values.

EXAMPLE 3.6 *Quartiles and Interquartile Range: Earthquake Magnitudes*

Suppose, on the one hand, we want to find out how small an earthquake can be and still be considered "major," and, on the other, how large it has to be before we consider it unusually large. The quartiles provide possible answers. Also, suppose we want to find the difference—the range—between the boundaries of "small" and "large" major earthquakes. This could be the interquartile range.

Recall the 15 earthquake magnitudes, which had a median rank of $(15 + 1)/2 = 8$. We find the quartile rank to be $(8 + 1)/2 = 4.5$. Thus, the quartiles are halfway between the data values ranked 4 and 5, counting in from either end of the ordered data. (See Table 3.5.)

Table 3.5

	DATA VALUES (IN ORDER)	RANK (FROM THE TOP)	RANK (FROM THE BOTTOM)
	8.1	1	15
	7.8	2	14
	7.7	3	13
	7.7	4	12
Upper quartile	→ ←	→ ←	
	7.7	5	11
	7.3	6	10
	7.3	7	9
	7.1	8	8
	6.9	9	7
	6.8	10	6
	6.8	11	5
Lower quartile	→ ←		→ ←
	6.5	12	4
	6.5	13	3
	6.3	14	2
	5.5	15	1

DISPLAY 3.3

5.5 6.3 6.5 6.5 6.8 6.8 6.9 7.1 7.3 7.3 7.7 7.7 7.7 7.8 8.1

| lower | low middle | upper middle | upper |

LQ = 6.65 Md = 7.1 UQ = 7.7

As we see in Display 3.3, the upper quartile value is 7.7, which is the average of the 7.7 ranked fourth from the top and the 7.7 ranked fifth from the top. So an earthquake of 7.7 or more would be unusually large, even for a "major" earthquake. The lower quartile value is 6.7, which is the (rounded) average of the 6.5 ranked fourth from the bottom and the 6.8 ranked fifth from the bottom. So an earthquake of 6.7 or less would be unusually small for a "major" earthquake.

Because the interquartile range is the difference between the quartiles, we have

$$\begin{aligned}\text{Interquartile range} &= \text{upper quartile value} - \text{lower quartile value}\\ &= 7.7 - 6.7\\ &= 1.0\end{aligned}$$

So there is a spread of 1.0 among Richter scale values of the medium "major" earthquakes.

This entire situation can be seen in Figure 3.2 (page 80), which displays the data, the median, the quartiles, and the interquartile range, arranged along a measured line. Notice how the median and quartiles divide the data set into four parts of approximately equal numbers of data points (about four points in each part) and see how the interquartile range measures the distance between the quartiles.

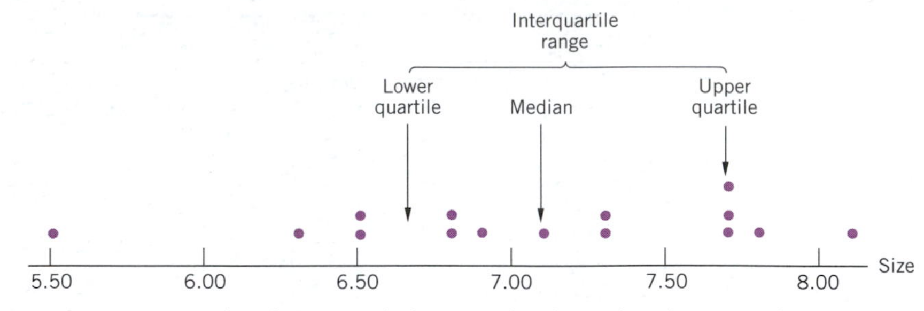

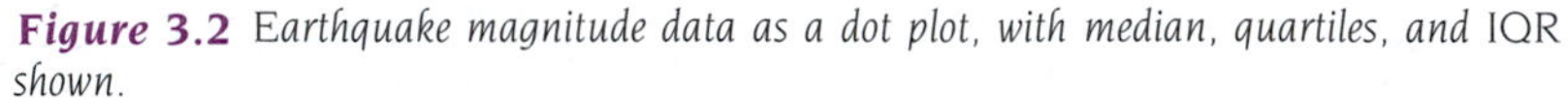

Figure 3.2 *Earthquake magnitude data as a dot plot, with median, quartiles, and IQR shown.*

EXAMPLE 3.7 *Interquartile Range with Skewed Data: Earthquake Deaths*

Suppose we want to find not only a typical number of deaths in a major earthquake, but also boundaries for unusually low, medium, and unusually large numbers of deaths. The quartiles and the interquartile range again provide possible answers.

Here are the reported numbers of deaths in major earthquakes, originally presented in Table 2.7. We have arranged them in order.

8	62	81	114	146	250	1,000	1,000
1,200	1,300	1,621	4,000	4,200	40,000	55,000	

This is clearly a skewed data set. Working out the median and quartile values (from the original data), we get

median rank = 8
median value = 1,000
quartile rank = 4.5
lower quartile value = (114 + 146)/2 = 130
upper quartile value = (1,621 + 4,000)/2 = 2,810
interquartile range = 2,810 − 130 = 2,680

Thus, 130 or fewer deaths is low, 2,810 or more deaths is high, and the medium numbers of deaths can differ by as much as 2,680.

Exercise Set 3.4

13. Consider these numbers:

12 14 16 18 20

Recall that you have already found their median rank (3) and median value (16).

a. What is the quartile rank?

b. What are the quartile values?

c. What is the interquartile range?

14. Consider these numbers:

12	14	16	18	20	22

Recall that their median rank is 3.5 and their median value is 17.

a. What is the quartile rank?

b. What are the quartile values?

c. What is the interquartile range?

15. Here are the EPA gas mileage figures for 1992 two-seat car models:

8	9	10	13	13	15	15	15	16	17
17	17	18	21	22	24	25	28	32	

Recall that their median rank is 10 and their median value is 17.

a. Find a number that marks the boundary between low gas mileage and medium gas mileage.

b. Find a number that marks the boundary between high gas mileage and medium gas mileage.

c. How many miles per gallon difference can there be between one medium-mileage car and another?

16. Here are the elevations (in feet) of the first 15 mountains mentioned in the Appalachian Mountain Club's *White Mountain Guide*:

6,288	5,532	5,715	5,798	5,363	2,404	5,585	5,470
3,450	5,385	5,004	4,761	4,310	4,052	3,919	

a. What is the boundary between a low-elevation mountain and a medium-elevation one?

b. What is the boundary between a high-elevation mountain and a medium-elevation one?

c. How much difference can there be in the elevation of two medium-elevation mountains?

17. Here are the high elevations (in feet) of the first 15 hikes listed in *Best Hikes with Children in Western Washington*:

6,000	5,699	5,200	6,250	5,200	4,550	4,300	4,800
900	5,537	5,392	5,475	6,200	6,254	7,000	

a. What is the boundary between a low-elevation hike and a medium-elevation one?

b. What is the boundary between a high-elevation hike and a medium-elevation one?

c. How much difference can there be in the elevation of two medium-elevation hikes?

18. Recall the ages of the students in the hypothetical statistics class in Exercise Set 2.3:

16	17	17	17	18	18	18	18	19	19	19	19	19	20	20	20	20
20	21	21	21	21	22	22	22	23	23	23	24	24	25	29	30	

Find the:

a. median

b. quartiles

c. interquartile range

d. range

19. Suppose you had a list of ages identical to the list in the previous exercise, except the oldest student is 65 rather than 30. How would this difference affect the median, the quartile values, and the interquartile range? How would it affect the range?

Discussion: Interpreting the Interquartile Range

A good way to think about the interquartile range is as the "width of the middle half of the data." The median and the quartiles divide the data set into four pieces, with about one quarter of the data values in each piece. From this, it is immediately clear that the interquartile range, as a measure of variability, is primarily dependent on the middle of the data set and is not likely to be overly dependent on extreme values and outliers. (Recall that the division of the data set into four pieces by the median and the quartiles was illustrated in Figure 3.2.)

The interquartile range must always be either a positive number or zero because the upper quartile must always be at least as large as the lower quartile. If you find a negative value for the interquartile range, it is wrong and you should recompute it. A small value of the interquartile range indicates only a small amount of variability, whereas larger values indicate more variability in the data set.

The interquartile range can be zero even when there is some variability in the data. This happens when all the data values in the middle of the distribution are the same. In such a case, we might say that most of the data set shows no variability, although some values (outliers, perhaps, or a small cluster of data values) are different.

3.5 Grouped Data

Sometimes we want to calculate medians and quartiles from data that have already been organized into groups. For example, perhaps all we have is a report that includes a summary table but not the raw data. Or we may have already organized our own data into a histogram and do not want to go back and do hand calculation with the raw data. The following example will show the correspondence between a list of numbers and its representation as grouped data.

EXAMPLE 3.8 *Grouped Data: Productivity Measures*

Suppose a group of measurements of productivity in 24 different manufacturing plants was as follows:

20	50	30	30	30	70	30	50	30	40	30	30
40	30	30	50	20	30	20	50	40	50	30	20

Because only a few different values are shown here, this data set may be represented more economically. Another way to record these data values is to count the number of times the value 20 occurred, the number of times 30 occurred, and so on, as shown in Table 3.6.

There are two items to notice here. First, although the data are presented very differently in the two displays (the first is simply a list of data values, whereas the second has two different columns of numbers), they, in fact, represent exactly the same data. Although data sets with two different columns often do represent more complicated data structures (which we will study later on), in this case they still represent a single group of numbers, but in a different way.

The second item of note is that the grouped-data display is shorter and simpler: 24 numbers have been summarized by two columns of five numbers each. Moreover, if we double the size of the sample by including data on an additional 24 plants, the grouped-data representation will still be quite compact. The counts in the right-hand column will increase, but the size of the

Table 3.6 Productivity measures, recorded as grouped data

PRODUCTIVITY MEASURE	NUMBER OF TIMES IT OCCURRED
20	4
30	11
40	3
50	5
70	1

display will stay about the same (and will increase only if a new productivity value, such as 60, occurs with the additional data, forcing inclusion of a new row in the display).

If this grouped-data format reminds you of a histogram, there is a good reason for it. The column of counts on the right corresponds to the height that a histogram would have at the corresponding productivity value (see Figure 3.3).

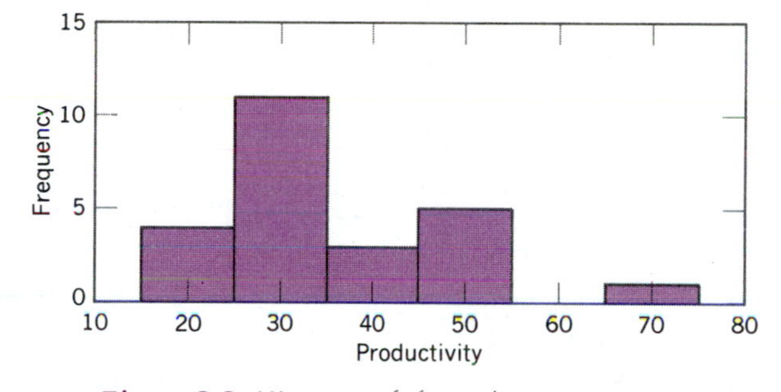

Figure 3.3 *Histogram of the productivity measures*

We will work directly with the grouped format and add a few additional columns that will help us compute the median and the quartiles. (See Table 3.7.) The new columns will include *cumulative* running totals of the number of times the productivity values have occurred, both from the top and from the bottom.

Table 3.7

X PRODUCTIVITY VALUE	N NUMBER OF TIMES	CUMULATIVE N UP	CUMULATIVE N DOWN
20	4	4	24
30	11	4 + 11 = 15	20
40	3	18	6 + 3 = 9
50	5	23	1 + 5 = 6
70	1	24	1
	Total 24 (Total number of data values)		

Finding the Median and Quartiles

To find the median and quartiles, note that the "cumulative N" column in Table 3.7 corresponds to the ranks of the data. For example, by looking at the "cumulative up" column, we see an entry of 4 for the first data value, 20. This says that the productivity value of 20 has ranks 1, 2, 3, and 4 from the smallest value. The next cumulative N value, 15, for the next data value, 30, says that the productivity value of 30 has ranks 5 (one more than 4, the number above it) through 15.

The rank of the median is $(1 + 24)/2 = 12.5$, which tells us to average the data values at ranks 12 and 13. Since the data value 30 has ranks 5 through 15 (which includes both ranks 12 and 13), the median is 30 averaged with itself, and we have found the median:

$$\text{Median} = 30$$

The quartile rank is

$$\frac{1 + 12}{2} = 6.5$$

so we will average the data values of ranks 6 and 7. For the lower quartile, we will use ranks from the smallest, for which we know that the data value of 30 represents ranks 5 through 15, which includes both 6 and 7. Thus, the lower quartile averages 30 with itself, so

$$\text{Lower quartile} = 30$$

This result is slightly strange: The median and the lower quartile are the same number! Although this is surprising, it is not impossible and is, in fact, neither an error nor a problem. This can happen whenever the same number is repeated often enough in the data set that it is located both in the middle and a quarter of the way in (in terms of ranks) from the smallest value.

For the upper quartile, we will average the data values with ranks 6 and 7, but using ranks from the largest value this time. From the second column of "cumulative N," we see that the data value of 50 represents ranks 2 (one more than the 1 below it) through 6 and that the data value of 40 represents ranks 7 through 9. Thus, the upper quartile will be found by averaging 50 (the data value of rank 6 from the largest) with 40 (the data value of rank 7 from the largest). The upper quartile is therefore

$$\text{Upper quartile} = \frac{50 + 40}{2} = 45$$

Finally, the interquartile range, our robust measure of variability, can be computed in the usual way from the two quartiles:

$$\text{Interquartile range} = 45 - 30 = 15$$

Exercise Set 3.5

20. Suppose that subjects in a taste-preference study were asked to rate a new food on a 7-point scale, where 1 = awful, 4 = okay, and 7 = great. The resulting data are shown in Table 3.8.

a. Find the median rating and the quartile ratings.

b. What is the interquartile range?

Table 3.8

RATING		NUMBER
Great	7	3
	6	5
	5	10
Okay	4	8
	3	2
	2	2
Awful	1	

Computer Analysis of Median, Extremes, and Quartiles

The extreme values, quartile values, and median value are easily found by computer. Display 3.4, which was produced by the Minitab command DESCRIBE, includes these values.

DISPLAY 3.4

```
MTB > Retrieve 'equakes.mtw'.
MTB > Describe 'size'.

              N       MEAN      MEDIAN     TRMEAN     STDEV     SEMEAN
Size         15      7.067       7.100      7.108      .696       .180

            MIN        MAX         Q1         Q3
Size      5.500      8.100       6.500      7.700
```

The MEDIAN, which is the third data column, is shown as 7.100. The extreme values are given as the MIN and MAX, with values of 5.500 and 8.100, respectively. The lower quartile is given as Q1 with a value of 6.500, and the upper quartile as Q3 with a value of 7.700.

Notice that, except for the lower quartile value, these are the same values as those we found by hand. The lower quartile value is different because it falls between two observed values. When we found the lower quartile by hand, we averaged the two values around it. The computer program uses a slightly different procedure. It does a calculation based on the total number of data values and then calculates a quartile value that may fall one quarter, one half, or three quarters of the way between the neighboring observed values. It may even, as in this case, declare one of those neighboring values to be the quartile. Such differences between the procedures are usually noticeable only with small data sets. When there are differences, the computer program produces a larger interquartile range.

Exercise Set 3.6

21. Display 3.5 (page 86) shows output for the Minitab "describe" procedure for the number of deaths in the major earthquake data set. According to the display:

a. What is the fewest number of deaths?

b. What is the lower quartile value?

c. What is the median value?

d. What is the upper quartile value?

DISPLAY 3.5

```
MTB > Describe 'deaths'.

             N        MEAN      MEDIAN     TRMEAN     STDEV     SEMEAN
Size         15       7332      1000       4229       16605     4287

             MIN      MAX       Q1         Q3
Size         8        55,000    114        4000
```

e. What is the highest value reported?

f. What is the interquartile range?

22. We entered the data from Exercise Set 3.5 into MINITAB and ran Describe. Here are the results:

```
MTB > Describe C1.

           N         MEAN       MEDIAN     TRMEAN     STDEV      SEMEAN
C1         30        4.767      5.000      4.808      1.305      0.238

           MIN       MAX        Q1         Q3
C1         2.000     7.000      4.000      6.000
```

a. What are the median and quartile values?

b. What is the interquartile range?

c. Do these results agree with the ones you got when you analyzed the data by hand?

3.6 Box Plots and the Five-Number Summary

A more complete description of the distribution of a single group of numbers is provided by the **five-number summary**, which consists of five "landmark" numbers in the distribution: the median, the two quartiles, and the two extremes (these are the smallest and the largest numbers). These numbers provide a condensed description of even a very large data set, while preserving some of the most important information. We will adjoin the size of the data set (the number of data values) and report these five summary values in order from smallest to largest in the following format:

Five-number summary (plus one)

size of data set (lower extreme, lower quartile, median, upper quartile, upper extreme)

EXAMPLE 3.9 *Five-number Summary: Earthquake Magnitudes*

Let's reconsider the data on earthquake magnitudes. There are 15 data values, and, from our earlier calculations, the median is 7.1, the quartiles are 6.7 and 7.7, and the extremes are 5.5 and 8.1. The five-number summary of this data set is

15 (5.5, 6.7, 7.1, 7.7, 8.1)

Exercise Set 3.7

23. Suppose that we say a certain data set can be described by this five-number summary:

132 (10, 50, 75, 100, 900)

True or false:

a. The authors of this book cannot tell the difference between five numbers and six numbers.

b. There were 132 values in the data set.

c. The smallest value was 10 and the largest 900.

d. The median value was 75.

e. About half the values were from 50 to 100.

24. Regarding the gas mileage data set, we have already calculated the median and quartile values, and we have identified the extreme values (Exercise 15 in Exercise Set 3.4). Show these five numbers, along with the size of the data set, in a five-number summary.

25. What is the five-number summary for the earthquake deaths data set (Example 3.7)?

26. Suppose you have this five-number summary: 33 (16, 19, 21, 22, 30)

a. How many numbers are in the original data set?

b. What is the median of these numbers?

c. What is their interquartile range?

d. What is their range?

27. Suppose you have this five-number summary: 33 (16, 19, 21, 22, 65)

a. How many numbers are in the original data set?

b. What is the median of these numbers?

c. What is their interquartile range?

d. What is their range?

Discussion: Importance of the Five-number Summary

The five-number summary is useful because it summarizes, in a few numbers, enough information to approximately describe a distribution. First, the median summarizes the center of the data. Second, the quartiles give an indication of variability. (In fact, the interquartile range is easily computed as the difference of the two quartiles.) Third, the location of the median with respect to the quartiles gives us an indication of symmetry (if it is about the same distance from each of them) or skewness (if the median is much closer to one quartile than to the other). Finally, the extremes give bounds for the data set because all data values must lie between them. Perhaps more important than that, if there are any extreme outliers, they wil—l show up in the extremes. Therefore, the extremes would bring to our attention the presence of strange, unlikely, or impossible data values.

Box Plots

Plotting data with pictures is one of the best ways to use the abilities of our eyes and minds to detect, recognize, and classify structure in data. For a single group of numbers, we would like to display information about center, variability, symmetry, and outliers, but without the distractions that are present in a plot of all the individual data values. After all, statistics concerns itself primarily with general facts about a data set, and the finer details can often distract from this more general goal.

Also, we would like the plot to reflect the data itself and not be influenced by any preconceived ideas or "models" about it. In particular, the display should work well whether the data distribution is normal or not and whether outliers are present or not. Two plots based on the five-number summary take this approach: the **quick box plot,** which displays four equal-sized groups of the data, and the **full box plot,** which identifies outliers.

The Quick Box Plot

The quick box plot is a simple display of the five-number summary, drawn such that it is easy to see these important features of the data set.

PROCEDURE: CONSTRUCTING A QUICK BOX PLOT

1. Draw a scale below where the plot will go. We will use a horizontal scale, but a vertical scale can also be used.
2. Draw a rectangular box with ends at the quartiles (the height is not important).
3. Draw a vertical line through the box at the median value.
4. Draw two lines ("whiskers") connecting the ends of the central box with the corresponding extreme value.

A quick box plot looks like the one in Figure 3.4 (except the five values would not ordinarily be labeled as they have been).

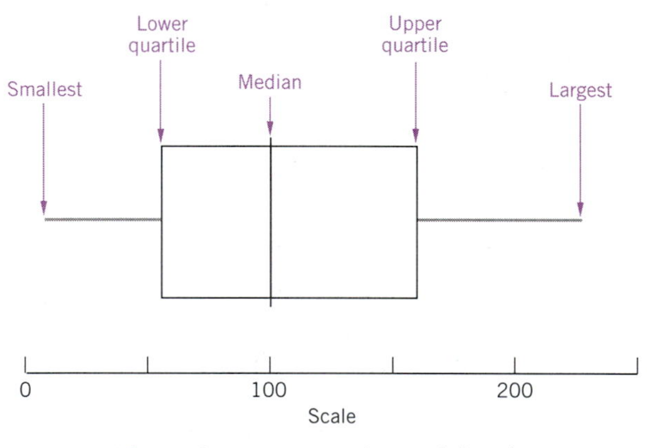

Figure 3.4 *How to read a quick box plot*

Note that the quick box plot indicates where the data set would divide into four groups of approximately equal sizes, as shown in Figure 3.5. The box itself highlights the middle half of the data, appropriately bringing it to our attention and emphasizing the central part of the data set. The extensions to the extreme values are noticeable, but are downplayed somewhat. This is where the tail of the distribution is.

Recognizing Symmetry and Skewness. A quick box plot for a perfectly symmetric data set is illustrated in Figure 3.6. The median line is exactly in the middle of the box formed by the quartiles, and the extremes are at equal distances from the corresponding quartiles. The plot is an exact mirror image of itself, as if a mirror were placed at the median. Of course, this perfection is unrealistic. Real data sets will have a random aspect to them so we could expect to see only approximate symmetry. The median location within the box (without regard to extreme values) is somewhat more reliable as an indication of symmetry because each extreme value is heavily

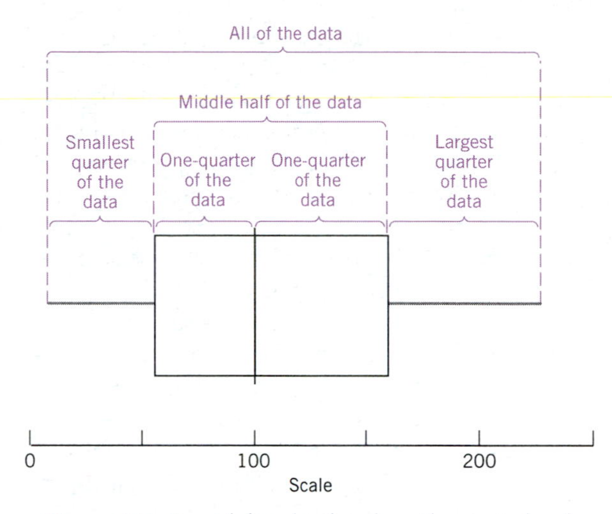

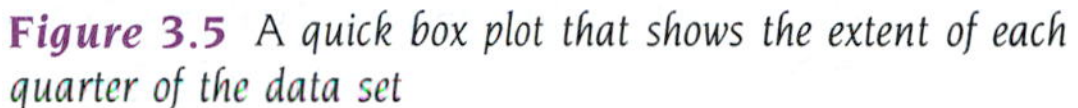

Figure 3.5 *A quick box plot that shows the extent of each quarter of the data set*

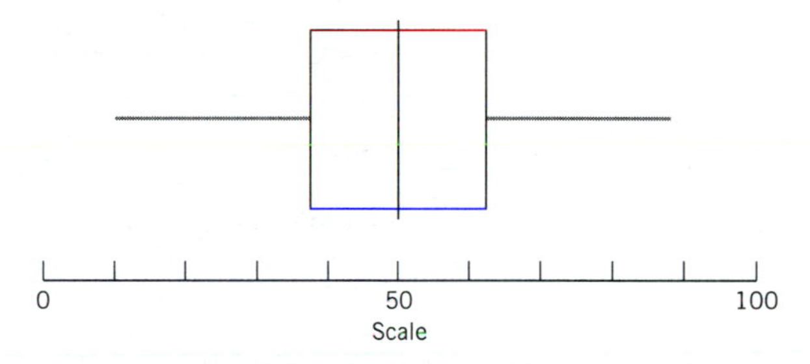

Figure 3.6 *Quick box plot for a perfectly symmetric data set*

dependent on one single extreme data value, which may or may not be representative of the data set as a whole.

Figure 3.7 indicates two quick box plots for skewed data sets. One is skewed toward low values, and one is skewed toward high values. Real data examples can be expected to show more variability than these, especially in the location of the extremes.

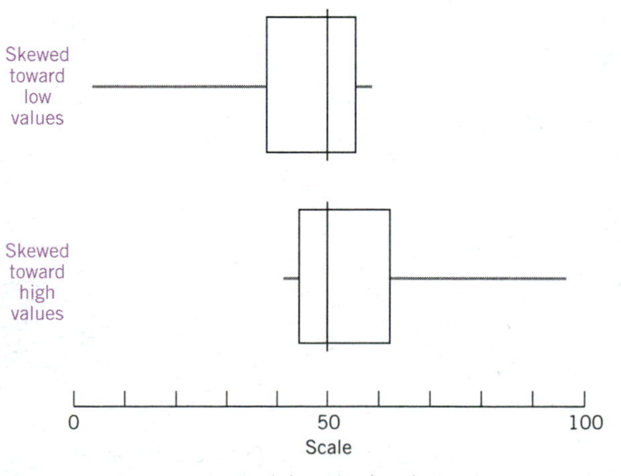

Figure 3.7 *Quick box plot for skewed data*

For the earthquake magnitude data, the five-number summary is 15 (5.5, 6.7, 7.1, 7.7, 8.1). The quick box plot based on this is in Figure 3.8. Looking at this plot, we can determine that the center of the data set is close to 7.0. We know this because the vertical line in the middle of the box indicates the median value, and it is between 7.0 and 7.5, but closer to 7.0. We can also determine that the upper quartile value, which separates the top quarter of the data from the rest, is about halfway between 7.5 and 8.0, say at 7.7. We know this because that is where the right edge of the box is. Similarly, we know that the lower quartile value, which separates the bottom quarter of the data from the rest, is between 6.5 and 7.0, say at 6.6. Combining these two observations, we now know that half of the data fall in the interval from 6.6 to 7.7. The smallest value in this data set is 5.5. We know this because the left whisker extends down to 5.5. The largest value in the data set is a little over 8.0—about 8.1—and we know this because that is how far up the right whisker extends. This approximately reconstructs the five-number summary. In a sense, then, the two presentations are equivalent, but we gain some extra information from the box plot. For example, notice that the whiskers are not the same length. This suggests that the data may be skewed. Although not extreme, this slight skew is easy to see, yet was not at all clear from the numbers in the five-number summary.

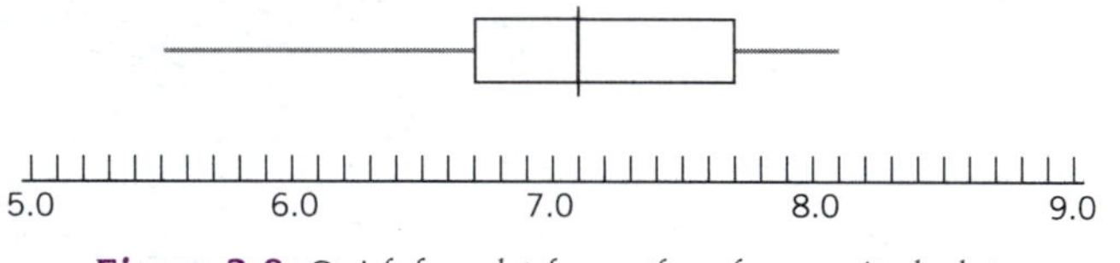

Figure 3.8 *Quick box plot for earthquake magnitude data*

The data on the heights (in feet) of the first 15 mountains in the *AMC Guide* provide another example of how skewed distributions are shown in box plots. First, look at the histogram of the data shown in Figure 3.9. The histogram shows a distribution that seems to center at around 5,500 feet, with a clear skew toward low values. Lets us see how this skew is reflected in the box plot.

The five-number summary for this data set is 15 (2,404, 4,181, 5,363, 5,558, 6,288), and the box plot based on the five-number summary is in Figure 3.10. Notice that the median is not

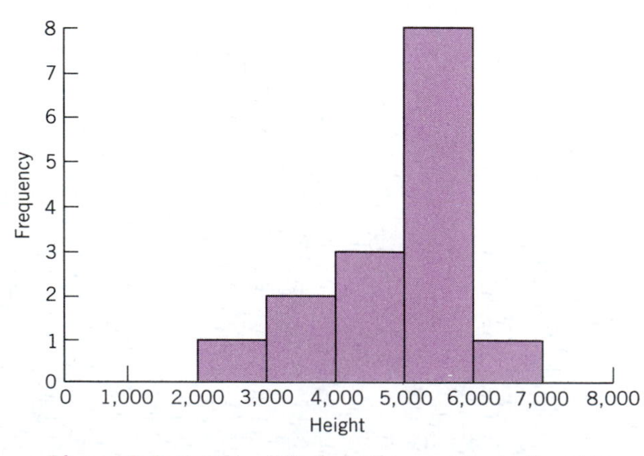

Figure 3.9 *Heights of the first 15 mountains in the* AMC Guide

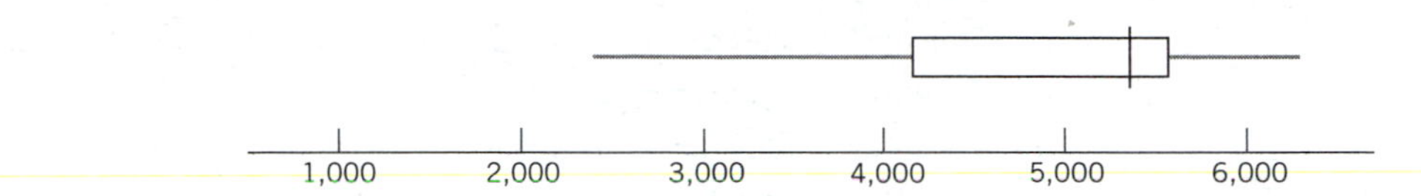

Figure 3.10 *Box plot based on the five-number summary of mountain heights*

centered in the box. Because the median is closer to the upper quartile, we know that data values are more densely concentrated here. They are sparser and more spread out between the median and the lower quartile, which is the direction of the skew. This tendency to trail off more slowly toward the low values is confirmed in this case by the fact that the line to the lower extreme value is much longer than the line to the upper extreme value.

The all-terrain ski prices provide another example of a box plot for a skewed data set, but this time the skew is toward high values. The five-number summary is 22 (400, 450, 475, 495, 825), and a quick box plot is shown in Figure 3.11. The whisker from the upper quartile to the highest value is very long compared to the lower whisker, indicating that the top quarter of the values is more spread out.

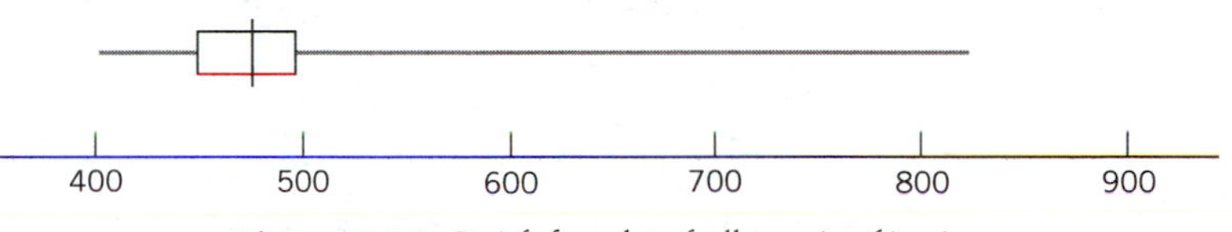

Figure 3.11 *Quick box plot of all-terrain ski prices*

Exercise Set 3.8

28. Using the box plot shown in Figure 3.12, determine (approximately):

- **a.** the lowest value
- **b.** the lower quartile value
- **c.** the median value
- **d.** the upper quartile value
- **e.** the highest value
- **f.** the interquartile range
- **g.** whether this distribution is symmetric (yes or no)?

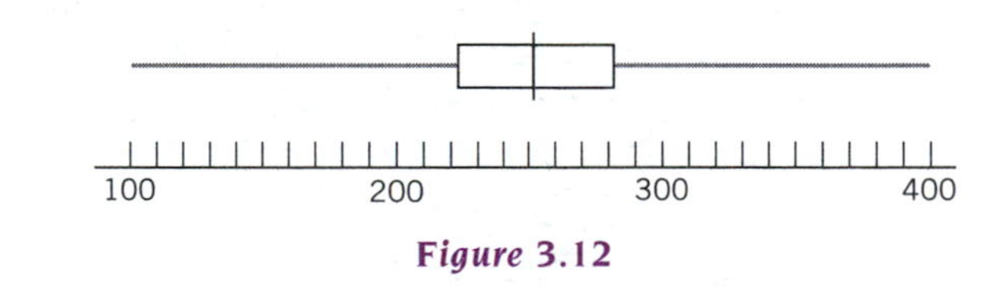

Figure 3.12

29. Construct a box plot of each of these sets of data:

- **a.** a rectangular distribution: (10, 20, 30, 40, 50)
- **b.** a skewed distribution: (10, 15, 20, 35, 70)

30. Here is the five-number summary for the earthquake deaths data set: 15 (8, 130, 1,000, 2,810, 55,000). Present this as a box plot.

31. Using the three hypothetical box plots shown in Figure 3.13, answer the following questions.

a. Which shows a symmetric distribution?

b. Which shows a distribution skewed to the right?

c. Which shows a distribution skewed to the left?

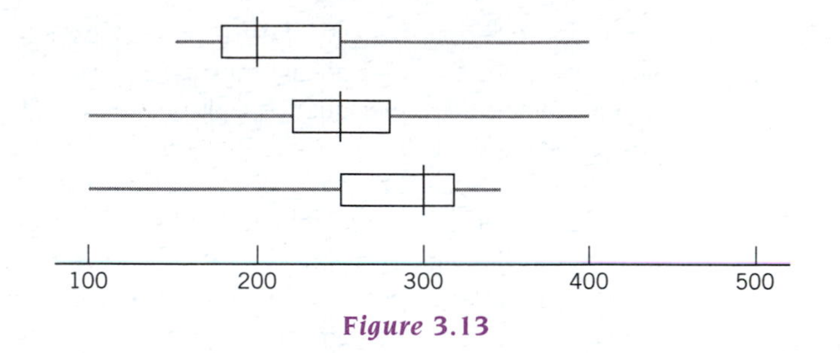

Figure 3.13

32. Construct a box plot of the ages of the students in the hypothetical statistics class:

16 17 17 17 18 18 18 18 19 19 19 19 19 20 20 20 20
20 21 21 21 21 22 22 22 23 23 23 24 24 25 29 30

33. Construct a box plot of the ages of the students in a hypothetical statistics class, where the ages are the same as before except for the oldest:

16 17 17 17 18 18 18 18 19 19 19 19 19 20 20 20 20
20 21 21 21 21 22 22 22 23 23 23 24 24 25 29 65

The Full Box Plot

The full box plot is very similar to the quick box plot, except that it distinguishes extreme outliers in the data. This requires some additional calculation beyond finding the five-number summary of the data, but the extra work often provides important information about the data.

Outliers. The first problem to address is to identify precisely what an outlier is. We know that it is a data value that is well separated from most of the rest of the data, but different people may judge "well separated" differently. We need a rule to decide exactly how extreme a data value must be before it is declared an outlier.

Tukey, the founder of exploratory data analysis, suggested the rule of thumb that we will use.[4] It relies on the interquartile range as a measure of the basic variability in the data and sets a threshold as follows.

PROCEDURE: OUTLIER CRITERION

1. Define one *step* as 1.5 times the interquartile range.
2. Define the *upper outlier threshold* to be the *upper* quartile *plus* one step. The *lower outlier threshold* is defined correspondingly as the *lower* quartile *minus* one step.
3. Any data value that is outside the outlier thresholds (either larger than the upper outlier threshold or smaller than the lower outlier threshold) will be declared an *outlier.*

FORMULA

To express this procedure using a mathematical formula, we would consider a data value X to be an outlier if either

$$X > \text{upper quartile} + \text{step}$$

or

$$X < \text{lower quartile} - \text{step}$$

where

$$\text{step} = 1.5 \times (\text{interquartile range})$$

The step and the outlier criterion are shown in Figure 3.14. Note that there need not be any outliers in a given data set. Only if one or more extreme data values exceeds these criteria will there be any outliers.

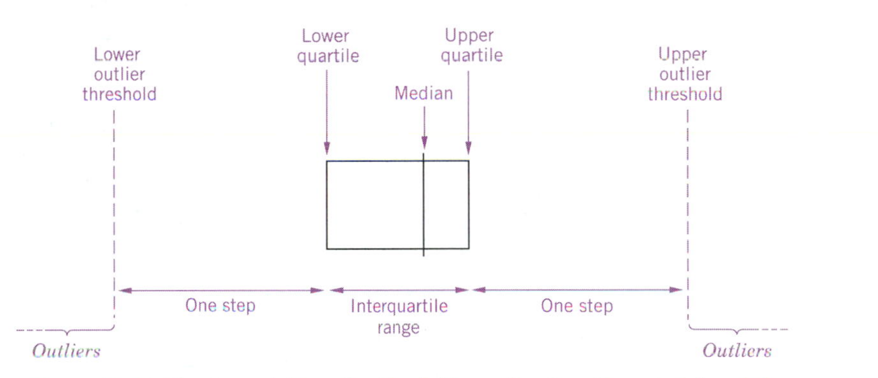

Figure 3.14 *The step and the outlier thresholds are based on the central box of the quick box plot.*

Adjacent Values. How far out do we draw the whiskers? We draw them out to the last values that are not outliers, which are called the **adjacent values.** We do *not* draw the whiskers all the way to the thresholds. This helps the basic part of the box reflect the majority of the data, while outliers will be shown separately but in the same display.

PROCEDURE: CONSTRUCTING A FULL BOX PLOT

1. Draw a scale below where the plot will go.
2. Draw a rectangular box with sides at the quartiles.
3. Draw a line through the box at the median value.
4. Draw a line connecting each end of the central box to the corresponding adjacent value. Draw a small crossbar there.
5. Label the adjacent values (this is optional, but recommended). The label should be a brief informative description of the data value that is represented there.
6. Place a mark at each outlier value. The outliers may also be labeled.

EXAMPLE 3.10 *Constructing a Full Box Plot: All-terrain Skis*

For the all-terrain ski data, the construction of a full box plot proceeds as follows. Preliminary calculations are based on the five-number summary—that is, 22 (400, 450, 475, 495, 825). The interquartile range is therefore

$$\text{Interquartile range} = 495 - 450 = 45$$

and the step is

$$\text{Step} = 1.5 \times 45 = 67.5$$

The outlier thresholds are now easily computed:

$$\text{Lower outlier threshold} = 450 - 67.5 = 382.5$$

$$\text{Upper outlier threshold} = 495 + 67.5 = 562.5$$

From the five-number summary, we know that there are no low outliers: The lowest value (400) is well above 382. But the highest value—825—is above the upper threshold of 562. To identify this outlier, and any other high outliers, we need to go to the original data values (given in Table 3.4). We find that there is only the one high outlier:

La Croix Reference, $825

There are two adjacent values, the most extreme data values that are not outliers. Because there are no lower outliers, the lower adjacent value is simply the smallest value (400) for this data set. The upper adjacent value is 550, which is the largest data value that is not an outlier. The adjacent values are therefore:

Lower adjacent value: *RD Locomotion,* $400

Upper adjacent value: *Elan MBX 16,* $550

We are now ready to draw the box plot. If we follow the procedure given earlier, the only remaining question is how to do the labels. For this data set, it is natural to use the brand name of the ski. Therefore, our completed box plot looks like the one in Figure 3.15.

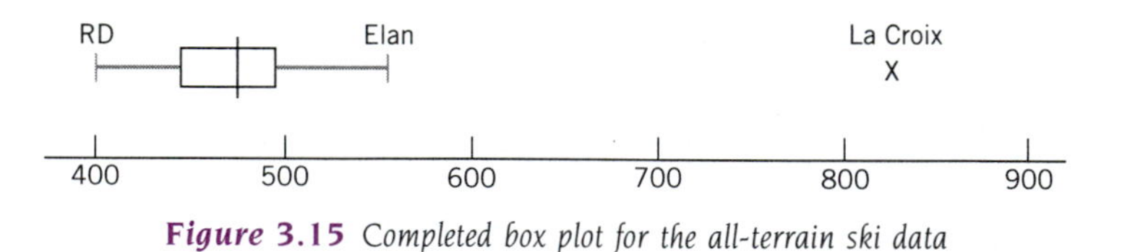

Figure 3.15 *Completed box plot for the all-terrain ski data*

Comparing this to the quick box plot (Figure 3.11), we see that additional information has been gained, especially in the upper tail of the distribution. Whereas the quick box plot merely showed us that the largest value was something around $825, the full box plot shows us that there are no other ski prices between it and the adjacent value of $550. The existence of this gap between the most expensive skis could not have been noticed from the quick box plot and provides useful information about the distribution.

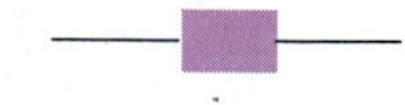

Discussion

There is something arbitrary about any rule that decides which data values are outliers and which are not. But a definite decision has to be made even in ambiguous and uncertain situations. Our choice of outlier thresholds depended on the choice of the constant 1.5, used to determine the size of the step in relation to the interquartile range. A larger constant would result in fewer outliers being identified, possibly causing us to miss some important ones. A smaller constant would result in more data values being classified as outliers, possibly requiring attention to details that are not important.

The choice of 1.5 has been confirmed as a useful one by the experience of people who analyze data routinely. One consequence of this choice is that in a large group of data values, if the data are normally distributed, we expect to see a bit less than 1% (in fact, .7% on the average) of the data values classified as outliers. In smaller samples from a normally distributed population the fraction of outliers is larger. For samples of only five numbers from a normal distribution, about 8% of the data will be classified as outliers (on the average). This technical issue is considered in more detail by Hoaglin, Mosteller, and Tukey (1983).[5] Naturally, if the data are not normally distributed (which is very often the case), we might see a larger number of outliers, especially when the distribution is skewed or long-tailed.

Exercise Set 3.9

34. Construct a full box plot of the ages of the students in the hypothetical statistics class. Be sure to indicate that you have checked for outliers.

16 17 17 17 18 18 18 18 19 19 19 19 19 20 20 20 20
20 21 21 21 21 22 22 22 23 23 23 24 24 25 29 30

35. Construct a box plot of the ages of the students in a hypothetical statistics class, where the ages are the same as before except for the oldest. Be sure to indicate that you have checked for outliers.

16 17 17 17 18 18 18 18 19 19 19 19 19 20 20 20 20
20 21 21 21 21 22 22 22 23 23 23 24 24 25 29 65

3.7 Multiple Box Plots and Computer Analysis

As we said before, the great virtue of box plots is that they let us use our eyes to perceive structure. They really come into their own when used to present several groups of data simultaneously.

Table 3.9 (page 96) lists EPA estimated highway mileage for the different base models of several car makers. The table also includes a code number for the car make; this is arbitrary.

Using Minitab to draw box plots of the data, we get the results shown in Figure 3.16. From these box plots, several things are immediately obvious. For instance, it is clear that on the whole the cars of maker number 6—which is BMW—get the worst mileage. Indeed, the median value for BMW is lower than any value for any other car maker. It is also clear that the best mileage is obtained by a model of car maker number 15, which is Ford. Many more detailed comparisons can also be made. For instance, makers 7 and 9—Buick and Chevrolet—have median mileages that appear to be the same, but the Chevrolet mileages are more variable.

Computer box plots may be slightly different from hand-drawn box plots. There are usually three differences: choices of quartiles, labeling, and accuracy of plotting.

Table 3.9 EPA *estimated highway mileage for selected automobile makers*

CODE	MAKE	MILES PER GALLON							
1	Acura	28	31	24	26				
6	BMW	27	23	21	18				
7	Buick	31	28	27	30	27	25	31	
9	Chevrolet	34	27	26	34	28	25	28	
12	Dodge	36	32	26	26	34	32	24	
15	Ford	36	42	25	29	31	29	33	27

Notice that the Minitab boxes seem a bit more spread out than the ones we drew by hand. This is because the program uses a slightly different procedure for finding quartile values. If the quartile value falls between observed values, the program puts the quartile value between them, but it may be one-fourth or three-fourths of the way out rather than in the middle.

For instance, the BMW mileage figures are 18, 21, 23, and 27. Using our procedures to find quartiles, we get a lower quartile of 19.5 and an upper quartile of 25. The computer program, on the other hand, gets a lower quartile value of 18.75 and an upper quartile value of 26. Neither is right or wrong—they are just different ways of splitting the difference when the quartile value is between two observed values. Differences between the procedures are usually apparent only with very small data sets.

The labeling of the computer output is not all that informative. We could label the image ourselves—and if we were preparing this for a presentation about gas mileage, we would—but for now we want to show you what to expect from a computer program.

Finally, the accuracy of the *plotting* of computer output is limited by the screen or printer it is being sent to. Many printers are limited, essentially, to the accuracy you can get with a typewriter. The take-home lesson is this: Unless you know the accuracy of the printer, do not make life-and-death decisions based on *small* features of computer printouts.

In spite of these differences, however, the computer is certainly faster, making it reasonable to do a lot of exploratory analysis quickly.

Exercise Set 3.10

36. Refer to Figure 3.16, showing 1992 gas mileage ratings by car make.

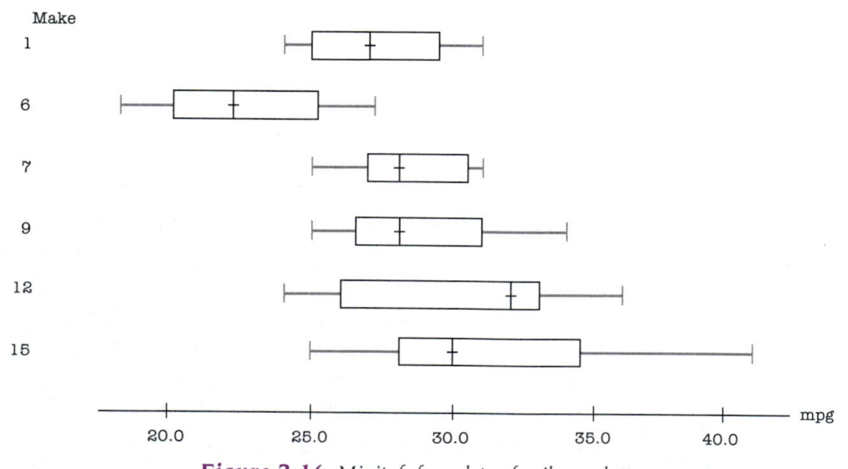

Figure 3.16 *Minitab box plots of mileage data*

a. Is there more variability among Fords (15) or Dodges (12)? What do you base your answer on?

b. How does the worst Buick mileage (7) compare with the median BMW mileage (6)?

c. Is the mileage distribution for Fords (15) perfectly symmetric or at least a little skewed? If skewed, is the skew toward high values or low values?

3.8 Percentiles

For much of this chapter we have been concerned with locating values that let us give a summary description of the overall distribution. Thus we have asked which value falls in the middle, which value cuts off the lowest 25%, and which value cuts off the top 25%. Now we will discuss percentiles, which give a way to indicate where *any* given value falls in the distribution.

When we say that a given value is the 75th percentile, we mean that 75% of the distribution is lower than or equal to this value. Thus the median value is the 50th percentile, because 50% of the values in the distribution are smaller (or equal), the lower quartile is the 25th percentile because 25% of the values are smaller (or equal), and the upper quartile value is the 75th percentile because 75% of the values in the distribution are smaller (or equal).

Exercise Set 3.11

37. Suppose that the median on a certain test is 100. What is the 50th percentile?

38. Suppose that 100 students take a test, and 80% of them score 73 points or lower. What is the 80th percentile?

39. Suppose that 130 is the 98th percentile on a test. What percent of the people taking this test score *over* 100?

Summary

Summarization is the process of going from extensive details to simple, useful generalizations about a situation. We are primarily concerned with summarizing the shape, the center, and the variability of a distribution. Summarizing the shape was discussed in the previous chapter. In this chapter, numerical summaries and plots based on the *five-number summary* (the median, quartiles, and extremes) were considered as representatives of a data set.

There is no unique measure of center that is best for all situations, especially if the data are not symmetric, but the arithmetic average (considered in detail in the next chapter), the mode, and the median can be useful. The *mode* is useful for categorical data. The *median* value splits the distribution in half, so that half the values are above it and half are below it. It can be used with ordered categorical as well as measured numerical data, and it is not sensitive to skewness or outliers.

Similarly, there are several measures of variability. The standard deviation will be considered in the next chapter. The *range* is the width of the distribution from smallest to largest. The *interquartile range* is the width of the middle half of the data—that is, the distance between the *quartiles* (which are the data values about one quarter of the way in from each end of the distribution). Because the range is very sensitive to just two data values (the extreme smallest and largest, which may be outliers) the interquartile range is often a better choice of summary.

We have seen two simple plots that summarize basic information about a single group of numbers. The *quick box plot* displays the five-number summary only (median, quartiles, and extremes). The *full box plot* displays individual outliers, in addition to the five-number summary. Outliers are determined using upper and lower outlier thresholds, which are 1.5 times the interquartile range beyond the quartiles. Outliers represent data values, if any, that stray far enough from the middle part of the data to warrant further attention.

The quick box plot is very easy to draw, whereas the full box plot takes some effort (but is worth the extra time if it is available). These plots are designed to help the observer quickly assess the center of the data, the amount of variability, skewness, and (for the full box plot) the existence and location of outliers. Both of these plots are robust in the sense that they provide a realistic display even if the data set does not follow a normal distribution due to skewness or the presence of outliers.

Box plots are especially useful when looking at and comparing several data sets simultaneously, with one box plot per data set (all drawn on the same numerical scale). The quartiles are examples of *percentiles*, which are data values that represent a known percentage of the data values. For example, the lower quartile corresponds to the 25th percentile, the median to the 50th percentile, and the upper quartile to the 75th percentile.

Problems

Basic Reviews

1. Consider this data set: 12, 23, 34

a. What is the median *rank*?

b. What is the median *value*?

2. Consider this data set: 12, 35, 23, 27, 50, 21, 6. (Notice that the numbers have not yet been arranged in order.)

a. What is the median rank?

b. What is the median value?

c. What is the quartile rank?

d. What are the quartile values?

3. Consider this data set: 45, 37, 28, 32, 42, 22, 35, 36.

a. What is the median rank?

b. What is the median value?

c. What is the quartile rank?

d. What are the quartile values?

e. What are the extreme values?

f. Construct a quick box plot of the data.

g. What is the interquartile range of these data?

h. What is the value of the step used to calculate outlier thresholds?

i. What are the outlier thresholds?

j. Are there any outliers?

k. Create a full box plot of the data, identifying any outliers.

4. Here are three sets of ten random numbers each, drawn from a normal distribution.

Set 1:	120	85	79	75	109	96	97	126	84	80
Set 2:	96	112	122	82	116	115	80	103	98	86
Set 3:	120	101	153	111	92	113	99	91	114	90

a. Prepare a five-number summary of each.

b. Draw a box plot of each, displaying the three box plots over the same scale, as we did in Figure 3.16.

Analysis

5. There are 50 measured petal lengths in the *I. setosa* data given in Display 3.6. Notice that we can count entries in the rows to find the quartile values and the median.

DISPLAY 3.6

STEM-AND-LEAF PLOT OF PETAL LENGTH FOR ONE SPECIES OF IRIS, *I. SETOSA*. SAMPLE SIZE, n = 50. LEAF UNIT = .10 CENTIMETER.

Stem	Leaves	
10	0	10 | 0 = 10.0 centimeters
11	0	
12	0 0	
13	0 0 0 0 0 0 0	
14	0 0 0 0 0 0 0 0 0 0 0 0 0	
15	0 0 0 0 0 0 0 0 0 0 0 0 0	
16	0 0 0 0 0 0 0	
17	0 0 0 0	
18		
19	0 0	19 | 0 = 19.0 centimeters

a. Construct a quick box plot of the data in Display 3.6.

b. Construct a full box plot of the data in Display 3.6.

6. There are 50 measured petal lengths in the *I. versicolor* data given in Display 3.7 (page 100).

a. Construct a quick box plot of the data in Display 3.7.

b. Construct a full box plot of the data in Display 3.7.

7. There are 50 measured petal lengths in the *I. virginica* data given in Display 3.8 (page 100).

a. Construct a quick box plot of the data in Display 3.8.

b. Construct a full box plot of the data in Display 3.8.

8. Compare the box plots you prepared in Problems 5, 6, and 7. Which variety of iris typically has the longest petals? In what sense are they typically the longest? Which typically has the shortest? In what sense are they typically the shortest? Which shows the greatest typical variation in petal length? How are you determining this?

DISPLAY 3.7

STEM-AND-LEAF PLOT OF PETAL LENGTH FOR IRIS SPECIES *I. VERSICOLOR*; SAMPLE SIZE, n = 50. LEAF UNIT = .10 CENTIMETER.

Stem	Leaves	
3	0	3 | 0 = 3.0 centimeters
3	3 3	
3	5 5	
3	6 7	
3	8 9 9 9	
4	0 0 0 0 0 1 1 1	
4	2 2 2 2 3 3	
4	4 4 4 4 5 5 5 5 5 5 5	
4	6 6 6 7 7 7 7 7	
4	8 8 9 9	
5	0 1	5 | 0 = 5.0 centimeters

DISPLAY 3.8

STEM-AND-LEAF PLOT OF PETAL LENGTH FOR IRIS SPECIES *I. VIRGINICA*; SAMPLE SIZE, n = 50. LEAF UNIT = .10 CENTIMETER.

Stem	Leaves	
4	5	4 | 5 = 4.5 centimeters
4		
4	8 8 9 9 9	
5	0 0 0 1 1 1 1 1 1 1	
5	2 2 3 3	
5	4 4 5 5 5	
5	6 6 6 6 6 6 7 7 7	
5	8 8 8 9 9	
6	0 0 1 1 1	
6	3	
6	4	
6	6 7 7	
6	9	6 | 9 = 6.9 centimeters

9. Figure 3.17 shows mileage ratings for the base models of a series of car makers.

a. What is the lowest mpg rating for make 26?

b. Three makes appear to have the same median mpg rating. What are they?

c. About what is the median value that these three makes share?

d. Which manufacturer has the model with the lowest mpg rating?

e. Which manufacturer has the greatest interquartile range in mileage ratings?

10. Here are the measurements of nine killer bee wing lengths (in millimeters).[6]

8.39 8.51 8.51 8.54 8.56 8.56 8.62 8.82 8.96

a. Prepare a quick box plot of these data.

b. Does this distribution appear to be symmetric or skewed? If skewed, in which direction?

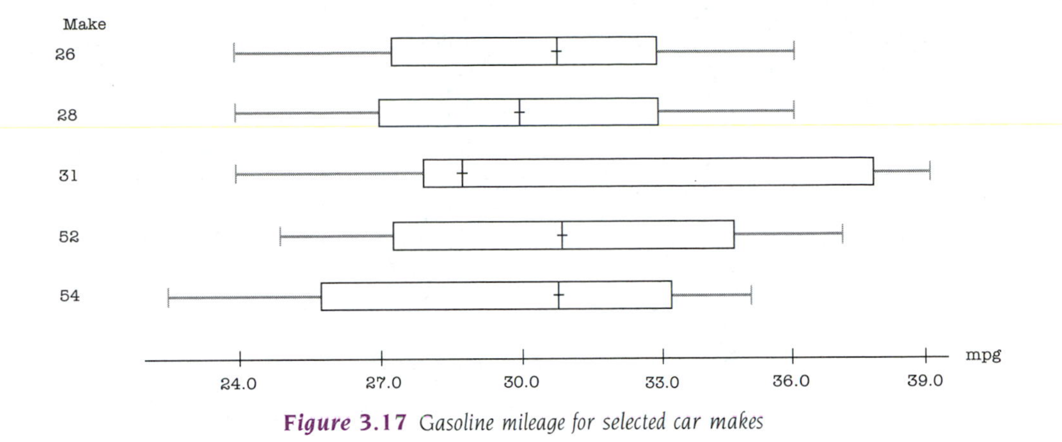

Figure 3.17 *Gasoline mileage for selected car makes*

c. By eye, do there appear to be any outliers?

d. Calculate the outlier thresholds. Are there any outliers?

e. Prepare a full box plot of the data, showing outliers separately.

11. Michelson's measurements of the speed of light are shown in Table 3.10.[7]

Table 3.10 *Determinations of the speed of light by Michelson, in kilometers per second*

299,960	299,940
299,960	299,940
299,880	299,800
299,850	299,880
299,900	299,840

a. What is the median value for the measured speed of light?

b. What are the quartile values?

c. Prepare a quick box plot of these data.

d. What is a typical value for the speed of light, based on these data?

12. The debt positions of the telephone companies resulting from the Bell Telephone breakup are shown in Table 3.11 (page 102).[8]

a. Prepare a box plot of these data.

b. Describe the distribution in terms of its shape, center, variability, and any unusual features.

13. The region of Appalachia, defined by the Appalachian Regional Development Act of 1965, represents low-income portions of 13 states. The average yearly income per person for this portion of each state is shown in Table 3.12 (page 102).[9]

a. Prepare a box plot of these data.

b. Describe the distribution in terms of its shape, center, variability, and any unusual features.

Table 3.11 Debts of the Bell System's local operating companies in billions of dollars at the end of 1980

Bell Telephone of Pennsylvania	$1.5
Chesapeake and Potomac Telephone (DC)	.2
Chesapeake and Potomac Telephone (MD)	.7
Chesapeake and Potomac Telephone (VA)	.7
Chesapeake and Potomac Telephone (WV)	.2
Cincinnati Bell	.2
Diamond State Telephone	.1
Illinois Bell Telephone	1.7
Indiana Bell Telephone	.5
Michigan Bell Telephone	1.2
Mountain States Telephone	2.0
New England Telephone	1.7
New Jersey Bell Telephone	1.0
New York Telephone	3.4
Northwestern Bell Telephone	1.3
Ohio Bell Telephone	.9
Pacific Northwest Bell Telephone	1.1
Pacific Telephone	5.8
South Central Bell Telephone	2.7
Southern Bell Telephone	3.2
Southern New England Telephone	.6
Southwestern Bell Telephone	4.6
Wisconsin Telephone	.5

Table 3.12 Per capita income, Appalachian portion of each state, in thousands of dollars (1965 dollars)

Alabama	4.1	Ohio	3.8
Georgia	3.8	Pennsylvania	4.4
Kentucky	3.2	South Carolina	4.1
Maryland	4.1	Tennessee	3.8
Mississippi	3.1	Virginia	3.7
New York	4.1	West Virginia	4.0
North Carolina	4.0		

14. Data on heights of boys who participated in a study of the effect of diet are shown in Table 3.13.

a. Prepare a box plot of these data.

b. Describe the distribution in terms of its shape, center, variability, and any unusual features.

15. Consider the data, shown in Table 3.14, on the total deposits of the ten largest U.S. commercial banks from about 1979.[10] (You may want to work with units of billions of dollars and will probably not want or need to keep all of the digits of accuracy listed here.)

a. Prepare a box plot of these data.

b. Describe the distribution in terms of its shape, center, variability, and any unusual features.

Table 3.13 *Heights of boys on macrobiotic diets, in centimeters*

718	883	853
630	768	840
1,002	830	820
560	810	862
715	813	785
1,010	1,105	1,132
1,085	860	1,084
750	1,090	1,152
510	615	490
1,112	912	781
1,093	684	1,000

Table 3.14 *Deposits, in dollars, of the ten largest commercial banks in the United States, about 1979*

Bank of America, San Francisco	85,069,944,000
Citibank, New York	72,429,528,000
Chase Manhattan Bank, New York	56,173,381,000
Manufacturers Hanover Trust Co.	39,000,618,000
Morgan Guaranty Trust Co.	35,485,994,000
Chemical Bank, New York	30,521,757,000
Continental Illinois National Bank	25,566,850,000
Bankers Trust Co., New York	22,121,919,000
First National Bank, Chicago	20,784,969,000
Security Pacific National Bank	19,590,576,000

16. Consider the data, shown in Table 3.15, on assets (primarily deposits) of the ten largest U.S. commercial banks in 1991. (Amounts are rounded to the nearest billion.)

Table 3.15 *Deposits, in billions of dollars, of the ten largest commercial banks in the United States, 1991*

Citicorp, New York	217
Chemical Banking, New York	139
BankAmerica, San Francisco	116
NationsBank Corp., Charlotte, North Carolina	110
J. P. Morgan & Co., New York	103
Chase Manhattan, New York	98
Security Pacific Corp., Los Angeles	76
Bankers Trust, New York	69
Wells Fargo & Co., San Francisco	54
First Chicago Corp., Chicago	49

a. Prepare a box plot of these data.

b. Describe the distribution in terms of its shape, center, variability, and any unusual features.

17. Discuss the changes in the sizes of the largest U.S. banks from 1979 to 1991, as reflected in Problems 15 and 16. Comment on the relative sizes of (a) the very largest banks, (b) the "typical" large bank, and (c) the variation in the size of the largest banks.

18. The two graphic displays in Display 3.9 present the same data: cholesterol levels for a group of heart attack patients 2 and 4 days after their heart attack (file *Cholesterol,* Minitab data set). The first graphic is a dot plot, and the second is a full box plot produced using a typewritten character set. The "+" sign marks the median, the "I" marks the edges of the box, and dashes mark the whiskers. There is no specific indication that outliers have been looked for (we used a crossbar when we did this by hand); when outliers are found they are marked with an asterisk (*).

DISPLAY 3.9

CHOLESTEROL LEVELS FOR A GROUP OF HEART ATTACK PATIENTS
2 AND 4 DAYS AFTER THEIR HEART ATTACK

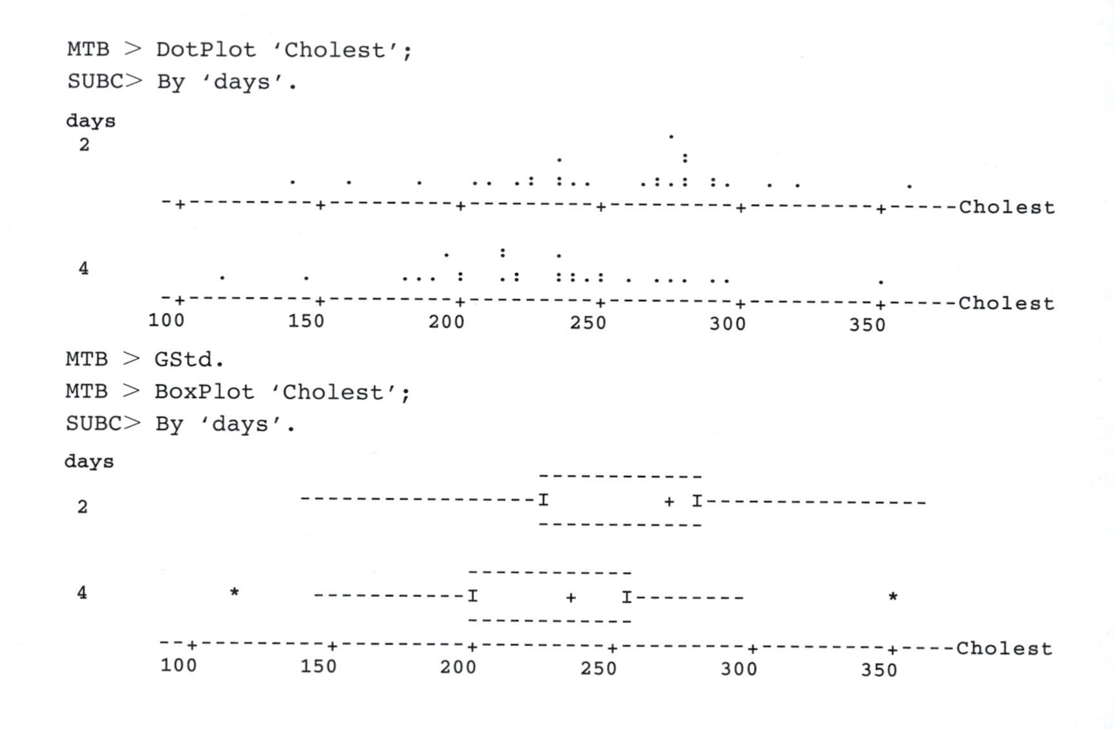

a. Looking at the dot plot, and without doing any calculations,
- Are cholesterol levels typically higher 2 days or 4 days after a heart attack? Why do you say this?
- Does it seem that there is more variability in the cholesterol levels 2 days or 4 days after a heart attack? Why do you say this?

b. Looking at the box plot, and without doing any calculations,
- Are cholesterol levels typically higher 2 days or 4 days after a heart attack? Why do you say this?

- Does it seem that there is more variability in the cholesterol levels 2 days or 4 days after a heart attack? Why do you say this?

c. Which display do you prefer? Why? Is there anything you see in one display but not the other?

19. Data on professional golf prizes are shown in Display 3.10.

DISPLAY 3.10

A DATA SET WITH TWO MODES: PRIZE MONEY (IN THOUSANDS OF DOLLARS) FOR PROFESSIONAL GOLF TOURNAMENTS IN 1991.
LEAF UNIT = $10,000.

Stem	Leaves
0	4 5 5 5
0	6 6 6 6 6 6 6 6 6 6 6 7 7 7 7
0	8 9
1	0 1 1
1	2 2
1	4 4 5
1	6 6 6
1	8 8 8 8 8 8 8 8 8 8 8 8 8 8 8 8 8 8 9 9 9 9 9
2	1 1
2	2 2 3 3
2	5
2	
2	8

a. Prepare a box plot of these data.

b. Comment on whether the box plot is an effective way to present bimodal distributions. Why or why not?

20. a. Construct a full box plot of the data on the number of seats on the various lifts of Whistler Mountain, presented in Table 2.8, page 52.

b. What is the interquartile range?

c. What is the outlier?

21. By computer, and using the Whistler Mountain data set, prepare a box plot of the vertical rise of the lifts. Are there any outliers? If so, identify them. (You may have to look at Table 2.8 to do this.) Might there be something fundamentally different about the outlier lift(s) compared to the others? (*Hint:* A chair lift is basically an open bench with back rests, suspended from a cable. A gondola lift is a small cabin suspended from a cable.)

22. Here, in chronological order, are the ages of U.S. presidents (Washington through Clinton) at the time of their first inauguration.[11]

57	61	57	57	58	57	61	54	68	51	49	64	50	48
65	52	56	46	54	49	51	47	55	55	54	42	51	56
55	51	54	51	60	62	43	55	56	61	52	69	64	47

a. Find a typical age at inauguration. Say how you found it, and give the technical name for the measure of central tendency you used.

b. Describe the variability in age at inauguration. Say how you found it, and give the technical name for the measure of variability you used.

23. Data on the savings for education by high school seniors employed in 1981 and 1990 are found in Table 3.16.[12] The figures in the body of the table are percentages; each column adds to 100%. Much more detail is given for the 1990 seniors than the 1981 seniors.

Table 3.16 *Savings for education by high school seniors employed in 1981 and 1990*

EXPENSE AND SPENDING PATTERN	PERCENT OF 1981 SENIORS	PERCENT OF 1990 SENIORS						
		TOTAL	SEX		RACE		PLANNING TO GO TO 4-YEAR COLLEGE	NOT PLANNING TO GO TO 4-YEAR COLLEGE
			MALE	FEMALE	WHITE	BLACK		
None or only a little	70.2	73.1	73.8	72.7	73.0	69.9	70.5	80.9
Some	12.4	10.7	9.9	11.5	11.1	6.2	10.9	9.8
About half	6.4	7.5	8.2	6.8	7.5	10.4	8.1	5.4
Most	6.4	4.5	4.5	4.3	4.9	5.0	5.4	2.1
All or almost all	4.6	4.2	3.7	4.5	3.6	8.5	5.2	1.8

a. For 1981, what is the modal savings category?

b. In 1990, what percentage of those planning to go to a 4-year college saved all or almost all of their earnings for education? How does this compare with those not planning to go to a 4-year college?

c. Prepare two graphic displays of educational savings, one for white students and the other for black students. Write a brief report of what these show, commenting on any obvious similarities and differences.

24. Use the median, the quartile, and the interquartile range to summarize the list of prices you compiled for Problem 3, Chapter 1. How do they compare with the typical value you made up for that exercise? How do the quartiles compare to the low typical and high typical values you made up?

25. Recall the data from Table 2.20 (Minitab data file Cal_VT.mtw), page 62, showing changes in the amount of exercise a person can do before being out of breath and reaching their ventilatory threshold. Describe the distribution of changes in ventilatory threshold and prepare a graphic display.

26. *Computer problem:* Data file SailSmll.Mtw (also available as Excel file SailSmll.xls) contains a subset of the boats listed for sale in Puget Sound. It is limited to the boats of five different makers: Benteau (1), Catalina (2), Choey Lee (3), J/Boat (4), and San Juan (5). Construct box plots of the prices of these different manufacturers. Comment on the results.

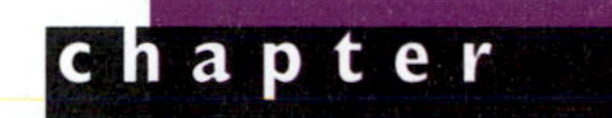

chapter 4

Describing a Normal Distribution and Summarizing Binary Data

So far we have considered the problem of describing a distribution in terms of its shape, its central value, and its variability. We now want to introduce and focus on two particular measures: the *mean*, as a measure of central tendency, and the *standard deviation,* as a measure of variability. These two measures are widely used. They are particularly useful when the underlying distribution is normal. It turns out that if we know the mean and standard deviation of a normal distribution, we can calculate the percentage of cases above or below any given number.

The average and standard deviation can also be used to describe binary (yes/no) data, which often result from sample surveys, among other places. We show how to do these calculations.

Sometimes we have data that have already been grouped into a table. We show how to calculate the mean and standard deviation for such data.

4.1 The Average or Mean

Among the many procedures for finding a central value, the **arithmetic average,** also called the **mean,** is most often used.[1] It has the satisfying property that if the total of all data values were to be divided equally among all cases, then each case would receive the average. That is, if we replaced each data value by the average, the total would remain unchanged. No other summary measure has this property.

For the rest of this book, we will use the term *average* to mean *arithmetic average.*

PROCEDURE: FINDING THE MEAN OR ARITHMETIC AVERAGE OF A SINGLE GROUP OF NUMBERS

1. Add all the numbers together.
2. Divide by the number of numbers.

The Formula for the Mean

This procedure can also be expressed as a mathematical formula. To write the formula, we first represent the list of data values by the following symbols:

$$X_1, X_2, X_3, \ldots, X_n$$

Here, n represents the number of data values. For example, the data set

157 269 198

would be represented with

$$X_1 = 157$$
$$X_2 = 269$$
$$X_3 = 198$$
$$n = 3$$

The formula for the mean can then be written in three different but completely equivalent ways.

FORMULA

$$\text{Mean} = \frac{X_1 + X_2 + X_3 + \cdots + X_n}{n}$$

or

$$\text{Mean} = \frac{\text{SUM}(X)}{n}$$

or

$$\bar{X} = \frac{\Sigma X}{n}$$

The first expression uses the familiar plus sign to represent the addition. The second and third expressions use a **summation operator.** We wrote it first using the English word "SUM," which is consistent with computer programming traditions (and which you will see if you use a spreadsheet program like Excel. We have in fact used function names from Excel in writing the English versions of the formulas.) Then we rewrote the equation using the standard mathematical symbols.

The X with a bar over the top, $\bar{X}$, usually read as "X bar," is the traditional notation for the sample mean. The capital Greek sigma, Σ, is the traditional notation for the summation operator. The traditional notation is compact and easy to manipulate algebraically. You will see it again if you take another course, and it is used to label the keys of statistical calculators. We will use both notations in this text.

These formulas are simply shorthand notes outlining the procedure for calculating the mean. The formula is interpreted and applied by using the steps outlined in detail earlier.

EXAMPLE 4.1 *Calculating a Mean*

The mean of the group of numbers 1, 8, 4, 6, 8 is calculated as follows:

$$\begin{aligned} \text{Mean}(1, 8, 4, 6, 8) &= \frac{1 + 8 + 4 + 6 + 8}{5} \\ &= \frac{27}{5} \\ &= 5.4 \end{aligned}$$

We have divided the sum by 5 here because the group contains 5 numbers; the value 8 must be counted twice because it occurs twice in the data set.

Checking Your Answer. Always look at your answer and ask yourself whether it seems reasonable. The first check is to make sure that the answer falls somewhere between the smallest value and the largest value. (If your answer is smaller than the smallest value or larger than the largest value, it is clearly wrong.) More subtly, if you have a symmetric data distribution, the mean should be fairly close to what looks like the middle by eye. It is always reasonable to repeat the calculation to make sure you get the same answer.

Exercise Set 4.1

1. Consider these numbers: 4, 5, 6.
   - **a.** What is the value of n?
   - **b.** What is the value of X_1?
   - **c.** What is the value of X_2?
   - **d.** What is the value of SUM(X)?
   - **e.** What is the value of ΣX_i?
   - **f.** What is the value of their mean?

2. Consider these numbers: 10, 20, 30, 40, 50.
   a. What is the value of n?
   b. What is the value of X_1?
   c. What is the value of X_2?
   d. What is the value of X_n?
   e. What is the value of SUM(X)?
   f. What is the value of ΣX_i?
   g. What is the value of their mean?
3. Suppose we have five numbers whose mean is 40. Four of the numbers are 10, 20, 30, and 40. What is the fifth number?
4. What is the average age of the students in this hypothetical statistics class?

   16 17 17 17 18 18 18 18 19 19 19 19 19 20 20 20 20
   20 21 21 21 21 22 22 22 23 23 23 24 24 25 29 30
5. What is the average age of the students in this hypothetical statistics class?

   16 17 17 17 18 18 18 18 19 19 19 19 19 20 20 20 20
   20 21 21 21 21 22 22 22 23 23 23 24 24 25 29 65

Discussion

The mean can be thought of as a single number with the property that if every number in the group were changed to the mean, then the sum of all the numbers would still be the same. For example, if the numbers 1, 8, 4, 6, 8 represent dollar amounts that five people put into a hat, then each person could reclaim the mean value of $5.40 and there would be nothing left in the hat.

This is why the mean works as a way of settling a restaurant check. If each person contributes the mean, the total of the contributions will equal the check. Because people pay the same amount regardless of how much they ordered, this is not really fair to the people who ordered the less expensive items on the menu and is too kind to the people who ordered the more expensive items. But it does work out in the sense that if each person contributes the mean, then the entire check will be paid exactly. The mean represents the best (in fact, the only) single-number summary of the individual costs that always works out the total in the sense of adding up to the original check.

More generally, the mean is very appropriate in situations when total amounts matter, such as when resources are to be allocated, projected, or accounted for. This happens often in public policy and other budgetary matters. In cases like this, it can be helpful to require that the mean value multiplied by the number of things gives an approximation to the total amount of expenditure.

However, the mean is not always appropriate. When things are not divided up evenly (at least approximately) to begin with, the mean can give a distorted picture. Rather than use a summary that preserves the total (as the mean does), we might prefer to use a summary like the median that is closer to most of the data. (We discuss this further in Chapter 6.)

4.2 The Standard Deviation

The mean or arithmetic average provides us with a way to describe the center of a distribution. But like the median, it does not tell us anything about how close together or how spread out the numbers are.

Now we want to construct a measure of variability to go with the mean. The basic idea is to calculate a number that tells us approximately how close to the mean most of the data values are. The bigger the number, the bigger the spread around the mean.

We will use a two-stage approach. First, we see how far each number is from the mean. This distance from the mean is called a *deviation from the mean* or simply a *deviation.* (It is sometimes called a *residual.*) Second, we find a typical or standard value for these deviations. (See Figure 4.1.)

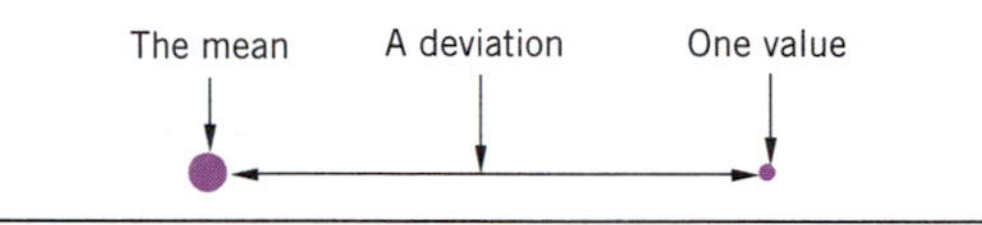

Figure 4.1 *A deviation from the mean*

EXAMPLE 4.2 Deviations

The numbers 10, 20, 30, 40, and 50 have a mean value of 30. Some are farther away from the mean than others; in each case the difference is their deviation. We have these deviations:

$10 - 30 = -20$;	the first item is 20 below the mean
$20 - 30 = -10$;	the second item is 10 below the mean
$30 - 30 = 0$;	the third item is just the mean
$40 - 30 = 10$;	the fourth item is 10 above the mean
$50 - 30 = 20$;	the fifth item is 20 above the mean

If we combine the deviations from the mean according to a special averaging process, we obtain a traditional measure of variability called the **standard deviation.** This is the best accepted and most widely used of all variability measures. It is more sensitive to all of the data than the range because the deviations of all data values from the mean enter equally into the computation. When it is computed from a sample of data, it is often called the *sample standard deviation* or the *standard deviation of the sample.*

PROCEDURE: COMPUTING THE STANDARD DEVIATION

1. Subtract the mean from each data value to obtain the *deviations from the mean* (these are also called the *residuals from the mean*).
2. Square these deviations. (This ensures that the deviations that were negative now become positive. It also increases the effect of large deviations.)
3. Add up the squared deviations.
4. Divide this sum by 1 less than the original number of data values in the group. The result is called the **variance**.
5. Take the square root of the variance. This result is the standard deviation.

EXAMPLE 4.3 *Calculating the Standard Deviation*

Using the numbers 10, 20, 30, 40, and 50 from Example 4.2, with mean 30, we:

1. Find the deviations:

$$10 - 30 = -20$$
$$20 - 30 = -10$$
$$30 - 30 = 0$$
$$40 - 30 = 10$$
$$50 - 30 = 20$$

2. Square the deviations:

$$(-20)^2 = 400$$
$$(-10)^2 = 100$$
$$0^2 = 0$$
$$10^2 = 100$$
$$20^2 = 400$$

3. Sum the squared deviations:

$$\text{SUM} = 400 + 100 + 0 + 100 + 400 = 1{,}000$$

4. Divide SUM by $n - 1$ to find the variance:

$$\text{Variance} = \frac{1{,}000}{4} = 250$$

5. Take the square root of the variance to find the standard deviation:

$$\text{Standard deviation} = \text{SQRT}(250) = 15.8$$

So, for the group of numbers 10, 20, 30, 40, and 50, we can say they have a mean of 30 and a standard deviation of 15.8. More informally, we say that a typical value is about 30, plus or minus 15.8.

The formula for the standard deviation summarizes the step-by-step approach just outlined, expressing the procedure in mathematical form. Here is the formula written in two equivalent ways, without and with the sigma summation operator.

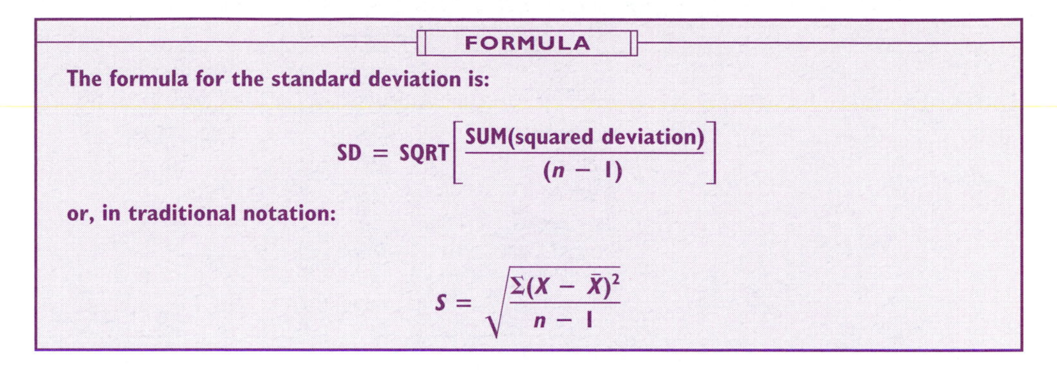

FORMULA

The formula for the standard deviation is:

$$\text{SD} = \text{SQRT}\left[\frac{\text{SUM(squared deviation)}}{(n-1)}\right]$$

or, in traditional notation:

$$S = \sqrt{\frac{\Sigma(X - \bar{X})^2}{n - 1}}$$

Note how the sigma summation operator says that we should sum "the square of each data value minus the mean." Compare this to the way sigma was used to define the mean

$$\bar{X} = \frac{\Sigma X}{n}$$

which says to sum "each data value." Generally speaking, the sigma summation operator says to add up whatever comes right after it.

EXAMPLE 4.4 Standard Deviation: Frozen Orange Juice Prices

Suppose we want to quantify the variability in the price of a product. We might want to use this measure of variability to compare the price structure of different commodities, or we might want to use it to tell whether a price is far out of line. For example, consider the prices of frozen orange juices that were rated "good" by *Pacific Magazine*,[2] as shown in Table 4.1.

Table 4.1 *Prices, in cents per ounce, for frozen orange juices*

Minute Maid	2.6
Sunkist	2.4
Treesweet	2.5
Lady Lee	2.1
Donald Duck	2.3
Bel-Air	2.5

We can compute the standard deviation to see by how much these brands tend to differ from the mean price. First we find that the mean price for all six brands is 2.4 cents per ounce (the total of 14.4 divided by 6). Then we compute a column of deviations by subtracting this mean value from each price in the list. We then square the deviations to obtain a column of squared deviations (see Table 4.2).

Because there are six data values (one per brand, in this case), we will divide the sum of squared deviations by 6 − 1 = 5 to obtain the variance:

Table 4.2 *Frozen orange juices*

BRAND	PRICE/ OUNCE	DEVIATIONS FROM THE MEAN	SQUARED DEVIATIONS
Minute Maid	2.6	.2	.04
Sunkist	2.4	.0	.00
Treesweet	2.5	.1	.01
Lady Lee	2.1	−.3	.09
Donald Duck	2.3	−.1	.01
Bel-Air	2.5	.1	.01
		Sum of squared deviations	.16

$$\text{Variance} = \frac{.16}{6 - 1} = \frac{.16}{5} = .032$$

Finally, the standard deviation emerges as the square root of this variance:

$$\begin{aligned} \text{SD} &= \text{SQRT(variance)} \\ &= \text{SQRT}(.032) \\ &= .18 \text{ cent per ounce} \end{aligned}$$

That is to say, orange juice prices typically deviate by about .18 cent per ounce from the mean price.

Exercise Set 4.2

6. Table 4.3 shows calculations for the standard deviation of the data set 1, 3, 5. Find:

a. the sum of squared deviations.

b. the variance (by dividing the sum of the squared deviations by $n - 1$).

c. the standard deviation (by taking the square root of the variance).

Table 4.3

DATA	DEVIATION = DATA − MEAN	SQUARED DEVIATION
1	1 − 3 = −2	4
3	3 − 3 = 0	0
5	5 − 3 = 2	4
	SUM(dev²) =	?

7. Find the standard deviation of 11, 12, 13. (The steps parallel those in Exercise 6, but you have to fill out a deviations table yourself.)

8. Suppose a group of numbers has a mean of 100. One individual value is 102.

a. How great is its deviation from the mean?

b. How great is its squared deviation from the mean?

9. Find the standard deviation of this group of numbers: 10, 30, 50. How does it compare with the standard deviation you calculated in Exercise 6?

10. What is the standard deviation of the ages of the students in this hypothetical statistics class?

16 17 17 17 18 18 18 18 19 19 19 19 19 20 20 20 20
20 21 21 21 21 22 22 22 23 23 23 24 24 25 29 30

11. What is the standard deviation of the ages of the students in this hypothetical statistics class?

16 17 17 17 18 18 18 18 19 19 19 19 19 20 20 20 20
20 21 21 21 21 22 22 22 23 23 23 24 24 25 29 65

Discussion

For the orange juice example, how should we interpret the standard deviation of .18 cent per ounce? It says that, typically, a brand is likely to cost within .18 cent per ounce of the mean price of 2.4 cents per ounce. To find these likely prices, we want to construct the interval that extends one standard deviation to each side of the mean. Subtracting from and adding the standard deviation to the mean, we find that

$$\text{one standard deviation below the mean} = 2.4 - .18 = 2.22$$

and

$$\text{one standard deviation above the mean} = 2.4 + .18 = 2.58$$

This provides a reasonable interval of values, from 2.22 to 2.58, inside which we see that the prices of four out of the six brands happen to be located. Another way to describe this is to say that in this particular case, two-thirds of the data values were within a distance of one standard deviation from the mean.

This is a particular example of a general property of the standard deviation in the case of a normal distribution. For normally distributed data, about two-thirds (68%) of the data values will be within a distance of one standard deviation on either side of the mean. We were quite lucky in this example to find two-thirds of the data values to be within one standard deviation. Generally (due to randomness), this will not be as exact, and slightly more or fewer than two-thirds of the data values will be within this range of values.

Three convenient properties of the mean and standard deviation with normal distributions may be summarized as follows (see also Figure 4.2 on page 116):

About 68% of the data values will be within a distance of one standard deviation on either side of the mean.

About 95% of the data values will be within a distance of two standard deviations on either side of the mean.

About 99.7% of the data values will be within a distance of three standard deviations on either side of the mean.

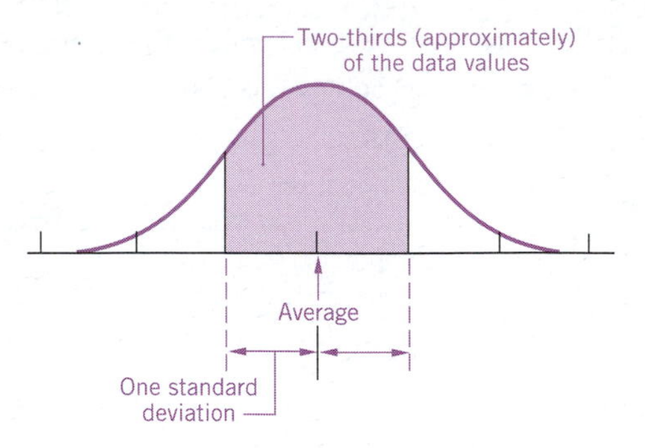

Figure 4.2
(a) One-standard-deviation interval around the mean for normal distribution

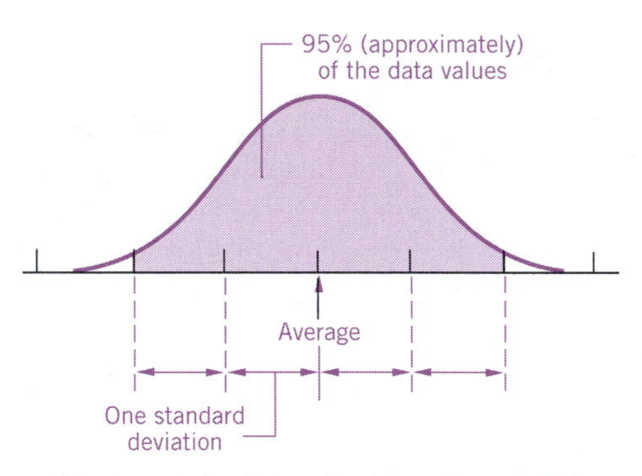

(b) Two-standard-deviation interval around the mean for normal distribution

Checking Your Calculation of the Standard Deviation. There are several ways you can check your calculations of the standard deviation. If any of the following checks fails, your answer is probably wrong! Find the mistake and correct it. Never be too proud to check an answer—even experienced statisticians make mistakes—and double-checking is one of the best ways to eliminate problems.

PROCEDURE: CHECKING THE STANDARD DEVIATION CALCULATION

1. It must be a positive number or zero (no negative numbers allowed) and will be zero only when all of the data values are the same.
2. It should seem reasonable in the sense that some of the deviations should be smaller in size (ignoring minus signs) than the standard deviation, but some should be larger. The standard deviation should certainly be smaller than the range (largest minus smallest).
3. All of the squared deviations should be positive or zero (no negative values are possible).
4. If you add up the deviations themselves (subtracting when they are negative so that positive values and negative values can cancel), you should always get zero (except possibly for a small amount of rounding error).

Exercise Set 4.3

12. Here are the asking prices of several used motor homes, in dollars:

47,950 32,500 15,500 17,500 30,000

a. Find their mean.

b. Find their standard deviation.

c. About how much is a typical asking price, plus or minus how much?

d. Calculate a low typical price and a high typical price, using your answers to part (c) above.

13. Here are the numbers of whaling trips made by Eskimo boats from Gambell, Alaska, in the spring of 1976:[3]

10 7 5 5 4 4 4 3 3 3 2 2 1 1 1 1

a. Find their mean.

b. Find their standard deviation.

c. About how many trips did a typical boat make, plus or minus how many?

14. Here are the crew sizes, where known, of the active whaling boats in Gambell, Alaska, in the spring of 1976:[4]

5 5 7 5 7 7 5 5 5

a. Find their mean.

b. Find their standard deviation.

c. What is a typical crew size, plus or minus how many? (*Note:* Given that there are only two different values in this data set, there is a strong temptation to just say "five," which is the most common value, or mode. For this exercise, use the mean as the typical value.)

Statistical Calculators and the Standard Deviation

Look at the manual for your calculator to find out exactly how to use it to calculate the standard deviation of a group of numbers. There may be a "mode" selection switch, in which case you want the "statistics" mode. The relevant keys may have the statistical functions marked *above* the keys, as second functions: Look for keys with second labels like "data," "ΣX," "$\overline{X}$," and "σ" or "S."

Your calculator may well offer you a choice of two slightly different standard deviation calculations. In one, it will use $n - 1$ in the denominator for the variance, and in the other it will use n. If your calculator offers both formulas, it probably uses one of these notations:

DIVIDE BY $n - 1$	DIVIDE BY n
S	σ
σ_{n-1}	σ_n

The use of n is appropriate only under certain limited conditions (for the population standard deviation), which we are not yet ready to talk about. Please use the $n - 1$ formula.

If you have only one key, or if you are unsure of which key uses which divisor, enter this data set: 1, 2, 3. If you get SD $= 1$ you are dividing by $n - 1$ (which we want); if you get SD $= .816\ldots$ you are dividing by n (which we do not want). If you have only one key, it probably uses $n - 1$. If your calculator only divides by n, you can multiply the answer by the square root of $n/(n - 1)$ to get the answer we want.

Computer Analysis

All commercially available statistics packages calculate means and standard deviations, usually as part of a basic set of descriptive statistics. In Minitab, the procedure "Describe" includes calculations of the mean and the standard deviation. Display 4.1 shows the results for the orange juice example. The number of data values is shown as N and is equal to 6. The mean (MEAN) is 2.4000 (which is spuriously precise), and the standard deviation (STDEV) is .1789, which is within rounding of our value of .18.

DISPLAY 4.1

```
MTB > Retrieve 'OJ.MTW'.
MTB > Describe 'Price'.

           N         MEAN       MEDIAN     TRMEAN     STDEV      SEMEAN
Price      6         2.4000     2.4500     2.4000     0.1789     0.0730

           MIN       MAX        Q1         Q3
Price      2.1000    2.6000     2.2500     2.5250
```

Exercise Set 4.4

15. Display 4.2 is the Minitab output on the all-terrain ski prices from Chapter 3.

DISPLAY 4.2

```
MTB > Retrieve 'A_TSKIS.MTW'.
MTB > Describe 'Price'.

           N         MEAN       MEDIAN     TRMEAN     STDEV      SEMEAN
Price      22        486.0      475.0      473.4      83.7       17.8

           MIN       MAX        Q1         Q3
Price      400.0     825.0      448.7      495.8
```

a. How many data values are in the set?

b. What is their mean value?

c. What is the standard deviation of their values?

16. Recall the data set with percent of vote in New Jersey counties for each presidential candidate in the 1992 presidential election. In Display 4.3 we used Minitab to calculate the mean vote percent for each candidate in the 21 New Jersey counties.

a. From how many counties do we have data?

b. What was the mean percent of the vote for Clinton?

c. What was the standard deviation of the percent voting for Clinton?

d. Who had the most variability in their vote percentage? Who had the least?

DISPLAY 4.3

```
MTB > Retrieve 'ELECTION.MTW'.
MTB > Describe 'Vote %';
SUBC>By 'Candidat'.
```

	Candidat	N	MEAN	MEDIAN	TRMEAN	STDEV	SEMEAN
Vote %	1	21	40.78	42.46	40.70	8.17	1.78
	2	21	41.39	42.03	41.32	5.81	1.27
	3	21	17.83	17.94	17.89	4.86	1.06

	Candidat	MIN	MAX	Q1	Q3
Vote %	1	26.00	57.01	35.48	45.88
	2	32.05	52.14	36.89	45.67
	3	8.27	26.30	14.42	21.42

Candidate 1 = Clinton, 2 = Bush, and 3 = Perot.

17. From Display 4.3, what is the lowest percentage in any county voting for

- **a.** Clinton
- **b.** Bush
- **c.** Perot

18. From Display 4.3, what is the highest percentage in any county voting for

- **a.** Clinton
- **b.** Bush
- **c.** Perot

19. From the data in Display 4.3, what is the interquartile range for each candidate?

Residuals

Deviations from the mean are a special case of a general concept called **residuals.** Generally speaking, residuals are the numbers that remain when a summary is subtracted from the data:

Residual = *observation* − *summary*

Because there are different summaries that can be used, there are also different types of residuals. In the case of computing the standard deviation, the summary used is the mean.

The residuals from the mean are like the original data set, except that the values are now centered at a standard place (zero) instead of some other value. Residuals can be interpreted in a simple way as the number of units above the summary value (if positive) or below it (if negative). In the orange juice example (Figure 4.3, page 120), the residual of 0 for Sunkist says that the price of Sunkist was exactly equal to the mean price. The residual of .2 for Minute Maid expresses that the price for Minute Maid is .2 cent higher than the mean price per ounce for this group. The residual of −.1 for Donald Duck says that this price is .1 cent lower than the mean.

As illustrated in Figure 4.3, if a price is equal to the mean, it has a residual of zero and falls on the horizontal line. If it is more than the mean, it is above the line; if it is less, it is below the line.

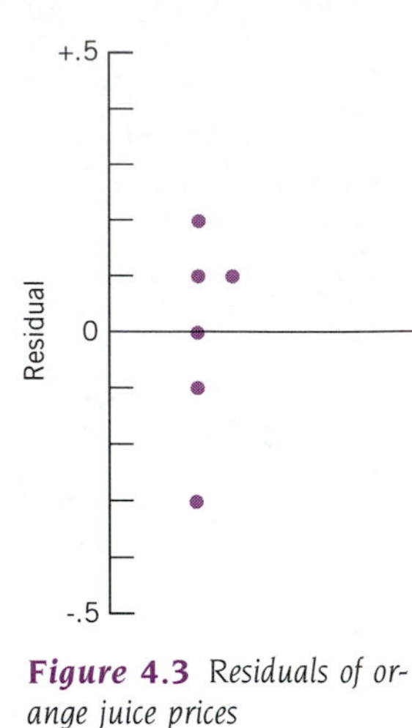

Figure 4.3 Residuals of orange juice prices

Sidebar 4.1

Why is the normal curve as important as it is? Part of the answer lies in the way that statistics calculated from samples behave, which we take up in Chapter 8. Another part has to do with the way that many data distributions behave.

In the mid-1800s a French mathematician and social researcher, Adolphe Quetelet, proposed two ideas (which generated both heat and light): The first is the notion of an "average man" who can serve as an archetype or model for a population, and from whom all others are deviations. The second—the one we are concerned with here—is the notion "that all naturally occurring distributions of properly collected and sorted data follow a normal curve."[5]

Although this is not true—many data distributions follow other forms—there are enough data distributions that are normal or nearly normal that the techniques developed for normal distributions can be very useful.

Table 4.4 Quetelet's data on Scottish soldiers

CHEST SIZE	NUMBER
33	3
34	18
35	81
36	185
37	420
38	749
39	1,073
40	1,079
41	934
42	658
43	370
44	92
45	50
46	21
47	4
48	1

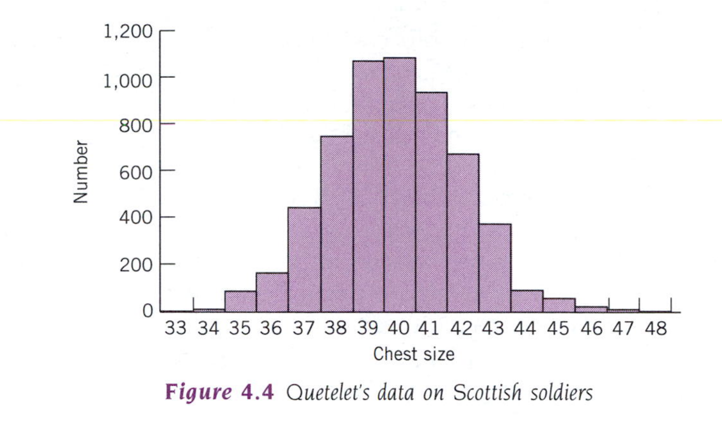

Figure 4.4 *Quetelet's data on Scottish soldiers*

One of the data sets that Quetelet analyzed and presented in support of his thesis consisted of the chest measurements of 5,738 Scottish soldiers. He extracted the data from the *Edinburgh Medical and Surgical Journal* (1817) and presented his analysis in his 1846 book, *Lettres à S.A.R. Le Duc de Saxe-Cobourg et Gotha, sur la Théorie des Probabilités, appliquée aux sciences morales et politiques*. We reproduce the data in Table 4.4 and show the corresponding histogram in Figure 4.4.

Discussion

There are two aspects of the standard deviation calculation that may seem a bit puzzling at first glance.

First, why square the deviations? The standard answer is that this turns any negative-valued deviations into positive numbers so that deviations can add up instead of canceling each other. This is certainly very reasonable, but you might wonder why we do not just erase the minus sign and average the magnitudes of the deviations instead of the squares. There is no perfectly convincing reason not to. In fact, this approach leads to a sensible but different way of doing statistics that statisticians have been exploring and that may be used more often in the future.

Perhaps people have decided that it is best to square the deviations because (1) squaring is easier to deal with when manipulating mathematical formulas and (2) the variance (the square of the standard deviation) has special properties when the data are at least approximately normally distributed. In fact, an entire set of statistical methods, called the *analysis of variance*, is based on these special properties.

For direct interpretation, the standard deviation is easier to work with than the variance (although they can be computed easily from one another) because it is a number in the same units as the original data values. That is, if the measurements are in years, centimeters, or dollars then the standard deviation will be in years, centimeters, or dollars, respectively. The variance, however, will be in units of squared years, square centimeters, or squared dollars, respectively.

Many traditional statistical methods were developed before powerful computers existed. Consequently, some methods became widely used for the simple reason that they required less use of precious computing resources. Back in the days when doing a computation meant using long division or a mechanical adding machine, and a "computer room" might have been a roomful of people good at arithmetic, computing costs were much higher than they are now!

Second, why divide by $n - 1$, one fewer than the number of terms added up in the standard deviation computation? Again, there are answers, some of which are good, but none of which are entirely satisfying without more math. The most convincing answers rely on mathematical reasoning to show that the process of subtracting the mean from each data value (to obtain the deviations) results in the loss of one piece of information from the original n numbers. Thus, we average the squared deviations by dividing by the reduced number, $n - 1$.

In fact, it is not difficult to see that a piece of information has indeed been lost when the mean is subtracted. The argument begins with the fact that the deviations themselves (not their squares) add up to zero, a fact that can be verified by straightforward algebra. This says that the deviations are not free to be just any group of n numbers. Whenever $n - 1$ of them are specified, the last one is completely determined by the condition that they all sum to zero. Thus, the last one must be the sum of the others but with the opposite sign so that they will cancel.

Statisticians sometimes refer to this phenomenon as the loss of a *degree of freedom*. We say that the original data have n degrees of freedom because there are n numbers that are free to have any value. The deviations from the mean (and also anything computed from them) have only $n - 1$ degrees of freedom because once $n - 1$ numbers have been freely specified, the last one is not free but is determined by the others.

Table 4.5 shows how with five data values, there are really only four pieces of information (degrees of freedom) in the deviations from the mean. The sum of the last four deviations (column on the right) gives away the value of the missing deviation. The sum is 3, forcing the first deviation to be -3 so that the total sum is zero. Because all five deviations are determined when we know only four of them, there are really only four pieces of information in these five deviations: $n = 5$ to begin with, but only 4 degrees of freedom remain for the standard deviation.

Table 4.5

DATA	DEVIATIONS	LAST FOUR DEVIATIONS
2	−3	(omitted)
9	4	4
4	−1	−1
6	1	1
4	−1	−1
Sum = 25	Sum = 0	Sum = 3
Mean = 5		

4.3 Percentiles and the Normal Curve

One really important feature of the normal distribution is this: If we know the mean and the standard deviation of a normal distribution, we can calculate the percent of cases that should be above or below any specified value. We do this by converting the value of interest into a standard score (sometimes called a Z score), which is the number of standard deviations it is from the mean. We then look up this standard score in a table.

EXAMPLE 4.5 Standard Scores of Aptitude Tests

Suppose an aptitude test has a mean of 500 points and a standard deviation of 100 points. Suppose Sally scored 533 points. We want to find the percent of students whose scores are

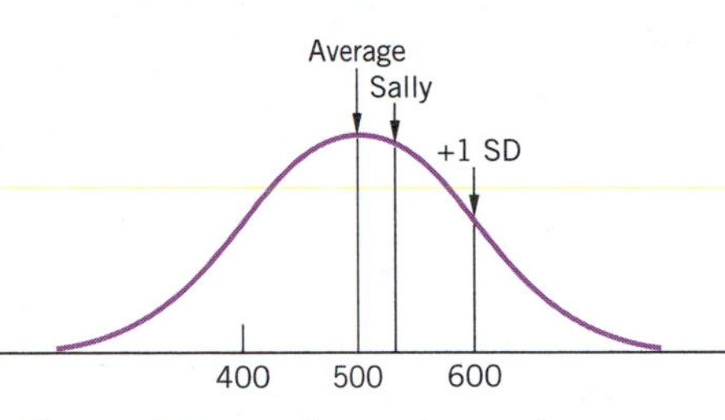

Figure 4.5 *Distribution of aptitude test scores*

lower than Sally's. (We need not worry about the difference between "equal to or lower than" or "lower than" when we are talking about a normal distribution.) A sketch such as the one drawn in Figure 4.5 helps us visualize the situation.

Now we convert Sally's 533 points into a standard score, saying how many standard deviations it is from the mean:

$$\text{Standard score} = \frac{\text{Sally's result} - \text{mean}}{\text{standard deviation}}$$

$$= \frac{533 - 500}{100}$$

$$= \frac{33}{100}$$

$$= .33$$

In words, Sally's score was one-third of a standard deviation above the mean.

Now, to find the percent of all scores that are no more than one-third of a standard deviation above the mean, we look in Table 4.6. (Table 4.6 is highly edited. A more complete normal

Table 4.6 *Edited normal table*

SCORE	PERCENT BELOW	SCORE	PERCENT BELOW
.00	50.0	.00	50.0
.25	59.9	−.25	40.1
.33	62.9	−.33	37.1
.50	69.1	−.50	30.9
.67	74.9	−.67	25.1
.75	77.3	−.75	22.7
1.00	84.1	−1.00	15.9
1.25	89.4	−1.25	10.6
1.33	90.8	−1.33	9.2
1.50	93.3	−1.50	6.7
1.67	95.3	−1.67	4.7
1.75	96.0	−1.75	4.0
2.00	97.7	−2.00	2.3
2.33	99.0	−2.33	1.0
2.50	99.4	−2.50	.6
3.00	99.87	−3.00	.13
3.50	99.98	−3.50	.02

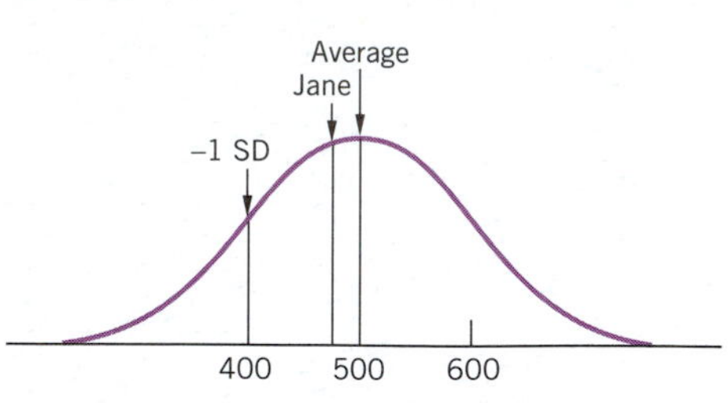

Figure 4.6 *Distribution of aptitude test scores*

table, expressed in proportions ranging from 0 to 1.0, is in the back of the book.) We see that 62.9%, say 63%, of the students would score 533 points or lower. Rounding to the nearest percent, we may also say that Sally's score was the 63rd percentile.

Now, suppose that Jane scores 475 points on the same normally distributed aptitude test, with mean 500 and standard deviation 100. What percent of the test takers would score at or below Jane's score of 475?

Looking at the sketch in Figure 4.6, we see immediately that since Jane's score is below the mean, we should get an answer that is less than 50%. We also see that it is not very far from the mean—less than one standard deviation—so the answer should not be very far from 50%.

Now, we calculate Jane's standard score:

$$\text{Standard score} = \frac{\text{Jane's result} - \text{mean}}{\text{standard deviation}}$$

$$= \frac{475 - 500}{100}$$

$$= \frac{-25}{100}$$

$$= -.25$$

Notice that because Jane's score was below the mean, we get a negative value for the standard score.

Looking up $-.25$ in Table 4.6, we find that 40.1% of the test results should be below this value. Rounding to the nearest percent, we say that Jane's score of 475 was the 40th percentile.

PROCEDURE: FINDING THE PERCENTILE RANK OF A NUMBER THAT IS PART OF A NORMAL DISTRIBUTION

0. *Optional:* Make a sketch of a normal curve, showing the mean, the value of interest, and one standard deviation from the mean.
1. Find the standard score.

$$\text{Standard score} = \frac{\text{value of interest} - \text{mean}}{\text{standard deviation}}$$

2. Look up this standard score in the normal table to find the percentage or percentile.

Exercise Set 4.5

20. Suppose the scores on another aptitude test are normally distributed with a mean of 100 and a standard deviation of 16. Suppose a student scores 108 points.

a. What is that student's standard score?

b. What percentile is the score of 108?

21. Suppose another student scores 96 on the aptitude test described in Exercise 13.

a. What is that student's standard score?

b. What percentile is the score of 96?

c. What percentage of students score over 96?

22. Suppose that in the FaXimly galaxy, the Freons (who are the resident intelligent race) have heights that are normally distributed with mean 50 hecadecibars and standard deviation 8 hecadecibars. What percent of the Freons are below 48 hecadecibars in height?

23. Suppose that one strain of chimpanzees averages 100 points on a certain aptitude test, with a standard deviation of 15 points.

a. What percentile is a score of 130 points?

b. What percentage of this strain of chimpanzees score over 130?

24. Suppose that another strain of chimpanzees also averages 100 points on the aptitude test, but with a standard deviation of 20 points.

a. For this strain of chimpanzees, what percentile is a score of 130 points?

b. What percentage of this strain of chimpanzees score over 130?

Discussion

The calculations we have done in this section on the normal distribution are the basis of many testing procedures. They are used in aptitude tests in schools and in quality control testing in factories. In each case the basic idea is that we know that the test results should be normally distributed, with a known mean and standard deviation, and we are asking where a particular result would fit in this larger distribution.

Further uses of the normal curve will be developed in Part Two, when we start looking at sampling theory.

4.4 Binary (Yes/No) Data

What Are Yes/No Data?

Recall our discussion of categorical data in Chapter 2. In the special case where there are only two categories, we can meaningfully do calculations of the mean and standard deviation. Like many other parts of statistics, this will be even more meaningful after we have discussed sampling in Part Two.

The following situations are examples of two-category cases:

1. When data represent answers to a "yes" or "no" question.

"Would you vote for so-and-so?"

"Do you believe in such-and-such?"

"Do you think the Congress should . . . ?"

2. When data represent a natural classification into two groups.

   male/female
   heads/tails
   taxable/tax-free

3. When data represent the presence or absence of something.

   For census data: presence of a telephone in a household
   In chemistry: formation of the desired end-product
   In meteorology: presence of rain on a given day

These situations are extreme examples of grouped data in which there are only two groups. The entire data set can be described by reporting the number of values in each group. For example, "There were 128 rainy days and 237 nonrainy days last year." In fact, this represents a single group of 365 numbers, with each number representing the outcome (rain or no rain) on each day.

Representation Using 0 and 1

To put many different examples into the same numerical context, we will represent data values from one group as 1's and data values from the other group as 0's.

Thus, for example, a rainy day would be recorded as a 1 and a nonrainy day as a 0. Sometimes, the data arrive already labeled as 1's and 0's; other times, we have to decide which is which. To guide this choice, the outcome that is of greater interest is conventionally denoted as the 1's group. If both groups are of equal importance, as in the case of male/female, then the choice is arbitrary and can be made either way.

Summarizing the Data

Now that we know a yes/no data set can be treated as numbers, we can do familiar computations such as summarizing its center and variability. Because the entire data set can be summarized as two numbers (the number of 0's and the number of 1's) even if it represents thousands of actual data values, the best computing methods will be based directly on these group totals. To keep our formulas small, let us use letters X, Y, and n for the numbers in each group and the total, respectively:

$$X = \text{number of 1's}$$

$$Y = \text{number of 0's}$$

$$n = X + Y = \text{total number of values}$$

The Mean

Although the mean could be computed in the standard way by adding all the data values, the direct formula is simpler and quicker.

$$\text{Mean} = \frac{X}{n}$$

This works because X, the number of 1's, is the same as the sum of the data values (because the 0's do not contribute anything to the sum). The natural interpretation of this mean is a proportion (or percentage) of the total that were 1's.

EXAMPLE 4.6 *Yes/No Data: Rainy Days*

For the rainy day example, we have

$$X = 128 \text{ rainy days}$$

$$Y = 237 \text{ nonrainy days}$$

$$n = 128 + 237 = 365 \text{ total days in the year}$$

$$\text{Mean} = \frac{X}{n} = \frac{128}{365} = .351 = 35.1\% \text{ rainy days in the year}$$

The natural interpretation of the mean here is as a percentage of the total days that were rainy. In this case, the percentage is 35.1%, which tells us that just over one-third of the days were rainy.

The Variance and Standard Deviation

There is also a shortcut for computing the variance (and its square root, the standard deviation) for yes/no data. It is based on the group totals, as follows:[6]

$$\text{Variance} = \frac{X \times Y}{n \times (n - 1)}$$

For the rainy day example,

$$\text{Variance} = \frac{128 \times 237}{365 \times 364}$$

$$= \frac{30{,}336}{132{,}860}$$

$$= .22833$$

Taking the square root of the variance, we find the standard deviation:

$$\text{Standard deviation} = .48$$

On the other hand, it may be that the data are reported as a percent and an n: "It rained 35.1% of the 365 days last year." We could, of course, use this information to find the number of rainy days and the number of nonrainy days, or we could use the next formula, which is written in terms of proportions and n.

Let us use p for the proportion of rainy days, which equals the percent divided by 100, and $(1 - p)$ for the proportion of nonrainy days. Then

$$\text{Variance} = \frac{n \times p \times (1 - p)}{n - 1}$$

For the rainy day example, we have

$$\text{Variance} = \frac{365 \times .351 \times (1 - .351)}{365 - 1}$$

$$= \frac{365 \times .351 \times .649}{364}$$

$$= \frac{83.1466}{364}$$

$$= .2284$$

which is the same as our original answer, except for rounding error introduced in the calculation of percent of rainy days.

Discussion

How should we interpret a standard deviation of .48 for rainy days in a year? As a concept, it represents the variability in "raininess" from day to day. The larger it is, the more difficult it would be to predict the weather on a random day. In a climate where it almost never rains, there is little variability in raininess, so the standard deviation would be much smaller than .48.

To interpret this as a statistical summary, our definition of standard deviation would say that this represents a kind of mean of deviations from the mean value of .351. Because a data value can only be either 0 or 1, the deviations from the mean are all either $0 - .351 = -.351$ or else $1 - .351 = .649$. The standard deviation represents a kind of mean of these values (found by squaring, combining, dividing, and taking a square root). The computed standard deviation, .48, is comfortably within the range .351 (leaving off the minus sign) and .649, which provides us with a rough check of our answer.

Any further interpretation of the standard deviation is difficult because this kind of yes/no data is not even approximately normally distributed. This means that we cannot rely on the rule for the normal distribution (that roughly two-thirds of the data will be within one standard deviation of the mean for normally distributed data). To see just how nonnormal the distribution is, note that a histogram of yes/no data will have only two vertical bars, each representing one of the two possible outcomes. (The histogram for the rainy day data is illustrated in Figure 4.7.) This is very different from the normal distribution, which is nicely symmetric with (if the scale is appropriately chosen) many bars trailing away from a central peak.

Exercise Set 4.6

25. Suppose of 100 people polled, 50 say they think the current president is doing a "good" or "excellent" job; the other 50 say he is doing a "fair" or "poor" job.

a. What proportion think the president is doing a "good" or "excellent" job?

b. Treating answers of "good" or "excellent" as 1's and "fair" or "poor" as 0's, what is the mean of these answers?

c. Still treating the answers as 1's or 0's, what is the standard deviation of these answers?

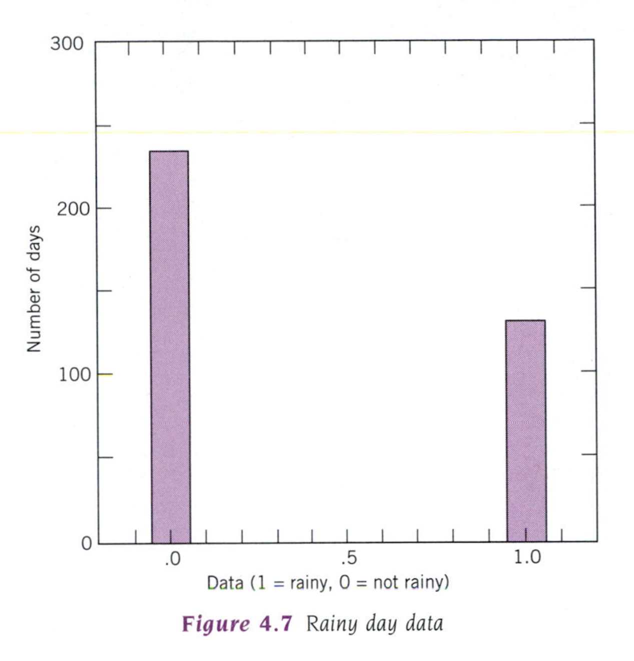

Figure 4.7 *Rainy day data*

26. Suppose of 40 people polled, 20 say they plan to vote for Smith; the other 20 say they are planning to vote for Brown.

a. What proportion are planning to vote for Smith?

b. Treating Smith voters as 1's and Brown voters as 0's, what is the mean of these answers?

c. Still treating the answers as 1's or 0's, what is the standard deviation of these answers?

27. Suppose of 100 people polled, 80 say they think the government should spend less; the other 20 say the government should spend more.

a. What proportion think the government should spend less?

b. Treating answers of "spend less" as 1's and "spend more" as 0's, what is the mean of these answers?

c. Still treating the answers as 1's or 0's, what is the standard deviation of these answers?

4.5 Grouped Data

In Chapter 3 we described how to calculate medians and quartiles from grouped data, such as those in a data table. We now show how to calculate the mean and standard deviation of grouped data.

EXAMPLE 4.7 Mean and Standard Deviation of Grouped Data: Productivity Measures

Table 4.7 shows the basic information on productivity measures introduced in Chapter 3. Now, instead of extra columns for the cumulative counts, we have columns for the calculation of the deviation. The calculations are done much the same way as for ungrouped data, except that we now multiply each data value and each deviation by the number of times they occur in the table, rather than adding them to the sum each time they occur.

Table 4.7 *Productivity measures*

X PRODUCTIVITY VALUE	n NUMBER OF TIMES	X TIMES n	DEVIATION	SQUARED DEVIATION TIMES n
20	4	20 × 4 = 80	20 − 35.42 = −15.42	237.78 × 4 = 951.12
30	11	330	30 − 35.42 = −5.42	29.38 × 11 = 323.18
40	3	120	40 − 35.42 = 4.58	20.98 × 3 = 62.94
50	5	50 × 5 = 250	50 − 35.42 = 14.58	212.58 × 5 = 1,062.90
70	1	70	70 − 35.42 = 34.58	1,195.78 × 1 = 1,195.78
Totals	24 Total number of data values	850 Sum of data values		3,595.92 Sum of squared deviations
		Mean = 850/24 = 35.42		

Dividing the sum of the squared deviations by $n - 1$, we find the variance:

$$\text{Variance} = \frac{3{,}595.92}{23}$$

$$= 156.34$$

Taking the square root of the variance, 156.34, we find the standard deviation to be

$$\text{Standard deviation} = 12.50$$

which we round to 12.5 for reporting.

Summary

In this chapter we used the arithmetic average or mean to describe the center of a normal distribution and the standard deviation to describe the amount of variability around the center.

The *average* or *mean* has the satisfying property that if the total of all data values were to be divided equally among all cases, each case would receive the average. No other summary measure has this property. We found the mean by adding all the numbers and dividing the sum by the number of numbers. The standard notation for this is

$$\bar{X} = \frac{\Sigma X}{n}$$

The *standard deviation* tells us approximately how close to the mean most of the data values are. We found the standard deviation by finding the difference, called the deviation, between each data value and the mean. We then square each deviation and find their mean, using $n - 1$ as the divisor. This mean squared deviation is the variance, and it is in squared units. We take its square root to get the standard deviation, which is back in the original units. The standard notation for this is

$$S = \sqrt{\frac{\Sigma(X_i - \bar{X})^2}{n - 1}}$$

The procedure for calculating the standard deviation may seem like a roundabout way of finding a typical value for the deviations. However, *if* we are dealing with a normal distribution, we can use the standard deviation to tell us where any given value should be in that distribution. We do this by calculating a *standard value, Z,* for the observation. The standard value tells us how many standard deviations the observation is above or below the mean. The formula for finding the standard value is:

$$Z = \frac{X - \bar{X}}{S}$$

We can then place this observation in the distribution by looking up the standard score in a table.

For the special case of *binary (yes/no) data,* there are formulas that provide a shortcut to finding the mean and standard deviation. Labeling the data values as 1 or 0 according to whether a particular category happened, we let X be the sum of these data values (this will be the number of times the particular category happened). We then define the number of times this category did not happen as $Y = n - X$, where n is the total number of cases. Then the average of these 0's and 1's is X/n. The variance of the list of 0's and 1's is

$$\text{Variance} = \frac{X + Y}{n \times (n - 1)}$$

As usual, the standard deviation is the square root of the variance.

Special procedures for computing the mean and standard deviation are also available for grouped data.

Problems

Basic Reviews

1. a. What does the formula $\frac{\text{SUM}(X)}{n}$ calculate?

b. What does the formula $\frac{\Sigma X}{n}$ calculate?

2. Find the mean of the following group of numbers.

81 86 96 97 103 111 114 115 121

3. When we speak of a deviation from the mean, we mean the difference between the mean and a particular observation. For the number group in Problem 2, you should have found a mean of 102.67.

a. What is the value of the deviation of the first number, 81, from the mean of 102.67?

b. What is the value of the deviation of the last number, 121, from the mean of 102.67?

4. Set up a table with three columns. Put the group of numbers from Problem 2 in the first column.

a. In the second column, put the deviation of each number from the mean (you have to do the arithmetic). What is the sum of the deviations?

b. Square each deviation and put the result in the third column of the table. What is the sum of the squared deviations?

c. Divide the sum of the squared deviations by $n - 1$. What is the variance?

d. Take the square root of the variance. What is the standard deviation of the numbers?

5. Calculate the mean and standard deviation of the numbers below.

89 91 92 99 101 111 112 113 120 153

6. Calculate the standard deviation of the numbers below.

74 79 79 84 84 96 96 109 119 125

7. a. What does the formula

$$\text{SQRT}\left[\frac{\text{SUM}((x - \text{mean})^2)}{n - 1}\right]$$

calculate?

b. What does the formula

$$S = \sqrt{\frac{\Sigma(X_i - \bar{X})^2}{n - 1}}$$

calculate?

8. Suppose an aptitude test has normally distributed scores with a mean of 100 and a standard deviation of 15.

a. A student scores 130. How many standard deviations from the mean is this score? What is its standard score?

b. About what percent of the students taking the test would be expected to score 130 or lower?

c. By subtraction, what percent of the students taking the test would be expected to score over 130?

9. Suppose an aptitude test has normally distributed scores with a mean of 100 and a standard deviation of 15.

a. A student scores 85. How many standard deviations from the mean is this score? What is its standard score?

b. About what percent of the students taking the test would be expected to score 85 or lower?

c. By subtraction, what percent of the students taking the test would be expected to score over 85?

10. Suppose an aptitude test has normally distributed scores with a mean of 100 and a standard deviation of 15.

a. A student scores 100. How many standard deviations from the mean is this score? What is its standard score?

b. About what percent of the students taking the test would be expected to score 100 or lower?

c. By subtraction, what percent of the students taking the test would be expected to score over 100?

11. Figure 4.8 shows the stem-and-leaf plots of three sets of random numbers. Each set contains 100 numbers; the first set is 73, 76, 77, 77, and so on. Each set was drawn from a normal distribution with a mean of 100 and a standard deviation of 15.

Sample A

```
 7  3
 7  677
 8  000033344
 8  5899
 9  00011112333344
 9  555566677999
10  000011123333334
10  555566666899
11  0122233344
11  557799
12  0002334
12  5788
13  13
13
14  4
```

Sample B

```
 6  0
 6  5679
 7  4
 7  668999
 8  000113344
 8  566677778888999
 9  01111222334
 9  5567789999
10  111233344
10  5555666777777778999
11  01223
11  557
12  0113
12  555
```

Sample C

```
 4  5
 5  7
 6  58
 7  0144467888
 8  122444455666789
 9  01222224566678889999
10  00002223345555667788999
11  00122335588889
12  00111334568
13  045
```

Figure 4.8

a. About what percent of the numbers should be over 130?
b. Since we have 100 numbers in each sample, about how many of these should be over 130?
c. How many numbers in sample A are over 130? Is this about right?
d. How many numbers in sample B are over 130? Is this about right?
e. How many numbers in sample C are over 130? Is this about right?

Analysis

12. Here are the killer bee wing length measurements (from Chapter 2).[7]

8.56 8.51 8.51 8.56 8.96 8.82 8.39 8.54 8.62

a. Calculate the mean wing length.
b. Calculate the standard deviation of the wing length.

13. Repeated here are Michelson's measurements of the speed of light (Table 2.11).[8]

299,960	299,940	299,960	299,940	299,880
299,800	299,850	299,880	299,900	299,840

a. What is the mean value of these measurements?
b. What is their standard deviation?
c. Subtract 299,000 from each of the measurements, so that you are dealing with numbers like 960 and 940. What are the mean and standard deviation of these numbers?
d. Add 299,000 to the mean you calculated in part (c). How does this compare to the answer you got in part (a)?

14. Table 4.8 shows the debt positions of telephone companies established in the breakup of the Bell Telephone system (Table 2.12).[9]

a. Calculate the mean debt (in billions of dollars).
b. Calculate the standard deviation of the debt.
c. Calculate the amount of debt that would be one standard deviation below the mean.
d. Calculate the amount of debt that would be one standard deviation above the mean.
e. *How many* companies had debts falling in the interval from one standard deviation below the mean to one standard deviation above the mean?
f. What *percent* of the companies had debts falling in the interval from one standard deviation below average to one standard deviation above average?

15. Table 4.9 shows data on heights of boys who participated in a study of the effect of diet on their growth (from Table 2.14).

a. Calculate the mean height, in centimeters.
b. Calculate the standard deviation of height, in centimeters.
c. Calculate the height that would be one standard deviation below the mean.
d. Calculate the height that would be one standard deviation above the mean.
e. *How many* of these boys had heights falling in the interval from one standard deviation below average to one standard deviation above average?

Table 4.8 *Debts of the Bell System's local operating companies in billions of dollars at the end of 1980*

Bell Telephone of Pennsylvania	$1.5
Chesapeake and Potomac Telephone (DC)	.2
Chesapeake and Potomac Telephone (MD)	.7
Chesapeake and Potomac Telephone (VA)	.7
Chesapeake and Potomac Telephone (WV)	.2
Cincinnati Bell	.2
Diamond State Telephone	.1
Illinois Bell Telephone	1.7
Indiana Bell Telephone	.5
Michigan Bell Telephone	1.2
Mountain States Telephone	2.0
New England Telephone	1.7
New Jersey Bell Telephone	1.0
New York Telephone	3.4
Northwestern Bell Telephone	1.3
Ohio Bell Telephone	.9
Pacific Northwest Bell Telephone	1.1
Pacific Telephone	5.8
South Central Bell Telephone	2.7
Southern Bell Telephone	3.2
Southern New England Telephone	.6
Southwestern Bell Telephone	4.6
Wisconsin Telephone	.5

Table 4.9 *Heights of boys on macrobiotic diets, in centimeters*

718	883	853
630	768	840
1,002	830	820
560	810	862
715	813	785
1,010	1,105	1,132
1,085	860	1,084
750	1,090	1,152
510	615	490
1,112	912	781
1,093	684	1,000

f. What *percent* of these boys had heights falling in the interval from one standard deviation below average to one standard deviation above average?

16. The numbers of seats on the various lifts at Whistler Mountain (Table 2.8) are shown again in Table 4.10 (page 136). Recall that one of these values is an outlier.

a. Calculate the mean, including the outlier.

b. Calculate the standard deviation, including the outlier.

c. Calculate the mean, *not* including the outlier.

Table 4.10 *Whistler Mountain ski area data*

LIFT	NUMBER OF SEATS
Peak Chair	3
Blue Chair	2
Green Chair	4
Black Chair	3
Olympic Chair	3
Express Gondola	10
Quicksilver Quad	4
Alpine T-Bar Two	2
Alpine T-Bar One	2
Red Line Quad	4
Orange Chair	2
Platter Lift	1
Scampland Handle	
Tow 1	1
Tow 2	1

d. Calculate the standard deviation, *not* including the outlier.

e. Describe the effect of the outlier on the mean and standard deviation.

17. The data on prices of exotic skis from Table 2.10 are reproduced in Table 4.11.

Table 4.11 *Exotic ski prices*

MODEL	PRICE ($)
Atomic Arc 837 Exir	450
Authier Zurbriggen GS	500
Blizzard V20 ATV R30	475
Dynamic VR27 Equipc SLT	475
Dynastar Mondial 5.0	475
Dynastar Pantara 5.0	475
Elan MBX 16	550
Fisher Vacuum Technic SLS	425
Hart Extreme Ultimate SL	498
Head CE6	450
Kastle Extreme	425
Kneissl White Star Steep	465
K2 CdS	475
La Croix Reference	825
Olin OTSL	445
PRE 860	495
RD Locomotion	400
Rossignol DV6S	475
Salomon Force 9 35	490
Tyrolia Comp 30X	460
Volant FX-1	425
Volkl Renntiger R	540

From *Ski Magazine*, Nov. 1991, "The All-Terrain Gang," by Seth Masin.

a. Calculate the mean, including the outlier.

b. Calculate the standard deviation, including the outlier.

c. Calculate the mean, *not* including the outlier.

d. Calculate the standard deviation, *not* including the outlier.

e. Describe the effect of the outlier on the mean and standard deviation.

18. Recall the age at first inauguration of the U.S. presidents from Washington to Clinton.[10]

57	61	57	57	58	57	61	54	68	51	49	64	50	48
65	52	56	46	54	49	51	47	55	55	54	42	51	56
55	51	54	51	60	62	43	55	56	61	52	69	64	47

a. What is the typical age, expressed as mean age? How does this compare with the median age calculated in Chapter 3?

b. Calculate an interval within which about two-thirds of the ages can be expected to fall, if the ages are normally distributed.

c. Does the interval you calculated in part (b) appear to contain about two-thirds of the data?

19. Retrieve the list of prices you compiled for Problem 3, Chapter 1. Calculate their mean and standard deviation. Now find the number that is one standard deviation below the mean. Think of this as a low medium number. Now find the number one standard deviation above the mean. Think of this as a medium high number. How do you think these numbers compare with your eyeball estimates for the Chapter 1 problem?

20. Recall the study of voice training mentioned in Problem 4 of Chapter 1, where six raters were asked to grade students' singing voices before and after a special course. A rating of 6 was best. Display 4.4 lists the before and after ratings for the first student, followed by a set of basic statistical summaries.

DISPLAY 4.4

```
MTB > Retrieve 'C:\MTBWIN\STUDENT1\VOICE.MTW'.

MTB > Print '1 BEFORE' '1 AFTER'.

 ROW     1 BEFORE     1 AFTER
   1          5           3
   2          4           5
   3          4           5
   4          5           6
   5          4           5
   6          5           5

MTB > Describe '1 BEFORE' '1 AFTER'.

                N         MEAN      MEDIAN     TRMEAN     STDEV     SEMEAN
1 BEFORE        6         4.500     4.500      4.500      0.548     0.224
1 AFTER         6         4.833     5.000      4.833      0.983     0.401

                MIN       MAX       Q1         Q3
1 BEFORE        4.000     5.000     4.000      5.000
1 AFTER         3.000     6.000     4.500      5.250
```

a. What level of measurement do the ratings have?

b. What are the average "before" and the average "after" rating for this student?

c. Is there anything going on here that we should be careful about? What is it, and why should we be careful?

21. The data set GENETICS.MTW distributed by Minitab includes the phenotypic classification—the appearance—of 200 randomly chosen peas. The codes are 1 = smooth-yellow, 2 = smooth-green, 3 = wrinkled-yellow, and 4 = wrinkled-green. The phenotypic codes for the first ten peas in the data set are: 2, 3, 1, 1, 1, 1, 2, 1, 3, 4. A computer analysis of the phenotypes produced the results shown in Display 4.5.

DISPLAY 4.5

```
MTB > Retrieve
     'C:\MTBWIN\STUDENT8\DATA\GENETICS.MTW'.

MTB > Describe 'PHENO'.
```

	N	MEAN	MEDIAN	TRMEAN	STDEV	SEMEAN
PHENO	200	1.8700	2.0000	1.8000	0.9580	0.0677

	MIN	MAX	Q1	Q3
PHENO	1.0000	4.0000	1.0000	3.0000

a. What level of measurement do the ratings have?

b. What is the mean phenotype code?

c. Is there anything going on here that we should be careful about? What is it, and why should we be careful?

22. Wisconsin Power and Light measured energy consumption in a group of homes. The energy consumption figures are adjusted for house size and outside temperature. The consumption figures, along with a variety of other data, are in Minitab data file FURNACE.MTW. In Display 4.6 we show energy consumption with an automatic chimney damper installed. Write a brief description of these energy consumption figures, referring as necessary to the material in the display.

DISPLAY 4.6

```
MTB > DotPlot 'BTU.IN';
SUBC> Same.

                          .
                     ..   :..         .
                :: : :: ..::: . . :.  :
   .  .   . .:.:::::.::.:::::.::: :::.: :. ..  . .    .
---+---------+---------+---------+---------+---------+---BTU.IN
 3.0       6.0       9.0      12.0      15.0      18.0

MTB > Describe 'BTU.IN'.
```

	N	MEAN	MEDIAN	TRMEAN	STDEV	SEMEAN
BTU.IN	90	10.038	9.835	9.978	2.868	0.302
	MIN	MAX	Q1	Q3		
BTU.IN	2.970	18.260	7.915	12.167		

23. Describe the body length and the jaw (mouth) length of male bowhead whales (see Table 2.21 and data file Whales.mtw).

24. Describe the body length and the jaw (mouth) length of female bowhead whales, shown in Table 4.12 (also in data file Whales.mtw).[11]

Table 4.12 *Total body length and mandible (mouth) length of female bowhead whales, in centimeters*

TOTAL	MOUTH	TOTAL	MOUTH
750	292	884	267
757	230	927	333
784	243	975	285
795	295	1,111	421
808	262	1,135	263
825	267	1,321	445
846	255	1,600	580
848	285	1,620	562
854	245		

25. *Computer problem:* Using the data in the file Runaway.mtw, calculate the mean and the standard deviation of the ages of these runaways.

chapter 5

Basics of Data Transformation

Sometimes the values in a data set are very different, with some values hundreds or thousands of times as large as others. In such cases we may have to modify our techniques to take these huge differences into account. The first modification is to plot different sections of the data distribution at different scales. The second is to transform the data.

Transformation is changing the system of measurement of the data so that things become easier to interpret. It will often be worthwhile to search for a transformation (sometimes also called a *reexpression*) of the data to simplify its description and analysis. For example, many standard statistical procedures require that the data follow a normal distribution; transformation can help satisfy this requirement by at least helping the distribution be symmetric instead of skewed. The main goals that can be achieved by transforming a single group of numbers are:

Effective display of the data
Symmetry of the distribution

5.1 Orders of Magnitude

Data sets that include numbers very different from one another pose special problems. We will first define what we mean by "very different from one another" and then explain how to deal with the problem.

The group of numbers 15, 428, and 3 spans more than one **order of magnitude.** Here, an order of magnitude means a power of 10. This means that the *most significant digits* (the nonzero digits that are the farthest to the left) do not represent the same units in every case. The most significant digit of 15 is the 1 in the "tens" place, whereas that of 428 is the 4 in the "hundreds" place. The numbers 258, 635, and 479 are of the same order of magnitude, whereas .00627, .237, and 121,000 are of different orders of magnitude. The number of orders of magnitude may be defined as the number of digits (to the left of the decimal point) you get when you divide the largest by the smallest.

EXAMPLE 5.1 Data Spanning Orders of Magnitude: Earthquake Deaths

The data from Chapter 2 on deaths due to major earthquakes are repeated in Display 5.1. These numbers range from 8 to 55,000, spanning four orders of magnitude. If we construct a stem-and-leaf plot using tens of thousands as our stem values, we get the results shown in Display 5.2.

We see that 13 of the 15 data points are in the row with the 0 stem, the numbers less than 10,000. There are only two earthquakes in this data set with 10,000 or more deaths. This tells us that death totals of 10,000 or more are very rare, but we have lost all detail on earthquakes with fewer than 10,000 deaths.

If we want to display all the data in such a way that more details of the smaller events are visible, we can divide the stem-and-leaf plot into four scales representing tens, hundreds, thousands, and tens of thousands, respectively, as shown in Display 5.3.

DISPLAY 5.1

EARTHQUAKE DEATHS

250	81	1,300	146	4,200	4,000
1,000	1,000	55,000	62	115	8
40,000	1,621	1,200			

DISPLAY 5.2

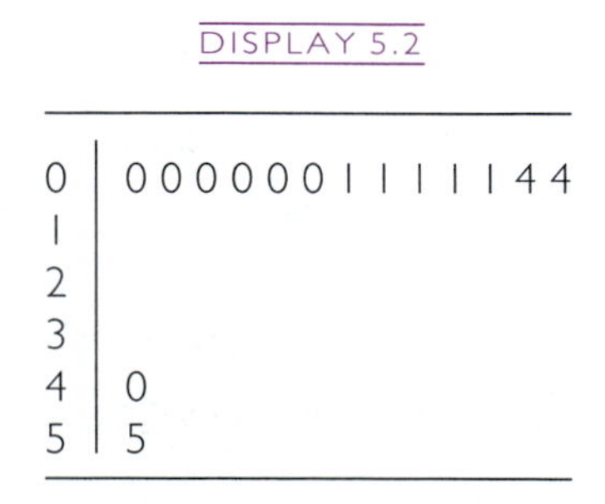

Stem	Leaves
0	0 0 0 0 0 0 1 1 1 1 1 4 4
1	
2	
3	
4	0
5	5

DISPLAY 5.3

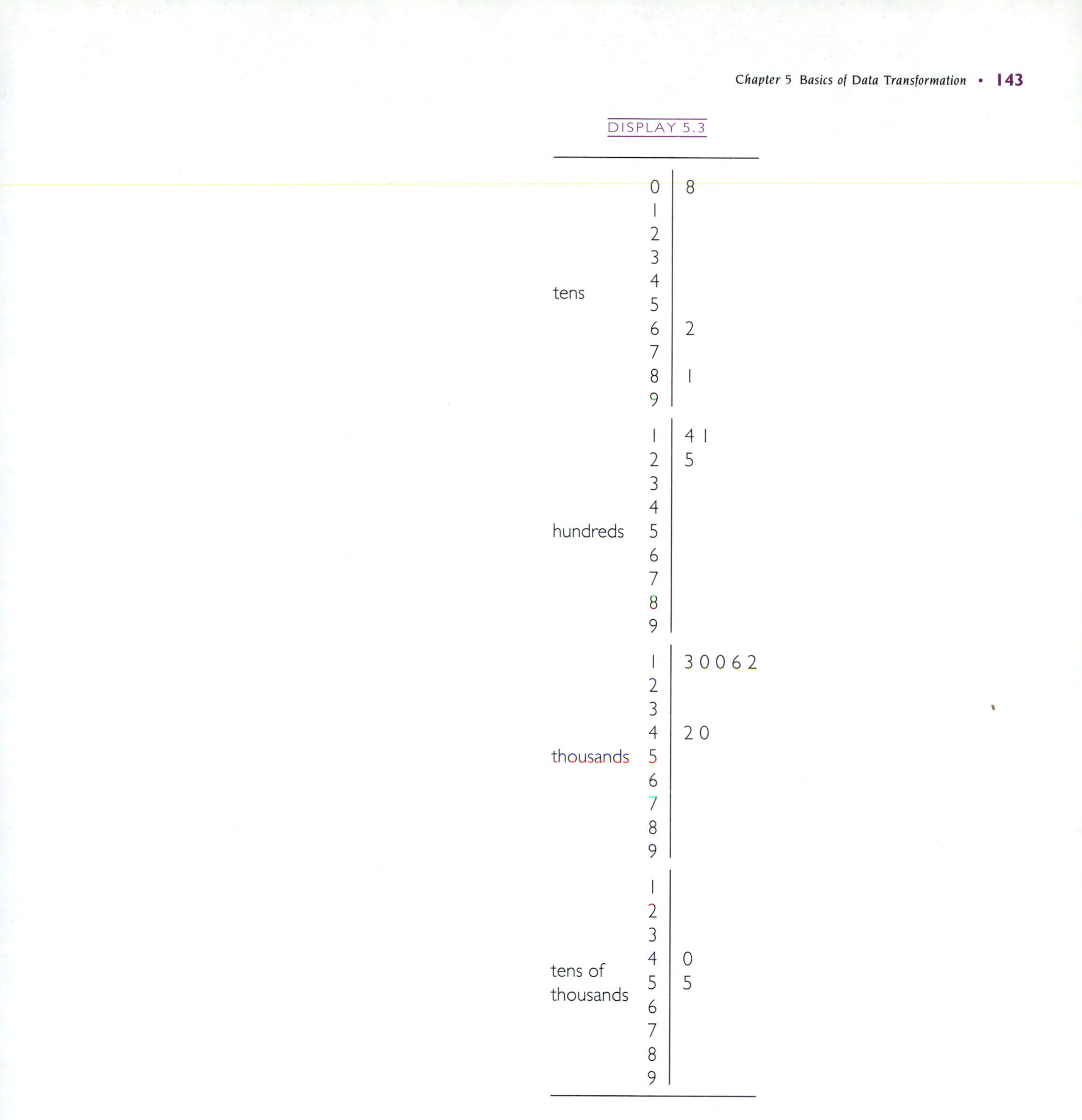

The advantage of breaking the scale into several components is that more detail is visible in the low range of data values. The disadvantage is that the scales have been distorted. We must be careful in interpreting a display like this. For example, we would have to stretch the middle scale (hundreds) vertically to be ten times as high as it is here to make it comparable to the upper scale (tens).

In the next section, we consider another way to deal with data spanning orders of magnitude.

Exercise Set 5.1

1. The first entry in Display 5.4 is 0 | 2. Which of the following does it represent?

a. 2 deaths

b. 20 to 29 deaths

c. 200 to 299 deaths

DISPLAY 5.4

STEM-AND-LEAF PLOT ON A SINGLE SCALE. STEM VALUES ARE HUNDREDS; LEAF VALUES ARE TENS. THE DATA ARE THE NUMBERS OF DEATHS IN MAJOR PEACETIME EXPLOSIONS FROM 1910 TO 1956.[1]

	Stem	Leaves
	0	2 3 3 4 4 4 4 4 5 5 5 6
	1	3 3 8 9
	2	9
	3	2
	4	
	5	
	6	6
	7	0
hundreds	8	
	9	
	10	
	11	0
	12	
	13	
	14	
	15	
	16	5

2. Which of the following does the entry 7 | 0 in Display 5.4 represent?

a. 70 deaths

b. 695 to 705 deaths

c. exactly 700 deaths

d. 700 to 709 deaths

e. 700 to 799 deaths

f. The length of time, in hours, before the train from Topeka passes the train from Spokane

3. The last two entries in Display 5.4 are 11 | 0 and 16 | 5, with stem values in the hundreds. Which of the following do these represent?

a. 11.0 and 16.5 deaths

b. 110 and 165 deaths

c. (1,100 to 1,109) and (1,650 to 1,659) deaths

4. The first entry on the tens scale in Display 5.5 is 2 | 1. Is this the same number that was represented as 0 | 2 in Display 5.4?

DISPLAY 5.5

THE SAME DATA AS IN DISPLAY 5.4, SHOWN AS SPLIT STEM-AND-LEAF PLOTS. STEM VALUES AS SHOWN.

	Stem	Leaf
tens	0	
	1	
	2	1
	3	0 0
	4	0 2 4 7 9
	5	0 1 5
	6	4
	7	
	8	
	9	
hundreds	1	3 3 8 9
	2	9
	3	2
	4	
	5	6
	6	
	7	0
	8	
	9	
thousands	1	16
	2	

5. Which of the following does the entry 7 | 0 on the hundreds scale in Display 5.5 represent?
   - **a.** 70 deaths
   - **b.** 695 to 705 deaths
   - **c.** exactly 700 deaths
   - **d.** 700 to 709 deaths
   - **e.** The length of time, in hours, before the train from Topeka passes the train from Spokane
6. The last set of entries in Display 5.5, with stem values in the thousands, is 1 | 16. Which of the following does this represent?
   - **a.** Two entries, the first being 1,100 to 1,199 and the second 1,600 to 1,699
   - **b.** A single value of 116,000
7. Consider this data, modeled after the data set shown in Display 5.4.

10	15	20	20	50	86
15	120	193	200	220	240
250	265	270	300	400	600
900	1,100	1,600			

   - **a.** Create a stem-and-leaf plot using a single scale.
   - **b.** Now construct a stem-and-leaf plot using three separate subscales: tens, hundreds, and thousands.

5.2 Transforming Data

In general, a mathematical transformation is any mathematical operation that turns one group of numbers into a different group of numbers in a systematic way. We will consider only transformations that keep the numbers in the same order before and after transforming. We have already seen several examples of simple transformations, although we did not call them transformations at the time. For instance, when we calculated percentiles for normal distributions, we transformed raw test scores into standard scores measured in standard deviations.

The transformations that we have used so far have been *linear* transformations, which shift the group of numbers up or down, but do not change the shape of the distribution. Now we want to talk about some transformations that change the shape of the data distribution.

Accommodating Many Orders of Magnitude

As we saw in the last section, a data set spanning many orders of magnitude must be transformed so that we may comprehend it in a practical way. When data values range over a wide range of magnitudes, it may be difficult to work with them, comprehend them, and properly compare them to others without the aid of transformation. This is because the large values dominate the situation and make it difficult to see detail in the rest of the data set. For the purpose of a quick look at the data, this is not a problem because we clearly see how skewed the data distribution is. However, for a more careful look at the details, transformation will be helpful.

EXAMPLE 5.2 *Two Orders of Magnitude: Distribution of Car Prices*

Consider the list prices of the two-seat cars available in the United States in 1992, shown in Table 5.1.[2] These prices span two orders of magnitude, from 11 thousand to 450 thousand

Table 5.1 *List prices of two-seat cars*

CAR MODEL	PRICES ($)
Honda CRX	11,900
Nissan NX	13,800
Mazda MX-5 Miata	21,000
Toyota MR2	25,000
Alfa Romeo Spider	26,325
Mazda RX-7	27,000
Lotus Elan SE	39,900
Morgan	48,000
Dodge Viper	50,000
Maserati Spyder	51,000
Porsche 968	51,000
Cadillac Allante	59,000
Acura NSX	65,000
Chevrolet Corvette	65,000
Lotus Esprit Turbo SE	87,000
Mercedes-Benz 300SL & 500SL	92,700
Porsche 911	96,900
Ferrari 348	108,000
Ferrari Testarossa	174,800
Lamborghini Diablo	239,000
Vector W8 Twinturbo	398,000
Ferrari F40	450,000

dollars. A stem-and-leaf plot with tens of thousands of dollars as stems has to be split a couple places to avoid using ridiculous amounts of space, as shown in Display 5.6.

DISPLAY 5.6

Stem	Leaves
1	1 3
2	1 5 6 7
3	9
4	8
5	0 1 1 9
6	5 5
7	
8	7
9	2 6
10	8
11	
12	
13	
14	
15	
16	
17	4
18	
19	
20	
21	
22	
23	9
24	
25	
26	
27	
28	
29	
39	8
45	2

If we try drawing it with split 100 thousand dollar stems, we get the results shown in Display 5.7 (page 148). This is a tolerable graphic presentation; the spread was not so extreme as to make this impossible. But now let's look at what happens when we compute measures of center and variability.

Doing the arithmetic (by computer), we get a *mean* price of $100,015 compared to a *median* price of $55,000. These are very different numbers. The difference arises because the numbers skewed out to the high end pull the mean up, but they do not affect the median.

The quartile values are $27,000 and (rounding slightly) $97,000, so the interquartile range is $70,000. The standard deviation, on the other hand, works out to $117,935.

DISPLAY 5.7

0	1 1 2 2 2 2 3 4
0	5 5 5 5 6 6 8 9 9
1	0
1	7
2	3
2	
3	
3	9
4	
4	5

Recall that for a normal distribution, about two-thirds of the data will be within plus or minus one standard deviation of the mean. If normal curve theory fits our data, our data should come reasonably close to this. Let's see if they do. We have

$$\text{Mean} - S = 100{,}015 - 117{,}935 = -17{,}920$$

and

$$\text{Mean} + S = 100{,}015 + 117{,}935 = 217{,}950$$

About one-third of the cars should be outside this range. Since there are 22 cars, seven cars should be outside this range, but only three are. Notice also that the normal curve calculations make some clearly wrong predictions: Car prices are calculated as extending down into negative numbers, which suggests that in many cases dealers will pay you to take new cars away from them. This is wrong; they don't do that.

Clearly, to do normal curve calculations we have to find a way to make this data more symmetrical. We have to *transform* it.

EXAMPLE 5.3 *Areas of Atlantic Islands: Six Orders of Magnitude*

Many data sets based on natural phenomena show strong asymmetries. For one example, consider the data in Table 5.2 representing the areas of islands in the Atlantic Ocean.[3]

This data set spans six orders of magnitude. Thus, any attempt to directly display all the numbers at once may well cause the smaller numbers to crowd together at one end of the scale, resulting in a loss of detail, as seen in Display 5.8. This stem-and-leaf plot is really not a very useful display of the data. The largest value, Greenland, at 840,000 square miles, is clearly an outlier and prevents us from seeing any details in the rest of the data set. However, to show some detail in the other data values, we can break the scale and display the outlier separately, as shown in Display 5.9 (page 150).

A classic histogram of the island areas is shown in Figure 5.1. This is better, but not yet good enough. The break in the scale makes it clear that the last value is actually much farther out

Table 5.2 *Areas in square miles of important islands in the Atlantic Ocean*

Anticosti	3,066	Iceland	39,769
Ascension	34	Long Island	1,396
Azores	902	Madeira	307
Bahamas	5,380	Marajo	15,528
Bermuda	20	Martha's Vineyard	91
Bioko	785	Mount Desert	108
Block	10	Nantucket	46
Canary	2,808	Newfoundland	42,030
Cape Breton	3,981	Prince Edward	2,184
Cape Verde	1,750	St. Helena	47
Faeroe	540	South Georgia	1,450
Falkland	4,700	Tierra del Fuego	18,800
Fernando de Noronha	7	Tristan da Cunha	40
Greenland	840,000		

DISPLAY 5.8

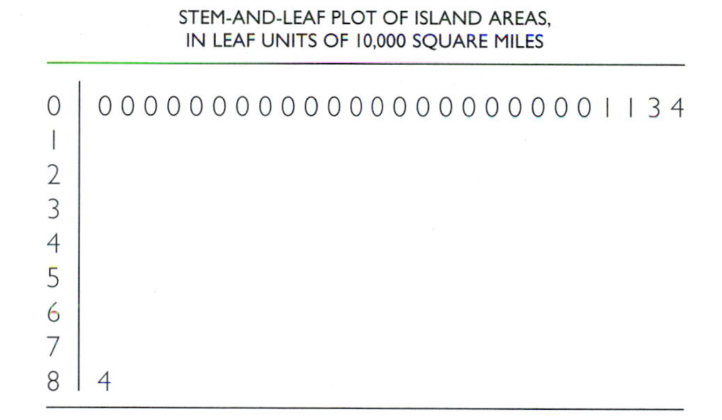

than it might appear to be. Much more detail is now displayed, but it still seems that a few large outliers are forcing the smaller data values to be bunched together at the low end of the scale.

Which Data Sets Should Be Transformed?

Not every data set should be transformed. In particular, we will usually transform only data sets consisting of positive values and possibly zeros. We will also require that the distribution be strongly skewed (that is, clearly in need of transformation). In addition, transformation will generally be useful only when the data set spans at least two orders of magnitude.

Although we will not transform negative values directly, it may be that such a data set consists of differences (such as measures of profit, which is gross income minus expenses). In such a case, we might consider transforming the more basic quantities (namely, income and expenses themselves) before subtracting them.

For now, we will suppose that the data set contains only positive numbers and that there is no natural fixed upper limit on the possible values. This excludes measurements such as percentages, which cannot exceed 100%.

DISPLAY 5.9

STEM-AND-LEAF PLOT OF ISLAND AREAS,
IN LEAF UNITS OF 1,000 SQUARE MILES

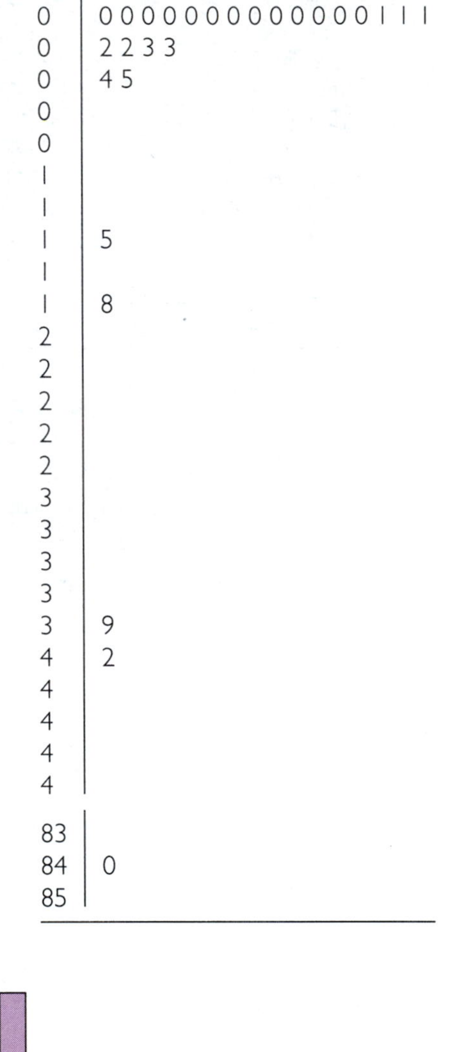

Stem	Leaves
0	0 0 0 0 0 0 0 0 0 0 0 0 0 0 1 1 1
0	2 2 3 3
0	4 5
0	
0	
1	
1	
1	5
1	
1	8
2	
2	
2	
2	
2	
3	
3	
3	
3	
3	9
4	2
4	
4	
4	
4	
83	
84	0
85	

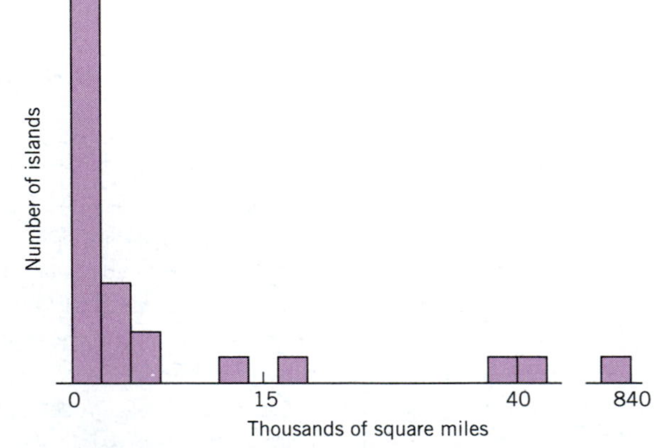

Figure 5.1 *Histogram of island areas, not transformed*

5.3 Shape Changing: More Than Just Adding or Multiplying

When we refer to transformation in this context, we are talking about changing the shape of the distribution by stretching the scale in some places and compressing it in others. Tukey called such transformations *reexpressions*. This cannot be accomplished simply by adding a constant to or subtracting a constant from each data value, as is illustrated in Figure 5.2. This would simply shift the distribution without changing its shape and is what happens, for example, when a measure of center such as the median or the average is subtracted from the data to form residuals.

Nor can a shape-changing transformation be accomplished by simply multiplying or dividing each data value by a constant. This would only change the scale of the data, stretching or contracting it uniformly everywhere, again without changing the actual shape of the distribution. (See Figure 5.3.)

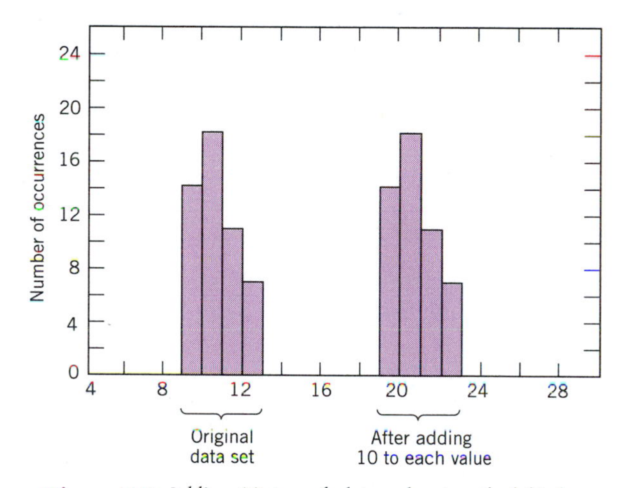

Figure 5.2 *Adding 10 to each data value (on the left) does not change the shape of the distribution.*

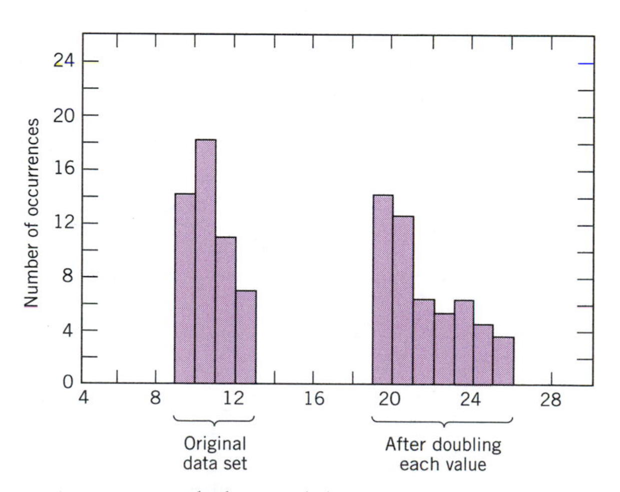

Figure 5.3 *Multiplying each data value (on the left) by 2 does not change the shape of the distribution.*

To accomplish a shape-changing transformation such as changing skewness into symmetry, we must distort the scale differently in the two tails of the distribution. For positive data values, the most useful transformations are either the *square roots* or the *logarithms* of the data. These transformations have the ability to pull in large values, compressing the scale where data values are likely to be sparse. They also tend to spread out small values by expanding the scale in the other tail of the distribution, where the data values are likely to be overconcentrated.

Square Roots

The **square root** of a data value is the number that, when multiplied by itself, gives the data value back again. For example, 3 is the square root of 9 because $3 \times 3 = 9$. Although there are tables where square roots can be looked up, that process is prone to errors. The best way to compute square roots these days is to use a calculator or computer. Square roots of some numbers are given in Table 5.3.

From these, we see that very large data values are made much smaller, condensing and shrinking the scale there, whereas very small data values are made larger, expanding the scale there. This ability of the square root transformation to change the expression of the scale can be seen in Figure 5.4.

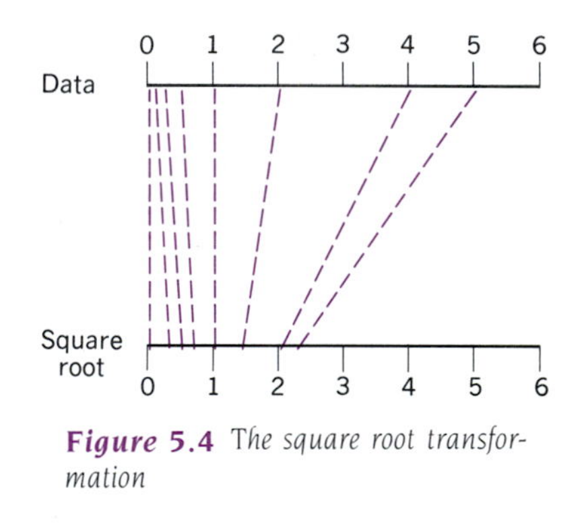

Figure 5.4 *The square root transformation*

Exercise Set 5.2

8. Use your calculator to check some of the figures in Table 5.3. Make sure you can go from the numbers on the left to those on the right.

Table 5.3 *Square roots of some numbers*

DATA VALUE X	SQUARE ROOT $\sqrt{X}$
0	0
.01	.1
.49	.7
1	1
10	3.1623 . . .
100	10
1,000,000	1,000

9. Use your calculator to find:
   a. The square root of 64
   b. The square of 8
   c. The square root of 225
   d. The square root of 1,600
   e. The square of 15
   f. The square of 40
   g. The square root of 160,000
   h. The square of 400

10. Suppose you know that an island has an area of 225 square miles. If the island happened to be a perfect square, how long would its sides be?

Logarithms

Logarithms (called *logs* for short) can be calculated using any fixed positive number as the *base*. The **logarithm** of a data value is the number such that, when the base is raised to the log value, the result is the original data value. For example, the log of the data value 1,000 to base 10 is 3 because 10 raised to the power 3 gives the data value of 1,000.

There are three kinds of logarithms in common use: base 2, base 10, and base e (where e is a special mathematical number that is approximately 2.718281828459045... and is not a repeating decimal).

Logs to Base 2. Let's start with logs to base 2; they are the easiest to explain. Here are the first few powers of 2:

$$
\begin{aligned}
2^1 &= 2 \\
2^2 &= 2 \times 2 = 4 \\
2^3 &= 4 \times 2 = 8 \\
2^4 &= 8 \times 2 = 16 \\
2^5 &= 16 \times 2 = 32 \\
2^6 &= 32 \times 2 = 64 \\
2^7 &= 64 \times 2 = 128 \\
2^8 &= 128 \times 2 = 256 \\
2^9 &= 256 \times 2 = 512 \\
2^{10} &= 512 \times 2 = 1{,}024
\end{aligned}
$$

With each increase in power, the size of the answer is doubled. These powers of 2 are common in computer sciences and in biology.

Now let's use this set of calculations to find the log of some numbers to base 2. The log of 4 to base 2 is 2, because 2 raised to the power 2 is 4. The log of 8 to base 2 is 3, because 2 raised to the power 3 is 8. The log of 16 to base 2 is 4, because 2 raised to the power 4 is 16. There are no deep tricks or mysteries here; all the definition of logs requires is a simple rearrangement of equations like

$$2^4 = 16$$

to

$$\log_2 16 = 4$$

The same three numbers appear, but now the 2 is a subscript on log, indicating the base, the 16 is the number whose log we are finding, and the 4 is the answer.

Exercise Set 5.3

11. **a.** What is the log of 32 to base 2?

b. What is the log of 64 to base 2?

c. What is the log of 128 to base 2?

d. What is the log of 256 to base 2?

12. Suppose the number of rabbits on an island doubles every year. Starting with 2 rabbits (of opposite sex), how many rabbits are on the island after 3 years?

13. Suppose you know that rabbit populations double every year, at least for the first few years, after a pair of rabbits is introduced to a new island. You go to an island and find eight rabbits there. How many years has it been since the first pair was introduced?

***Logs to Base* e *and Base* 10.** Logarithms to base e and to base 10 are widely used in mathematics and engineering. Logarithms to base 10 are called *common logarithms* and are usually written as "log" with no base indicated. Logarithms to the base e are sometimes called *natural logarithms* and are usually written as "ln."

Fortunately, the three kinds of logarithms differ only by constant multiplicative factors. For instance, the logarithm to the base e of any number is 2.3026... times its logarithm to the base 10. This means that in terms of their abilities to change the shape of a distribution, their effects are the same. The constant factor affects the scale but not the shape of a distribution.

Scientific and statistical calculators usually have function keys that calculate logarithms to base e or base 10. Table 5.4 lists some logarithms to both bases so you can see which one your calculator uses. Usually, a calculator key labeled "log" uses base 10, and one labeled "ln" uses base e.

Looking at the transformed values, we see that the very large data values are made much smaller, moving them closer together, whereas very small data values are made larger, spreading them apart from each other. This is a lot like the effect of the square root, but much stronger, as is evident from what happens to very large numbers (compare the last entry in each table). The ability of the log transformation to change the expression of the data is shown in Figure 5.5.

Table 5.4 *Logarithms (logs) of some numbers*

DATA VALUE X	LOG TO BASE e $\ln X$	LOG TO BASE 10 $\log X$
0	Nonexistent	Nonexistent
.01	−4.605 . . .	−2
.1	−2.303 . . .	−1
1	0	0
2.718 . . .	1	.434 . . .
10	2.303 . . .	1
100	4.605 . . .	2
1,000,000	13.816 . . .	6

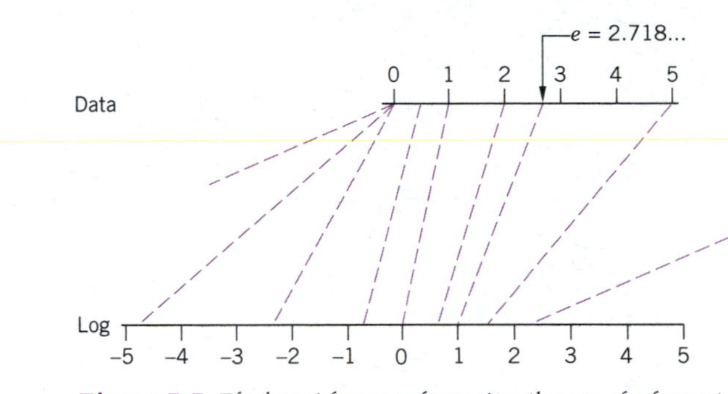

Figure 5.5 *The logarithm transformation (logs to the base* e*)*

EXAMPLE 5.4 Transforming Skewed Data: Car Prices

To see how these two transformations actually work, let us return to the data set of car model prices. Table 5.5 shows the prices in dollars, the square roots of these prices, and their logarithms. From just a quick glance, we see that the transformations have helped compress the numbers. But because it is hard to judge distribution shape by looking at lists of numbers, let's look at the stem-and-leaf-plots shown in Display 5.10.

In general, the square root transformation tends to pull in the long tail of the distribution on the right, but stretch it out on the left. Looking at Display 5.10, we see that this has happened.

Table 5.5 *High list prices of 1992 two-seat cars. Hi$ = raw prices; root(hi$) = square root of hi$; log(hi$) = log to base 10 of hi$; ln(hi$) = log to base* e *of hi$*

CAR MODEL	HI$	ROOT(HI$)	LOG(HI$)	LN(HI$)
Honda CRX	11,900	109.0871	4.075547	9.38429
Nissan NX	13,800	117.4734	4.139879	9.53242
Mazda MX-5 Miata	21,000	144.9138	4.322219	9.95228
Toyota MR2	25,000	158.1139	4.397940	10.12663
Alfa Romeo Spider	26,325	162.2498	4.420368	10.17827
Mazda RX-7	27,000	164.3168	4.431364	10.2036
Lotus Elan SE	39,900	199.7498	4.600973	10.5941
Morgan	48,000	219.0890	4.681241	10.7790
Dodge Viper	50,000	223.6068	4.698970	10.8198
Maserati Spyder	51,000	225.8318	4.707570	10.8396
Porsche 968	51,000	225.8318	4.707570	10.8396
Cadillac Allante	59,000	242.8992	4.770852	10.9853
Acura NSX	65,000	254.9510	4.812913	11.0821
Chevrolet Corvette	65,000	254.9510	4.812913	11.0821
Lotus Esprit Turbo	87,000	294.9576	4.939519	11.3737
Mercedes-Benz SLs	92,700	304.4667	4.967080	11.4371
Porsche 911	96,900	311.2876	4.986324	11.4814
Ferrari 348	108,000	328.6335	5.033424	11.5899
Ferrari Testarossa	174,800	418.0909	5.242541	12.0714
Lamborghini Diablo	239,000	488.8763	5.378398	12.3842
Vector W8 Twinturbo	398,000	630.8724	5.599883	12.8942
Ferrari F40	450,000	670.8204	5.653213	13.0170

DISPLAY 5.10

LIST PRICES FOR 1992 TWO-SEAT CARS.
RAW DATA STEM UNITS = $100,000.

RAW DATA		SQUARE ROOTS		LOG	
0	1 1 2 2 2 2 3 4	1	0 1 4	4	0 1
0	5 5 5 5 6 6 8 9 9	1	5 6 6 9	4	3 3
1	0	2	1 2 2 2 4	4	4 4
1	7	2	5 5 9	4	6 6 6 7 7 7
2	3	3	0 1 2	4	8 8 9 9 9
2		3		5	0
3		4	1	5	2 3
3	9	4	8	5	5
4		5		5	6
4	5	5			
		6	3		
		6	7		

The higher values *have* been pulled in, and the lower values *are* more spread out. The distribution is much less skewed. Still, it is not really symmetrical.

The log transformation is generally stronger than the square root transformation. Indeed, looking at the stem-and-leaf plot of the log values in Display 5.10, we see that something truly remarkable has occurred: We now have a nice symmetrical distribution.

The average value of the logs is 4.790032, and they have a standard deviation (SD) of .422312. Finding limits of one SD on either side of the mean log, we have

$$\text{Mean log} + \text{SD log} = 5.212344$$

and

$$\text{Mean log} - \text{SD log} = 4.36772$$

But what do these numbers mean? What do they have to do with car prices, anyway? To get back to something that looks like a price, we have to reverse our transformation.

Let's take the average value of 4.790032 first. Recall, from the definition of the logarithm, that this is a power to which the *base* is raised. Our base here was 10, so

$$\text{Average log-transformed price} = 10^{4.790032}$$
$$= \$61{,}664$$

This answer still seems high, but it's a lot more reasonable than the $100,000 we got using the untransformed prices. (Notice, by the way, that we are not trying to compute the average price of cars *sold.* A lot more Honda CRXs are sold than Ferrari F40s, but each appears only once in our list. We are computing the average list price of the different models *available* on the U.S. market.)

Now watch what happens when we back-transform the values for one SD above the mean and one SD below the mean. We have

$$\text{Log-transformed price 1 SD above mean} = 10^{5.212344} = \$163{,}058$$

and

$$\text{Log-transformed price 1 SD below mean} = 10^{4.36772} = \$23{,}319$$

For a normal curve, we would typically see about two-thirds of our cases between these values and one-third outside them. Checking our list of prices, we see that we have 3 prices below the lower bound, 4 above the upper bound, and 15 in between. This is perfect. And, even nicer, we are not predicting that there are models where the dealer will pay us to take them away!

Exercise Set 5.4

14. Table 5.5 shows the Honda CRX listing at $11,900 in 1992.

a. Find the square root function key on your calculator. (Probably labeled $\sqrt{\ }$, it may be a second function above another key.) Use it to find the square root of $11,900. Your answer should match the value in the table.

b. If your calculator has an "X^2" key (or second function), use it on the square root value you just calculated. If your calculator does not have an "X^2" function, multiply the square root by itself. In either case, you should get back the original $11,900. (If you had to reenter the square root, there may be some rounding error.)

c. Your calculator should have either a "log" key, an "ln" key, or both. Use one to calculate the log (or ln) of $11,900. Your answer should match the value in the table. (*Note:* The tables do not necessarily show all the digits to the right of the decimal place.)

d. If your calculator has a log button, it probably also has a second function on it labeled "10^x." If it has an "ln" button, it probably has a second function on it labeled "e^x." Use the appropriate one on the log (or ln) value you just calculated. You should get back the original $11,900 value.

15. Pick two more list prices from Table 5.5.

a. Use your calculator first to find the square root, then to square that answer to get back the original price.

b. Use the log (or ln) key to find log transformations, then use the 10^x (or e^x) key to back-transform to the original price.

EXAMPLE 5.5 *Transforming Very, Very Skewed Data: Island Areas*

Let us now see how the square root and log transformations actually work for the data set of areas of important islands in the Atlantic Ocean. Table 5.6 (page 158) shows the areas (in square miles), the square roots of these areas, and their logarithms.

The second column represents the group of numbers consisting of the square roots of the areas, a transformation that we know tends to pull in the long tail of the distribution on the right, but stretch it out on the left. A histogram for these transformed values looks like that shown in Figure 5.6.

Table 5.6 *Raw (untransformed) areas and transformed areas for important islands in the Atlantic Ocean*

ISLAND NAME	AREA	SQUARE ROOT	LN (base *e*)	LOG (base 10)
Anticosti	3,066	55.37	8.03	3.49
Ascension	34	5.83	3.53	1.53
Azores	902	30.03	6.80	2.96
Bahamas	5,380	73.35	8.59	3.73
Bermuda	20	4.47	3.00	1.30
Bioko	785	28.02	6.67	2.89
Block	10	3.16	2.30	1.00
Canary	2,808	52.99	7.94	3.45
Cape Breton	3,981	63.10	8.29	3.60
Cape Verde	1,750	41.83	7.47	3.24
Faeroe	540	23.24	6.29	2.73
Falkland	4,700	68.56	8.46	3.67
Fernando de Noronha	7	2.65	1.95	.85
Greenland	840,000	916.52	13.64	5.92
Iceland	39,769	199.42	10.59	4.60
Long Island	1,396	37.36	7.24	3.14
Madeira	307	17.52	5.73	2.49
Marajo	15,528	124.61	9.65	4.19
Martha's Vineyard	91	9.54	4.51	1.96
Mount Desert	108	10.39	4.68	2.03
Nantucket	46	6.78	3.83	1.66
Newfoundland	42,030	205.01	10.65	4.62
Prince Edward	2,184	46.73	7.69	3.34
St. Helena	47	6.86	3.85	1.67
South Georgia	1,450	38.08	7.28	3.16
Tierra del Fuego	18,800	137.11	9.84	4.27
Tristan da Cunha	40	6.32	3.69	1.60

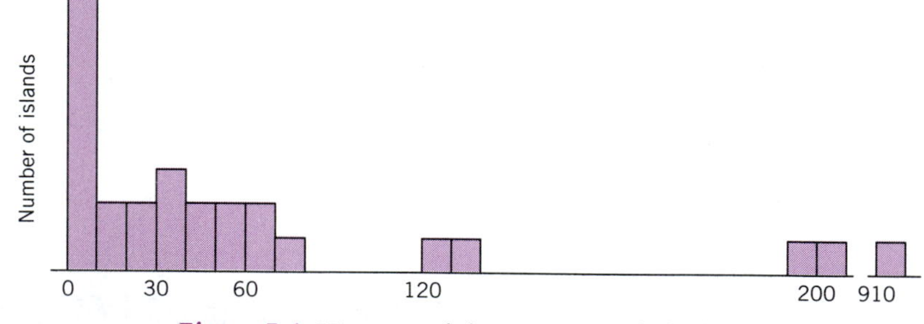

Figure 5.6 *Histogram of the square roots of island areas*

The stem-and-leaf plot in Display 5.11 shows more detail.

This transformation has certainly done something to spread out the distribution near the smaller islands and to display some information about their sizes. This is much better than the uninformative 0's visible on the stem-and-leaf plot of raw areas that we saw in Display 5.8. However, the distribution is still strongly skewed toward high values, although not nearly so much as before. Even ignoring the extreme outlier (916) at the end (which is actually much

DISPLAY 5.11

STEM-AND-LEAF PLOT OF THE SQUARE ROOTS OF ISLAND AREAS. LEAF UNIT = 1 MILE.

0	2 3 4 5 6 6 6 9
1	0 7
2	3 8
3	0 7 8
4	1 6
5	2 5
6	3 8
7	3
8	
9	
10	
11	
12	4
13	7
14	
15	
16	
17	
18	
19	9
20	5
91	6

farther out than it appears, as indicated by the break in the scale), this distribution is far from symmetric. This suggests that an even stronger transformation than square roots will be needed to symmetrize the distribution.

The third column of Table 5.6 represents the group of numbers consisting of the natural logarithms (to the base e) of the areas, a transformation we know to be stronger than square roots. A histogram for these logarithmically transformed values is shown in Figure 5.7 and the stem-and-leaf plot in Display 5.12.

Compared to the previous displays, these are very nice indeed. They are fairly symmetric, with neither end trailing off much faster than the other. Greenland, which had been a huge outlier (way off the scale in the other displays), has been tamed by the logarithmic scale and now is only slightly (though noticeably) more extreme than the other large islands. In fact,

Figure 5.7 Histogram of the logarithms of the island areas

DISPLAY 5.12

STEM-AND-LEAF PLOT OF THE LOGARITHMS OF THE ISLAND AREAS

Stem	Leaves
1	9
2	3
3	0 5 6 8 8
4	5 6
5	7
6	2 6 8
7	2 2 4 6 9
8	0 2 4 5
9	6 8
10	5 6
11	
12	
13	6

Greenland is now (just barely) inside the outlier cutoffs as we would compute them for a box plot.

Discussion: Interpreting the Log Scale

Logarithms were invented to turn multiplication into addition. Addition is simpler. No matter which base is being used (so long as the same one is used throughout), it is true for all positive numbers x and y that $\log(xy) = \log(x) + \log(y)$. For example, in base 10 with $x = 10$ and $y = 100$, we have $1{,}000 = 10 \times 100$, so the following are all equal:

$$\begin{aligned} 3 &= \log(1{,}000) = \log(10 \times 100) \\ &= \log(10) + \log(100) = 1 + 2 = 3 \end{aligned}$$

Similarly, division is converted to subtraction so that $\log(x/y) = \log(x) - \log(y)$. Thus, in base 10 with $x = 1{,}000$ and $y = 10$, we have $10 = 1{,}000/100$, so the following are all equal:

$$\begin{aligned} 1 &= \log(10) = \log(1{,}000/100) \\ &= \log(1{,}000) - \log(100) = 3 - 2 = 1 \end{aligned}$$

These facts about logarithms are useful in the interpretation of an analysis of logarithms of data values. For instance, consider the deviations from the center, working with the logs of the data. These can be interpreted as a *proportionate* increase or decrease in the original units of the data (before taking logs). For example, using logs to base 10, a difference of 2 units means that one of the original values is 100 times the others because 100 is 10 raised to power 2.

Logarithms are so well suited to some situations that the transformed values have actually become the standard units. One example of this is the measurement of the intensity of sound in standard units called *decibels.* These units represent the logarithm of the physical power in the sound waves (the energy per unit of time). The use of the logarithm is practical here because it allows sounds ranging from a quiet whisper across an empty hallway to a loud rock concert from the front row, to be expressed on the same scale, while allowing distinctions to be made at the quiet end of the scale. Logarithms here also correspond nicely to the human ear's ability to distinguish subtle differences in the loudness of sound.

Another example of a standard use of the logarithmic transformation is the Richter scale for measuring the intensity of earthquakes. The logarithmic units easily accommodate the different energies released by earthquakes of different orders of magnitude.

EXAMPLE 5.6 *Interpreting the Log Transformation: Areas of Atlantic Islands*

Returning to the Atlantic island data, how can we interpret the fact that the distribution seems so symmetric and well behaved on the log scale? What does it mean that Greenland, an extreme outlier on the original scale, is no longer an outlier on the log scale? The key to interpreting logarithms is proportionality.

On the untransformed scale, comparisons are made by seeing how many square miles larger one island is than another. For example, on the original scale, comparing the islands of Bioko and Bermuda, we see that

Bioko is $785 - 20 = 765$ square miles larger than Bermuda.[4]

On the log scale, however, the comparison corresponds to seeing how many *times* larger one island is than another. Thus,

Bioko is $785/20 = 39.25$ times larger than Bermuda.[5]

To see why the log scale indicates proportions we look at the histogram of the logarithms and see that visual comparisons correspond to subtraction of the logarithms because subtraction measures how far one log value is from another. And we know that subtraction of logarithms corresponds to division of original values, just as addition of logarithms corresponds to multiplication of original values.

In this example, the distribution seems well behaved under the log scale. This says that if we use a measurement scale that reflects proportionate size rather than actual size, we can think of the island sizes as being at least roughly symmetric and normally distributed. This is helpful because the normal distribution is much more easily understood than a skewed distribution with outliers.

With this in mind, why is Greenland no longer an outlier? On the one hand, it is undeniably true that Greenland is much larger than all of the other islands in terms of the number of square miles. On the other hand, when we consider how many times larger one island is than another, Greenland is within (albeit just barely) the usual variation in size from one island to another. To see that Greenland is not so terribly extreme in this proportionate sense, consider the following two comparisons:

Greenland is $840{,}000/4{,}700 = 179$ times larger than the Falkland Islands.

The Falkland Islands are $4{,}700/20 = 235$ times larger than Bermuda.

Proportionately speaking then, the comparison between the Falkland Islands and Bermuda is more extreme than the comparison between Greenland and the Falkland Islands. Thus, Greenland is no longer so extremely different from the others.

Exercise Set 5.5

Suppose that a researcher invents a scale of artistic value, based on the average price of the last ten works of an artist sold on the open market (providing that at least ten works had been sold on the open market in the last 10 years). In the process of developing the scale, the researcher first tries to use just the average price in dollars. The data for a sample of nine artists look like this:

10 100 200 300 400 500 1,000 5,000 10,000

The distribution is skewed toward high values, so the investigator tries a log transformation, to base 10, and gets the following values:

1.0000 2.000 2.3010 2.4771 2.6020 2.6990 3.0000 3.6990 4.0000

The histogram of the log values is symmetric, so the investigator decides to use this transformation.

16. Suppose another artist has an average sales price of 100,000. What is the artist's value on the log-transformed scale?

17. Suppose artist A has a value of 1 and artist B a value of 2 on the log scale.

  a. How many *more* dollars does artist B make on an average sale?

  b. How many *times* as much does artist B make on an average sale?

18. Suppose artist C has a value of 3 on the log scale.

  a. How many *more* dollars does artist C make compared to artist B on an average sale?

  b. How many *times* as much does artist C make compared to artist B on an average sale?

Computer Analysis

Once raw data are stored on a computer file, it is very easy to use the computer to transform the data values. The transformed values can then be stored in different columns, and analysis can be done on either the transformed values or the raw values.

In the island area example, the data set (shown in Display 5.13) was stored as a variable named "area" in column c1. The following three commands were used to calculate and store the square roots, logs to base 10, and log to base e, in columns c2, c3, and c4, respectively.

```
MTB > Sqrt 'area' c2.
MTB > LogTen 'area' c3.
MTB > LogE 'area' c4.
```

We then renamed the columns "root," "Log10," and "Ln_e." We can now use the computer for preliminary graphic analysis (done here with dot plots) as well as for calculating various numerical summaries.

DISPLAY 5.13

COMPUTER-TRANSFORMED ISLAND AREAS

```
MTB > DotPlot 'area'-'Ln_e'.

:
:
:
:
:
:
:
:
:
:: :                                                    .
+---------+---------+---------+---------+---------+-------area
0     150000    300000    450000    600000    750000

:
: .
::::
::::. ..  :                                  .
+---------+---------+---------+---------+---------+-------root
0        200       400       600       800      1000

 . .  . ..:  :    . . ...:....:    .. :               .
---+---------+---------+---------+---------+---------+---Log10
 1.0       2.0       3.0       4.0       5.0       6.0

            .
    ..  . .:  ..    . . : :..:.:    :  ..              .
-------+---------+---------+---------+---------+---------Ln_e
     2.5       5.0       7.5      10.0      12.5      15.0

MTB >

MTB > Describe 'area'-'Ln_'e'.
```

	N	MEAN	MEDIAN	TRMEAN	STDEV	SEMEAN
area	27	36510	1396	5831	160965	30978
root	27	82.0	37.4	51.8	175.9	33.8
Log10	27	2.930	3.145	2.894	1.244	0.239
Ln_e	27	6.747	7.241	6.663	2.864	0.551

	MIN	MAX	Q1	Q3
area	7	840000	47	4700
root	2.6	916.5	6.9	68.6
Log10	0.845	5.924	1.672	3.672
Ln_e	1.946	13.641	3.850	8.455

Summary

Transformation is changing the system of measurement of the data so that description and analysis are simpler. The main goals possibly achieved by transforming a single group of numbers are (1) an effective display of the data (especially when data values span several orders of magnitude), and (2) approximate symmetry of the distribution shape (when the untransformed data are skewed). To change the shape of the distribution, a transformation must do more than merely add a constant to each data value (which would only shift the distribution) or multiply each data value by a constant (which would only stretch or contract the distribution in the same way overall).

We will usually transform only data sets consisting of positive values and possibly zeros. If data contain negative values representing the difference between two positive quantities, then we might transform the original positive quantities before taking differences. To correct for skewness toward high values, the most commonly used transformations are *square roots* and *logarithms*. The logarithm transformation is stronger than the square root transformation. When logarithms are used, the scale shows proportionate differences, indicating how many times larger one data value is than another (instead of how much larger in absolute measurement units) because logarithms turn multiplication into addition. Log transformations are often appropriate for *highly* skewed data.

Problems

Basic Reviews

1. Consider these numbers: 2, 4, 8, 16, 32.

a. Find their logs to base 2.

b. Find their logs to base 10 or to base e.

2. Find the square roots of these numbers: 4, 16, 25, 36.

3. Find the logarithms to base e (natural logs) of the following numbers.

a. .02

b. 1.0

c. 2.72

d. 10

e. 100

f. 50,000

4. Find the logarithms to base 10 (common logs) of the numbers in Problem 3.

5. Find the square roots of the numbers in Problem 3.

6. Just for fun, find the squares of the numbers in Problem 3.

Analysis

7. Suppose we were studying bacteria growth and got these estimated numbers in a series of tests:

100 198 412 786 1,698 3,250

a. Construct a quick box plot of the numbers. Does it appear symmetric or skewed?

b. Take logs of the original numbers.

c. Construct a quick box plot of the logs. Does it appear symmetric or skewed?

8. Here are the numbers of deaths in major earthquakes from Display 5.1. They are clearly skewed toward high values.

8	62	81	115	146	250	1,000	1,000
1,200	1,300	1,621	4,000	4,200	40,000	55,000	

a. How many orders of magnitude do they span?

b. Would a transformation make their distribution more symmetric?

9. Table 5.7 gives the areas of major world lakes in square miles.

Table 5.7 *Areas, in square miles, for major world lakes*

Aral Sea	25,300	Maracaibo	5,217
Athabasca	3,064	Michigan	22,300
Balkhash	7,115	Nicaragua	3,100
Baykal	12,162	Nyasa	11,430
Caspian Sea	148,550	Onega	3,710
Chad	6,300	Ontario	7,550
Erie	9,910	Superior	31,700
Eyre	3,600	Tanganyika	12,700
Great Bear	12,028	Titicaca	3,200
Great Slave	11,031	Victoria	26,828
Huron	23,100	Volta	3,276
Ladoga	6,835	Winnipeg	9,417

a. By eye, does it appear that this data set might be a good candidate for transformation? Why or why not?

b. Make a stem-and-leaf plot or (by computer) a histogram or dot plot of the data. Describe the shape of the distribution.

c. Do a square-root transformation of the lake areas. Create a stem-and-leaf plot or (by computer) a histogram or dot plot of the square roots of the lake areas. Describe the shape of the distribution.

d. Do a log transformation of the lake areas. Construct a stem-and-leaf plot or (by computer) a histogram or dot plot of the logs of the lake areas. Describe the shape of the distribution.

e. Which transformation makes the distribution most nearly symmetric?

10. Discuss (using graphic displays or numerical analysis as appropriate) whether the data set in Table 5.8 (page 166) is a good candidate for transformation. If it is, perform an appropriate transformation and describe the results.

11. Suppose that you are able to measure the amount of energy released in an earthquake. Most earthquakes are quite small and never even noticed except locally, but some of the larger ones do a great deal of damage and are widely reported, becoming "major" earthquakes in popular reference works. Still others are quite large but take place in remote areas, or deep under the ocean, and only frighten the wildlife. In any case, you obtain the following energy-release figures for a sample of major earthquakes.

316,000	50,118,000	12,589,000	63,095,000	125,892,000
19,952,000	3,162,000	19,952,000	6,309,000	7,943,000
1,995,000	3,162,000	50,118,000	50,020,000	6,309,000

Table 5.8 *Debts owed to the United States arising from World War I, in millions of dollars*

Armenia	11	Hungary	2
Austria	26	Italy	2,044
Belgium	423	Latvia	7
Cuba	10	Liberia	.026
Czechoslovakia	185	Lithuania	6.6
Estonia	16	Nicaragua	.14
Finland	9	Poland	213
France	4,128	Romania	68
Germany	1,059	Russia	192
Great Britain	4,933	Yugoslavia	63
Greece	34		

Discuss (using graphic displays or numerical analysis as appropriate) whether the data set in Table 5.8 is a good candidate for transformation. If it is, perform an appropriate transformation and describe the results.

12. Suppose you decide to investigate some refinements of Pavlovian conditioning. Using laboratory rats, you pair electric shocks with flashing lights on several occasions, then you allow the rats to approach a water bottle. As the rat approaches the bottle, the flashing lights are activated. You measure how much longer than 5 seconds it takes the rat to complete a total of 5 seconds of licking at the water bottle. These data are shown in column 1, "lights," of Table 5.9,[6] as well as in data file RatLicks.mtw. [Data are loosely

Table 5.9 *Time over 5 seconds required by rats to complete 5 cumulative seconds of licking in the presence of three different conditioned stimuli, each of which has been previously paired with electric shocks*

LIGHTS	NOISE	ODOR
5.15	13.24	3.36
2.28	.79	6.81
1.11	1.62	.76
13.54	3.31	7.01
2.88	12.06	1.42
18.09	6.51	1.26
4.47	20.09	3.95
12.92	3.96	3.03
6.04	9.04	1.86
3.98	4.48	2.39
4.70	5.64	1.05
3.14	15.66	1.40
5.40	1.20	3.65
48.35	4.81	3.53
10.36	58.51	3.81

modeled after Yin et al. (1994).] Describe this distribution and, if appropriate, transform the data and describe the transformed distribution.

13. In a variation of the experimental situation described in Problem 12, you pair the electric shocks with the sound of a buzzer, then test for the effects of the buzzer. These data are shown in column 2, "noise," of Table 5.9 and in data file RatLicks.mtw. Describe this distribution and, if appropriate, transform the data and describe the transformed distribution.

14. In another variation of the experimental situation described in Problem 12, you pair the electric shocks with the smell of wintergreen, then test for the effects of this odor. These data are in column 3, "odor," of Table 5.9 and in data file RatLicks.mtw. Describe this distribution and, if appropriate, transform the data and describe the transformed distribution.

15. Does it appear that a transformation would be helpful in describing the GL− data in Table 5.10?[7] Why or why not? What might be an appropriate transformation, if any? [*Note:* G is a particular gene. In GL+ and GL−, this gene has mutated (L stands for "lost"). GW+ and GW− are the unmutated "wild" types. The "+" and "−" refer to the presence or absence of dextrose sugar in the growth solution.]

Table 5.10 *Numbers of cells in growth solution on a series of days, in hundreds of thousands*

DAY	GL−	GL+	GW−	GW+
1	1.0	1.0	1.0	1.0
3	1.2	1.1	2.7	2.8
5	3.7	4.0	6.8	11.0
7	6.9	7.7	15.0	29.5
9	5.5	5.5	12.0	56.5
11	6.8	8.0	10.5	210.0
13	12.2	12.2	5.0	550.0

16. Does it appear that a transformation would be helpful in describing the GL+ data in Table 5.10? Why or why not? What might be an appropriate transformation, if any?

17. Does it appear that a transformation would be helpful in describing the GW− data in Table 5.10? Why or why not? What might be an appropriate transformation, if any?

18. Does it appear that a transformation would be helpful in describing the GW+ data in Table 5.10? Why or why not? What might be an appropriate transformation, if any?

19. Table 5.11[8] (page 168) contains data on the length of time it takes junior high and high school students to respond to a graphing problem. Some of the subjects are given multiple choice (MC); others are given a pencil and a sheet of graph paper for use in free response (FR).

a. Graph each distribution. Note that the intervals of the table are not equal.

b. Are transformations likely to be helpful? Why or why not?

Table 5.11 *Number of subjects responding to a graphing problem in a given time interval, under multiple-choice* (MC) *and free-response* (FR) *conditions*

TIME	MC	FR
under 15 sec	55	5
15–29 sec	25	30
30–44 sec	100	60
45–59 sec	5	5
1 min	330	270
2 min	100	200
3 min	5	75
4 min	2	25
5 min	0	5
6 min	0	2

chapter 6

Choosing a Description

The mode, the median, and the mean all provide descriptions of the center of a distribution. The range, the interquartile range, and the standard deviation all provide measures of variability.

The *median* and the *interquartile range* are usually paired as one set of descriptors, and the *mean* and the *standard deviation* as another. In this chapter we discuss when one set is preferred over the other. We also explore some less commonly used descriptors that are useful in certain situations.

It may be that we want to do normal curve calculations using the mean and standard deviation. If the distribution is skewed, these calculations can be misleading. In such cases we may want to transform the data, as discussed in the previous chapter.

6.1 Measures of Center: The Mode, Median, and Mean

The mode, median, and mean will be exactly equal whenever the distribution is perfectly symmetric. In the presence of the randomness that we always expect to see in real data, they will be approximately equal when the distribution is fairly symmetric. If the distribution is not really symmetric, the measures can give systematically different values.

The Mode

The mode is the most common category for a categorical variable. In fact, it is the only measure of the center that we can meaningfully present for a categorical variable. If the variable consists of ordered categories or numerical measurements, we can compute the mode as the most common value or small range of values. This can be instructive, but the mode has no associated measure of variability and does not provide a good basis for further analysis. For numerical measurements, we are usually interested in the median or the mean of the distribution rather than the mode.

The Mean and Median

EXAMPLE 6.1 *Mean and Median of Symmetric Data: Time for a Ball to Roll*

Consider the stem-and-leaf plot of the times a ball took to roll down a hill, shown in Display 6.1. This distribution is not perfectly symmetric, but it is reasonably so if we consider that this is a fairly small amount of data.

DISPLAY 6.1

STEM-AND-LEAF PLOT OF THE TIME REQUIRED, IN HUNDREDTHS OF A SECOND, FOR A BALL TO ROLL DOWN AN INCLINED PLANE

Stem	Leaf
91	
92	5
93	
94	7 7 8 9
95	0 3 3 3 3 4 7
96	0
97	0
98	2
99	

The average value is .9534 second and the median value is .9530 second, two different summaries of the typical amount of time a ball required to travel along its course. Both summaries are located near the center of the distribution, as we see clearly from looking at the display, and both summaries are nearly identical in this case.

EXAMPLE 6.2 *Mean and Median of Skewed Data: Home Prices*

Now let's consider a case where the distribution really is skewed. Here, in thousands of dollars, are the listed prices for a group of homes advertised for sale:

94	95	96	98	99	101	105	108
110	115	118	130	130	135	135	137
140	140	142	144	145	150	150	150
154	155	156	165	170	170	179	180
210	218	230	299	475			

This is a real data set—they are all the houses advertised for sale one Thursday in a Seattle paper in 1993—and they are a good example of a general phenomenon. Often economic data are not normally distributed. Notice the 475 thousand dollars at the end. It is not a mistake, but it is far from the next highest number. This is even clearer when we look at the stem-and-leaf plot shown in Display 6.2.

DISPLAY 6.2

STEM-AND-LEAF PLOT OF HOUSE PRICES.
(n = 37; LEAF UNIT = 10)

```
0 | 99999
1 | 000111333334444 4
1 | 55555567778
2 | 113
2 | 9
3 |
3 |
4 |
4 | 7
```

Notice that the distribution now really appears skewed. To see how the shape of the distribution affects the performance of our measures of center, we used Minitab to compute measures of central tendency. The results are shown in Display 6.3. Notice that the mean is 154.8 and the median 142.0. The mean is close to $13,000 above the median, which is a substantial difference. Because the distribution is not symmetric, the tail on the high end pulls the mean up.

DISPLAY 6.3

```
Describe 'Prices'.

            N       MEAN     MEDIAN    TRMEAN    STDEV    SEMEAN
Prices      37      154.8    142.0     144.4     68.8     11.3

            MIN     MAX      Q1        Q3
Prices      94.0    475.0    112.5     167.5
```

Robustness to Outliers

In the presence of an outlier, the mean and the median will usually give different results. To see the effect of an outlier on each measure of center, consider the following simple data set.

Original data: 3 6 7 8 10

Now, suppose this is copied and an outlier is introduced by mistakenly copying 10 as 50.

Copied data: 3 6 7 8 50

Display 6.4 shows how each measure of center behaves under the introduction of this outlier. The mean has jumped from 6.8 to 14.8, whereas the median remained unchanged and unaffected at 7.

DISPLAY 6.4

	ORIGINAL DATA (NO OUTLIER) 3, 6, 7, 8, 10	COPIED DATA (ONE OUTLIER) 3, 6, 7, 8, 50
Mean	6.8	14.8
Median	7	7

We can visualize this situation using the number line in Figure 6.1. In this illustration, we see that the mean has been pulled to the right by the outlier, whereas the median has resisted this tendency. In fact, the mean has been driven to the right well beyond most of the data and is located somewhere in empty space between most of the data and the outlier.

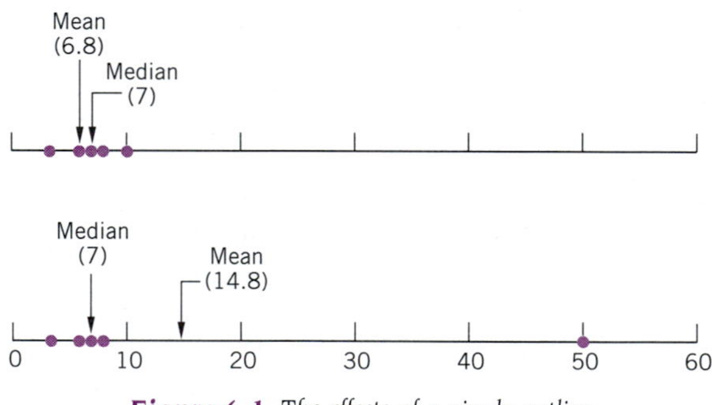

Figure 6.1 *The effects of a single outlier*

This expresses a fundamental difference between the two measures: The mean is much more sensitive to extreme values than the median. Thus, the median is more robust than the mean. (Recall that a *robust* statistic is not sensitive to outliers in the data.) Although we often want to use a method that is as sensitive as possible to all the data, especially when the data have been gathered at great trouble and expense, such a method can backfire if it pays too much attention to only a few of the data values.

Exercise Set 6.1

1. Consider this data set: 2, 4, 4, 4, 6. By eye, will the mean be different from the median? Why or why not?

2. Consider this data set: 2, 4, 4, 4, 60. Without doing any arithmetic, do you think the mean will be different from the median? If different, will it be higher or lower?
3. Consider this data set: 2, 40, 40, 40, 60. Without doing any arithmetic, do you think the mean will be different from the median? If different, will it be higher or lower?
4. Here, in thousands of dollars, are the prices of the first 30 homes listed for sale in *The Seattle Times* on Sunday, 30 July 1995:

79	83	85	106	110	115	116	119	129	129	130	139	140	143	143
150	152	167	179	182	185	189	195	200	200	235	290	349	369	374

   a. Do a stem-and-leaf plot or a dot plot of these data.
   b. By eye, is there strong reason to think that the mean and median will be substantially different?
   c. On the basis of experience with similar data sets, is there reason to think that the mean and the median will be substantially different?
   d. Calculate the mean and median, and discuss whether they are "substantially" different.
5. Here, in thousands of dollars, are the prices of all the homes on the east side of Lake Washington advertised by one particular broker in *The Seattle Times* on Sunday, 30 July 1995. (The last house is a waterfront home in a *very* nice area.)

140	145	150	170	184	185	190	240	250	252	259	279	295	299	300	315
320	349	349	349	382	394	399	425	450	475	515	540	565	639	899	2,975

   a. Do a stem-and-leaf plot or a dot plot of these data.
   b. By eye, is there strong reason to think that the mean and median will be substantially different?
   c. On the basis of experience with similar data sets, is there reason to think that the mean and the median will be substantially different?
   d. Calculate the mean and median, and discuss whether they are "substantially" different.

The Trimmed Mean

There is a compromise choice between the mean and the median. The **trimmed mean** removes observations (as one might trim branches from a tree) from both extremes of the distribution and then averages the remaining data.

If there are not too many outliers, they will be trimmed away, providing robustness. The averaging process then provides a summary that is quite sensitive to the remaining data. If there are no outliers, the trimming process will affect both ends of the distribution approximately equally. Then the trimmed mean will not differ very much from the ordinary untrimmed mean.

Look back at the computer analysis of house prices (Display 6.3). There is an entry labeled "TRMEAN," for "trimmed mean," whose value is 144.4. This is quite close to the value for the median (142.0), and substantially below the value of the untrimmed mean (154.8).

Notice that in the trimming process it is as though we were making an important assumption—that the outliers are mistakes and do not really belong in the data set.

PROCEDURE: FINDING THE TRIMMED MEAN OF A SINGLE GROUP OF NUMBERS

1. Put the numbers in order from smallest to largest.
2. Delete an equal number of the largest and smallest data values.
3. Calculate the average of the data that remain.

How many data values should be trimmed at step 2? It depends on how many outliers we need to protect against and how sensitive the summary measure needs to be. These two goals should be balanced against each other. If we do not trim at all, we are left with the mean. If we trim as much as possible, only the middle value (or values) remain, and we will find the median. Any trimming amount between these extreme cases therefore represents a compromise between the mean and the median.

One commonly used choice is called the **10% trimmed average.** We actually trim 10% of the data from each of the two ends, resulting in an overall total of 20% of the data being removed. If we have 10 data values, the largest and the smallest numbers will be removed. Thus a single outlier will not be able to greatly distort the computed summary value because it will be trimmed, regardless of whether it is a small or large outlier.

When the sample size is not evenly divisible by 10, the 10% trimmed mean is found by truncating (rounding down to the nearest smaller whole number). For example, for a group of 37 numbers, the 10% amount is 3.7, which is not a whole number. Therefore, we truncate 3.7 to 3 and trim three data values from each end of the distribution.

Minitab calculates a 5% (rounded) trimmed mean.

EXAMPLE 6.3 *Calculating a Trimmed Mean: Attention Spans*

Consider the following set of 11 numbers representing (hypothetical) attention spans, in minutes, for students in a class.

5 18 15 2 8 55 11 3 9 8 6

Proceed as follows to find the 10% trimmed mean:

1. Order: 2 3 5 6 8 8 9 11 15 18 55
2. Trim: 2 and 55
3. Average: 3 5 6 8 8 9 11 15 18

$$\begin{aligned}10\% \text{ trimmed mean} &= (3 + 5 + 6 + 8 + 8 + 9 + 11 + 15 + 18)/9 \\ &= 9.22 \text{ minutes}\end{aligned}$$

We have trimmed one number from each end because 10% of 11 is 1.1, which truncates to 1. There is an outlier, 55, that shows up as the largest observed time. (Perhaps 5 or 5.5 was meant instead? Or does some student have an extraordinary attention span?) The outlier 55 was trimmed away and could not affect the trimmed mean. The ordinary mean, 12.73, is larger than the trimmed mean, indicating the ability of the outlier to influence this summary when it is not trimmed away.

Discussion

Although the trimmed mean is indeed better than the ordinary mean when we suspect mistakes in the data, it is not always better. Trimming is good in that it provides some protection from errors while retaining much of the sensitivity of the mean. This robustness does not come free, however; a price must be paid. One cost of using trimmed means is that the summary is not the "best" when the data do come from a normal distribution with no extreme outliers. For normally distributed data, it can be proven mathematically that the average of all the data is the best possible summary of the center of the distribution. Even if the data are not normally distributed, trimming may discard real information.

Exercise Set 6.2

6. Consider these 11 numbers: 1, 3, 4, 5, 5, 6, 7, 7, 8, 9, 10.

a. To compute a 10% trimmed mean, how many numbers would you drop from each end?

b. What is the 10% trimmed mean of these numbers?

7. Consider these 12 numbers: 1, 3, 4, 5, 6, 6, 7, 7, 8, 9, 10, 33.

a. What is their 10% trimmed mean?

b. Is the untrimmed mean different?

8. Suppose you have been collecting data on the heights of second graders. You are not doing the measurements yourself, but are using data reported by school nurses to a review panel at the school board. By the time you get the data, all identifying information has been stripped off, including the individual school where the data were collected, so there is no way to go back and check individual measurements. On doing a stem-and-leaf plot, you notice some unusual observations: At least one student appears to be under 1 foot tall, and another appears to be over 30 feet tall. Describe a way to handle this situation.

6.2 Measures of Variability: The Standard Deviation Compared to the Interquartile Range

Generally speaking, a variability measure is basically a summary of the extent to which the data values differ from each other (or from a summary measure). This concept seems simple, but in practice there is disagreement on just how this should be done. Perhaps any method that seeks a single number to summarize a potentially complicated situation is asking for trouble, but there are several different reasonable opinions. Let us compare and contrast the two major variability measures we have studied so far.

The standard deviation is the traditional, universally accepted measure of variability, whereas the interquartile range is less common but has gained some acceptance. How do these two measures compare to one another? First, they both adhere to the following basic principle:

The location of the center is irrelevant to the variability and is ignored when measuring it.

This makes sense because the center has already been summarized (by the mean, the median, or something else) and we are supposed to be looking for some other (different) aspect of the data set now. Thus, if two distributions are the same except for being centered at different places, then their variabilities should be the same. Both the standard deviation and the interquartile range meet this requirement.

In particular, if a fixed value is added to (or subtracted from) each data value, then the variability should remain unchanged. For example, the two distributions in Display 6.5 (page 176) have the same variability because each data value in stem-and-leaf plot (a) (3, 16, 22, 24, . . . , 57) is 20 less than the corresponding value in stem-and-leaf plot (b) (23, 36, 42, 44, . . . , 77).

DISPLAY 6.5

TWO DATA SETS WITH DIFFERENT CENTERS, BUT OTHERWISE IDENTICAL

	Stem	Leaves
a	0	3
	1	6
	2	2 4 8 8 9
	3	2 6 7
	4	4 5 5 8
	5	1 7
	6	
	7	
b	0	
	1	
	2	3
	3	6
	4	2 4 8 8 9
	5	2 6 7
	6	4 5 5 8
	7	1 7

The standard deviation is unaffected by adding or subtracting constants because the deviations from the mean will be unchanged. These deviations represent the distance from each data value to the center, and this distance is unchanged by shifting the data set to the right (by adding) or the left (by subtracting) because the mean will also shift by the same amount. The interquartile range is also unaffected because it is also a relative distance: the distance between upper and lower quartiles. Again, this distance is not changed by shifting the data to the right or left because both quartiles will shift by the same amount.

Comparison for a Normal Distribution

Although they both adhere to this fundamental point, the two variability measures differ in many other respects. Even for a normal distribution, each would probably result in a different number for the variability summary. Fortunately, however, in the case of a normal distribution, they measure basically the same thing, and it is easy to convert one to the other. The basic relationship between the two measures of variability is illustrated in Figure 6.2.

First, note that the interquartile range is larger than the standard deviation. Each quartile is less than one standard deviation away from the center (mean and median are the same here), but the distance from one quartile to the other is larger than a standard deviation (but less than two standard deviations). Notice also that one-quarter of the data values (represented by one-quarter of the area under the bell-shaped curve) are located outside each of the two quartiles. One-quarter of the values are smaller than the lower quartile and another quarter are larger than the upper quartile.

By consulting a normal table (Chapter 4), we can say that the interquartile range is about one-third larger than the standard deviation (about 34.9% larger to be more exact). If we compare the other way, the standard deviation is smaller by a factor of about three-fourths, or about 74.1% as big as the interquartile range.

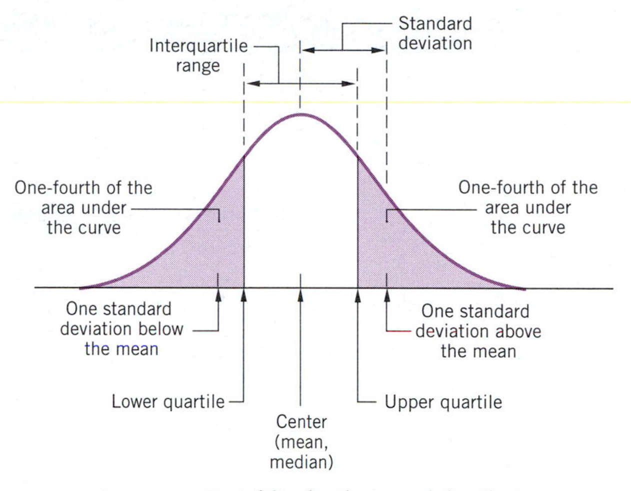

Figure 6.2 *Variability for the normal distribution*

If we summarize for an approximately normal distribution, it will be approximately true that

$$\frac{\text{Interquartile range}}{\text{Standard deviation}} = 1.3$$

In practice, this relationship will be roughly true whenever the data look somewhat normally distributed (a fairly symmetric histogram with a bell-shaped curve trailing off in the tails with no extreme outliers).

For the 15 earthquake magnitudes—5.5, 6.3, 6.5, 6.5, 6.8, 6.8, 6.9, 7.1, 7.3, 7.3, 7.7, 7.7, 7.7, 7.8, 8.1 (from Chapter 2)—the variability measures are easily computed, as follows:

$$\text{Standard deviation} = .696$$

$$\text{Interquartile range} = 1.05$$

The interquartile range is indeed bigger, and the ratio is

$$\frac{\text{Interquartile range}}{\text{Standard deviation}} = \frac{1.05}{.696} = 1.5$$

This is close to the ideal value of 1.3 for a perfectly normal distribution.

Comparison for a Skewed Distribution

When the data follow a distribution that is not normal, these two variability measures are not so nicely related to each other. For skewed distributions and those with outliers, the standard deviation can be larger (sometimes much larger) than the interquartile range. This is because the standard deviation responds quite readily to outliers and the interquartile range does not.

EXAMPLE 6.4 *Variability of Skewed Data: Revenues of States*

Consider the distribution of the revenues of states in the United States shown in Display 6.6.[1]

DISPLAY 6.6

REVENUES OF STATES IN THE UNITED STATES, IN BILLIONS OF DOLLARS, 1992

```
MTB > Stem-and-Leaf 'Billions'.

Stem-and-Leaf of Billions N = 50
Leaf Unit = 1.0

(30)  0 | 1 1 1 1 2 2 2 2 3 3 3 3 4 4 4 4 5 5 5 6 7 7 7 8 9 9 9 9 9 9
 20   1 | 0 0 2 2 3 3 4 4 5 6 8
  9   2 | 4 4 4 5 7
  4   3 | 1 3
  2   4 |
  2   5 |
  2   6 | 5
  1   7 |
  1   8 |
  1   9 | 0

MTB > Describe 'Billions'.

              N       MEAN     MEDIAN    TRMEAN    STDEV    SEMEAN
Billions      50      13.20    9.15      10.56     15.92    2.25

              MIN     MAX      Q1        Q3
Billions      1.63    90.78    4.15      14.71
```

This is clearly a nonnormal distribution, seen in the extreme, asymmetric skewness. The standard deviation of these values is $15.92 billion, which is larger than the interquartile range of $10.56 billion. Note that the ratio

$$\frac{\text{Interquartile range}}{\text{Standard deviation}} = \frac{10.56}{15.92} = .66$$

is about half the ideal value (1.3) that we would expect for normally distributed data. The standard deviation is large here because it is more sensitive to the large revenue values of a few states (represented as a long tail on the right), whereas the interquartile range is more responsive to the middle of the distribution.

General Comparison for a Skewed Distribution

Figure 6.3 shows how the mean, median, standard deviation, and interquartile range might compare for a skewed distribution. For completeness, the mode (the highest point, indicating the densest part of the distribution) is also illustrated. The long tail toward the right corresponds to the presence of a few large values in that region.

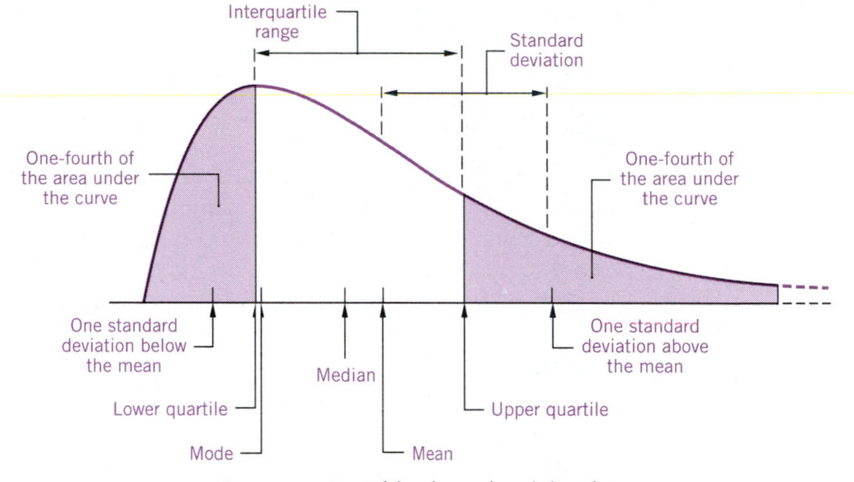

Figure 6.3 *Variability for a skewed distribution*

Discussion: Effects of Outliers

Even a single outlier can have profound effects on the standard deviation, but its influence over the interquartile range is limited. Display 6.7 shows how these variability measures change in response to an outlier. Five of the six data values are kept the same. The sixth value goes from reasonably small to reasonably large and then continues on to the ridiculously large value of 907, requiring a break in the scale on the histogram.

DISPLAY 6.7

HISTOGRAM	MEAN	MEDIAN	STANDARD DEVIATION	INTER-QUARTILE RANGE
7 6 5 2 4 5 0 1 2 3 4 5 6 7 8 9	21.5	25.0	11.0	15.0
7 6 2 4 5 6 0 1 2 3 4 5 6 7 8 9	30.00	26.5	14.7	11.0
7 6 2 4 5 4 0 1 2 3 4 5 6 7 8 9	36.3	26.5	29.2	11.0
7 6 2 4 5 7 0 1 2 3 4 5 6 7 8 90	171.8	26.5	360.2	11.0

Whereas the robust measures (the median and the interquartile range) change only somewhat, the nonrobust measures (the mean and the standard deviation) find themselves almost entirely at the mercy of this outlier. Notice how the mean goes from 21.5 to 171.8, while the median changes only from 25.0 to 26.5. A similar comparison can be made with the standard deviation and the interquartile range.

Generally speaking, robust measures summarize the "rule" (that is, the majority or bulk of the data observations), whereas the nonrobust measures accommodate everything, including any "exception(s)." Thus, nonrobust procedures may not provide an accurate reflection of the primary structure of the data when extreme outliers are present.

This suggests that a large standard deviation could be due to one of two things:

Possibility 1: *Generally large variability throughout the data set*
Possibility 2: *Generally small variability throughout the data set, with a few outlying data values*

To know which is the case we need to do more than just compute the standard deviation. We also need to *look at the data,* say, by using a histogram. Only then can we be sure that the statistical methods are accurately describing useful aspects of the data.

Exercise Set 6.3

9. Recall the prices, in thousands of dollars, of the first 30 homes listed for sale in *The Seattle Times* on Sunday, 30 July 1995:

79 83 85 106 110 115 116 119 129 129 130 139 140 143 143
150 152 167 179 182 185 189 195 200 200 235 290 349 369 374

a. If you have not already done a stem-and-leaf plot or a dot plot of these data, do one now.

b. By eye, is there strong reason to think that the ratio of the standard deviation to the interquartile range will be much different from what we expect for a normal distribution?

c. On the basis of experience with similar data sets, is there reason to think that the ratio of the standard deviation to the interquartile range will be much different from what we expect for a normal distribution?

d. Calculate the interquartile range and the standard deviation. Is their ratio substantially different from what we expect with a normal distribution?

10. Recall the prices, in thousands of dollars, of all the homes on the east side of Lake Washington advertised by one particular broker in *The Seattle Times* on Sunday, 30 July 1995. (The last house is a waterfront home in a *very* nice area.)

140 145 150 170 184 185 190 240 250 252 259 279 295 299 300 315
320 349 349 349 382 394 399 425 450 475 515 540 565 639 899 2,975

a. If you have not already done a stem-and-leaf plot or a dot plot of these data, do one now.

b. By eye, is there strong reason to think that the ratio of the standard deviation to the interquartile range will be much different from what we expect for a normal distribution?

c. On the basis of experience with similar data sets, is there reason to think that the ratio of the standard deviation to the interquartile range will be much different from what we expect for a normal distribution?

d. Calculate the interquartile range and the standard deviation. Is their ratio substantially different from what we expect with a normal distribution?

Summary

The process of choosing a way to describe a data set should be interactive. You should look at the data and ask questions about the data, and you should think about the audience to whom you will be presenting your analysis. We present an outline of the process here.

1. Determine what level of measurement has been used.
2. If you have categorical data, you are restricted to counts, percents, and the mode as descriptions. These can be presented as bar charts or pie charts.
3. If you have ordinal data, you can use histograms, medians, quartiles, and box plots to describe the data.
4. If you have numerical measurement data:
   a. Look at the data. Construct a stem-and-leaf plot, dot plot, or histogram.
   b. Are the data symmetrically distributed, with an approximately normal distribution?
   c. If the data are symmetric and approximately normally distributed, you have a choice. You can use the median and the interquartile range, or the mean and the standard deviation. If you are planning to do further computations (see the rest of this book), you will probably want to use the mean and standard deviation. If you are not planning further computations but are preparing data for presentation, the median and interquartile range, presented in box plots, are good choices. Medians are also often suitable if you are planning a nonparametric analysis.
   d. If the data are *not* symmetrically distributed, you have to decide whether the distribution is really skewed or whether the asymmetric shape results from errors in the data. If you think there are errors in the data, you can discard (if justified) or correct the mistakes and reexamine the data, or you can use a trimmed mean and forge ahead. If you think the distribution is really skewed, the mean and the standard deviation are bad choices because (1) the mean is pulled in the direction of the skew, away from the bulk of the cases, and (2) the standard deviation can be wildly inflated by the extreme values. Thus the mean and standard deviation not only poorly describe the data, but normal table calculations using them also give wrong answers. Therefore, you should use the median, interquartile range, and box plots to describe skewed data. This is precisely the situation for which they were designed.
5. If the data are skewed but you really want to do normal curve calculations, you may be able to *transform* the data so that they are normally distributed, as discussed in Chapter 5. This does not simplify the problem of describing the data to a nontechnical audience; indeed, it complicates the matter. But it does make it possible to do normal curve calculations, which can be tremendously helpful.

Problems

Basic Reviews

1. Consider the distribution shown and described in Display 6.8 on page 182.

   a. Does the distribution appear to be basically symmetric, or does there appear to be a skew toward high values?

   b. From looking at the stem-and-leaf plot, do you think the mean and median should be about the same?

DISPLAY 6.8

STEM-AND-LEAF PLOT AND BASIC STATISTICS COMPUTED ON A RANDOM SAMPLE OF 100

```
STEM-AND-LEAF OF C1 (N = 100)
Leaf Unit = 10

    1   2   7
    4   3   0 3 3
   21   3   5 6 6 6 6 7 7 7 7 7 8 8 9 9 9 9 9
   37   4   0 0 0 1 1 1 1 2 2 2 3 3 3 4 4 4
 (21)   4   5 5 5 5 6 6 6 6 7 8 8 8 9 9 9 9 9 9 9 9 9
   42   5   0 0 1 1 1 1 1 1 1 2 3 3 3 3 4 4 4
   25   5   5 5 6 6 6 6 7 7 7 7 8 9 9
   12   6   0 0 1 1 3 3
    6   6   5 5 6
    3   7   3
    2   7   5 8

MTB > Describe C1.

        N        MEAN      MEDIAN    TRMEAN    STDEV
C1      100      488.90    491.41    485.56    97.59

        MIN      MAX       Q1        Q3
C1      272.14   783.20    414.06    555.91
```

(*Note:* If you do hand calculations to check these values you will get slightly different answers. The original data had several more digits, and the computer used them all in its calculations. Only the first two digits are preserved in the stem-and-leaf plot.)

c. From looking at the basic statistics, are the mean and the median about the same, or are they substantially different?

d. In the basic statistics, Q1 is the lower quartile value and Q3 is the upper quartile value. Calculate the interquartile range for the data and compare it to the standard deviation. Is their ratio about right for a normal distribution?

e. Would calculations that assumed that the distribution was a normal distribution produce reasonable results?

2. Consider the distribution shown and described in Display 6.9.

a. Does the distribution appear to be basically symmetric, or does there appear to be a skew toward high values?

b. From looking at the stem-and-leaf plot, do you think the mean and median should be about the same?

c. From looking at the basic statistics, are the mean and the median about the same, or are they substantially different?

d. In the basic statistics, Q1 is the lower quartile value and Q3 is the upper quartile value. Calculate the interquartile range for the data and compare it to the standard deviation. Is their ratio about right for a normal distribution?

e. Would calculations that assumed that the distribution was a normal distribution produce reasonable results?

DISPLAY 6.9

STEM-AND-LEAF PLOT AND BASIC STATISTICS COMPUTED ON A RANDOM SAMPLE OF 100

```
STEM-AND-LEAF OF C5 (N = 100)
Leaf Unit = 1.0

  27   0 | 0 0 0 1 1 1 1 1 1 2 2 2 2 2 2 2 3 3 3 3 3 3 3 3 4 4 4 4
(32)   0 | 5 5 5 5 5 5 5 6 6 6 6 6 6 7 7 7 7 7 7 8 8 8 8 8 8 8 8 9 9 9 9 9
  41   1 | 0 0 0 0 0 1 1 2 2 2 2 2 2 2 3 3 3 4 4
  22   1 | 6 6 6 7 7 9
  16   2 | 0 0 2 3
  12   2 | 5 6 7 7 9
   7   3 | 1 1 2 3
   3   3 |
   3   4 |
   3   4 |
   3   5 |
   3   5 |
   3   6 |
   3   6 | 6
   2   7 | 2
   1   7 |
   1   8 | 0

MTB > Describe C5.

         N       MEAN     MEDIAN    TRMEAN    STDEV
C5       100     12.13    8.26      10.27     13.42

         MIN     MAX      Q1        Q3
C5       0.33    80.23    4.47      13.80
```

Analysis

3. Here are the prices of the houses listed for sale on June 4, 1992, in the Seattle morning paper, in thousands of dollars:

192	499	239	154	129	112	179	97	300	169	113	136
122	104	59	139	199	169	159	150	89	125	385	179
225	155	172	230	163	160	129	300	189			

Describe these data. Include a description of the distribution shape and measures of center and variability. Briefly indicate how the measures you use should be interpreted.

4. The Bell Telephone data are shown again in Table 6.1 (page 184). In Chapter 2 you did a stem-and-leaf plot of these data, in Chapter 3 you prepared a five-number summary and box plot of them, and in Chapter 4 you calculated their mean and standard deviation and compared the distribution to a normal distribution.

Table 6.1 *Debts of the Bell System's local operating companies in billions of dollars at the end of 1980*

Bell Telephone of Pennsylvania	$1.5
Chesapeake and Potomac Telephone (DC)	.2
Chesapeake and Potomac Telephone (MD)	.7
Chesapeake and Potomac Telephone (VA)	.7
Chesapeake and Potomac Telephone (WV)	.2
Cincinnati Bell	.2
Diamond State Telephone	.1
Illinois Bell Telephone	1.7
Indiana Bell Telephone	.5
Michigan Bell Telephone	1.2
Mountain States Telephone	2
New England Telephone	1.7
New Jersey Bell Telephone	1
New York Telephone	3.4
Northwestern Bell Telephone	1.3
Ohio Bell Telephone	.9
Pacific Northwest Bell Telephone	1.1
Pacific Telephone	5.8
South Central Bell Telephone	2.7
Southern Bell Telephone	3.2
Southern New England Telephone	.6
Southwestern Bell Telephone	4.6
Wisconsin Telephone	.5

a. Did the normal distribution come reasonably close to describing the shape of this distribution, or did it make some obviously wrong predictions?

b. Do you think this distribution would be more nearly normal if it were transformed, say with a log transformation?

c. *Open question:* Which description of the distribution do you prefer? Why? Which would you prefer to use to describe the data to, say, a meeting of the city council? Why? (You may assume that the members of the city council are reasonably bright, but that they have probably forgotten most of the little they learned in the one statistics course they took years ago.)

5. The Appalachian regional data are shown again in Table 6.2. In Chapter 2 you did a stem-and-leaf plot of these data and in Chapter 3 you prepared a five-number summary and box plot of them.

Table 6.2 *Per capita income, Appalachian portion of each state, in thousands of dollars (1965 dollars)*

Alabama	4.1	Ohio	3.8
Georgia	3.8	Pennsylvania	4.4
Kentucky	3.2	South Carolina	4.1
Maryland	4.1	Tennessee	3.8
Mississippi	3.1	Virginia	3.7
New York	4.1	West Virginia	4
North Carolina	4		

a. Would the mean and standard deviation be useful summaries of these data?

b. Calculate the mean and standard deviation.

c. Use the mean and standard deviation, together with your work from Chapters 2 and 3, to defend your answer to part (a).

6. The data on heights of boys on macrobiotic diets are shown again in Table 6.3. In Chapter 2 you did a stem-and-leaf plot of these data, in Chapter 3 you prepared a five-number summary and box plot of them, and in Chapter 4 you calculated their mean and standard deviation and compared the distribution to a normal distribution.

Table 6.3 *Heights of boys on macrobiotic diets, in centimeters*

718	883	853
630	768	840
1,002	830	820
560	810	862
715	813	785
1,010	1,105	1,132
1,085	860	1,084
750	1,090	1,152
510	615	490
1,112	912	781
1,093	684	1,000

a. Did the normal distribution come reasonably close to describing the shape of this distribution, or did it make some obviously wrong predictions?

b. Do you think this distribution would be more nearly normal if it were transformed, say with a log transformation?

c. *Open question:* Which description of the distribution do you prefer? Why? Which would you prefer to use to describe the data to a school dietitian? Why?

7. Consider the data on the total deposits of the ten largest U.S. commercial banks shown in Table 6.4.[2] (You may want to work with units of billions of dollars and will probably not want or need to keep all of the digits of accuracy listed here.) In Chapter 2 you did a stem-and-leaf plot of these data and in Chapter 3 you prepared a five-number summary and box plot of them.

Table 6.4 *Deposits, in dollars, of the ten largest commercial banks in the United States, about 1979*

Bank of America, San Francisco	$85,069,944,000
Citibank, New York	72,429,528,000
Chase Manhattan Bank, New York	56,173,381,000
Manufacturers Hanover Trust Co.	39,000,618,000
Morgan Guaranty Trust Co.	35,485,994,000
Chemical Bank, New York	30,521,757,000
Continental Illinois National Bank	25,566,850,000
Bankers Trust Co., New York	22,121,919,000
First National Bank, Chicago	20,784,969,000
Security Pacific National Bank	19,590,576,000

a. Would the mean and standard deviation be useful summaries of these data?

b. Calculate the mean and standard deviation.

c. Use the mean and standard deviation, together with your work from Chapters 2 and 3, to defend your answer to part (a).

8. Consider the seating capacities of baseball stadiums in the National League, shown in Table 6.5.[3] In Chapter 2 you did a stem-and-leaf plot of these data.

Table 6.5 *Baseball stadium seating capacities*

Atlanta Braves	52,003
Chicago Cubs	38,170
Cincinnati Reds	52,952
Colorado Rockies	approx. 50,000
Florida Marlins	47,226
Houston Astros	54,816
Los Angeles Dodgers	56,000
Montreal Expos	43,739
New York Mets	55,601
Philadelphia Phillies	62,382
Pittsburgh Pirates	58,727
St. Louis Cardinals	56,227
San Diego Padres	59,022
San Francisco Giants	58,000

a. Do you think the mean and standard deviation would be appropriate summaries of these data?

b. Calculate the mean and standard deviation.

c. Use the mean and standard deviation, together with your work from Chapter 2, to defend your answer to part (a).

9. Consider the number of deaths caused by the 15 major earthquakes:

8	62	81	115	146	250	1,000	1,000
1,200	1,300	1,621	4,000	4,200	40,000	55,000	

a. Will the mean or the median be higher?

b. Does it appear that a log transformation will make this data distribution more symmetric?

c. Transform the data and do a stem-and-leaf plot (you may want to do this by computer—the raw data are on the course data disk as "equakes.mtw").

d. Is the transformed data distribution more nearly symmetric?

10. The following questions are based on the data displayed in Table 6.6.[4] If you write out the answers to all of them you will have a nice report of a basic statistical analysis.

a. Construct a stem-and-leaf plot or a classic histogram of the three tree variables in the display above. Do they appear to be basically symmetric, badly skewed, or somewhere in between? Do they appear so skewed that transformation will be needed to make them more symmetric?

b. Compute the mean and standard deviation of each variable.

Table 6.6 *Three dimensions of a group of trees:* Diameter *(inches at* 4.5 *feet above ground),* height *(feet),* and volume *(cubic feet)*

DIAMETER	HEIGHT	VOLUME	DIAMETER	HEIGHT	VOLUME
8.3	70	10.3	12.9	85	33.8
8.6	65	10.3	13.3	86	27.4
8.8	63	10.2	13.7	71	25.7
10.5	72	16.4	13.8	64	24.9
10.7	81	18.8	14	78	34.5
10.8	83	19.7	14.2	80	31.7
11	66	15.6	14.5	74	36.3
11	75	18.2	16	72	38.3
11.1	80	22.6	16.3	77	42.6
11.3	79	24.2	17.3	81	55.4
11.4	76	21	17.5	82	55.7
11.4	76	21.4	17.9	80	58.3
11.7	69	21.3	18	80	51.5
12	75	19.1	18	80	51
12.9	74	22.2	20.6	87	77

c. Compute the median, the quartiles, and the interquartile range of each variable (preferably by computer).

d. Are the means and the medians about the same or substantially different?

e. What are the ratios of the standard deviations to the interquartile ranges? How do these compare with what we would see with a perfectly normal distribution?

11. Retrieve the list of prices you compiled for Problem 3, Chapter 1.

a. Does the distribution appear to be skewed or symmetric? (You may have prepared plots as a Chapter 2 problem.)

b. By eye, do you think the mean and the median will be about the same? Why or why not?

c. Calculate either (1) the median, upper and lower quartiles, and interquartile range, or (2) the mean, standard deviation, and the values one standard deviation above and below the mean. State why you chose the set you did.

d. Compare your answers in part (c) with the estimate you did by eye for Problems 3(b) and 3(c) of Chapter 1. About how close were your eyeball estimates to the calculated values?

12. Table 6.7 (page 188) shows data on the percent of time males and females of one species of fish spend "guarding" or "fanning" the eggs in their nest, under certain test conditions.[5] (Note that we are not suggesting these fish as models of human mating behavior.) Your answers to parts (a) and (b) should include measures of central tendency and variability, a brief indication of why you chose those measures, and a verbal comparison. Graphics would be nice, too.

a. Describe the percent of time males spend guarding the nest. Compare this to the percent of time females spend guarding the nest.

b. Describe the percent of time females spend fanning the eggs. Compare this to the percent of time males spend fanning the eggs.

Table 6.7 Percent of time males and females guard or fan eggs in their nests

MALES		FEMALES	
GUARD	FAN	GUARD	FAN
48.5	3	26.9	65.6
54.8	.7	24.2	62.8
50.9	6	32.4	52.1
58.1	7.5	35.9	44.8
49.7	5	36.8	49.7
42.5	5.5	42.4	42.1
41.4	5.9	35.6	49.1
35.4	11.5	32.9	49.7
46	14	29.2	51.1
40.7	18.8	32.3	49.8

13. Table 6.8 shows part of a table reporting total income for U.S. households in larger cities.[6] Household numbers are in thousands. The median income is reported at the bottom of the column of household counts.

Table 6.8 Household income

RANGE	NUMBER OF HOUSEHOLDS
Less than $5,000	2,280
$5,000 to $9,999	3,563
$10,000 to $14,999	3,056
$15,000 to $19,999	2,717
$20,000 to $24,999	2,688
$25,000 to $29,999	2,844
$30,000 to $34,999	2,114
$35,000 to $39,999	1,582
$40,000 to $49,999	2,593
$50,000 to $59,999	1,982
$60,000 to $79,999	2,136
$80,000 to $99,999	830
$100,000 to $119,999	485
$120,000 or more	817
Total households	29,687
Median income	$25,947

a. How many thousand households had incomes of less than $20,000?

b. What income category breakpoint (rounding to the nearest percent) is the 30th percentile?

c. There are two unexpected jumps in the number of households in each category—one as income goes over $40,000 and the other as it goes over $60,000. How do you explain these jumps? (*Hint:* Look first at how the table is constructed.)

d. Why does the census bureau report median rather than mean income?

14. *Computer problem:* Minitab file WGTHGT.MTW includes information on the weight (in pounds) and height (in inches) of 73 students in two first-grade classes. Stem-and-leaf-plots and basic statistics for the two variables are shown in Display 6.10.

DISPLAY 6.10

WEIGHT (IN POUNDS) AND HEIGHT (IN INCHES) OF 73 STUDENTS IN TWO FIRST-GRADE CLASSES

```
MTB > Stem-and-Leaf 'WEIGHT' 'HEIGHT'.

Stem-and-leaf of WEIGHT N = 73
Leaf Unit = 1.0

    8   3 | 0 1 2 2 4 4 4 4
   21   3 | 5 5 5 5 6 7 7 7 8 8 9 9 9
 (22)   4 | 0 0 0 0 0 1 1 1 1 1 2 2 3 3 4 4 4 4 4 4
   30   4 | 5 5 5 6 6 6 6 7 7 8 8 8 8 9 9 9
   14   5 | 0 0 1 3 3 3
    8   5 | 6 8 9
    5   6 | 0 0 2 4
    1   6 |
    1   7 | 2

Stem-and-leaf of HEIGHT N = 73
Leaf Unit = 1.0

    1   3 | 1
    1   3 |
    1   3 |
    1   3 |
    4   3 | 9 9 9
   15   4 | 0 0 0 0 0 0 0 1 1 1 1
   32   4 | 2 2 2 2 2 2 2 2 2 2 3 3 3 3 3 3 3
 (21)   4 | 4 4 4 4 4 4 4 4 4 4 5 5 5 5 5 5 5 5 5 5 5
   20   4 | 6 6 6 6 6 6 6 6 6 6 6 6 6 6 7
    5   4 | 8 9 9 9
    1   5 | 1

MTB > Describe 'WEIGHT' 'HEIGHT'.
```

	N	MEAN	MEDIAN	TRMEAN	STDEV	SEMEAN
WEIGHT	73	43.908	43.500	43.435	8.333	0.975
HEIGHT	73	43.551	43.750	43.596	2.977	0.348

	MIN	MAX	Q1	Q3
WEIGHT	30.000	72.000	38.250	48.000
HEIGHT	31.000	51.000	42.000	45.500

Write a brief description of each distribution (you may choose to supplement this with a graphic display). In describing the distribution, comment on the shape, the center, and the variability. Name the measures you present, say why you used them, and briefly indicate how they should be interpreted.

15. Describe the distribution of sailboat lengths in the data set in Table 2.19. Support your description with appropriate calculations and/or graphics.

16. Describe the distribution of sailboat prices in the data set in Table 2.19. Support your description with appropriate calculations and/or graphics.

17. Describe the distribution of sailboat motor types in the data set in Table 2.19. Support your description with appropriate calculations and/or graphics.

18. Describe the distribution of the *ages* of the sailboats in the data set in Table 2.19. Support your description with appropriate calculations and/or graphics. (Note that age, per se, is not one of the variables. These boats were listed for sale in the summer of 1992.)

19. Data on the gross wealth of the citizens of Tuscany (around Florence, Italy) in the years 1427–1430 are included in Herlihy and Klapisch-Zuber (1985, Figure 4.4).[7] Reading the numbers from their figure, we obtain approximately the data shown in Table 6.9. Describe this distribution and support your description with graphics and/or calculations.

Table 6.9 *Gross wealth of citizens of Tuscany, 1427–1430*

GROSS WEALTH (FLORINS)	PERCENT OF HOUSEHOLDS
0 to 1	11
1 through 24	33
25 through 49	20
50 through 99	16
100 through 199	10
200 through 399	7
400 plus	3

20. The distribution of household (as distinct from individual or family) income in the United States in 1991 is shown in Table 6.10.[8] Describe this distribution and support your description with graphics and/or calculations.

Table 6.10 *Distribution of U.S. household income in 1991*

INCOME	PERCENT OF HOUSEHOLDS
under $5,000	4.8
$5,000 to $9,999	10.1
$10,000 to $14,999	9.4
$15,000 to $24,999	17.4
$25,000 to $34,999	15.2
$35,000 to $49,999	17.3
$50,000 to $74,999	15.4
$75,000 plus	10.4

21. In 1989, the mean net worth of American families was $183,700, which is to say that if every family sold everything they owned (at fair market value) and paid off all their debts, the average amount left would be $183,700. At the same time, the median net worth of

American families was $47,200, which is to say that if every family sold everything and paid off all debts, the median amount left would be $47,200.

a. How can both these statements be true?

b. Given that they are both true, what can we conclude about the distribution of net worth among American families?

22. Describe the data distributions in Table 2.22, which gives the heart fill rate and maximum oxygen use for two age groups. Support your description with appropriate calculations and/or graphics.

23. *Computer problem:* Write a description of the distribution of number of foster home placements (variable "Nfoster") among the runaways in data file Runaway.mtw. Support this description with appropriate computations and/or graphics.

24. *Computer problem:* The data file CarsMajr.mtw contains information on car models available in the United States from manufacturers with four or more models. Describe the distribution of brake horse power ("bhp") among these models. Support your description with appropriate computations and/or graphics.

25. *Computer problem:* The data file CarsMajr.mtw contains information on car models available in the United States from manufacturers with four or more models. Describe the distribution of length ("length") among these models. Support your description with appropriate computations and/or graphics.

26. *Computer problem:* The data file PULSE.MTW, distributed by Minitab and used here with permission, contains data from a simple demonstration experiment in which students were asked to report certain background information and then record their pulse rates two times. Half of the students, randomly picked, ran in place between the two recordings.

a. Describe the distribution of pulse rates before the experiment (Pulse1). Support your description with appropriate calculations and/or graphics.

b. Describe the distribution of pulse rates at the second reading (Pulse2). Support your description with appropriate calculations and/or graphics.

27. Describe the distribution of calories spent in exercise per week, shown in Table 2.20, from the datafile Cal_VT.mtw. Briefly explain why you chose the description you used.

28. Describe the distribution of changes in exercise capacity, as indicated by changes in "ventilatory threshold," from the datafile Cal_VT.mtw (also shown in Table 2.20). Briefly explain why you chose the description you used.

29. *Computer problem:* For the data set WGTHGT.MTW:

a. Describe the distribution of weight in the sample. Briefly explain why you chose the description you used.

b. Describe the distribution of height in the sample. Briefly explain why you chose the description you used.

30. Table 6.11 (also included in data file State.mtw and State.xls) on page 192 contains data on the number of lawyers per state and in Washington, D.C.[9]

a. Describe the distribution of the number of lawyers.

b. Briefly explain why you chose the description you used.

Table 6.11 *Number of lawyers, density of lawyers in terms of nonlawyers per lawyer, and highest points (in feet) of the 50 states and the District of Columbia*

STATE	LAWYERS	CAP/LWYR	HIPOINT
Alabama	7,261	565	2,405
Alaska	1,940	270	20,320
Arizona	8,859	394	12,633
Arkansas	3,939	608	2,753
California	93,054	304	14,494
Colorado	12,253	269	14,433
Connecticut	13,144	246	2,380
Delaware	1,813	364	442
Washington, D.C.	32,114	19	410
Florida	33,259	371	345
Georgia	15,314	414	4,784
Hawaii	3,130	351	13,796
Idaho	2,242	447	12,662
Illinois	40,743	285	1,235
Indiana	10,091	551	1,257
Iowa	6,159	460	1,670
Kansas	5,978	417	4,039
Kentucky	7,517	496	4,139
Louisiana	11,515	383	535
Maine	2,922	412	5,267
Maryland	14,608	316	3,360
Massachusetts	24,218	243	3,487
Michigan	21,644	427	1,979
Minnesota	12,723	339	2,301
Mississippi	4,384	598	806
Missouri	13,463	382	1,772
Montana	2,161	373	12,799
Nebraska	4,473	358	5,424
Nevada	2,489	423	13,140
New Hampshire	2,613	415	6,288
New Jersey	27,308	283	1,803
New Mexico	3,503	430	13,161
New York	81,698	219	5,344
North Carolina	10,238	634	6,684
North Dakota	1,286	519	3,506
Ohio	28,290	384	1,549
Oklahoma	8,853	366	4,973
Oregon	7,743	357	11,239
Pennsylvania	30,319	396	3,213
Rhode Island	2,654	374	812
South Carolina	5,482	633	3,560
South Dakota	1,385	515	7,242
Tennessee	9,515	514	6,643
Texas	43,455	388	8,749
Utah	3,640	464	13,528
Vermont	1,537	362	4,393
Virginia	14,916	403	5,729
Washington	12,925	360	14,410
West Virginia	3,016	622	4,861
Wisconsin	10,829	448	1,951
Wyoming	1,168	410	13,804

31. Table 6.11 also contains data on the *density* of lawyers per state (the number of nonlawyers per lawyer) and in Washington, D.C.

a. Describe the distribution of the density of lawyers.

b. Briefly explain why you chose the description you used.

c. Are there any outliers? If so, is there justification for excluding this outlier?

d. How would you describe the lawyer density distribution excluding the outlier (if any)?

32. Table 6.11 also contains the *height* (in feet) of the highest point in each state and Washington, D.C.

a. Describe this distribution.

b. Briefly explain why you chose this description.

PART TWO

Probability, Sampling, and Tests of Statistical Significance

chapter 7

Probability

We live in a world of randomness. Indeed, uncertainty may be found in nearly every aspect of our lives. For example,

- *The weather:* Will it rain today? How cold or hot will it be?
- *Games of chance:* Will I win the lottery? Is a particular bet worthwhile?
- *Medical considerations:* Will the patient survive? How long will the illness last? Will a new experimental treatment prove to be successful? When will we cure cancer?
- *Economic forecasts:* Will the stock market go up tomorrow? Will the economy expand next year? Which investments will give the highest return?
- *Business activities:* Will the public like our new product? Will we meet the deadline? Will our customers pay their bills on time?
- *Personal considerations:* Will the boss be in a good mood today? How large a raise will I receive next year? Will the talent agent call back? Will the traffic be heavy on my way home? Will I run into an old friend?

All of these examples have one feature in common: The outcome cannot be predicted perfectly in advance. How can such an inexact situation be studied? The methods of *probability* make precise the degree of predictability and indicate the likelihood of occurrence of each possibility. Thus, even though a situation is unpredictable, we may be able to precisely describe its unpredictability. The methods of probability help us understand randomness and are therefore useful for planning our activities and making decisions in the face of uncertainty.

We introduce probability theory because the statistical procedures discussed in later chapters are based on it. For example, we might have test results on a new medical drug. The results might seem to show that the new drug is a little better than the already available drugs. But is this apparent difference real or a chance variation? This is a question of statistical significance (Chapter 11). Or we might want to use survey results to predict the range of likely results of an election. This is a question of statistical inference (Chapter 10).

7.1 Random Experiments

In the study of randomness, it is necessary to be precise. We begin by identifying the situation of interest to us, called a *random experiment*.

DEFINITION

A *random experiment* is any procedure or situation that produces a definite outcome that may not be predictable in advance.

This definition helps us to decide whether something is a random experiment or not. The following *are* random experiments:

- Count the number of trees in a particular forest grove.
- See how our portfolio of investments performs tomorrow.
- See who wins a particular baseball game.

The random experiment serves the purpose of focusing our attention on exactly what is being studied and exactly which kind of outcome is important to us. When a problem becomes difficult to solve or hard to understand, it is usually helpful to identify exactly what the underlying random experiment of interest is. For example, the U.S. census taken every 10 years is such a procedure. However, the census produces too many outcomes to be understood completely unless we further focus our attention on a specific kind of outcome.

The following is *not* a random experiment:

- Waiting for the bus.

Although this is a situation with randomness, no outcome has been specified and so it is not specific enough to be a random experiment. In fact, many different random experiments can be defined for this situation, such as

- Wait for the bus and record the time it takes to arrive.
- Wait for the bus and record its color.
- Wait for the bus and count the number of people on it.

Exercise Set 7.1

1. Which of the following are random experiments:
   - **a.** Put 49 numbered Ping-Pong balls in a vat, use a blower to mix them thoroughly, then have a blindfolded chimpanzee pick one using a pair of tongs. Record the number on the Ping-Pong ball chosen.
   - **b.** Make a batch of soup.
   - **c.** Organize a bus trip.
   - **d.** Take a bus trip and record the arrival time.

7.2 Outcomes and Events

Each time we observe a random experiment, it produces one outcome that specifies the result of the random experiment. In many cases, we are not interested in the particular outcome, but focus instead on some general property of the outcome. For example, it may be more useful for us to hear the event "It is hot" instead of the outcome "It is 85.342°F" for the purpose of deciding what to wear.

DEFINITION

An *outcome* is a single possible result of a random experiment.

DEFINITION

An *event* is any collection of outcomes.

DEFINITION

A *simple event* is an event consisting of just one outcome.

Just as there are many different outcomes that may be produced by a random experiment, there may also be several different events of interest in a given situation. When we focus on a particular event, we may ask "Did the event occur?" and receive a yes-or-no answer depending on the outcome of the random experiment.

EXAMPLE 7.1 Outcome and Events: Baseball Scores

Consider the random experiment of going to a baseball game and recording the score as the outcome. Here are some possible outcomes and an event:

A possible outcome:	3 to 9
Another possible outcome:	4 to 3
Another possible outcome:	2 to 6
An event:	One team's score is three times the other's.

To see that having one score three times the other really is an event, note that it stands for the collection of outcomes 1 to 3, 2 to 6, 3 to 9, and so on. Also note that the question "Is one score three times the other?" can be answered with a yes or a no based on the outcome of the random experiment. This yes-or-no criterion can be used to decide whether something is an event.

Another event: A shutout

This event (a shutout) is the collection of outcomes (scores) for which one team scored zero points.

A simple event: The score was 5 to 8.

This simple event consists only of one outcome: The score was 5 to 8.

EXAMPLE 7.2 Outcomes and Events: Income Levels

Consider the following random experiment: Interview a family at random and determine their annual income. In practice, not every family will tell you their income, so let us consider one random experiment to repeat the interview with a new family, if necessary, until we obtain an income. The list of possible outcomes is enormous. Literally, it would consist of 0, 1 penny, 2 pennies, 3 pennies, and so on—a list of all possible (nonnegative) dollar amounts. Here are some more typical outcomes:

A possible outcome:	$26,500
Another possible outcome:	$58,900
Another possible outcome:	$14,000

One of the important things that events can provide for us is groupings of incomes. Some events of this form would be:

An event:	The income is over $60,000.
Another event:	The income is under $10,000.
Another event:	The income is between $10,000 and $20,000.
Another event:	The income is between $25,000 and $40,000.
A simple event:	The income was exactly $34,965.02.

The first two events listed might correspond to the notions "rich" and "poor," respectively.

Be sure that you understand that each of these events defines a list of possible outcomes and that each time we run the random experiment, we can determine whether the event happened or not.

Exercise Set 7.2

2. If we toss two coins, say a nickel and a dime, we have four possible outcomes: (nickel H, dime H), (nickel H, dime T), (nickel T, dime H), (nickel T, dime T), where H = heads and T = tails.

a. Which outcome is in the event "no heads"?

b. Which outcomes are in the event "one head and one tail"?

c. Which outcomes are in the event "one or more heads"?

3. Suppose we have a large jar filled with marbles, some red (R) and some blue (B). We pick two, putting the first on our left and the second on our right. (That is to say, we distinguish between the first and second marble picked.) The four possible outcomes are: RR, RB, BR, and BB.

a. Which outcome is in the event "no reds"?

b. Which outcomes are in the event "one red and one blue"?

c. Which outcomes are in the event "one or more reds"?

4. Suppose a company has two positions it is going to fill from a single group of applicants. One position is division director and the other is administrative assistant to the division director. Looking only at the genders of the applicants, and pretending for the sake of argument that this is a random experiment, we have four possible outcomes: MM, MF, FM, and FF, where the division director is listed first and the administrative assistant second and M = male, F = female.

a. Which outcome is in the event "neither is male"?

b. Which outcomes are in the event "one male and one female"?

c. Which outcomes are in the event "one or more males"?

5. Suppose a couple plans to have two children and are wondering about the various boy–girl possibilities.

a. What are the possible outcomes? (First write the gender of the first born, then the gender of the second born, to distinguish between them.)

b. What outcomes produce the event "two children of the same gender"?

c. What outcomes produce the event "a boy and a girl"?

d. What outcomes produce the event "at least one girl"?

e. What outcomes produce the event "at least one boy"?

The Relative Frequency of an Event

There is very little that can be said about precisely what will happen on a single run of a random experiment. However, when the experiment is repeated many times, some patterns emerge involving the proportion of times that an event happens, which is measured by its *relative frequency*.

DEFINITION

When a random experiment is repeated, the *relative frequency* of an event is equal to the number of times the event happened divided by the number of runs of the random experiment.

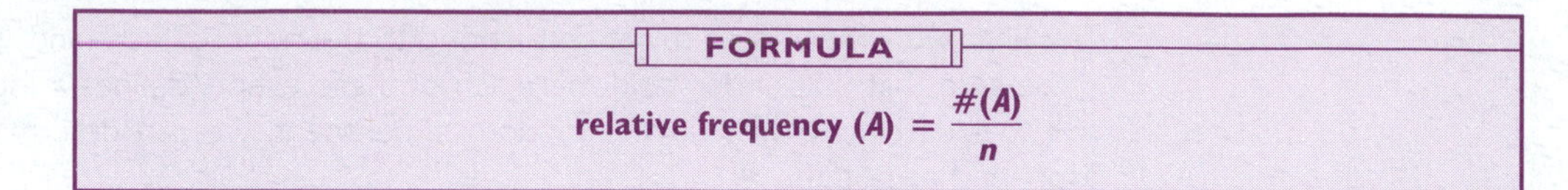
FORMULA

$$\text{relative frequency }(A) = \frac{\#(A)}{n}$$

For example, suppose the random experiment of tossing a coin and observing the outcome is repeated 20 times and that heads occurs 11 times out of those 20. Then, the relative frequency of heads is

$$\frac{11}{20} = .55$$

and the relative frequency of tails is

$$\frac{9}{20} = .45$$

Both of these relative frequencies are near $\frac{1}{2}$, indicating that heads and tails each occurred about half of the time.

Rare events—those not likely to occur very often—will tend to have small relative frequencies. For example, an event such as oversleeping might have a relative frequency of .02, indicating that it occurred in only 2% of the times that the random experiment (recording the wake-up time on weekdays) was repeated.

Common events—those that occur often—will have relative frequencies near 1. For example, the event of a warm day measured during the summer at a particular location might have a relative frequency of .993, indicating that 99.3% of summer days were warm.

Exercise Set 7.3

6. Find the relative frequency of heads when you get

a. 7 heads in 10 experiments

b. 9 heads in 20 experiments

c. 17 heads in 30 experiments

7. What would be the relative frequency of the event "the patient felt better" if you gave 100 patients 2 aspirins each, and 46 reported they felt better?

8. Suppose you are looking through the freshman yearbook at a school. You randomly pick ten students; six of the ten are female. What is the relative frequency of females in this sample?

The Law of Large Numbers

Although the relative frequency of an event is a useful indication of how likely that event is to occur, it is slightly unfortunate that the relative frequency is itself a random quantity because it depends on the particular outcomes of the repeated random experiment. Fortunately, the relative frequency is more stable (less random) if the random experiment is repeated many, many times under the exact same circumstances. In other words, if you and I each separately perform ten repetitions of a random experiment, then we may see very different relative frequencies. But if we each complete 10,000 repetitions of the random experiment, our relative frequencies will be nearly identical. This is a consequence of **the law of large numbers.**

LAW OF LARGE NUMBERS

Every event has a special number called its probability such that if the random experiment is repeated a large number of times, then the relative frequency of the event will be close to this probability. The more times the random experiment is repeated, the closer the relative frequency will tend to be to this probability.

Figure 7.1 illustrates the law of large numbers for three events. The probabilities of the events are .2, .5, and .8, as indicated by the horizontal lines. The three jagged, random lines indicate the approach of the relative frequency toward the probability as more and more runs of the random experiment are performed. Notice how the relative frequencies get closer to their respective probabilities, but observe also that they move somewhat randomly. Even after 100 runs, they have not yet settled down and are not quite equal to the probability.

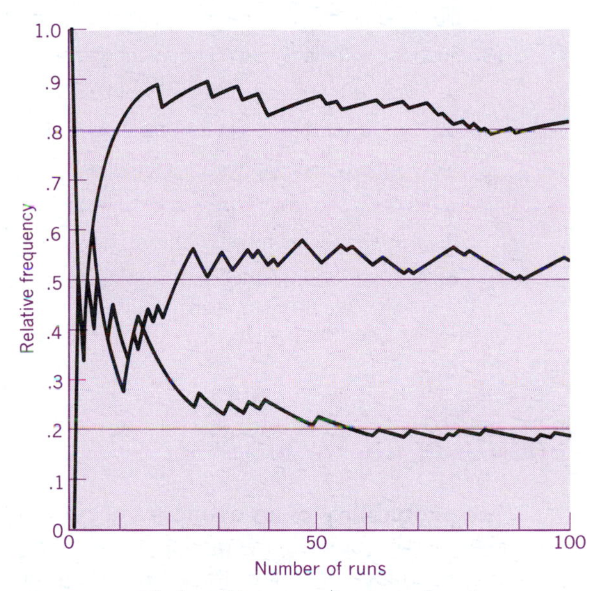

Figure 7.1 *The law of large numbers says that after many runs, the relative frequency tends to stabilize and be close to the probability of the event. This is illustrated for three events, with probabilities .2, .5, and .8, respectively.*

EXAMPLE 7.3 *Law of Large Numbers: Coin Flips*

If you flipped a coin ten times, the relative frequency of heads might easily be .6. If you flip it 100 times, the relative frequency of heads should be closer to .5.

The law of large numbers is what people are thinking of when they talk about "the law of averages" and "things evening out in the long run." Notice that this law does not say anything about things evening out in the short run. If you have flipped three heads in a row, the chance of a head on the next flip is still .5. Things even out only as you compute relative frequencies for a larger and larger number of flips. To a statistician, the "long run" can easily mean 2,000 or more coin flips.

Exercise Set 7.4

9. Does a relative frequency of .75 for heads seem unreasonable if you flip an honest coin

a. 4 times?

b. 1,000 times?

10. *True or false:* After you have flipped a coin 1,000 times the relative frequency of heads has to settle down to exactly .500, and stay there. Exactly. Forever.

11. Get the list of 20 coin flips that you did for the Chapter 1 problem set. Compute the relative frequency of heads after

a. 10 flips.

b. 20 flips.

c. *Yes or no:* Does the relative frequency get closer to .5 as the number of flips increases? (Because this is a small sample, there will be different answers to this. There are no guarantees.)

12. Flip a coin 20 more times, recording the number of heads. Add these to the number of heads you got in the 20 tosses for Exercise 11. What is the relative frequency of heads in these 40 flips? Is it closer to .5 than was the relative frequency for the first 20 flips? (Again, there are no guarantees.)

13. Suppose you flip a coin 10 times and get seven heads.

a. What is the relative frequency of heads?

b. If you expected to get 5 heads, how many "extra" heads do you have?

Now suppose you flip the coin 100 times and get 60 heads.

c. What is the relative frequency of heads?

d. If you expected to get 50 heads, how many "extra" heads do you have?

e. In what sense (if any) are things evening out with the larger number of coin flips? In what sense (if any) are they not evening out?

7.3 The Probability of an Event

The **probability of an event** is a number between 0 and 1 that indicates how likely it is an event will occur when the random experiment is performed. It is an exact number (not random at all) and it is the limit of the relative frequency if we could perform the experiment over and over an unlimited number of times.

Probability may be thought of as a precise description of randomness because an event may be quite unpredictable and yet have a probability of, say, exactly .7. Thus, the event would occur exactly 70% of the time if we could observe the random experiment over and over forever. It would occur *around* (rather than exactly) 70% of the time in a large sample of observations. We can state nothing very exact about a single observation of the random experiment—just that the probability is .7, indicating that it is somewhat more likely to happen than not. We cannot be very certain about what will happen during a single occurrence.

The interpretation of probability is straightforward, as shown in the following list.

- A probability of 0 indicates an essentially impossible event.
- A probability *near* 0 indicates a rare, but possible event.
- A probability *near* .5 indicates an event that happens about half the time.
- A probability *near* 1 indicates an event that happens most, but not all, of the time.
- A probability of 1 indicates an event that essentially always occurs.

Other probabilities have similar interpretations (a probability of .25 indicates an event that occurs about one-quarter of the time, and so on).

The standard notation for a statement like

"The probability of event A is .25."

is:

$$P(A) = .25$$

Exercise Set 7.5

14. Match these probability values with the best interpretation for each from parts (a)–(g).

PROBABILITY	INTERPRETATION
0	
.25	
.50	
.60	
.95	
1.00	
2.75	

a. This should happen just about half the time.
b. This should happen about a quarter of the time.
c. This should happen almost all of the time.
d. Whatever this number is, it is not a probability.
e. This should happen essentially all the time.
f. This should essentially never happen.
g. This should happen a little over half the time.

Probability When All Outcomes Are Equally Likely

The easiest kind of random experiment is one for which all outcomes are equally likely. (We are also assuming that the number of possible outcomes is limited, so that—at least in principle—we could list each of them.) In this case, the probability of an *outcome* (or a *simple event*) is equal to 1 divided by the number of possible outcomes.

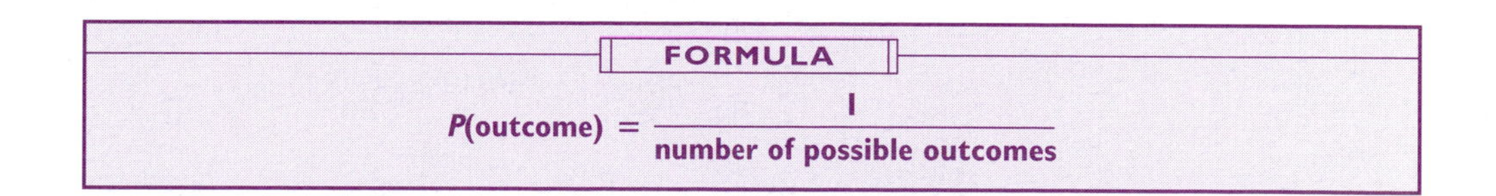

So the probability of an *event*, which is made up of one or more *outcomes*, is the number of outcomes in the event divided by the total number of possible outcomes.

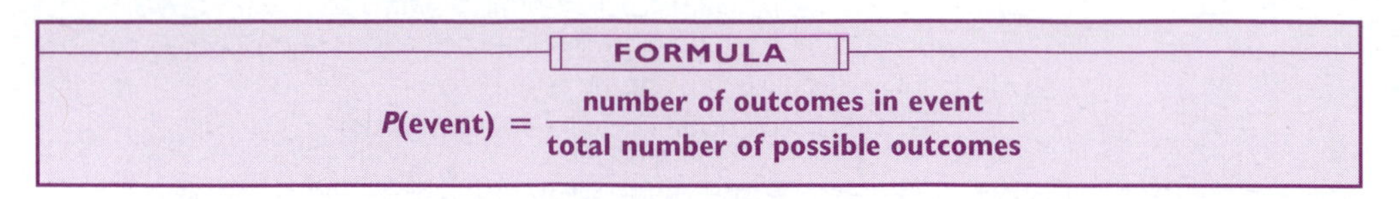

EXAMPLE 7.4 *Equally Likely Outcomes: The Six-Sided Die*

Consider a six-sided die of the sort used by many games in which "tossing the dice" introduces a random element to the play. The die has six sides that, by the symmetry of the cube, are all equally likely. This seems reasonable because the die does not "know" which face has which number on it, and all faces would look alike without their numbers. It is easy to compute the probability of obtaining an even number (2, 4, or 6) with one toss of a die because this event, "evens," has three outcomes in it:

$$\text{Probability of an even number} = \frac{3}{6} = .5$$

What is the probability of a large number? More precisely, say that the event "a large number" means a 5 or 6. Then the probability would be $\frac{2}{6}$ or $\frac{1}{3}$. In fact, this is the probability of obtaining one of any two numbers specified in advance.

EXAMPLE 7.5 *Equally Likely Outcomes: A Raffle*

Consider a drawing in which a TV set is to be given away to one of 653 people, each of whom bought one $1 ticket. The probability of any one ticket being drawn is then $\frac{1}{653}$ or .0015, which we recognize as quite a rare event. If there are 314 women and 339 men in the group of 653 ticketholders, then the probability that a woman wins will be $\frac{314}{653} = .48$.

EXAMPLE 7.6 *Equally Likely Outcomes: Who Collects the Papers?*

Suppose there are 52 students in a class. Each fills out an index card with his or her name and his or her year in school (freshman, sophomore, junior, senior). It turns out that there are 13 students in each of the four school years. If the instructor shuffles the stack of cards thoroughly and picks one at random to determine who collects papers, then the probability of obtaining any particular card is the same as for any other particular card. For example, the

$$\text{Probability of picking Jane Doe} = \frac{1}{52} = .019$$

The probability of obtaining one of two particular cards does not depend on which two are specified. For example, the

$$\text{Probability of picking Jane Doe or Jack Smith} = \frac{2}{52} = .038$$

There are 13 cards of each of the four years (freshman, sophomore, junior, senior), so the probability of drawing a card of a particular year is:

$$\text{Probability of picking a sophomore} = \frac{13}{52} = \frac{1}{4} = .25$$

$$\text{Probability of picking a junior} = \frac{13}{52} = \frac{1}{4} = .25$$

Exercise Set 7.6

15. Recall that there are four possible outcomes when we toss two coins, say a nickel and a dime. There are also four possible outcomes if we toss one coin twice and keep track of the order:

HH, HT, TH, TT

Suppose each of the outcomes is equally likely. What is the probability of getting

a. two heads?

b. two tails?

c. a head, and then a tail?

d. a tail, and then a head?

e. the same side on both flips?

f. no heads?

g. at least one head?

16. There are four possible outcomes for the gender of two siblings (with the oldest child written first):

BB, BG, GB, GG

This is not exactly true, but for now suppose that these four outcomes are equally likely. For a two-child family, then,

a. what is the probability of two boys?

b. what is the probability of two girls?

c. what is the probability of a boy first, then a girl?

d. what is the probability of a girl first, then a boy?

e. what is the probability of one boy and one girl (regardless of who is older)?

f. what is the probability of the oldest child being a girl?

g. what is the probability of the family including at least one girl?

h. what is the probability of the family including at least one boy?

17. Suppose we are doing a study of friendship formation. We pick pairs of people and (privately) ask each individual if he or she likes (L) or does not like (D) the other. List all the possible L and D combinations.

Probability When Outcomes May Not Be Equally Likely

As we just learned, probabilities are easily calculated when outcomes are equally likely and we know how many outcomes there are to consider. However, this approach will not give correct answers when the various outcomes are not equally likely. For example, consider this random experiment with two possible outcomes.

Outcome 1: I arrive at class on time.

Outcome 2: I arrive at class late.

It would be foolish to suppose that these are equally likely; probably outcome 2 is a rare possibility (at least we would like to think so!).

Even when outcomes are not equally likely, if we can make a list assigning a probability to each simple event, then it will be easy to find any probability of interest.

The probability of an event is the sum of the probabilities of occurrence of each of the simple events it represents.

There are only two requirements to check to be sure a table giving a probability value for each simple event is valid.

Requirement 1: All probabilities must be between 0 and 1, inclusive.

Requirement 2: The sum of all probabilities must be 1.

Requirement 1 states that each simple event has some chance of occurring between the bounds of 0 and 1, which represent the two extremes of "never happening" and "always happening," respectively. Requirement 2 states mathematically that the probability of *something* happening is 1; something always happens.

EXAMPLE 7.7 Events that Are Not Equally Likely: Vehicle Sizes

Consider the random experiment of observing the size of the next motorized vehicle to pass through an intersection. Based on past experience, a probability table has been constructed that lists the probability of each simple event (Table 7.1). Notice that this table is similar to the frequency table in percents for a categorical variable (Section 2.3). Notice also that from this table, we can find not only the probabilities of simple events (such as the probability of a medium car being .24), but also the probabilities of compound events (consisting of more than one outcome).

For example, to find the probability that the next vehicle is a truck, we first ask ourselves exactly which event we want to determine the probability of. The event "truck" is the collection of two simple events (as we have listed them in the table)—namely, "large truck" and "small truck." Adding the probabilities of these simple events, we have

$$\text{Probability that the next vehicle is a truck} = .07 + .11 = .18$$

Table 7.1

SIMPLE EVENT	PROBABILITY
Large truck	.07
Small truck	.11
Large car	.19
Medium car	.24
Compact car	.21
Subcompact car	.18

Similarly, we find

$$\text{Probability that the next vehicle is a car} = .19 + .24 + .21 + .18 = .82$$

And finally,

$$\text{Probability that the next vehicle is large (either car or truck)} = .07 + .19 = .26$$

Exercise Set 7.7

18. Using the hypothetical data on planned number of children in Table 7.2, what are the probabilities of a couple planning:

Table 7.2 *Planned number of children for a sample of newly married couples (hypothetical)*

NUMBER OF CHILDREN PLANNED (SIMPLE EVENTS)	PROBABILITY
0	.347
1	.209
2	.191
3	.125
4	.067
5 or more	.061

a. 1 or 2 children?

b. 3 or 4 children?

c. 4 or more children?

19. Suppose that we have the data shown in Table 7.3 on the age distribution of doctorate holders employed by 4-year colleges and universities.[1] What is the probability that a randomly selected doctor at a 4-year college or university is

a. under 40?

b. 40 or over?

c. under 55?

Table 7.3

AGE	PROBABILITY
under 40	.263
40–55	.510
55 or over	.225
no report	.002

7.4 Not: The Complement of an Event

To every event there corresponds a **complementary event** that consists of all other outcomes. Table 7.4 shows complementary event pairs. Because an event together with its complement make up all possibilities, the probabilities of the two add up to 1. Thus it is simple to find the probability of the complement of an event when the event's probability is known.

Table 7.4

EVENT	COMPLEMENT OF THAT EVENT
Red	Not red (all other colors)
Large	Not large (all other sizes)
Tie score	Win or lose (not a tie score)

COMPLEMENT RULE

The probability of the complement of an event is equal to 1 minus the probability of the event itself.

$$P(\text{not } A) = 1 - P(A)$$

Thus, if the probability of winning a game is .29, then we find

$$\begin{aligned} \text{Probability of not winning} &= 1 - \text{probability of winning} \\ &= 1 - .29 \\ &= .71 \end{aligned}$$

There are three main reasons that complements are important. First, when the probability of the event of interest is unavailable but the probability of its complement is known, then the needed probability can be calculated using the complement rule. Second, the complement rule provides a useful check against errors in a complicated problem. If the probability of an event and its complement are separately available but do not add up to 1, then a mistake has been made somewhere (perhaps the two probabilities are from different random experiments or surveys and therefore are not really complements of one another).

Third, in many cases it is difficult to *calculate* the probability of an event, but fairly easy to calculate the probability of the complementary event. Then we can get the probability we want using the complement rule. For example, we might want to find the probability of "at least one head" in five coin flips. This is a fairly tedious calculation. But calculating the probability of "no heads" in five flips is fairly simple (see Section 7.10). So we calculate the probability of "no heads" and subtract that from 1 to get the probability of "at least one head."

Exercise Set 7.8

20. For each of the following events, specify its complement:

a. A plane arrives at the gate at its scheduled arrival time.

b. A plane arrives at the gate before its scheduled arrival time.

c. A plane arrives at the gate after its scheduled arrival time.

21. Suppose you are told that "somebody has to win" the state lottery and that it might be you . . . if you enter. If your chances of winning are 1 in 4 million, what are your chances of losing?

22. Suppose the probability of a family with two children having two boys is $\frac{1}{4}$. What is the probability of a family with two children having at least one girl?

23. Suppose the probability of a family with three children having three boys is $\frac{1}{8}$. What is the probability of a family with three children having at least one girl?

24. Referring again to the hypothetical data on planned family size (Table 7.2), use the complement rule to find the probability of a couple planning to have any children at all.

7.5 Picturing the Probability of an Event

In this section we look at three different ways of presenting probability information: probability trees, Venn diagrams, and joint probability tables. Each way has advantages in particular situations, and some people are more comfortable with one approach than another.

Probability Trees

We use **probability trees** to help us visualize a problem, especially when we are interested in the order in which things happen. The simplest tree shows an event and its complement, as in Figure 7.2. This tree has only one branch, starting from the certain event that something happened with probability 1 (circled at left) then dividing into the two cases "bought the car" and its complement "did not buy the car," with their probabilities circled at the endpoints to the right. The complement rule says that the two probabilities on the right must add up to 1 (which they do). Probability trees will be very important to us in the next few topics.

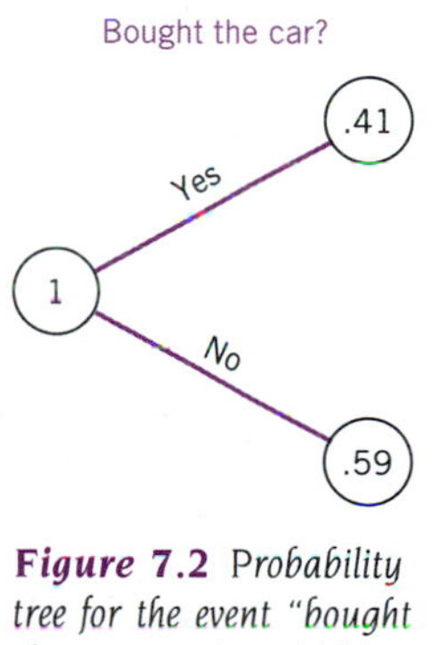

Figure 7.2 *Probability tree for the event "bought the car" (with probability .41) and the complement of this event "did not buy the car" (with probability .59)*

A probability tree for two events is shown in Figure 7.3 (page 212) for the events "read a textbook" and "read a novel," indicating two possible behaviors for a student on a weekend. The four possible combinations of occurrence of these two events are represented by the four branches at the far right of the figure. At the top is the event "novel and text," meaning both events happened, with probability .30. Next is "text, but no novel" with probability .10, then "no text, but novel" with probability .12, and finally "neither" with probability .48. To find the probability of the event "read text," we simply add up the two ending branches corresponding to this event, as indicated in Figure 7.4. The probability of the event "read novel" may be determined similarly.

Note the two fundamental properties of a probability tree. First, the probabilities at the final ends of the branches (at the far right) add up to 1.

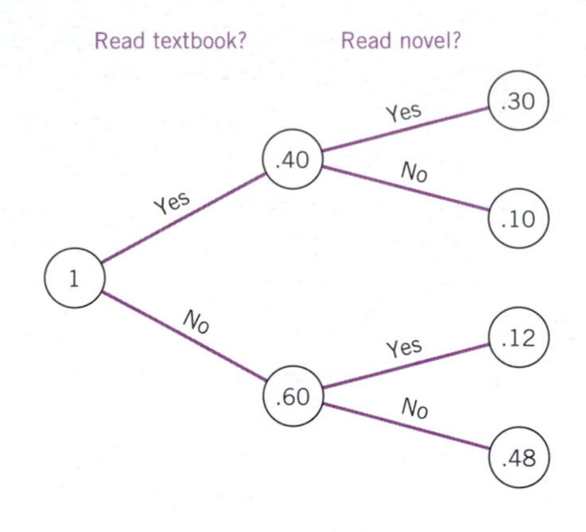

Figure 7.3 *Probability for the two events "read textbook" and "read novel"*

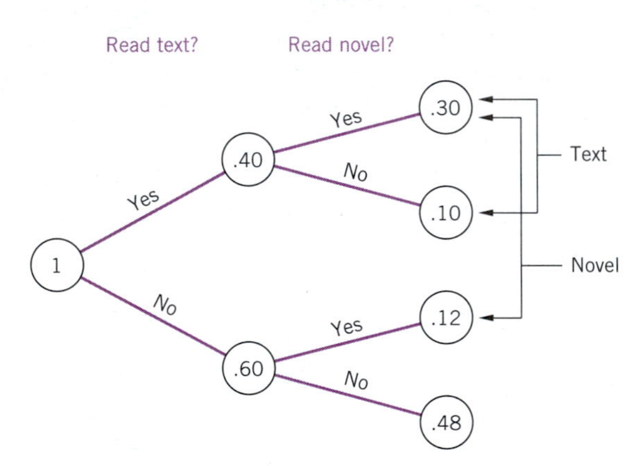

Figure 7.4 *Probability tree indicating that the probability of "read text" is .30 + .10 = .40 and that the probability of "read novel" is .30 + .12 = .42*

$$\text{Sum of ending branches} = .30 + .10 + .12 + .48 = 1$$

Second, the probability at the beginning of each branch point is equal to the sum of the probabilities at the ends of the branches coming from that point. For example, from the probability .40, there are two branches with probabilities .30 and .10 that do indeed add up to .40. Similarly, at the lower branch point we have .60 = .12 + .48.

Exercise Set 7.9

25. Table 7.5 shows a probability tree for the gender of two children (about 51% of the children born are male).

a. What is the probability that the first born is a boy?

b. What is the probability that both the first and second born are boys?

c. What is the probability that the second born is a boy?

d. What is the probability of *at least* one boy?

Table 7.5

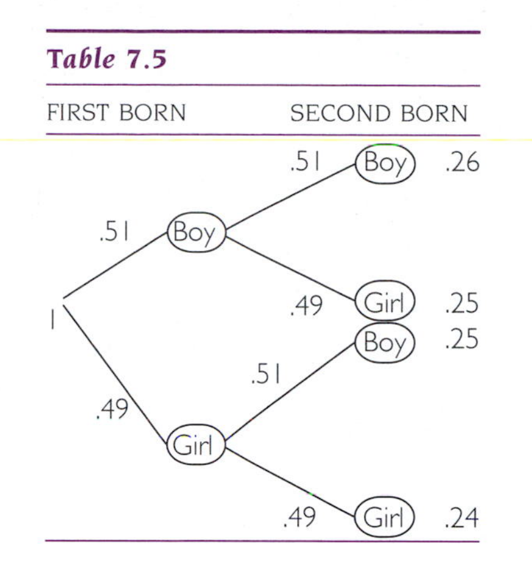

Venn Diagrams

Another technique that helps us visualize a problem is to use a circle (or other closed shape) for each event. The result is called a **Venn diagram.** Venn diagrams are particularly useful when we are interested in logical possibilities of things happening together. The simplest Venn diagram may be drawn to show an event and its complement, as in Figure 7.5. The outer box represents all possible outcomes. The circle divides the box into the two events "bought the car" (inside the circle, with its probability .41 noted there) and "did not buy the car" (outside the circle but in the box, with its probability .59 noted there). Venn diagrams will also be important in understanding the next few topics.

A Venn diagram is shown in Figure 7.6 (page 214) for the two events "read text" and "read novel." This is a different representation of the same situation depicted in the probability tree of Figure 7.3. The four possible combinations of events are displayed here as the four regions formed by two intersecting circles (do not forget the outer region corresponding to "read neither text nor novel"). The probability of the event "read text" is the sum of the numbers within its circle on the left: .10 + .30 = .40. Similarly, the probability of the event "read novel" is the sum of the numbers within its circle on the right: .30 + .12 = .42.

Probabilities for a Venn diagram follow two basic rules. First, there is a probability indicated for each separate region into which the enclosing rectangle is divided. Second, the probabilities for all of these (nonoverlapping) regions must add up to 1.

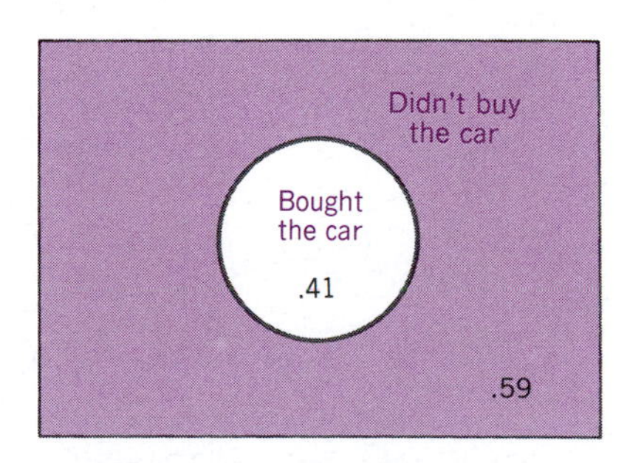

Figure 7.5 *Venn diagram for the event "bought the car" (with probability .41) and the complement of this event "did not buy the car" (with probability .59)*

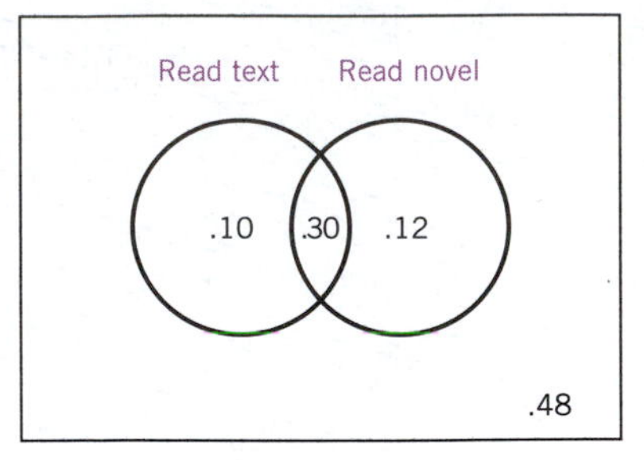

Figure 7.6 *Venn diagram for the two events "read text" and "read novel"*

Exercise Set 7.10

26. Figure 7.7 is a Venn diagram for some (fictional) data relating baldness and gender.

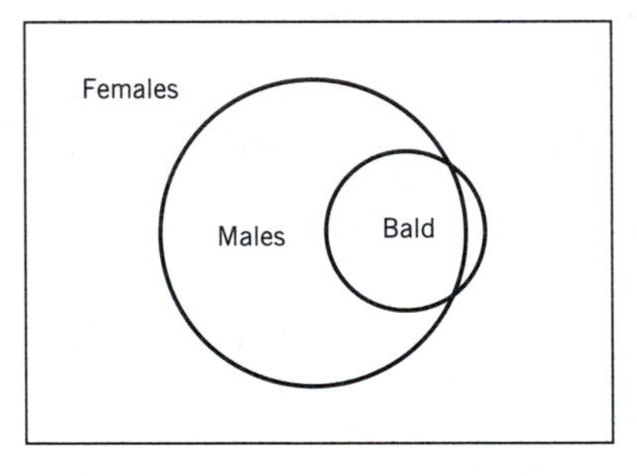

Figure 7.7 *Gender and baldness (hypothetical)*

Which of these statements is true, according to this diagram?

a. All males are bald.

b. All bald people are male.

c. Most, but not all, bald people are male.

27. Figure 7.8 is the same diagram, with some probabilities filled in.

a. What part of the diagram holds the not-bald females?

b. What is the probability that a randomly selected individual is a not-bald female?

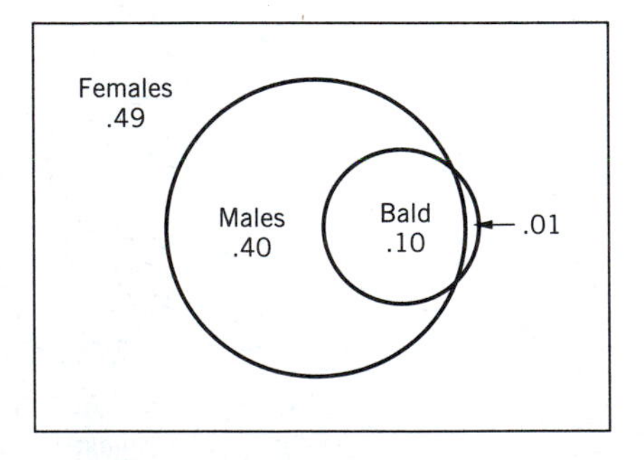

Figure 7.8 *Gender and baldness, with (hypothetical) probabilities*

c. What part of the diagram holds the not-bald males?

d. What is the probability that a randomly selected individual is a not-bald male?

e. What part of the diagram holds the bald males?

f. What is the probability that a randomly selected individual is a bald male?

g. What part of the diagram holds the bald females?

h. What is the probability that a randomly selected individual is a bald female?

Joint Probability Tables

When we want "just the facts" to summarize the probabilities of two events, instead of a picture, we would use a joint probability table. Joint probability tables are particularly useful for organizing numerical information. Such a table shows the probabilities of each event separately, with their complements, as well as the probabilities of each possible combination of the two events and their complements.

Table 7.6 is the joint probability table for the two events "read text" and "read novel." Within the body of the table, we recognize the four probabilities from the final right-hand endpoints of our probability tree (Figure 7.4) representing each of the four possibilities for one weekend, combining both events. For example, the .30 at the top left indicates the probability of "read text and novel." To the right of that, the .10 indicates the probability of "read text and not novel."

Table 7.6

		READ NOVEL		
		YES	NO	TOTAL
READ TEXT	YES	.30	.10	.40
	NO	.12	.48	.60
	TOTAL	.42	.58	1.00

The extra row and column indicate separately the probabilities for each event. The last column at the right of the table shows the probabilities for the event "read text" (.40) and its complement "did not read text" (.60); these are the sums of the two probabilities just to their left in the same row. The bottom row of the table indicates the probabilities for the event "read novel" (.42) and its complement "did not read novel" (.58); these are the sums of the two probabilities just above them.

Finally, the 1.00 in the lower right corner expresses the fact that *something* has to happen. Consequently, the two numbers to the left in the bottom row must add up to 1, the two numbers above in the rightmost column must also add up to 1, and so must the four numbers in the main body of the table.

Exercise Set 7.11

28. Table 7.7 (page 216) is a joint probability table for gender and baldness, using the same fictional probabilities as in Exercise Set 7.10. Which of these statements is true, according to this table?

a. All males are bald.

b. All bald people are male.

c. Most, but not all, bald people are male.

Table 7.7

		BALD		
		YES	NO	TOTAL
GENDER	MALE	.10	.40	.50
	FEMALE	.01	.49	.50
	TOTAL	.11	.89	1.00

29. Suppose we have a class with 20 students. Three are male and blue-eyed, 5 are male and brown-eyed, 6 are female and blue-eyed, and 6 are female and brown-eyed. We plan to select one student at random. The joint probability table is shown in Table 7.8.

Table 7.8

		EYE COLOR		
		BLUE	BROWN	TOTAL
GENDER	MALE	3/20 = .15	5/20 = .25	.40
	FEMALE	6/20 = .30	6/20 = .30	.60
	TOTAL	.45	.55	1.00

What is the probability of selecting:

a. a male?

b. a female?

c. a person with brown eyes?

d. a person with blue eyes?

e. a brown-eyed female?

f. a blue-eyed female?

30. Suppose you are observing wildlife at a watering hole in Africa. You classify each animal approaching the hole as a "carnivore" or "herbivore," and note whether it seems "bold" or "cautious" in its approach. At the end of the study, you have: 6 bold carnivores, 54 cautious carnivores, no bold herbivores, and 140 cautious herbivores.

a. Construct a joint probability table based on these observations.

What is the probability that a randomly selected observation is of:

b. a bold herbivore?

c. a cautious herbivore?

d. an herbivore (either bold or cautious)?

e. a bold carnivore?

f. bold (whether herbivore or carnivore)?

7.6 Combining Two Events Using OR and AND

One Event OR Another

When we speak of something like having a king or an ace in a card game, we are creating a single new event from two other events (having a king; having an ace) using the logical connector **OR.** The effect of OR is to combine all outcomes that were in either one event, the other event, or perhaps both. The logical OR includes the "or both" case. Be careful you do not mistake it for the *exclusive or,* which means one or the other, but not both, in ordinary English.

Remember that an event corresponds to a yes-or-no question about the result of a random experiment. Thus, we might construct a "truth table" for a student's activities, representing the two events "read text" and "read novel," and also the combined event "read text OR read novel." (See Table 7.9)

EXAMPLE 7.8 Combining Events: Read Text OR Read Novel

From Table 7.9, we see that the only way *A* OR *B* can fail to occur is when neither *A* nor *B* happens. In this case, the only way a student could have failed to read a text OR read a novel would be if she had done neither.

The event "read text OR read novel" is illustrated in a probability tree in Figure 7.9 as the first three events, excluding only one of the four combinations. Figure 7.10 (page 218) depicts "read text OR read novel" in a Venn diagram as the shaded region defined by the two circles,

Table 7.9

EVENT *A* (TEXT)	EVENT *B* (NOVEL)	EVENT *A* OR *B* (TEXT OR NOVEL)
Yes	Yes	Yes
Yes	No	Yes
No	Yes	Yes
No	No	No

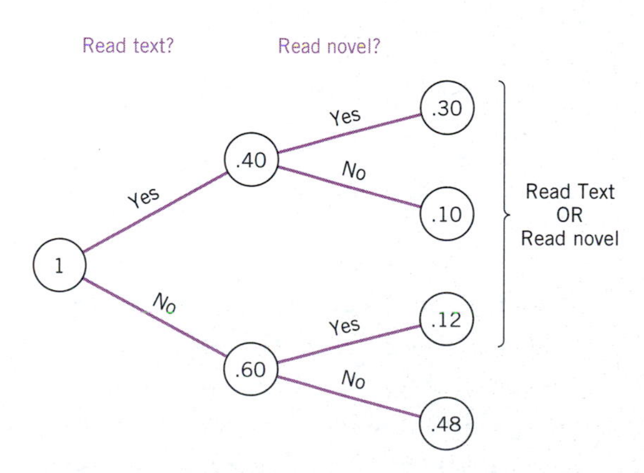

Figure 7.9 *Probability tree indicating the event "read text OR read novel"*

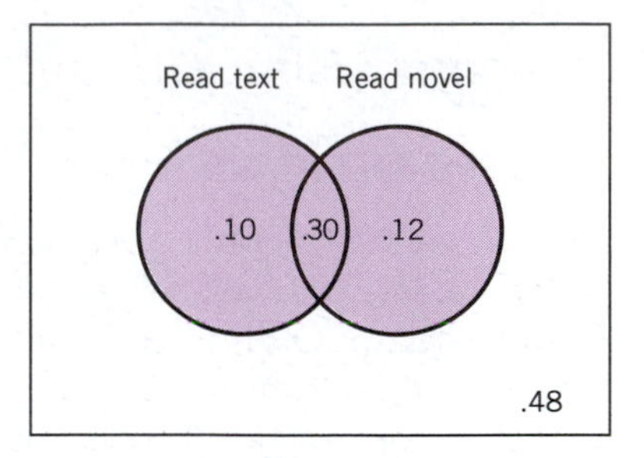

Figure 7.10 *Venn diagram indicating the event "read text OR read novel"*

one for each event. In either case, the probability is found by adding up the indicated probabilities in the figure.

$$\text{Probability of "read text OR read novel"} = .30 + .10 + .12 = .52$$

In a joint probability table, like Table 7.6, the OR condition is satisfied by a number if it is in a row or column that meets one of the conditions. Specifically, the "read text OR read novel" condition is satisfied by the circled numbers in Table 7.10. Notice that they form an "L" shape.

Table 7.10

		READ NOVEL		
		YES	NO	TOTAL
READ TEXT	YES	.30	.10	.40
	NO	.12	.48	.60
	TOTAL	.42	.58	1.00

Exercise Set 7.12

31. Consider the following conversation.

Mom: "Johnny, would you like (chocolate OR vanilla) ice cream?"
Johnny: "Yes."

a. Might Johnny mean he wants both?

b. Is Johnny's answer a legitimate interpretation of Mom's logical OR?

32. Harry has green hair and a red jacket. Does Harry have (green hair OR a red jacket)?

33. You roll two dice. You get two aces. Did you get (ace on first die OR ace on second die)?

34. Table 7.11 is a joint probability table for the gender of two children. What is the probability that

a. the first is female OR the second is female?

b. the first is male OR the second is male?

c. the first is male OR the second is female?

Table 7.11 *Joint probability for the gender of two children*

		FIRST		
		MALE	FEMALE	TOTAL
SECOND	MALE	.26	.25	.51
	FEMALE	.25	.24	.49
	TOTAL	.51	.49	1.00

Mutually Exclusive Events

Two events are said to be **mutually exclusive** if one excludes the other—that is, if they cannot both happen simultaneously. What is nice about mutually exclusive events is that if their individual probabilities are known, then their OR can easily be computed by adding up the individual probabilities.

FORMULA

For mutually exclusive events *A* and *B*, the probability of *A* OR *B* is equal to the probability of *A* plus the probability of *B*.

Note that mutually exclusive events need not cover all possibilities. For example, I might have no shoes on at all; what tells us that brown shoes and blue sneakers are mutually exclusive is that I cannot simultaneously be wearing both shoes and sneakers (at least not under the usual conventions of dress). So if the probability of my wearing brown shoes is .23 and the probability of blue sneakers is .64, then

Probability of my wearing brown shoes OR blue sneakers

= probability of brown shoes *plus* probability of blue sneakers
= .23 + .64
= .87

Mutually exclusive events may be represented in a Venn diagram either by using nonoverlapping circles or by indicating that the probability of the overlap is 0, as shown in Figure 7.11 (page 220).

■ Exercise Set 7.13

35. Are the following events mutually exclusive? (Assume the usual meanings of the events.)

- **a.** dead, alive
- **b.** male, female
- **c.** right, wrong

36. You flip one coin. Are "heads" and "tails" mutually exclusive on this one flip?

37. You flip two coins, a nickel and a dime. Are "heads on nickel" and "tails on dime" mutually exclusive?

38. Suppose that in a certain species of fish it is impossible to tell the sex of the immature ones, but easy to tell for the adults. In one tank we have 20 adult males, 30 adult females, and 50 immature fish. We randomly select one fish.

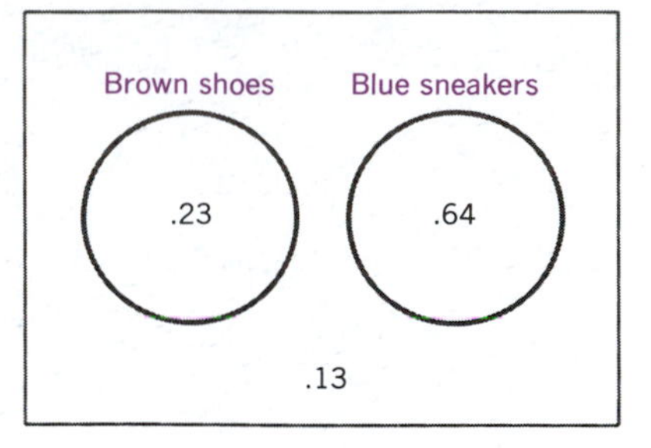

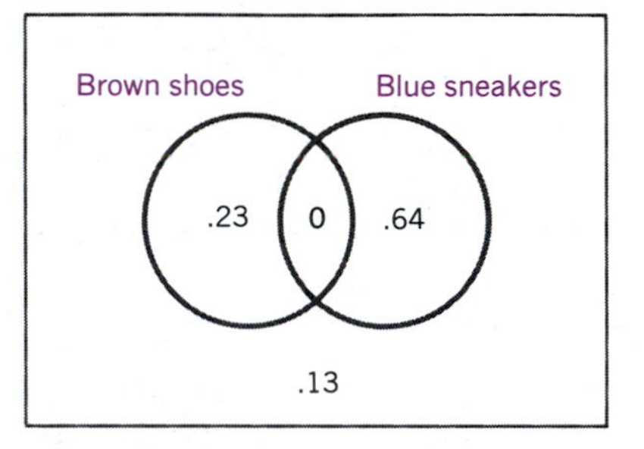

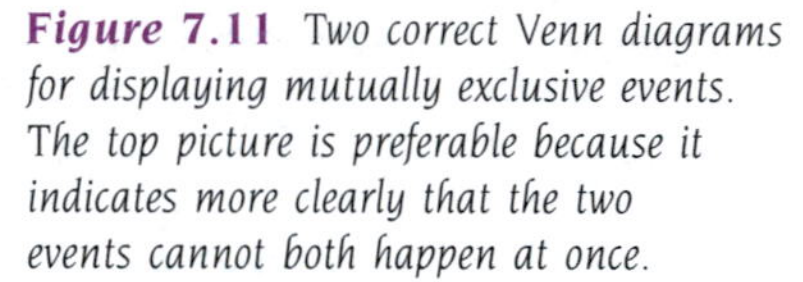

Figure 7.11 *Two correct Venn diagrams for displaying mutually exclusive events. The top picture is preferable because it indicates more clearly that the two events cannot both happen at once.*

a. What is the probability of selecting an adult male?

b. What is the probability of selecting an adult female?

c. On this one selection, are "adult male" and "adult female" mutually exclusive?

d. What is the probability we get an adult male OR an adult female?

e. Are the events "adult female" and "immature" mutually exclusive?

39. Suppose we have another very large tank with these fish in it, and that the probability a randomly selected fish is an adult male is .20, the probability it is an adult female is .30, and the probability it is immature is .50. We randomly select *two* fish.

a. What is the probability that the first fish is an adult male?

b. What is the probability that the second fish is an adult female?

c. Are "first fish is male" and "second fish is female" mutually exclusive?

d. If (for some reason) we want to find the probability of "the first fish is male OR the second fish is female," can we just add their separate probabilities?

One Event AND Another

When we speak of something like a person wearing a brown suit AND brown shoes, we are again creating a single new event from two other events (wearing a brown suit; wearing brown shoes), this time using the logical connector AND. The effect of AND is to restrict our attention to outcomes that are in both events. Because an event corresponds to a yes-or-no question about the result of a random experiment, we can again construct a truth table for a weekend's activities with the two events "read text" and "read novel." This time we will also list the combined event "read text AND read novel." (See Table 7.12.)

Table 7.12

EVENT *A* (TEXT)	EVENT *B* (NOVEL)	EVENT *A* AND *B* (TEXT AND NOVEL)
Yes	Yes	Yes
Yes	No	No
No	Yes	No
No	No	No

EXAMPLE 7.9 Combining Events: Read Text AND Read Novel

From Table 7.12, we see that the only way that *A* AND *B* can happen is for *both A* and *B* to occur. For all other cases *A* AND *B* fails to happen.

The event "read text AND read novel" is indicated in a probability tree in Figure 7.12 as the single ending branch on the top right, for which both events happened. Figure 7.13 depicts "read text AND read novel" in a Venn diagram as the small region common to both circles. In either case, the probability is seen to be .30.

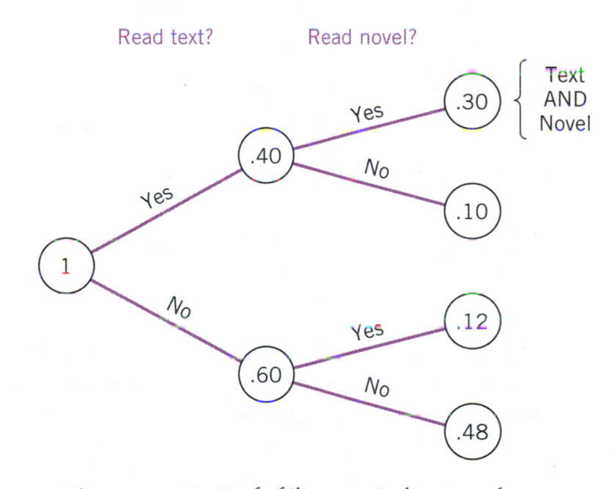

Figure 7.12 *Probability tree indicating the event "read text AND read novel"*

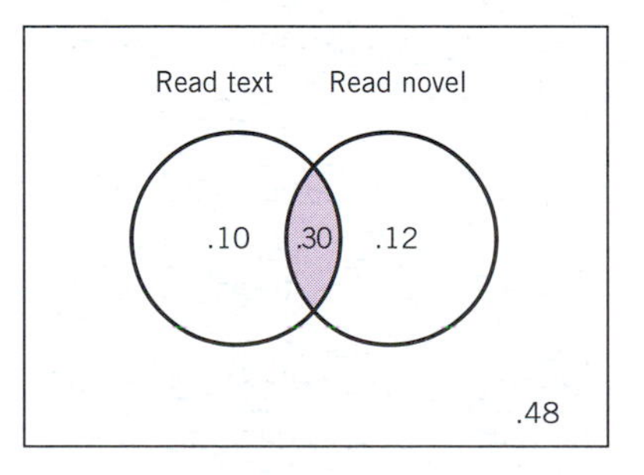

Figure 7.13 *Venn diagram indicating the event "read text AND read novel" (shaded region)*

Relationship Between AND and OR

There is a basic relationship between the probability of the AND of two events and the probability of the OR of those events that allows us to solve for one probability when the other is known and when we know the probability of each event. This rule has two forms.

1. The probability of one event OR another is equal to the sum of the probabilities of the two events minus the probability of one event AND the other.

$$P(A \text{ OR } B) = P(A) + P(B) - P(A \text{ AND } B)$$

2. The probability of one event AND another is equal to the sum of the probabilities of the two events minus the probability of one event OR the other.

$$P(A \text{ AND } B) = P(A) + P(B) - P(A \text{ OR } B)$$

The first rule, illustrated in Figure 7.14, shows that the last term needs to be subtracted because the overlapping region would otherwise be counted twice, when it should be counted only once. The second rule follows from the first rule by straightforward algebra, solving for the proper quantity. In any one problem we only use one version of the rule, picking the version that solves for the probability we are trying to find.

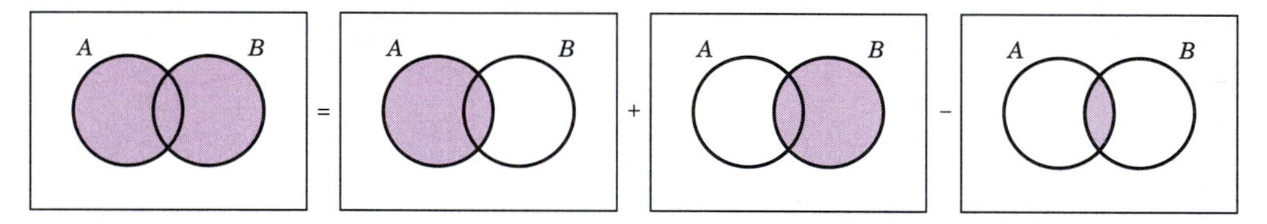

Figure 7.14 *Venn diagrams to illustrate the relationship between OR and AND: Probability of A OR B = probability of A + probability of B − probability of A AND B. Note that the last subtraction is needed so that this region is not counted twice.*

EXAMPLE 7.10 *Combining Events: Looking for a Rare Book*

Suppose we are looking for a particular rare book and there are two places we can look: the bookstore and the antique store. Assume the following probabilities:

$$P(\text{bookstore has it}) = .20$$
$$P(\text{antique store has it}) = .30$$
$$P(\text{bookstore has it AND antique store has it}) = .12$$

Based on this information, we can find the probability that we will be able to locate the book at one place or the other:

$$\begin{aligned}&\text{Probability that we will be able to find it}\\ &\quad = P(\text{bookstore has it OR antique store has it})\\ &\quad = P(\text{bookstore has it}) + P(\text{antique store has it}) - P(\text{both have it})\end{aligned}$$

$$= .20 + .30 - .12$$
$$= .38$$

Thus, there is a 38% chance that we will be able to find the book.

Exercise Set 7.14

40. You flip an honest coin twice. Suppose

P(head on first flip) = .50
P(head on second flip) = .50
P(head on first AND head on second) = .25

What is the probability of a head on the first flip OR a head on the second flip?

41. Suppose that you are in a poorly maintained boat with two engines. What is the probability that at least one engine fails, when:

P(left engine fails) = .40
P(right engine fails) = .30
P(both engines fail) = .12

42. This is not quite true, but for simplicity suppose that

P(first child is female) = .50
P(second child is female) = .50
P(both children are female) = .25

What is the probability that the first child is female OR the second child is female?

7.7 Conditional Probabilities

Information can be valuable. In particular, the information available affects our assessment of a situation and helps to guide our decision making. The effects of partial information in uncertain situations are expressed by the notion of **conditional probability.**

EXAMPLE 7.11 Conditional Probability: Job Interview

Consider a job interview situation to be a random experiment and focus on two events: "the applicant had good eye contact" and "the applicant got the job." Suppose the probabilities for this situation are summarized in the probability tree and Venn diagram of Figure 7.15 (page 224). From that figure, we see that the

Probability of "getting the job" = .12 + .08 = .20

This may be interpreted as saying that 20% of typical, or random, applicants get the job.

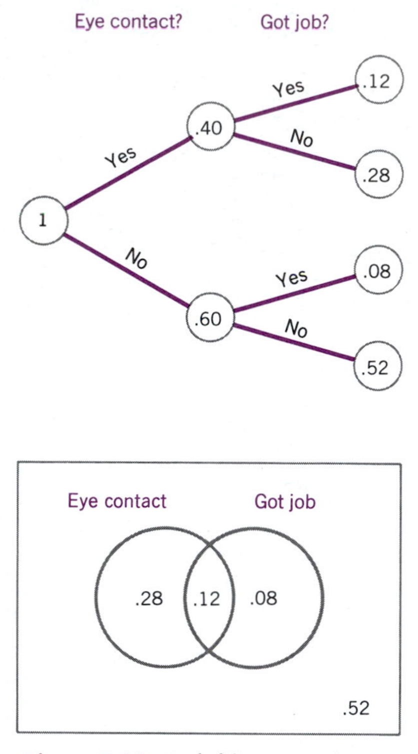

Figure 7.15 *Probability tree and Venn diagram for the job interview situation*

Does good eye contact enhance the chances of success? That is, does the additional information that "an applicant had good eye contact" have any effect on the probability of his or her getting the job? To answer this, we have to restrict ourselves to only those applicants who had good eye contact (rather than to consider all applicants). The size of this restricted group is 40% of all applicants because

$$\text{Probability of "good eye contact"} = .28 + .12 = .40$$

Within this group (40%), how many get the job? The answer is found by restricting the group to those who both had good eye contact AND got the job:

$$\text{Probability of "good eye contact" AND "got the job"} = .12$$

Finally, we can compute the probability that a person with good eye contact will get the job. This is found by dividing these last two numbers to answer the question "What percentage of .40 is .12?" This is the conditional probability of getting the job GIVEN the additional information that a person had good eye contact. In standard notation, a vertical bar between two events is read GIVEN, as shown below.

Conditional probability of getting the job GIVEN that the applicant had good eye contact

$$= P(\text{got the job} \mid \text{good eye contact})$$

$$= \frac{\text{probability of "got the job" AND "good eye contact"}}{\text{probability of "good eye contact"}}$$

$$= \frac{.12}{.40}$$
$$= .30$$

So we see that the chances of getting the job have increased from 20% (for typical applicants) to 30% (for applicants with good eye contact). That is, 30% of those job applicants who have good eye contact will get the job, as compared to only 20% for job applicants as a whole (including both those with and without good eye contact). Thus, it appears that good eye contact does indeed enhance the chances of success, and we have measured precisely how large that effect is.

There are always two events involved in a conditional probability computation: (*A*) the event whose conditional probability we wish to calculate and (*B*) the event that provides the additional GIVEN information. The general rule for computing the conditional probability of an event is as follows:

CONDITIONAL PROBABILITY OF *A* GIVEN *B*

$$P(A \mid B) = \frac{\text{probability of } A \text{ AND } B}{\text{probability of } B}$$

To apply this rule correctly, always be sure you know which of the two events provides the additional GIVEN information and which is the event whose probability is actually being computed. Do not make the mistake of dividing by the wrong probability: Always divide by the probability of the event that gives the additional information.[2]

Conditional probability can be represented with a Venn diagram by restricting our attention to the event that provides the extra information and enlarging this event to make up 100% or the entire region we are interested in. This is shown in Figure 7.16.

We can also use a probability tree, "decorating" it by inserting additional information about conditional probabilities along the branches, as illustrated in Figure 7.17 (page 226). At each place where branches divide, the conditional probabilities along the dividing branches add up to 1, indicating how things divide up once they get there. At the endpoints, the usual (unconditional) probabilities are displayed and circled as before. Note that the circled probability at the left of any branch times the conditional probability along the branch always gives the resulting circled probability to the right of the branch.

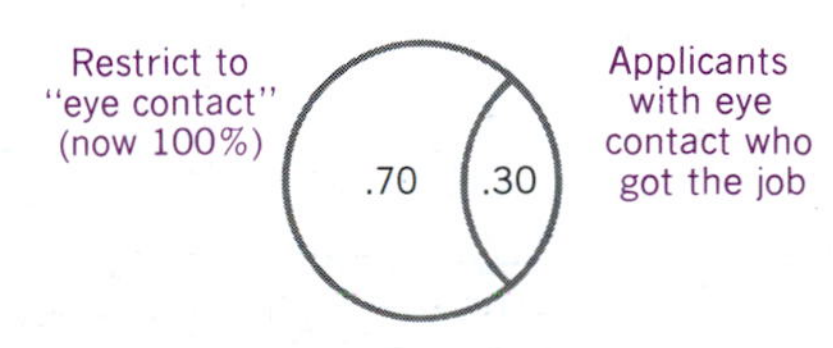

Figure 7.16 *Venn diagram for the conditional probability of an applicant's getting a particular job GIVEN that he or she had good eye contact. The given information "the applicant had good eye contact" restricts our attention to this event's circle, which now becomes 100% of the cases we are interested in.*

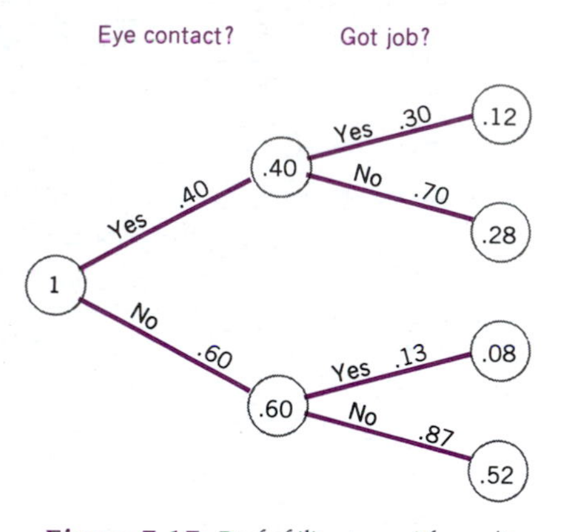

Figure 7.17 *Probability tree with conditional probabilities indicated along the branches (not circled). As before, the ordinary unconditional probabilities are circled at the ends of the branches.*

From Figure 7.17, we can see other conditional probabilities of interest in the situation. For example, the third one down on the right (.13 or 13%) represents the conditional probability of getting the job, given that the applicant did not have good eye contact. It certainly is a reasonable figure, expressing the lowered chances (down from both 20% for all applicants and 30% for those with good eye contact) resulting from a failure to maintain good eye contact. The other conditional probabilities on the right can be interpreted similarly. The conditional probabilities on the left, which branch from the initial point of the tree, are not conditional on anything (there is nothing to their left) and will be equal to the unconditional probabilities .40 and .60, respectively.

Exercise Set 7.15

43. From the information in Table 7.7 (p. 216), find the probability that an individual is bald, given that she is female. You need to find *P*(female AND bald) and *P*(female), then calculate

$$P(\text{bald GIVEN female}) = \frac{P(\text{female AND bald})}{P(\text{female})}$$

44. Also from Table 7.7, find the probability that an individual is male, given that he is bald.

45. You are told that

$$P(\text{rich AND old}) = .10$$
$$P(\text{old}) = .20$$

Find *P*(rich GIVEN old).

46. Recall the (hypothetical) wildlife study, with observations of 6 bold carnivores, 54 cautious carnivores, no bold herbivores, and 140 cautious herbivores (Exercise 30). Suppose you randomly pick one observation record for review. What is the probability you pick a record of

a. a cautious animal GIVEN that you picked an herbivore?

b. a bold animal GIVEN that you picked a carnivore?

c. a carnivore GIVEN that you picked a bold animal?

47. Recall the (hypothetical) data from the class with 20 students: 3 blue-eyed males, 5 brown-eyed males, 6 blue-eyed females, and 6 brown-eyed females (Exercise 29). Suppose you randomly select one person from this class. What is the probability of selecting

a. a female GIVEN blue eyes?
b. blue eyes GIVEN a female?
c. a male GIVEN blue eyes?
d. blue eyes GIVEN a male?

7.8 Solving Problems Using Joint Probability Tables

We can solve probability problems with a straightforward procedure using joint probability. The layout of a joint probability table forces us to consider all the possibilities and makes any required calculations fairly easy.

EXAMPLE 7.12 Filling in a Joint Probability Table: Effectiveness of Advertising

Suppose you are given the following information about the success of an ad campaign:

$$P(\text{heard ad}) = .35$$
$$P(\text{bought product}) = .23$$

We can use this information to fill in two of the totals on the joint probability table shown in Table 7.13. Since both the row totals and the column totals must sum to 1.00, we can now use the complement rule to fill in the two missing totals.

$$P(\text{did not hear ad}) = 1 - P(\text{heard ad}) = 1 - .35 = .65$$
$$P(\text{did not buy product}) = 1 - P(\text{bought product}) = 1 - .23 = .77$$

Filling these in the table, we have the results shown in Table 7.14. This is as far as we can go without information on one of the other values.

Table 7.13

		HEARD AD		
		YES	NO	TOTAL
BOUGHT PRODUCT	YES			.23
	NO			
	TOTAL	.35		1.00

Table 7.14

		HEARD AD		
		YES	NO	TOTAL
BOUGHT PRODUCT	YES			.23
	NO			.77
	TOTAL	.35	.65	1.00

Now suppose we are given the additional information that

$$P(\text{heard ad AND bought product}) = .15$$

We fill that probability in the appropriate cell of the table, as shown in Table 7.15. The remaining three cells can now be filled in by subtraction. The empty cell in the first row is

$$.23 - .15 = .08$$

since the row totals .23 and the first cell is .15 (see Table 7.16). The empty cell in the first column is

$$.35 - .15 = .20$$

since that column totals .35 and .15 of the total is in the top cell. The empty cell in the second column is

$$.65 - .08 = .57$$

since that column totals .65 and .08 is in the already filled cell. The completed table is shown in Table 7.17.

Table 7.15

		HEARD AD		
		YES	NO	TOTAL
BOUGHT PRODUCT	YES	.15		.23
	NO			.77
	TOTAL	.35	.65	1.00

Table 7.16

		HEARD AD		
		YES	NO	TOTAL
BOUGHT PRODUCT	YES	.15	.08	.23
	NO			.77
	TOTAL	.35	.65	1.00

Table 7.17

		HEARD AD		
		YES	NO	TOTAL
BOUGHT PRODUCT	YES	.15	.08	.23
	NO	.20	.57	.77
	TOTAL	.35	.65	1.00

Calculation of conditional probabilities from such a table is gloriously direct. For instance, in calculating the conditional probability that someone heard the ad GIVEN that that person bought the product, both required intermediate probabilities can be read directly from the table.

$$P(\text{heard ad} \mid \text{bought product}) = \frac{P(\text{heard ad AND bought product})}{P(\text{bought product})} = \frac{.15}{.23} = .65$$

Exercise Set 7.16

48. Using the joint probability table in Table 7.14, find the probability of "bought the product" GIVEN "heard ad."

49. **a.** Fill in the missing values in the joint probability table in Table 7.18.

b. What is the conditional probability of "Yes on *B*" GIVEN "Yes on *A*?"

Table 7.18

		A YES	*A* NO	TOTAL
B	YES	.30	.30	.60
	NO			.40
	TOTAL	.30	.70	1.00

50. Fill in the missing values in the joint probability table in Table 7.19.

Table 7.19

		C YES	*C* NO	TOTAL
D	YES	.05		
	NO			.75
	TOTAL	.20		1.00

51. Suppose that 100 students are surveyed about their courses. Thirty of them are taking a foreign language and 40 are taking a math course. Of these, 15 are taking both a foreign language and a math course.

a. Fill in a joint probability table using the above data.

b. What is the probability that one of these students selected randomly is taking a foreign language OR a math course?

c. What is the probability that one of these students selected randomly is taking a foreign language GIVEN that he or she is taking a math course?

7.9 Independent Events

We are often interested in the relationship between two events. In particular, based on the probabilities of occurrence of the various outcomes, are the events at all related? We say two events are related if information about the occurrence of one is useful in making a decision about how likely the other event is to occur. Two events are **independent** if information about one of them does *not* help us determine how likely the other event is. But this is a statement about conditional probabilities, so we may use the following definition.

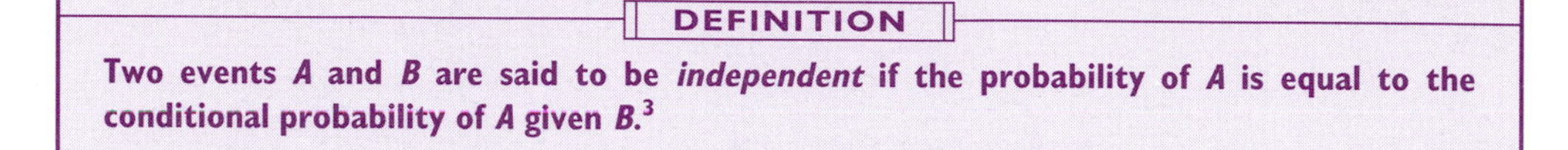

DEFINITION

Two events *A* and *B* are said to be *independent* if the probability of *A* is equal to the conditional probability of *A* given *B*.[3]

This definition simply says that two events are independent if information about one does not affect the (conditional) probability of the other. The next definition always gives identical results and is sometimes easier to work with.

EQUIVALENT DEFINITION

Two events *A* and *B* are said to be *independent* if the probability of *A* AND *B* is equal to the product of the probability of *A* times the probability of *B*.[4]

If we know—or are willing to assume—that the two events are independent, we can *calculate* the probability that they occur together. We do this by multiplying their separate probabilities, as shown in the following formula.

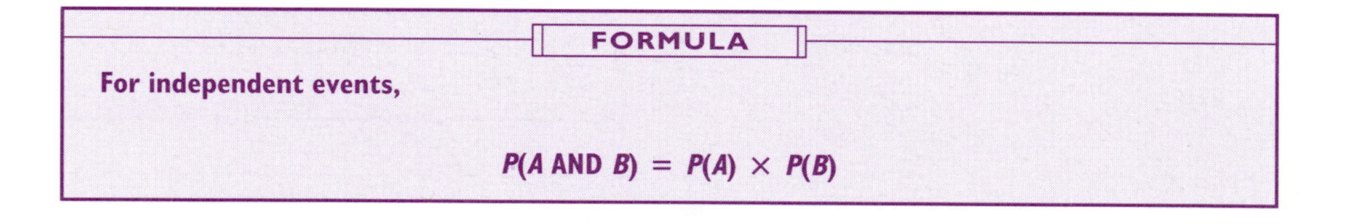

FORMULA

For independent events,

$$P(A \text{ AND } B) = P(A) \times P(B)$$

EXAMPLE 7.13 *Independent Events: Coin Tossing*

Consider a random experiment consisting of two tosses of an ordinary coin. Let the first event be "heads on the first toss" and the second event be "heads on the second toss." It is well established that the probabilities would be as given in Figure 7.18.

The conditional probability of "heads on the second toss" given that we had "heads on the first toss" is computed to be .5, which is .25 (the probability that both events occur) divided by .5 (the probability of "heads on the first toss"). The ratio of these gives the conditional probability .25/.5 = .5, which is indeed equal to the unconditional probability of "heads on the second toss."

Thus, these two events representing the results of two different coin tosses are independent. Note that this seems reasonable: The coin cannot "remember" whether it came up heads the first time; regardless of its past history, it will be heads or tails with probability .5 for each toss.

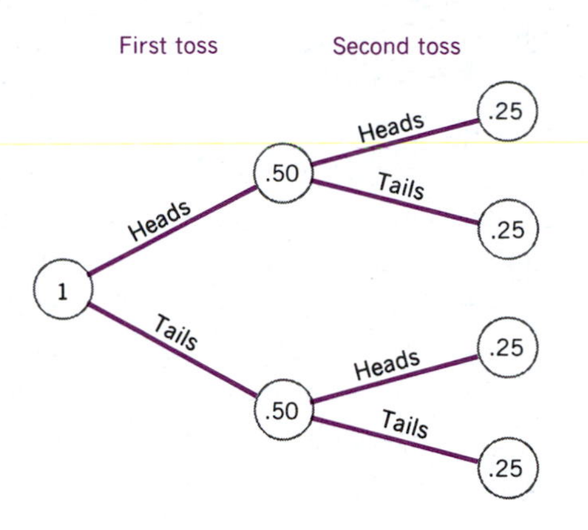

Figure 7.18 *Probability tree for two tosses of an ordinary coin*

Now, let's turn this example around. Suppose we know that the probability of a head on a single toss is .5, but that we *do not* know the probability of two heads in a row. Looking at the physical situation (coins have no memories), we decide that it is reasonable to *assume* that repeated tosses are independent. Then we can apply the formula for independent events to get

$$\begin{aligned} P(\text{two heads}) &= P(\text{heads on a single toss}) \times P(\text{heads on a single toss}) \\ &= .5 \times .5 \\ &= .25 = \frac{1}{4} \end{aligned}$$

We can continue this argument for more than two tosses.

$$\begin{aligned} P(\text{three heads}) &= P(\text{heads on two tosses}) \times P(\text{heads on a single toss}) \\ &= .25 \times .5 \\ &= .125 = \frac{1}{8} \end{aligned}$$

and

$$\begin{aligned} P(\text{four heads}) &= P(\text{three heads}) \times P(\text{heads on a single toss}) \\ &= .125 \times .5 \\ &= .0625 = \frac{1}{16} \end{aligned}$$

and so on.

Exercise Set 7.17

52. You randomly and independently select two angelfish from a (very large) population that is half female and half male.

a. What is the probability of selecting two males?

b. What is the probability of selecting two females?

53. Suppose, for simplicity, that the probability that a newborn child is female is .50. Now also suppose that the gender of each child born to a couple is independent of the gender of the last child born.

a. What is the probability of two females in a row?

b. What is the probability of three females in a row?

c. What is the probability of four females in a row?

54. Now suppose that the probability that a newborn child is female is .60. Continue the assumption that the gender is independent from one child to the next.

a. What is the probability of two females in a row?

b. What is the probability of three females in a row?

c. What is the probability of four females in a row?

55. What is the probability of five heads in a row (assuming independent flips of an honest coin)?

7.10 Dependent Events

If the probability of one event changes when another event has happened, the two events are said to be **dependent** because the probability of one depends on occurrence of the other. An extreme case of dependence is mutually exclusive events, which cannot both happen; in this case the probability of one event is reduced to 0 whenever the other happens.

If two events are not independent, they are dependent. In fact, events are very often dependent. However, we cannot reliably resolve the issue of independence or dependence by thinking about whether the events are connected or not. Instead, we must always work with the definition.

For example, consider the advertising example given earlier. Both the manufacturer and the advertising company would probably be upset if the two events "heard the ad" and "bought the product" were found to be independent. The independence of these events would suggest that advertising has no effect on purchases. To find out whether they are independent or not, we compare

Probability of "bought product" = .23

to the conditional probability of this event given the other

Probability of "bought product" GIVEN "heard ad" = .43

Because .23 is different from .43, we conclude that these are dependent events.

We cannot, however, conclude from this that one event causes the other because the probability numbers do not "know" the nature of the events they describe. When events are dependent, we know that they are related but need more knowledge (more than just probability) to decide the nature of that relationship. In this case, it seems reasonable to conclude that seeing the ad affects purchasing decisions (and to reject as nonsensical the possibility that buying the product causes people to view the ad).

Had we used the second equivalent definition to determine dependence, the conclusion would have been the same.

Probability of "heard ad" AND "bought product" = .15
Probability of "heard ad" times the probability of "bought product"
= (.35) (.23)
= .08

Because .15 is different from .08, we again conclude that these are dependent events.

Two events may be related for different possible reasons. One event might have a direct role in causing the other (for example, the event "drove too fast" can cause the event "got a ticket"). Or the events may be related by some third factor (for example, the events "wearing a wool suit" and "wearing a silk tie," neither of which causes the other, but which are related because people who wear one of the two are more likely to be wearing the other than are people in general). The reason for the relationship does not enter into the picture; only the probabilities matter in the definition of dependent and independent events. Naturally, the reasons would be considered in interpreting and explaining the resulting conclusion (dependence or independence).

Sidebar 7.1

THE THREE-DOOR PROBLEM

Here is a problem originally posed by Marilyn vos Savant in *Parade* magazine:

> Suppose you are on a game show, and you are given a choice of three doors. Behind one door is a car, behind the others, goats. You pick a door, say #1, and the host, who knows what's behind the doors, opens another door, say #3, which has a goat. He says to you, "Do you want to pick door #2?" Is it to your advantage to switch your choice of doors?

Her solution to the problem is as follows:

> Yes, you should switch. The first door has a $\frac{1}{3}$ chance of winning, but the second door has a $\frac{2}{3}$ chance. Here's a good way to visualize what happened. Suppose there are a million doors, and you pick door #1. Then the host, who knows what's behind the doors and will always avoid the one with the prize, opens them all except #777,777. You'd switch to that door pretty fast, wouldn't you?[5]

The answer provoked literally thousands of letters—most disagreeing, arguing that since two doors were left, the probability of the car being behind door 1 was the same as the probability of it being behind door 2.

It turns out that the probability of winning by switching depends on what rule the host is using to choose a door. We think three of the many possible rules are particularly interesting.

Rule 1: The host always opens another door and shows the player a goat.

This is the rule that vos Savant is considering, and she is correct that in this case you can improve your chances from $\frac{1}{3}$ to $\frac{2}{3}$ by switching. But why? Let's consider the case where you pick door 1. Assume the car was "randomly" placed to start with. There are three possibilities:

1. The car is behind door 1. The host shows you a goat behind door 2 or 3. You win if you do not switch and lose if you do.
2. The car is behind door 2. The host shows you a goat behind door 3. You lose if you do not switch and win if you do.
3. The car is behind door 3. The host shows you a goat behind door 2. You lose if you do not switch and win if you do.

Parallel analysis applies if you pick door 2 or door 3 to start with. By counting, we see that you win 1 time in 3 if you do not switch, compared to 2 times in 3 if you do. You should switch.

Rule 2: The host randomly opens one of the other doors without considering where the car is.

Again, let's consider the case where you pick door 1.

1. The car is behind door 1. The host opens door 2. If you switch, you lose.
2. The car is behind door 1. The host opens door 3. If you switch, you lose.
3. The car is behind door 2. The host opens door 2. Game over.
4. The car is behind door 2. The host opens door 3. If you switch, you win.
5. The car is behind door 3. The host opens door 2. If you switch, you win.
6. The car is behind door 3. The host opens door 3. Game over.

Similar possibilities can be listed if you choose door 2 or door 3 at the start.

Now this gets interesting. In two of the six cases, or one-third of the time, the host opens a door and shows you the car and the game is over. In the other four cases you are in a situation that is indistinguishable, from your point of view on a single turn, from the earlier one. You picked a door and the host opened another door and showed you a goat. What now? In two cases you lose if you switch, and in two cases you win. The odds are even. There is no advantage in switching.

Rule 3: The host opens a door, shows you a goat, and offers you a chance to switch only if you have picked the winning door.

Here the analysis is simple enough that we will not enumerate cases. If the host is using this rule, you always lose if you switch, so you should never switch.

The real point we want to make is that the probabilities depend on the rule that the host is using. We also point out that getting to a solution depends on being clear about the problem we are trying to solve.

7.11 Discussion: Where Do Probabilities Come From?

So far, when we have needed probability values to work with, they have been given to us as part of the problem to be solved. However, in real life the necessary numbers may not be handed to us so easily. There are three different ways to obtain probability values for use in assessing and understanding the uncertainty of a situation. The three methods are

1. computed (theoretical) probabilities
2. empirical (observed) probabilities
3. subjective (educated guess) probabilities

Computed probabilities are derived theoretically, based on given or assumed conditions or other reliable information. We derived probabilities when we filled out a joint probability table based on limited (but sufficient) available information. There are many mathematical "tricks," studied in more advanced courses, that help derive probability values in different situations.

Empirical probabilities are based on observed experience. The observed relative frequency of an event, based on a large number of runs of a random experiment, could be used as an approximate but good indication of the probability.

Subjective probabilities are usually proposed by an expert in the subject, based on the available information. They are meant to provide a "best educated guess" of what the probability is based on all related experiences. Because they are subjective, they are not normally based on calculations. Subjective probabilities are an important and useful way to introduce imprecise information into the situation for which there is no completely reliable knowledge. A branch of statistics called *Bayesian statistics* addresses the problem of combining subjective probability information with observed experiences.

Fortunately, the rules for manipulating probabilities generally apply regardless of the source of information.

Summary

The *probability* of an event is a number from 0 to 1 that indicates the event's likelihood of occurrence, with 0 indicating that the event never happens and 1 indicating that the event always happens. The study of probability is concerned with understanding the uncertain outcomes that will be produced by a known random system. This complements the study of statistics, which is concerned with discovering the nature of an unknown system based on some observed outcomes from it.

A *random experiment* is any procedure that produces a definite result (or *outcome*) that may not be predictable in advance. An *event* is any collection of outcomes; a *simple event* consists of just a single outcome. An event either "happens" or "does not happen," depending on the actual outcome of the random experiment. The *relative frequency* of an event is the number of times the event occurs divided by the number of times the random experiment was repeated. The *law of large numbers for relative frequencies* says that if the random experiment is repeated a large number of times, then the relative frequency will be close to the probability of the event.

When all the possible outcomes of a random experiment are equally likely, the probability of any particular event is equal to the number of outcomes represented by the event divided by the total number of possible outcomes of the experiment.

If all simple events can be listed for a random experiment, then the probability of any event is equal to the sum of the probabilities for each simple event it represents.

The *complement* of an event is a new event consisting of the collection of all outcomes not represented by the original event. Two events may be combined to form a new event in different ways, including the OR operation and the AND operation. Two events are said to be mutually exclusive if they cannot both happen as a result of a single run of a random experiment.

The *conditional probability* of one event GIVEN another represents a revision of the probability of an event based on the knowledge that some other given event has occurred.

To solve probability problems, it is useful to work with pictures that represent the situation, such as a *probability tree diagram* or a *Venn diagram.* The *joint probability table,* a useful summary of the probabilities for two events, also provides a good way to develop solutions.

Two events are said to be *independent* if information about the occurrence of one event does not affect the likelihood of the occurrence of the other. Two events that are not independent are *dependent.* Events may be dependent with or without the direct causal influence of one event over the other.

There are three sources of probability numbers in applications: *computed* probabilities based on theory, *empirical* probabilities based on observed experience, and *subjective* probabilities based on educated guesses about the situation. The rules of probability will apply to any situation regardless of the source of the probability numbers.

Problems

Basic Reviews

1. A die has six faces, numbered 1 to 6 with dots. What are the simple outcomes that are possible when you roll a die once?

2. What are the possible outcomes of flipping a single coin once?

3. You have a nickel and a dime and flip them both. What are the possible *outcomes* that satisfy the *event* "one head and one tail"?

4. You have a nickel and a dime and flip them both. What are the possible *outcomes* that satisfy the *event* "at least one head"?

5. You flip a coin 10 times and get 6 heads. In this set of flips, what is the relative frequency of heads?

6. You flip a coin 100 times and get 55 heads. In this set of flips, what is the relative frequency of heads?

7. What would be the relative frequency of the event "the sum equals 7" if you rolled two dice 36 times and the sum of 7 occurred six times?

8. Suppose the probability of rain is 20% today. What is the probability that it will *not* rain?

9. A group consisting of 26 men and 31 women is choosing a representative. If the selection process is not discriminatory, so that each person has an equal chance of being selected, what is the probability that the chosen person is a woman?

10. There are three possible outcomes when tossing a coin: "heads," "tails," and "lands on its edge." Does it follow that each of these three simple events has probability $\frac{1}{3}$? Why or why not?

11. A roulette wheel has 38 possible outcomes: the numbers 1 through 36 and the special outcomes 0 and 00. Each of these 38 simple events is equally likely to result from a spin of the wheel.
    a. What is the probability that a number from 1 to 36 will result?
    b. What is the probability that the special outcome 00 will result?

Analysis

12. Suppose the probabilities of success for a new software project are as given in Table 7.20.

Table 7.20

OUTCOME	PROBABILITY
Make a profit beyond our wildest dreams	.02
Make a fairly nice profit	.47
Make back the initial investment only (break even)	.26
Lose some money	.21
Lose our shirts (lose a lot of money)	.04

    a. Consider the event "make a profit." Which outcomes does this event represent? What is the probability of this event?
    b. Consider the event "lose money." Which outcomes does this event represent? What is the probability of this event?

13. Consider the random experiment "give the medication to seasick patients and then observe any change in their condition" as part of a test of a new medication. Suppose that this random experiment is repeated for each of 200 seasick patients and that 132 were observed to improve, whereas the others did not.

a. Find the relative frequency of the event "the patient improved."

b. Is your answer to part (a) close to the probability of the event "the patient improved"? Why or why not?

c. Can you conclude that your answer to part (a) is exactly equal to the probability of the event "the patient improved"? Why or why not?

14. In a study of animal behavior, consider the events "the monkey was playing in the brook" and "the monkey was sleeping in the tree." Suppose that the probabilities are .23 and .37, respectively, for these two events.

a. With the understanding that these events represent behavior at one particular time for one particular monkey, would you say that these events are mutually exclusive or not?

b. What is the probability that the monkey was playing in the brook OR sleeping in the tree?

c. What is the probability that the monkey was playing in the brook AND sleeping in the tree?

15. A restaurant has collected data on its customers' orders and so has estimated empirical probabilities about what happens after the main course. It was found that 20% of the customers had dessert only, 40% had coffee only, and 30% had both dessert and coffee.

a. Do one of the following.

i. Draw a Venn diagram for this situation.

ii. Draw a probability tree in which the first branch represents the event "ordered coffee."

iii. Draw a probability tree in which the first branch represents the event "ordered dessert."

iv. Fill in a joint probability table for these events.

b. Find the probability of the event "had coffee." (*Hint:* Be careful; this event includes those who did as well as those who did not have dessert.)

c. Find the probability of the event "did not have dessert."

d. Find the probability of the event "neither coffee nor dessert."

e. Find the probability of the event "had coffee OR dessert."

f. Are the events "had coffee" and "had dessert" mutually exclusive? How do you know?

g. Find the conditional probability of having coffee GIVEN that the customer had dessert.

h. Are "had dessert" and "had coffee" independent events? How do you know?

i. Find the conditional probability of having dessert GIVEN that the customer had coffee.

j. Find the conditional probability of having dessert GIVEN that the customer did not have coffee.

k. To see if coffee and dessert seem to go well together, compare your answers to parts (i) and (j) above. In particular, who is more likely to order dessert: a customer who orders coffee or one who does not?

16. A patient with a failing heart has been told that the probability of finding a suitable transplant is .38, the conditional probability of surviving GIVEN that a transplant operation is performed is .85, and the conditional probability of surviving GIVEN that a transplant operation is not performed is .30.

a. Find the probability that the patient survives.

b. Find the probability that a suitable transplant was found AND the patient survived.

c. Find the conditional probability that a transplant was found GIVEN that the patient survived.

17. Chances are sometimes stated as "odds ratios," like "2 to 1" or "50:50." These can be converted to regular probabilities by adding them up and dividing, so that "2 to 1" becomes $\frac{2}{3}$ (or "2 out of 3") and "50:50" becomes $\frac{50}{100}$ or $\frac{1}{2}$. Convert these following odds ratios to probabilities.

a. 3 to 1

b. 4 to 1

c. 1 to 9

18. You toss an honest coin three times. (By convention, when we talk of an "honest coin," we mean one where the probabilities of heads and tails are both .50, and we ignore the small real-world possibility that the coin lands on an edge, rolls out the door, bounces down the stairs, and is never seen again.) What is the probability

a. of three heads?

b. of three tails?

c. of no heads?

d. of no tails?

e. of at least one head?

f. of at least one tail?

19. Suppose you have a box with the names of 600 Republicans and 400 Democrats written on slips of paper. You thoroughly shake the box to stir the slips, then draw a single slip, write the party membership on a piece of paper, and put it back. You shake and stir the slips again, draw a single slip, write down its party, and put it back. You do this 100 times. (By the way, it is very hard—maybe impossible—to mix slips of paper so thoroughly that the draws are really random, but for this problem assume the draws are random and independent.)

a. About how many times should you get a Republican name?

b. About how many times should you get a Democratic name?

20. Suppose you have a long list of names. You are told that if you pick a random name, there is a .60 probability that person is brown-eyed. Assuming this is true, if you draw two names,

a. what is the probability that they are both brown-eyed?

b. what is the probability that neither is brown-eyed?

c. what is the probability that at least one is brown-eyed?

d. what is the probability that one is brown-eyed and the other is not?

21. Suppose there are 1,000 first-year students at a certain university. Two hundred are taking French (but not statistics), one hundred are taking statistics (but not French), and another 50 are taking both French and statistics.

a. Do one of the following.

i. Draw a Venn diagram of this situation.

ii. Draw a probability tree.

iii. Fill in a joint probability table.

If you pick a student at random, what is the probability that this student

b. is taking French, but not statistics?

c. is taking French (and maybe or maybe not statistics)?

d. is taking French AND statistics?

e. is taking French OR statistics?

f. is not taking French?

g. is taking neither French nor statistics?

22. A six-sided die is tossed 100 times, with the results given in the following chart.

Face showing	1	2	3	4	5	6
Frequency	20	14	17	16	18	15

a. Find the relative frequencies of the following events.

i. a 1 appears

ii. an even number appears

iii. a number of 4 or more appears

b. Do you think these results are random, or does the die appear to be "loaded"? Explain your answer.

23. Suppose that you are playing a computerized coin-tossing game written by a 10-year-old computer whiz. You are playing against the child who programmed it and after a while you begin to notice that you seem to be losing more than half the time. You start to keep track of the number of times you win and the total number of plays. Every ten plays you calculate the relative frequency of your winning. The graph of these calculations is shown in Figure 7.19.

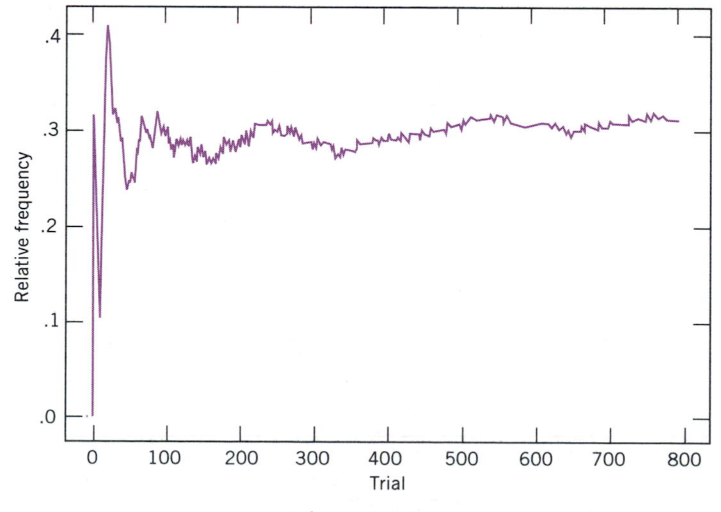

Figure 7.19

a. Does this graph look like one where your probability of winning is $\frac{1}{2}$?

b. What appears to be the probability that the relative frequency is fluctuating around? (Limit your possible answers to the fractions $\frac{3}{4}, \frac{3}{5}, \frac{3}{10}, \frac{1}{2}, \frac{1}{3}, \frac{1}{4}, \frac{1}{5}$ and $\frac{1}{6}$.)

24. Data on the living arrangements of 18- to 24-year-olds in 1989[6] are shown in Table 7.21. A "householder" is a person (or one of the persons) in whose name a home is owned or rented. Young adults living with their parents—or even in a college dorm—are counted as a "child of householder," not as co-householders, unless their name is on the papers.

Now, let's focus on the 1989 figures. There were a total of 25,629 thousand 18- to 24-year-olds in 1989; of these, 13,450 thousand (52.5%) are counted as "child of householder," 5,956 thousand (23.2%) had established their own family households (or were the spouse of family householders), 2,363 thousand (9.2%) had their own household alone or with nonrelatives, and 3,860 thousand (15.1%) had various other arrangements.

a. What is the probability in 1989 that a randomly selected 18- to 24-year-old was living at home or as an unmarried student in a college dorm?

b. What was the probability in 1989 that a randomly selected 18- to 24-year-old had established (or married into) their own family household?

c. What was the probability in 1989 that a randomly selected 18- to 24-year-old was living alone or as part of an unrelated household?

d. What was the probability in 1989 that an 18- to 24-year-old was living as a member of a family household, either that of their parents or a family they established themselves?

Table 7.21 *Living arrangements of 18- to 24-year-olds, by sex: 1960 to 1989*

LIVING ARRANGEMENTS OF 18- TO 24-YEAR-OLDS	NUMBER, IN THOUSANDS				PERCENTAGE DISTRIBUTION			
	1960	1970	1980	1989	1960	1970	1980	1989
Total	14,718	22,357	29,122	25,629	100.0	100.0	100.0	100.0
Child of householder[a]	6,333	10,582	14,091	13,450	43.0	47.3	48.4	52.5
Family householder or spouse	6,186	8,470	8,408	5,956	42.0	37.9	28.9	23.2
Nonfamily householder[b]	354	1,066	2,766	2,363	2.4	4.8	9.5	9.2
Other[c]	1,845	2,239	3,848	3,860	12.5	10.0	13.2	15.1
Male	6,842	10,398	14,278	12,574	100.0	100.0	100.0	100.0
Child of householder[a]	3,583	5,641	7,755	7,308	52.4	54.3	54.3	58.1
Family householder or spouse	2,160	3,119	3,041	2,055	31.6	30.0	21.3	15.9
Nonfamily householder[b]	182	563	1,581	1,345	2.7	5.4	11.1	10.7
Other[c]	917	1,075	1,902	1,916	13.4	10.3	13.3	15.2
Female	7,876	11,959	14,844	13,055	100.0	100.0	100.0	100.0
Child of householder[a]	2,750	4,941	6,366	6,142	34.9	41.3	42.7	47.0
Family householder or spouse	4,026	5,351	5,367	3,951	51.1	44.7	36.2	30.3
Nonfamily householder[b]	172	503	1,195	1,018	2.2	4.2	8.1	7.8
Other[c]	928	1,164	1,946	1,944	11.8	9.7	13.1	14.9

[a]Child of householder includes unmarried college students living in dormitories.
[b]A nonfamily householder is an unmarried person maintaining a household while living alone or with nonrelatives.
[c]Includes roomers, boarders, paid employees, and nonrelatives sharing a household but not classified as the householder.

Note: A householder is defined as a person (or one of the persons) in whose name the housing unit is owned or rented. There can be only one householder per household. This table excludes inmates of institutions and military personnel living in barracks.

Source: U.S. Department of Commerce, Bureau of the Census, Current Population Reports, Series P-20, Marital Status and Living Arrangements, nos. 410 and 445.

From *Youth Indicators 1991: Trends in the Well-Being of American Youth.* Office of Educational Research and Improvement, U.S. Department of Education. U.S. Government Printing Office, Washington, D.C., April 1991.

25. Notice that Table 7.21 also reports the living arrangements of males and females separately. From the 1989 data in this table, what is the probability that

a. a randomly selected male is living with his parents or in a college dorm?

b. a randomly selected female is living with her parents or in a college dorm?

c. a randomly selected male has established his own family ("Family householder or spouse")?

d. a randomly selected female has established her own family ("Family householder or spouse")?

26. Write a sentence or two describing these probabilities found in Problem 25. Speculate on why there may be differences.

27. From Table 7.21, can you see any changes from 1960 to 1989 in the probability that an 18- to 24-year-old is living with his or her parents or in a college dorm? If so, describe these changes and speculate on what may have caused them.

28. Variations on a technique of "inoculation" had been practiced for centuries in China, Turkey, Greece, and Russia. A healthy person was deliberately exposed to smallpox, in the hope that they would develop a mild case and then be protected from severe cases. (In one variation of inoculation, children were made to sleep in nightshirts smeared with pus from smallpox victims; in another, pus was rubbed into scratches on the arm.) A paper on inoculation was published by the Royal Society in 1714. Cotton Mather read it and vowed to try inoculation.

In April 1721 the British ship *Seahorse* entered Boston Harbor. After a sick sailor was taken ashore, it was discovered that he had smallpox. On May 21, Cotton Mather wrote, "The grievous calamity of smallpox has now entered the town."

On June 26, 1721, Dr. Zabdiel Boylston conducted the first inoculations on the American continent. His subjects were his own son (age 6), a slave, and the slave's son (age 2). Now, this was a very risky procedure—not a modern vaccination with a virus that has been rendered harmless, or even the vaccination procedure that Jenner later developed using the related, but less dangerous, cowpox. Boylston deliberately induced cases of smallpox in the three subjects, who luckily survived.

Boylston went on to inoculate 244 more people, six of whom died. By his count there were another 5,759 people in Boston who were susceptible to smallpox. Of these, 844 died.[7] For now, use the relative frequencies as (empirical) probabilities.

a. Based on the results of Boylston's inoculations, what was the (conditional) probability that an inoculated individual would die of smallpox?

b. Based on Boylston's data, what was the (conditional) probability that an uninoculated, but susceptible, individual would die of smallpox?

c. Irrespective of whether they were inoculated, what was the probability a susceptible Boston resident would die of smallpox during this epidemic?

d. Do "inoculation" and "death by smallpox" appear to be independent or dependent, based on this data?

29. *The Nation,* a weekly magazine of politics and the arts, occasionally sponsors poetry contests. The results of one were disputed by a reader, whose comments follow:

> Your poetry competition and its result were remarkable. Competitiveness is one of those traits of our present system I'd think you'd eschew. But what really irks me is your result: Four female poets win a competition cosponsored by a publication with a female poetry editor. Yes, there were males on the judging panel just as the queens of old had

eunuchs to attend them. Does *The Nation* mean to tell us that there were no entries by male poets that remotely approached the quality (however that's judged) of the winners? I can imagine the screams from the gallery if the results were as one-sided in the other direction—four male winners of a prize offered by a publication with a male poetry editor.

If progressive principles include freedom from gender bias, you're as regressive as anyone, only you've exchanged Neanderthal attitudes for Amazonian. You're as helplessly carried about by the raging beast of gender prejudice as anyone you've ever criticized.[8]

The editor responded that the contest had been judged anonymously and that none of the judges knew the names or genders of the authors they were judging. She went on to make a brief argument that sometimes things like that just happen.

Let's set up the problem. Suppose that equal numbers of males and females enter the contest and that the total number of entries is large enough so that there is no effective difference between sampling with and without replacement.

a. If we select four winners randomly with respect to gender, what is the probability they are all male?

b. If we select four winners randomly with respect to gender, what is the probability they are all female?

c. If we select four winners randomly with respect to gender, what is the probability they are all the same gender?

d. Suppose we ran this poetry contest annually for 80 years. In about how many years would we expect all the winners to be of the same gender?

30. Dinosaur fossils, including fossilized eggs, are relatively common in parts of the Gobi Desert. Unfortunately, eggs with identifiable embryos are rare, so figuring out which eggs go with which adults is sometimes a bit of a guessing game. Early fossil collections from the Flaming Cliffs region included 101 specimens of protoceretops, two velociraptors, one saurornithoides, and one oviraptor. At the same time, half the eggs collected were of a certain type, which we will call A.[9]

a. Just on the basis of the information above, which type of dinosaur probably laid type-A eggs? Explicitly state the assumptions that you are making in reaching this conclusion.

b. Now suppose that you are told that the first fossilized oviraptor was found lying on top of a clutch of type-A eggs. Would you change your conclusion? Why or why not? Again, explicitly state the assumptions you are making in reaching these conclusions.

31. Medical tests can be used for many purposes: to determine whether a person has a disease (say, AIDS), carries a disease-causing organism (say, HIV), has used certain drugs to enhance athletic performance, or has used an illegal recreational psychoactive substance. But these tests are sometimes wrong. They can be wrong in either of two ways: They can turn out negative when they should be positive, or they can turn out positive when they should be negative. If we are testing for drug use and the individual has not used drugs, a mistaken positive result is said to be a "false positive." If the individual has used drugs but the test result is negative, it is said to be a "false negative."

Suppose that random drug tests are given to 100 athletes. Suppose further that half these athletes have used a banned drug to enhance their performance. The tests indicate that half the athletes used drugs. Unfortunately, the tests are completely independent of truth: Half the drug users are found not to have used drugs, and half the nonusers are found to have used drugs.

a. Construct a joint probability table from these data.

b. Calculate the rates of false positives and false negatives.

32. Suppose that an HIV antibody test has, effectively, a 100% chance of detecting the presence of HIV antibodies, with a false positive rate of 1 in 20,000. (These are approximately the figures for the testing procedures used by the U.S. Army.[10])

a. Using this test, suppose you test 100,000 randomly selected people from a population where 20% of the individuals carry HIV. About how many true positive results do you expect? About how many false positive results do you expect? What share of the total number of positive results are false positives?

b. Now suppose you apply the test to 100,000 randomly selected individuals from a population where HIV infection is very rare, with only about 1 person in 10,000 infected. About how many true positive results do you expect? About how many false positive results do you expect? What share of the total number of positive results are false positives?

chapter 8

Random Variables, Probability Distributions, and the Central Limit Theorem

In the previous chapter we dealt with the probability that something random *would* or *would not* happen. In this chapter we extend the basic ideas to ask *how much* of some random quantity we are likely to see.

We first introduce the idea of a *random variable* that takes on numerical values. This leads to a discussion of *probability distributions,* especially *discrete probability distributions,* the *binomial distribution,* and the *normal distribution.* Something special happens when we compute the average of independent observations of a random variable. According to the central limit theorem, the probability distribution of this average becomes closer to a normal distribution as our samples get larger.

Much of the material in this chapter will already seem familiar. In particular, we will be using the mean and standard deviation to provide summary descriptions of these probability distributions, just as we used them to provide summary descriptions of data distributions.

8.1 Random Variables

If the outcome of a random experiment is a number (rather than, say, a category such as "green" or "raining"), then the random numerical result is called a **random variable.** Examples of random experiments and their random variables include:

- *Random experiment:* Pick a random classroom and count the number of students present. *Random variable:* Number of students.
- *Random experiment:* Randomly pick a patch of woods and count squirrels for 10 minutes. *Random variable:* Number of squirrels.
- *Random experiment:* Randomly assign patients who have volunteered for the study to "treatment as usual" or "experimental treatment." Collect data on recovery time. *Random variable:* Recovery time.

DEFINITION

A *random variable* is a numerical outcome (in the abstract) of a random experiment.

Discussion

It is important to recognize that the random variable is the idea or the process of generating some (partially) random number but it is not the number itself. The way we distinguish between the process and the number is to say that the random variable *generates* the number. For example, consider the random variable "the amount of rain today." After the day goes by, this total amount is found to be, say, 1.37 in. We say that this random variable has been observed and that it takes on the value 1.37 in. However, the random variable is *not* the number 1.37. This number is its observed value, but the random variable remains defined as the abstract idea "the amount of rain today."

Every random variable has a **probability distribution** that describes the possible values and the likelihood of occurrence of these values. Sometimes we can make a table to represent the probability distribution (for discrete distributions in general and for the binomial distribution in particular). For other situations (for example, the normal distribution), we cannot make a table because there are too many possible outcomes. Instead we work with a graph (for example, the normal bell-shaped curve) to represent the probability distribution.

Discrete Random Variables

If we are able to list all possible values that a random variable can take on, and there are no values between the ones listed, then the random variable is said to be **discrete.** If we list each possible value together with the probability that the value is observed, then we have the *probability distribution* of the random variable. From such a probability distribution, we can find nearly everything of interest about the random variable, including its mean, its standard deviation, and various probabilities.

DEFINITION

The *probability distribution* of a discrete random variable is a list of each possible value together with the probability that each value will be observed.

EXAMPLE 8.1 A Discrete Probability Distribution: Number of Tankers in Port

Consider the random variable "the number of tankers docked at a certain time at a certain seaport." Based on experience, suppose the probability distribution of the number of tankers is known to be as shown in Table 8.1.

Table 8.1

NUMBER OF TANKERS DOCKED	PROBABILITY
0	.17
1	.24
2	.30
3	.21
4	.08

A probability distribution can be represented as a picture and has an interpretation similar to a histogram (which represents the distribution of a set of data consisting of a single group of numbers). From Figure 8.1, we obtain an impression of the relative likelihoods of different outcomes. For example, we see that middle values are more likely than those at the endpoints of the range because the probabilities are higher in the middle.

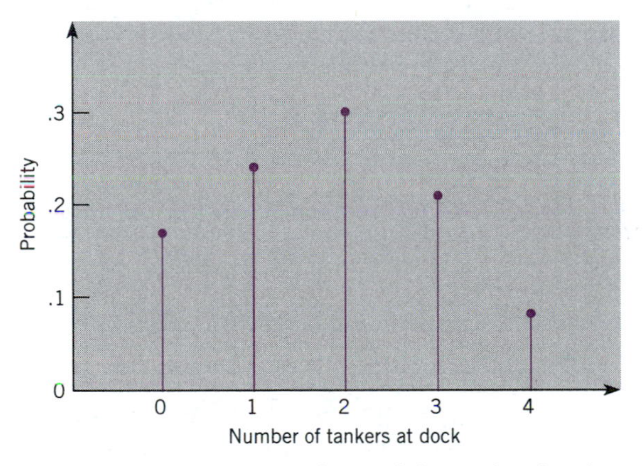

Figure 8.1 *Probability distribution of the number of tankers*

From this probability distribution, we can compute any probability of interest. For example, the probability that two or more tankers are docked is

$$.30 + .21 + .08 = .59$$

The Mean Value and the Standard Deviation

Just as we summarized a single group of numbers in an earlier chapter, we will now summarize the probability distribution of a random variable. The first summary number, the *mean* or *average*, describes a typical or central value of the random variable. It is computed using a *weighted average* as follows:

FORMULA

The mean of a discrete random variable is found by adding up the product of "value" times "probability" from the probability distribution:

$$\text{mean} = \Sigma[X \times P(X)]$$

For example, the average number of tankers at dock is easily determined as in Table 8.2. We see that on the average, there are about 1.79 tankers docked. Referring to Figure 8.1, we see that 1.79 is indeed a reasonable summary number, falling within the central region of the distribution. As was the case when we calculated the mean of a data distribution, the mean is the "balance point" of the histogram.

***Table* 8.2**

NUMBER OF TANKERS	PROBABILITY	TANKERS × PROBABILITY
0	.17	0 × .17 = 0
1	.24	1 × .24 = .24
2	.30	2 × .30 = .60
3	.21	3 × .21 = .63
4	.08	4 × .08 = .32
		Mean number of tankers: 1.79

The *standard deviation,* the second summary value, describes how far an observation will typically be from the mean and is a measure of variability or randomness of the distribution. It is computed as follows:

FORMULA

The standard deviation of a discrete random variable is found by squaring the deviation "value minus mean," multiplying by "probability," then summing these products and finally taking the square root.

$$\text{standard deviation} = \sqrt{\Sigma[(X - \text{mean})^2 \times P(X)]}$$

The *variance* is the sum just before the square root is taken.

For example, the standard deviation of the number of tankers at dock is computed as shown in Table 8.3 and the following equation:

$$\begin{aligned}\text{standard deviation} &= \text{square root of variance (from Table 8.3)}\\ &= \sqrt{1.4059} = 1.1857\end{aligned}$$

***Table* 8.3**

NUMBER OF TANKERS	PROBABILITY	TANKERS × PROBABILITY	DEVIATION: TANKERS − MEAN	SQUARED DEVIATION	PROBABILITY × SQUARED DEVIATION
0	.17	.00	0 − 1.79 = −1.79	3.2041	(.17)(3.2041) = .5447
1	.24	.24	1 − 1.79 = −.79	.6241	(.24)(.6241) = .1498
2	.30	.50	2 − 1.79 = .21	.0441	(.30)(.0441) = .0132
3	.21	.63	3 − 1.79 = 1.21	1.4641	(.21)(1.4641) = .3075
4	.08	.32	4 − 1.79 = 2.21	4.8841	(.08)(4.8841) = .3907
		1.79			Variance: 1.4059

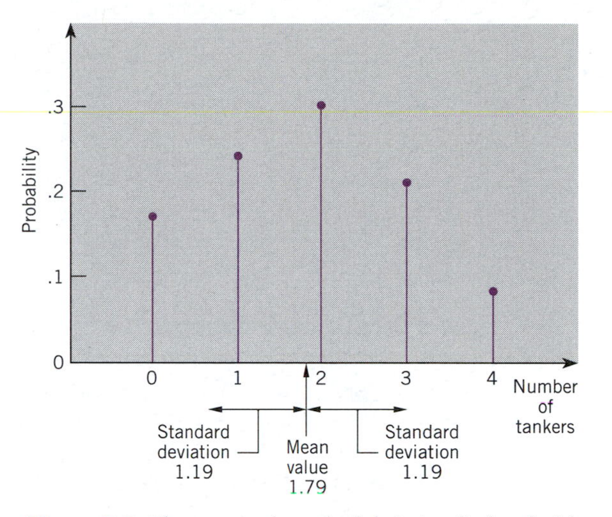

Figure 8.2 *The mean and standard deviation displayed with the probability distribution of the number of tankers*

This standard deviation, rounded to 1.19, indicates that typically there are about 1.19 tankers more than or less than the value of 1.79. Figure 8.2 shows the mean and standard deviation values displayed with the probability distribution itself. Note how the mean value summarizes a center for the distribution, whereas the standard deviation summarizes its variability.

The mean value of a probability distribution is sometimes called the *expected value*. It is "expected" in the sense that it is what we expect to average over the long run, not what we expect in any one experiment.

Exercise Set 8.1

1. a. What is the mean value of the discrete probability distribution shown in Table 8.4? (*Hint:* Set up a table like Table 8.4 and add a third column for the product of Value times Probability. Then find the average of these products.)

Table 8.4

VALUE	PROBABILITY
0	.20
1	.40
2	.20
3	.10
4	.10

b. What is the standard deviation of the probability distribution in Table 8.4?

c. Draw a picture (like Figure 8.2) of the probability distribution in Table 8.4. Indicate the mean and standard deviation.

2. a. What is the mean value of the discrete probability distribution shown in Table 8.5 (page 250)?

b. What is the standard deviation of the probability distribution in Table 8.5?

c. Draw a picture (like Figure 8.2) of the probability distribution in Table 8.5. Indicate the mean and standard deviation.

Table 8.5

VALUE	PROBABILITY
0	.33
1	.34
20	.33

3. What is the probability distribution for the number of heads in two tosses of an honest coin? What are the mean and SD of this distribution?
4. Suppose that your state runs a "game" where you pay a dollar (−$1) for a ticket. You then scratch a silver film off the ticket to see whether you win a prize. One ticket in ten pays $2 (which puts you one dollar ahead, since you paid a dollar for the ticket), and another one ticket in ten pays $5 (which puts you $4 ahead). The other eight tickets leave you behind by the dollar you paid for them.
   a. Construct a table showing the probability distribution of this game.
   b. What is the mean of this probability distribution?
   c. What is the standard deviation of this probability distribution?

8.2 The Binomial Distribution

Many processes result in binary data with only two outcomes: yes/no, heads/tails, male/female, and so on. We may want to make repeated independent trials—flipping a coin or surveying a random individual, for instance—where the probabilities do not change. (Technically, these are called *Bernoulli trials,* after James Bernoulli, who published *Ars Conjectandi* in 1713.) Our goal is to make a probability statement about the likelihood of seeing some particular outcome in a given number of trials (for instance, "What is the probability of eight heads in ten flips of an honest coin?" or "What is the probability that ten randomly selected voters will favor Smith, when 60% of all voters favor Brown?"). It turns out that the solution to this is very general and leads naturally to the binomial distribution.

Sidebar 8.1

Questions of the type that can be answered using the binomial theorem are very common. Are six heads in ten flips of a coin unusual? Seven flips? How about eight? If we select four winners in a poetry contest without regard to sex and are willing to assume that quality of submitted poems is the same for the two sexes and that half the contestants are male, what is the probability that two males win? Three males? Four males?

The form of the problem is common enough to provide the opening scene of Tom Stoppard's play *Rosencrantz and Guildenstern Are Dead.* As the play opens, Rosencrantz and Guildenstern are flipping coins. If it comes up heads, Rosencrantz wins. So far, it has come up heads 76 times in a row.

> *Guildenstern:* A weaker man might be moved to reexamine his faith, if in nothing else at least in the law of probability. . . . The law of averages, if I have got this right, means that if six monkeys were thrown up in the air for long enough they would land on their tails about as often as they would land on their [heads].[1]

EXAMPLE 8.2 *Binomial Probabilities: Financial Aid*

Suppose you are a student at a large university. You know from the registrar's office that 60% of the students get some financial aid (Y = Yes) and 40% get no financial aid (N = No). You are going to pick five students at random and ask whether they get financial aid. You want to calculate the probability of getting exactly three "yes" responses.

There are a lot of different orders in which you can get three financial aid students in the answers from five students. You could, for instance, pick the aid students first and then the two non-aid students: YYYNN. Or you could pick two aid students, a non-aid student, another aid student, and then another non-aid student: YYNYN. Listing all the different ways to get three aid students and two non-aid students, we have:

YYYNN, or YYNYN, or YNYYN, or NYYYN, or YYNNY, or
YNYNY, or NYYNY, or YNNYY, or NYNYY, or NNYYY

What is the probability of the first of these, YYYNN? Let's say the probability of Yes is .6, and call it p. Likewise, let's call the probability of No q. Since No is the complement of Yes, $q = 1 - .6 = .4$.

The probability of Yes to the first question is .6, and so is the probability of Yes on the second question. We assume the events are independent, so

$$P(YY) = .6 \times .6 = .36$$

or, more generally,

$$P(YY) = p \times p = p^2$$

The probability of Yes on the third question is also .6, so the probability of Yes to the first three questions is .6 times the probability of Yes to the first two questions:

$$P(YYY) = P(YY) \times P(Y) = .36 \times .6 = .216$$

or, more generally,

$$P(YYY) = p \times p \times p = p^3$$

Now, the probability of No on the fourth question is .4, so the probability of Yes on the first three questions and No on the fourth is

$$P(YYYN) = P(YYY) \times P(N) = .216 \times .4 = .0864$$

or, more generally,

$$P(YYYN) = p^3 \times q = p^3q$$

Making a parallel argument one more time, the probability of No on the fifth question is .4, so the probability of Yes on the first three questions and No on the last two is .4 times the probability of Yes on the first three questions and No on the fourth.

$$P(YYYNN) = P(YYYN) \times P(N) = .0864 \times .4 = .0346$$

or, more generally,

$$P(\text{YYYNN}) = p \times p \times p \times q \times q = p^3q^2$$

That takes care of YYYNN. Now how about the others, say YYNYN? We can work this out the same way, and we will get the same answer:

$$P(\text{YYNYN}) = ppqpq = p^3q^2 = .0346$$

Similarly, all the other arrangements of three Y's and two N's have the same probability.

Now, we were not originally concerned with the particular order, we just wanted to know the probability of getting three financial aid students. Since the ten possibilities are mutually exclusive, we can use the addition rule to find the probability of one OR another of them:

$$\begin{aligned} &P(\text{three Yes responses in five answers}) \\ &\quad = .0346 + .0346 + .0346 + .0346 + .0346 + .0346 + .0346 + .0346 + .0346 + .0346 \\ &\quad = 10 \times .0346 = .346 \end{aligned}$$

Let's stop and review what we have done: We have some number of trials, usually called n (equal to 5 in this example). We want the probability of three Yes (Y) responses in the five trials and are given that the probability of Yes, p, equals .6 on a single trial. We will call the number of Y's k. We calculated the probability of a particular sequence of three Yes responses and two No responses, then multiplied this by the number of different possible sequences. Writing this out in our notation, we have

$$\begin{aligned} P(\text{3 Y's in five answers}) &= (\text{number of sequences with 3 Y's}) \times .6^3 \times .4^2 \\ &= 10 \times .0346 = .346 \end{aligned}$$

or, more generally,

$$P(k \text{ Y's in } n \text{ tries}) = (\text{number of sequences with } k \text{ Y's}) \times p^kq^{(n-k)}$$

Exercise Set 8.2

5. Suppose a computer is programmed to play a dishonest game of roulette. It randomly picks red (R) with probability $p = .6$ and green (G) with probability $q = .4$. You have the computer "spin the wheel" three times. What is the probability of gettting

a. RRG?

b. RGR?

c. GRR?

d. two reds and one green (without worrying about their order)?

e. RGG?

f. GRG?

g. GGR?

h. one red and two greens?

i. RRR?

j. GGG?

k. Are there any other possibilities?

l. What do the probabilities of three reds, two reds and a green, one red and two greens, and three greens add up to?

6. You are going to buy four aquarium fish at the pet store. You cannot tell their sexes by looking at them. Assume that half of the fish in the tank are male and half female, and that there are so many that removing four of them does not change any of the probabilities measurably. You want to calculate the probability that you get two males and two females in picking your four fish.

   a. There are six different ways to get two males and two females. What are they? (By "different way," we mean different arrangements of MMFF. Think of them as different orders of being netted out of the tank.)

   b. What is the probability of getting two males and two females in one particular order (say MFMF)?

   c. What is the probability of getting two males and two females, without worrying about the order?

7. Suppose a couple plans to have three children, and that for each birth a boy is as probable as a girl. What is the probability of exactly two girls?

8. Suppose you flip an honest coin three times. What is the probability of exactly two heads?

Numbers of Ways: The Binomial Coefficients (Optional)

You will have noticed that a critical part of the procedure for finding binomial probabilities is counting the number of different ways we can arrange a given number of Yes responses (Y) and No responses (N). When we are dealing with only a few arrangements it is easy just to list them, but if we have to deal with a large number of them, the task quickly becomes intolerable. There is a better way (originally due to Isaac Newton).

Let's consider the case where $n = 5$ and $k = 3$, three Yes responses in five trials. We already know that there are ten ways to arrange 3 Y's and 2 N's; they are listed on page 251. How can we calculate this number?

Let's start by asking how many ways we can arrange five different letters, say ABCDE. At the start, we have five choices for the first letter (A or B or C or D or E). Having picked a first letter, we have four choices left for the second letter. (If we picked A for a first letter, the second can be B or C or D or E; if we picked B for a first letter, the second can be A or C or D or E; and so on.) That gives us $5 \times 4 = 20$ possibilities for the first two letters. Now, having picked the first two letters, we have three choices left for the third letter. (If our first two are AB, we can now pick C or D or E, and so on.) This gives us $20 \times 3 = 60$ possibilities for the first three letters. Now we have two possibilities left for the fourth letter, for $60 \times 2 = 120$ four-letter arrangements. Finally, having picked four letters, there is only one letter possible for the fifth letter, and $1 \times 120 = 120$.

We do not have our final answer yet, but we want to point something out before we go on. When we take a number and multiply it repeatedly by numbers that are smaller by one each time, we are calculating a *factorial.* For example, 5 factorial is $5 \times 4 \times 3 \times 2 \times 1 = 120$. An exclamation mark is used to indicate a factorial, so "5!" is read "5 factorial."

Back to the problem. The 120 we got when we calculated 5! is far higher than the 10 arrangements of three Y's and two N's we found by hand. The difference is that we distinguished each of the five letters when we developed the 5 factorial, but in the yes/no problem we are only distinguishing Y's from N's. Suppose that we somehow distinguish the three Y's ("Yes, I have a full scholarship," "Yes, I have a government grant," "Yes, I have a guaranteed loan"). We can then arrange the three Y's in $3! = 3 \times 2 \times 1 = 6$ ways. Each of these six ways is indistinguishable from the others in the original problem. Dividing the 120 by 6, we find that we are down to 20 ways we can arrange 5 items, three of which are indistinguishable.

Now, let's look at the two N's. If we distinguish the two N's ("No, my parents are paying," "No, I'm working my way through school"), there are $2! = 2 \times 1 = 2$ ways to arrange them. Dividing the 20 ways we got earlier by the 2 ways the N's can be arranged, we now get 10 ways to arrange three indistinguishable Y's and two indistinguishable N's.

Writing this out in a formula, we have

$$\text{Number of ways to get three Y's in five tries} = \frac{5!}{3!2!}$$

or, more generally, we have the formula for the binomial coefficient:

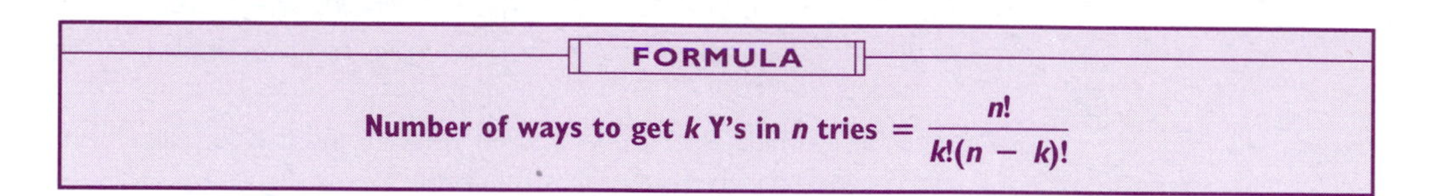

FORMULA

$$\textbf{Number of ways to get } k \textbf{ Y's in } n \textbf{ tries} = \frac{n!}{k!(n-k)!}$$

Two comments: (1) Zero factorial, 0!, is defined as equal to 1. (2) There is a standard notation for "number of ways to get k Y's in n trials,"

$$\binom{n}{k}$$

which is read as "n choose k" or "n combination k" and is also called a **binomial coefficient.**

Exercise Set 8.3

9. Suppose we toss a coin five times. How many different arrangements of three heads and two tails are there?

10. Suppose we toss a coin five times. How many different arrangements of four heads and one tail are there?

11. Suppose we toss a coin five times. How many arrangements of five heads and zero tails are there? (This is where the definition of 0! comes in.)

12. Suppose we toss a coin ten times. How many different arrangements of three heads and seven tails are there? (*Comment:* 10! is a big number. Rather than trying to evaluate it and then dividing by the product of 7! and 3!, you can write it out as $10 \times 9 \times 8 \times 7!$, put it over $7! \times 3 \times 2 \times 1$, and look for cancellations.)

Finding a Binomial Probability

Now let's put the two parts of our discussion together. To find the probability of getting k Y's in n independent trials, when the probability of Yes (Y) on a single trial is p, we use the following procedure.

PROCEDURE: FINDING A BINOMIAL PROBABILITY

1. Find the probability of No (N), q, by subtracting p from 1.
2. Calculate the probability, $p^k q^{(n-k)}$, of k Y's and $n - k$ N's in any one particular order by multiplying p by itself k times, then multiplying that by q times itself $n - k$ times.
3. Find the number $\binom{n}{k}$ of ways k Y's and $n - k$ N's can be arranged, either by counting them (which is error prone if n is even medium-sized), or by taking $n!$ and dividing it by $k!$ and then $(n - k)!$. (Be aware that there are a lot of cancellations available to simplify the arithmetic.)
4. Multiply the probability of k Y's in n trials in a particular order (which you found in step 2) by the number of ways this could happen (from step 3). This is the answer.

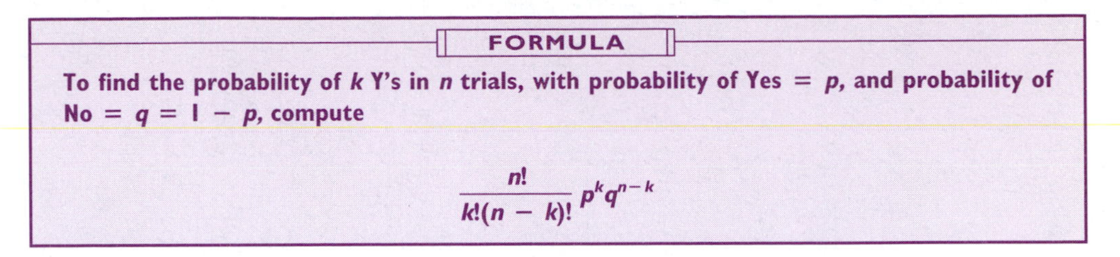

FORMULA

To find the probability of k Y's in n trials, with probability of Yes $= p$, and probability of No $= q = 1 - p$, compute

$$\frac{n!}{k!(n-k)!}p^k q^{n-k}$$

Discussion

Let's step back a moment and relate this to the rest of the chapter. First, notice that something like flipping a coin ten times and counting the number of heads is a classic random experiment. The possible results range from 0 to 10, and their probabilities form a probability distribution. We can use the formula developed above to calculate the probability of each value. Taken together, they will make up a binomial distribution. Binomial distributions were extraordinarily important in the development of probability and statistics. Indeed, the normal distribution was originally developed as a way to simplify binomial calculations when the numbers involved were large.

Sidebar 8.2

THE INTRODUCTION OF THE NORMAL CURVE

DeMoivre introduced what became known as the normal curve as a way to solve the binomial for large values of *n*. In his book *The Doctrine of Chances,* he wrote:[2]

> I shall here translate [from Latin] a Paper of mine which was printed November 12, 1733, and communicated to some Friends, but never yet made public, reserving to myself the right of enlarging my own Thoughts, as occasion shall require.
>
> Novemb. 12. 1733
>
> **A Method of approximating the Sum of the Terms of the Binomial $(a + b)^n$ expanded into a Series, from whence are deduced some practical Rules to estimate the Degree of Assent which is to be given to Experiments.**
>
> Altho' the Solution of Problems of Chance often require that several Terms of the Binomial $(a + b)^n$ be added together, nevertheless in very high Powers the thing appears so laborious, and of so great a difficulty, that few people have undertaken that Task; for besides *James* and *Nicolas Bernoulli,* two great Mathematicians, I know of no body that has attempted it; in which, tho' they have shown very great skill, and have the praise which is due their Industry, yet some things were farther required; for what they have done is not so much an Approximation as the determining of very wide limits, within which they demonstrated that the Sum of the Terms was contained. Now the Method which they have followed has been briefly described in my *Miscellanea Analytica,* which the reader may consult if he pleases, unless they rather chuse [*sic*], which perhaps would be the best, to consult what they themselves have writ upon the Subject: for my part, what made me apply myself to that Inquiry was not out of opinion that I should excel others, in which however I might have been forgiven; but what I did was in compliance to the desire of a very worthy Gentleman, and good Mathematician, who encouraged me to do it: I now add some new thoughts to the former; but in order to make their connexion the clearer, it is necessary for me to resume some few things that have been delivered by me a pretty while ago.

By way of example, let us work out the distribution for the problem of the number of heads in ten tosses. We do the first three in detail and then provide a table of all the answers.

$$P(0 \text{ in } 10, p = .5) = \frac{10!}{10!0!} \times .5^0 \times .5^{10}$$
$$= 1 \times 1 \times .0009765625 = .0009765625, \text{ or } .0010$$

$$P(1 \text{ in } 10, p = .5) = \frac{10!}{1!9!} \times .5^1 \times .5^9$$
$$= 10 \times .5 \times .001953125 = 10 \times .0009765625$$
$$= .009765625, \text{ or } .0098$$

$$P(2 \text{ in } 10, p = .5) = \frac{10!}{2!8!} \times .5^2 \times .5^8$$
$$= 45 \times .25 \times .00390625 = .0439453125, \text{ or } .0439$$

Table 8.6 shows all the probabilities. There are several things to notice here. First, the distribution is symmetric: The chances of getting all heads is the same as getting all tails, the chance of one head is the same as of nine heads, and so on. This is because the chances of heads and tails were the same to start with. Notice also the best chance is of getting five heads and five tails, which should happen about a quarter of the time. Notice that four heads and six heads are also pretty likely; each of them should happen about a fifth of the time. The three center values, then, should happen about 45% of the time. The least likely events—ten heads or ten tails—should each happen about 1 time in 1,000.

Table 8.6 Binomial probabilities, $n = 10$, $p = .500000$

k	p(of getting k heads)
0	.0010
1	.0098
2	.0439
3	.1172
4	.2051
5	.2461
6	.2051
7	.1172
8	.0439
9	.0098
10	.0010

Exercise Set 8.4

13. Suppose that there are ten students in a class and that the probability any one student misses a randomly picked class is $\frac{1}{5}$. (We are not suggesting that this is a reasonable attendance record; we are trying to keep the arithmetic simple.)

a. What is the probability that (exactly) four students miss class on one randomly picked day?

b. What is the probability that (exactly) five students miss class on a randomly picked day?

14. Suppose that a computer is programmed to generate a head with probability $p = .6$ and a tail with probability $q = .4$. You let the computer generate a string of 6 heads or tails. What is the probability of

a. five heads?

b. six heads?

c. more than four heads?

15. Using Table 8.6 showing the probability of k heads in ten trials, when $p = .5$, what is the probability of

a. ten heads?

b. nine heads?

c. eight heads?

d. more than seven heads?

e. at least seven heads?

f. fewer than five heads?

16. Suppose you flip an honest coin 10 times.

a. How many different ways can you get exactly 5 heads?

b. What is the probability that you get exactly 5 heads?

17. Suppose that an instructor grades quizzes by standing at the top of the stairs and throwing the quizzes down. Quizzes that land on the top third of the stairs are given A's, and the rest are given C's. There is a .4 chance of getting an A. Over the course of the term, ten sets of quizzes are graded this way.

a. How many different ways can a student get exactly 5 A's?

b. What is the probability of getting exactly 5 A's?

Computer Calculations

Most statistical programs allow us to calculate a binomial distribution. In Minitab, for instance, we use the PDF (for "probability density function") command with the Binomial subcommand. We have to specify two parameters, *n* and *p*.

Display 8.1 is a computer-generated table for the number of heads in 20 tosses of a fair coin. The program did not print probabilities for 0 heads and 20 heads. It did print the probabilities of 1 head and 19 heads as .0000, so we know that 0 heads and 20 heads are at least that unlikely.

DISPLAY 8.1

```
MTB > pdf;
SUBC> binomial 20 .5.
 BINOMIAL WITH N = 20 P = 0.500000
 K             P(X = K)
 1             0.0000
 2             0.0002
 3             0.0011
 4             0.0046
 5             0.0148
 6             0.0370
 7             0.0739
 8             0.1201
 9             0.1602
10             0.1762
11             0.1602
12             0.1201
13             0.0739
14             0.0370
15             0.0148
16             0.0046
17             0.0011
18             0.0002
19             0.0000
```

EXAMPLE 8.3 Binomial Probabilities: Hiring and Gender Representation

Suppose that a research project has openings for four interviewers and that 80% of the applicants for the jobs are female. Also suppose that the applicant pool is large enough so that the percent female does not change noticeably as people are offered jobs. If applicants are offered jobs randomly with respect to gender, how can we find the probabilities of getting various numbers of female interviewers?

Solution We set this up as a binomial distribution problem with $n = 4$ and $p = .8$. Generating the probability distribution by computer, we get the results shown in Display 8.2. We see that there is a 41% chance (with rounding) of hiring four female interviewers, and a 41% chance of hiring three female interviewers.

DISPLAY 8.2

```
MTB > pdf;
SUBC> binomial 4 .8.
 BINOMIAL WITH N = 4 P = 0.8.
K                P(X = K)
0                  0.0016
1                  0.0256
2                  0.1536
3                  0.4096
4                  0.4096
```

Exercise Set 8.5

18. Looking at Display 8.1, what is the probability of

a. 10 heads in 20 tosses?

b. exactly 15 heads in 20 tosses?

c. 15 or more heads in 20 tosses?

Display 8.3 is a Minitab printout of the binomial probabilities for $n = 4$ and $p = .6$. Use it to answer Exercises 19 and 20.

DISPLAY 8.3

```
MTB > pdf;
SUBC> binomial 5 0.6.
 BINOMIAL WITH N = 5 P = 0.6
K                P(X = K)
0                  0.0102
1                  0.0768
2                  0.2304
3                  0.3456
4                  0.2592
5                  0.0778
```

19. Suppose—to continue the example we used to develop the method for solving binomial problems—you ask five randomly selected students whether they are receiving financial aid, and the probability any one of them says yes is .6. What is the probability that

a. all five say yes?

b. none say yes?

c. three say yes?

d. most of them say yes?

20. Suppose you are eating evening meals at a random assortment of college dining halls over a 5-day period. Suppose the probability any given meal includes chicken is .6. What is the probability that

a. all five meals include chicken?

b. none include chicken?

c. three include chicken?

d. most of them include chicken?

21. Suppose you are doing quality-control work at an apple orchard. From previous years' work, you know that about 40% of the apples will be acceptable as table fruit and the rest will be processed into applesauce, juice, and pie filling. You randomly select four apples. You want to find the probability that three of the four will be acceptable ("yes") as table fruit.

a. For this problem, what is n?

b. What is k?

c. What is p?

d. What is q?

e. What is the probability the *first three* apples will be acceptable as table fruit and the fourth unacceptable?

f. How many ways can you arrange three acceptable apples and one unacceptable one?

g. What is the probability of getting three acceptable apples and one unacceptable one?

8.3 The Normal Distribution

The **normal distribution,** the most important probability distribution in statistics, is represented by the bell-shaped curve of Figure 8.3. We have already seen its use in describing

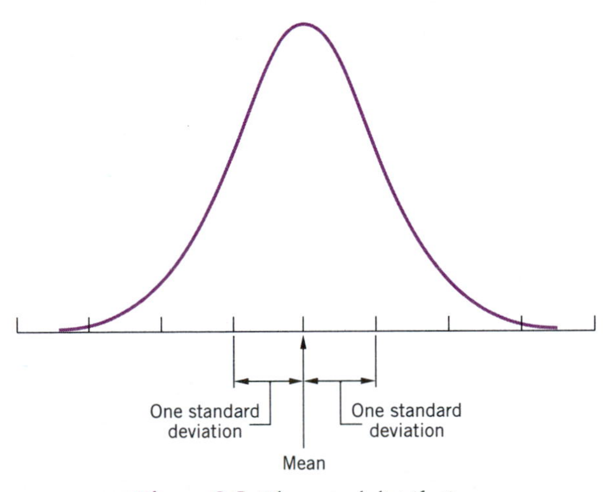

Figure 8.3 *The normal distribution*

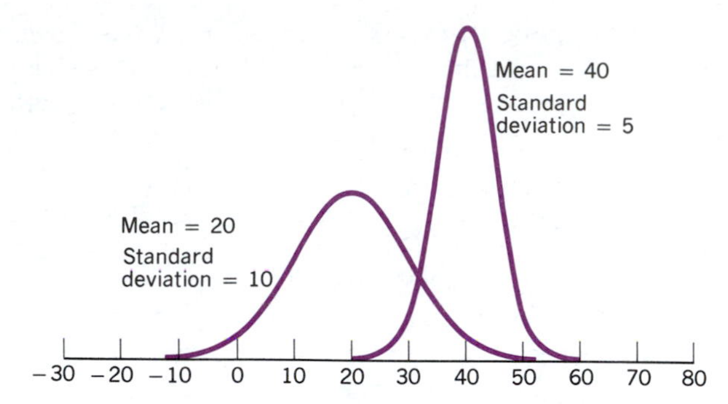

Figure 8.4 *Two different normal distributions. Normal distributions can differ only in the mean (which shifts it over) and the standard deviation (which stretches or compresses it overall). The basic shape remains the same.*

symmetric bell-shaped histograms of data (see Section 3.2). In this section, we will learn how to compute probabilities for a normal distribution.

The calculations parallel the ones we did in Chapter 4 to find percentile values for a normal data distribution. The major practical difference is that our normal curve table is now in probability values ranging from 0 to 1.0, rather than in percents ranging from 0 to 100.

There are many normal distributions. All have the same basic bell shape and differ only in their mean and standard deviation. The mean value determines the location of the center of symmetry, whereas the standard deviation sets the scale by indicating how wide the bell will be. Figure 8.4 shows two different normal distributions with different means and standard deviations. The height of the curve indicates the likelihood of the random value falling near that value. For example, the normal distribution represented by the taller curve on the right in Figure 8.4 would more likely produce values near 40, corresponding to the highest point on the curve, and most values would be between 30 and 50. The shorter curve on the left would produce values usually between 0 and 40, with values near 20 being the most likely.

The **standard normal distribution** has mean 0 and standard deviation 1, as shown in Figure 8.5. Instead of having a different table for each normal distribution (there would be too many of them), tables for the normal distribution usually give probabilities only for the standard normal distribution. We will first learn how to use this table and then will see how to compute probabilities for nonstandard normal distributions.

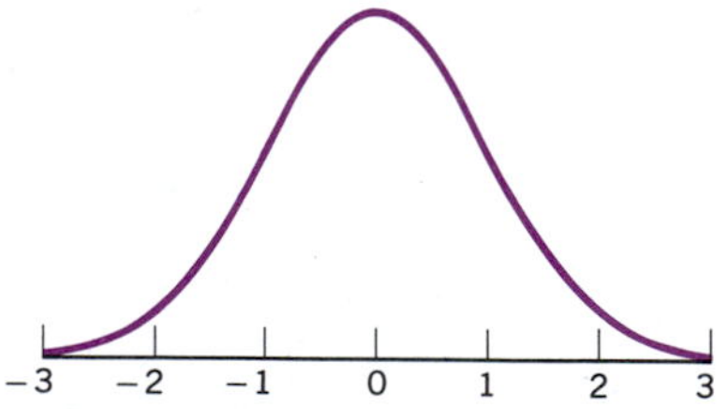

Figure 8.5 *The standard normal distribution with mean 0 and standard deviation 1*

Probability Is Area Under the Curve

For a standard normal random variable, the probability of falling within an interval is equal to the area under the standard normal curve within that interval, as indicated in Figure 8.6. This notion of "area under the curve" for the distribution of a random variable corresponds to the "proportion of data values" within that interval for a histogram. The area under the curve within an interval (for a random variable) is proportional to the probability of falling within that interval, whereas the area under the histogram within an interval (for a data set) gives the frequency (or relative frequency) of the interval.

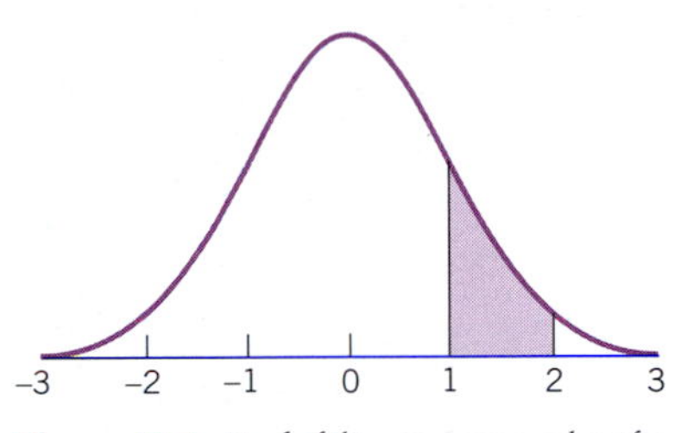

Figure 8.6 *Probability is area under the curve. The probability of being between 1 and 2 is equal to the shaded area. Note that the probability of being between 0 and 1 is larger because the curve is taller there.*

The standard normal probability table in the next section (Table 8.7) allows us to find the probability that a standard normal random variable is less than any particular number of interest. Figure 8.7 indicates this tabled probability value as the area under the curve to the left of the value of interest, 1.5. From the figure, we see that most of the area under the total curve is located to the left of 1.5, and so we expect the probability (represented by the shaded area) to be close to 1. In fact, from the table we will see that the probability that a standard normal random variable is less than 1.5 is .933. Thus, a standard normal random variable will be less than 1.5 about 93% of the time.

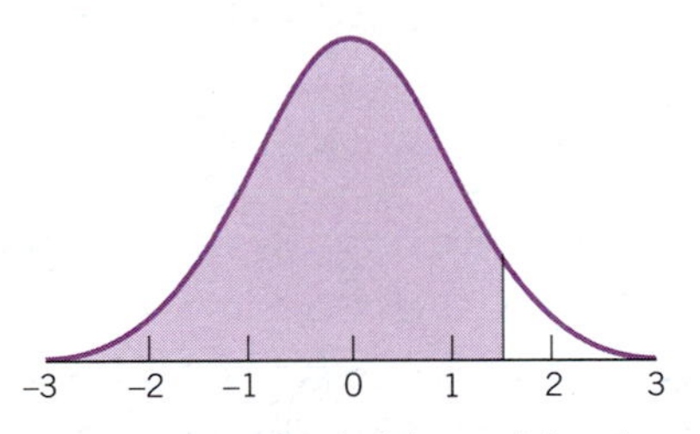

Figure 8.7 *The shaded area below the curve to the left of 1.5 indicates the probability that a standard normal random variable is less than 1.5. The exact amount of area that is shaded is found to be .933 from the table.*

The Standard Normal Probability Table

The standard normal probability table gives the probability that a standard normal random variable (with mean 0 and standard deviation 1) is less than a given value. An edited version of the table is given in Table 8.7 (page 262). For example, the probability of being less than 1.5 is

Table 8.7 *Short normal table*

VALUE	PROBABILITY	VALUE	PROBABILITY
.00	.500	− .00	.500
.25	.599	− .25	.401
.33	.629	− .33	.371
.50	.691	− .50	.309
.67	.749	− .67	.251
.674	.750	− .674	.250
.75	.773	− .75	.227
1.00	.841	− 1.00	.159
1.25	.894	− 1.25	.106
1.282	.900	− 1.282	.100
1.33	.908	− 1.33	.092
1.50	.933	− 1.50	.067
1.645	.950	− 1.645	.050
1.67	.953	− 1.67	.047
1.75	.960	− 1.75	.040
1.960	.975	− 1.960	.025
2.00	.977	− 2.00	.023
2.326	.990	− 2.326	.010
2.33	.990	− 2.33	.010
2.500	.994	− 2.500	.006
2.576	.995	− 2.576	.005
3.000	.9987	− 3.000	.0013
3.090	.9990	− 3.090	.0010
3.291	.9995	− 3.291	.0005
3.500	.9998	− 3.500	.0002
4.000	.99997	− 4.000	.00003

.933, and the probability of being less than −.5 is .309. For routine work, use the probability for the closest value in the table. For example, the probability of being less than .2948 (which is not listed exactly in the table) is approximately .614, the probability of being less than .29 (the closest value that is listed in the table).

Exercise Set 8.6

22. What is the probability that a standard value is less than 1.282?

23. What is the probability that a standard value is less than −2?

24. What is the probability that a standard value is between −1.0 and +1.0 ?

25. What is the probability that a standard normal score is less than 1.645?

26. What standard normal score has only 2.5% of the normal distribution above it?

27. What is the probability that a standard normal score is between −1.960 and +1.960?

Finding Normal Probabilities

To compute probabilities for a normal distribution that is *not* standard, we have to convert the numbers in the actual problem to those for a standard normal distribution so that the answer may be found in the standard normal table. This is done by converting a value to the *number of standard deviations it is from the mean* by subtracting the mean from the value and dividing by the standard deviation.

Probability of Being Less Than Some Amount. For example, a problem might begin with the statement that after 30 years, a certain kind of tree has a height that is normally distributed with a mean of 20 feet and a standard deviation of 5 feet. The problem might then ask for the probability that a randomly selected tree will be less than 25 feet tall. A good first step in solving this (or any) word problem is to summarize the important information.

- Normal distribution for height
- Mean = 20
- Standard deviation = 5
- Find probability that height is less than 25

A good second step for a problem like this is to draw and label a picture of the normal curve for the situation. It should be centered at the mean of 20 and extend a little more than two standard deviations on either side of the mean, as shown in Figure 8.8. Using this figure as a guide, with a little practice you can very roughly estimate probabilities from the area under the curve. The rough estimate can then be used to check your calculated answer.

Next, to solve the problem "What is the probability that the height is less than 25?", we ask "How many standard deviations above the mean is 25?" The answer is clear in this case: 1. This may be computed more generally by subtracting the mean and dividing by the standard deviation. We find that 25 is

$$\frac{25 - \text{mean}}{\text{standard deviation}} = \frac{25 - 20}{5}$$
$$= 1 \quad \text{standard deviation above the mean}$$

In Chapter 4, we would have referred to this as the tree's *standard score*. The convention in probability problems is to call it a *Z score* and to call the number of standard deviations from the mean *Z*. So, in standard notation, we have

$$Z = \frac{X - \mu}{S}$$

The problem now becomes "What is the probability of being less than 1 standard deviation above the mean?" that corresponds to the shaded area in Figure 8.9 (page 264). (Drawing a picture showing the area of interest helps you check whether your answer is reasonable.) We find in the standard normal probability table that the answer is .841. To summarize,

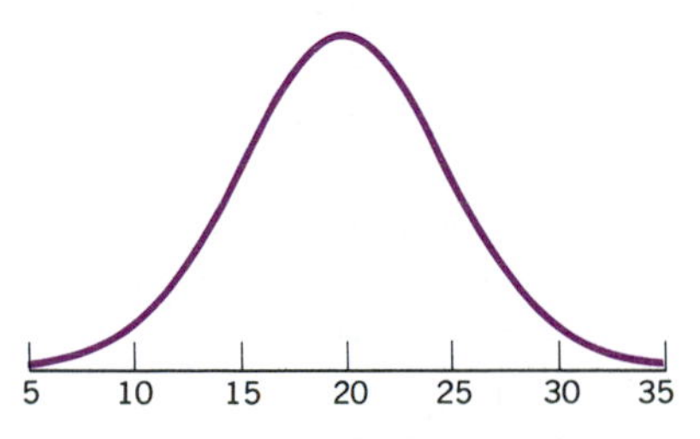

Figure 8.8 *The probability distribution of tree heights, centered at the mean value of 20 and having standard deviation 5*

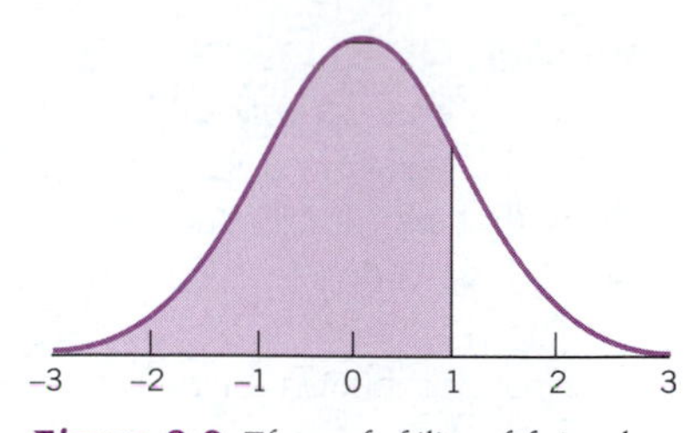

Figure 8.9 *The probability of being less than "1 standard deviation above the mean" is represented by the shaded area (.841 from the table) for a normal distribution.*

Probability that the height is less than 25
  = probability that a standard normal is less than (25 − 20) / 5
  = probability that a standard normal is less than 1
  = .841 (from the table)

Thus, about 84% of these trees will be under 25 feet tall after 30 years.

Probability of Being More Than Some Amount. In general, to find the probability that a random variable with a normal distribution is more than a certain amount, we use the complement rule to change it to less than a certain amount, then convert to standard deviations above the mean, and finally use the standard normal probability table.

Specifically, to find the probability that the height of a tree will be over 25 feet:

Probability that height is over 25
  = 1 − (probability that height is less than 25)
  = 1 − [probability that standard normal is less than (25 − 20)/5]
  = 1 − (probability that standard normal is less than 1)
  = 1 − .841
  = .159

For a second example, let us find the probability of a tree being more than 10 feet tall. The sequence of steps is as follows:

Probability that height is over 10
  = 1 − (probability that height is under 10)
  = 1 − [probability that standard normal is under (10 − 20)/5]
  = 1 − (probability that standard normal is under −2)
  = 1 − .023 (from the table)
  = .977

So we see that about 98% of the trees are over 10 feet tall. Note the negative value (−2), which expresses the fact that the value (10) is below rather than above the mean. Figure 8.10 shows the probability that was found in the table.

What is the probability that a tree is *exactly 10 feet or more?* The answer remains the same: .977. The only difference between the events "over 10 feet" and "10 feet or more" is the outcome "exactly 10 feet," which has probability 0 because the area is 0 above this single point. Another way to think about it is that, if measured with exact precision, a tree that appeared to

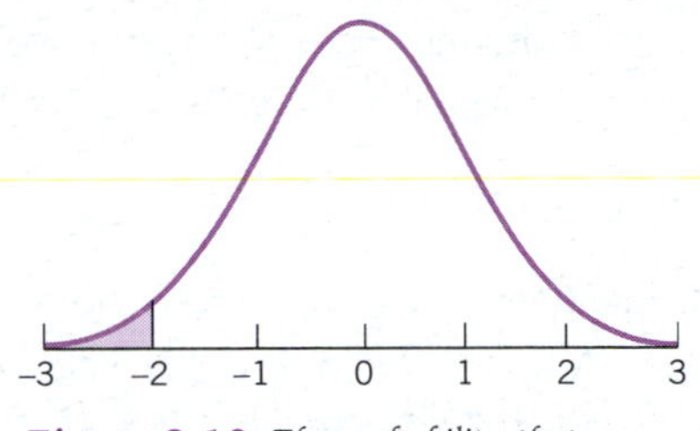

Figure 8.10 *The probability that a standard normal random variable (with mean 0 and standard deviation 1) is less than −2 is equal to the shaded area.*

be exactly 10 feet tall would, because of some slight randomness, actually have a height slightly over or slightly under 10 feet.

Probability of Being Between Two Values To find the probability that a random variable with a normal distribution falls between two given values, we

1. Convert both values to standard deviations from the mean.
2. Look up both Z values in the standard normal probability table.
3. Subtract the probabilities to find the answer (always subtract the smaller from the larger, so that the answer is a positive number).

Consider again the example of normally distributed heights of trees with mean 20 and standard deviation 5. What is the probability of a tree being between 12 and 26 feet tall?

Probability of height being between 12 and 26
= probability of a standard normal being between $(12 - 20)/5$ and $(26 - 20)/5$
= probability of a standard normal being between -1.6 and 1.2
= difference between .055 and .885 (from the table)
= $.885 - .055$
= .830

Thus, 83% of the trees are between 12 and 26 feet tall. Figure 8.11 indicates this probability as

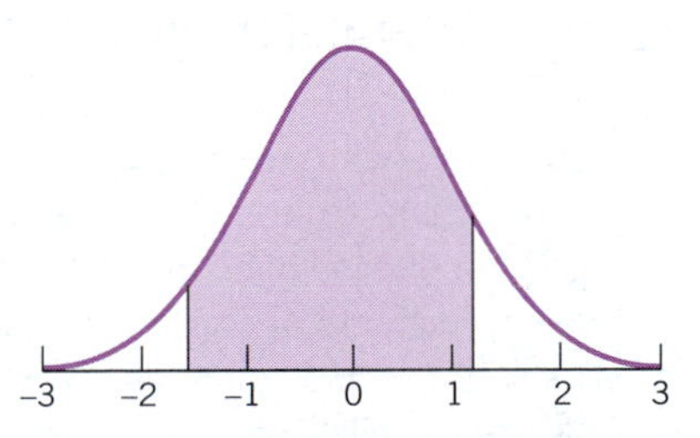

Figure 8.11 *The shaded region indicates the probability that a standard normal random variable is between −1.6 and 1.2. The answer, .83, is found by subtracting the two tabled probabilities.*

an area under the curve. The reason we subtract to find the answer can be explained as follows: The area between -1.6 and 1.2 is the area to the left of 1.2 minus the area to the left of -1.6.

Exercise Set 8.7

28. Find standard values or Z scores for each of the following observations.

a. A value of 130, from a distribution with mean = 100 and SD = 15. What is the probability of a value smaller than this? What is the probability of a value higher than this?

b. A value of 70, from a distribution with mean = 100 and SD = 15. What is the probability of a value smaller than this? What is the probability of a value higher than this?

c. A value of 550, from a distribution with a mean = 500 and SD = 100. What is the probability of a value smaller than this? What is the probability of a value higher than this?

d. A value of 450, from a distribution with a mean = 500 and SD = 100. What is the probability of a value smaller than this? What is the probability of a value higher than this?

e. A value of 665, from a distribution with a mean = 500 and SD = 100. What is the probability of a value smaller than this? What is the probability of a value higher than this?

f. A value of 335, from a distribution with a mean = 500 and SD = 100. What is the probability of a value smaller than this? What is the probability of a value higher than this?

29. What is the probability of each of the following observations, from a normal distribution with mean = 100 and SD = 16?

a. a value of 108 or less

b. a value of 92 or less

c. a value of 132 or less

d. a value of 84 or less

30. If we select a random value from a normal distribution with mean = 100 and SD = 16, what is the probability it is

a. between 92 and 108? (Do not worry about correcting for whether it is exactly 92 or exactly 108.)

b. between 84 and 132?

c. between 100 and 108?

d. between 100 and 132?

e. between 92 and 116?

f. outside the range 84 to 132 (that is, less than 84 or more than 132)?

g. outside the range 52 to 148?

8.4 Distributions That Are Not Normal

The probabilities we have just computed were found by looking up values in the standard normal probability tables. Thus, these answers apply only to a normal distribution, strictly speaking. If the distribution is not normal, then these probabilities will be in error.

However, many distributions are *approximately normal,* so methods for the normal distribution will still give *approximate* answers. One case in which this happens is when the random variable is actually the sum of many equivalent independent processes; the central limit theorem says that such a sum will approximately follow a normal distribution.

Nonetheless, some distributions are very far from being normal, making the probabilities from the standard normal table seriously in error. Care must be exercised in particular with discrete distributions taking on only a few values and with highly skewed distributions that are not symmetric enough to be like a normal distribution.

EXAMPLE 8.4 *Lottery Winnings*

For an example of a distribution that is not well approximated by a normal distribution, consider a lottery like those that many states are now using to help supplement their tax revenues. Suppose we pay $1 for a ticket that will pay us $100 with the small probability of .006, but otherwise nothing. If we win, we receive the net value of $99 after subtracting the cost of the ticket, but if we lose, we get −$1 (that is, we lose the dollar paid for the ticket). The probability distribution can be summarized as in Table 8.8.

Table 8.8

WINNINGS	PROBABILITY
$99	.006
−$1	.994

$$\text{Mean winnings: } (\$99)\,(.006) + (\$-1)\,(.994) = \$-.400$$

$$\text{Standard deviation} = \sqrt{(9{,}880)\,(.006) + (.36)\,(.994)}$$

$$= \sqrt{59.64} = 7.72$$

Consider now the probability of winning at least $5. The actual answer is easily found because the only way in which at least $5 can be won is to actually win the lottery. So,

The actual probability of winning at least $5 is .006.

To see that the normal distribution would not be very accurate as an approximation to this discrete, skewed distribution, let us find the probability that a normal distribution with this same mean (−.4) and standard deviation (7.72) is at least 5.

Probability that a normal distribution with mean −.4 and standard deviation 7.72 is at least 5
= probability that a standard normal is at least [5 − (−.4)] / 7.72
= probability that a standard normal is at least .699
= 1 − (probability that a standard normal is less than .699)
= 1 − .76 (from the table)
= .24

If the winnings distribution were normally distributed with the same mean and variance, then we would expect to win $5 or more about 24% of the time. However, this number is much too optimistic: The actual probability is less than 1%.

Exercise Set 8.8

31. Suppose we have a lottery where each ticket costs $1, and (as in Example 8.4) there is a small probability, .006, of winning $100.

a. What is the actual probability of winning $10 or more?

b. Using the normal approximation, what is the probability of winning $10 or more?

32. Suppose that your state has a game like the one in Exercise Set 8.1, where a ticket costs \$1, and one ticket in ten pays \$2 and another ticket in ten pays \$5.

- **a.** What is the actual probability that the purchaser of a single ticket will net one dollar or more?
- **b.** Using the normal approximation, what is the probability a player will net one dollar or more, on a single ticket?

8.5 The Normal Approximation to the Binomial

The normal curve can be used to provide satisfactory (if approximate) answers to questions involving the binomial distribution. We see this in Figures 8.12a to 8.12d, which show binomial sampling distributions with matching normal distributions overlaid on them.

The mean of the binomial distribution—and of the matching normal distribution—is the sample size (n) times the probability of the outcome of interest (p). (The outcome of interest is a Yes.) The standard deviation of the number of Yes answers is the square root of the sample size (n) times the probability of Yes (p) times the probability of No ($q = 1 - p$).

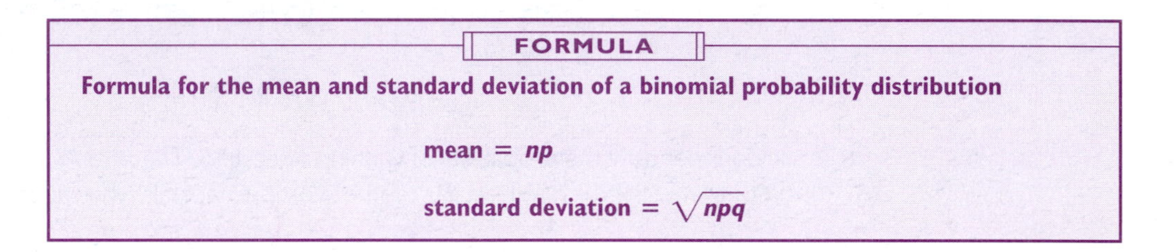

FORMULA

Formula for the mean and standard deviation of a binomial probability distribution

$$\textbf{mean} = np$$

$$\textbf{standard deviation} = \sqrt{npq}$$

As a rule of thumb, the normal approximation to the binomial gives acceptable answers when at least 5 Y's and 5 N's are expected in each sample. This is true when n times p is 5 or greater and n times q is 5 or greater.

EXAMPLE 8.5 *Using the Normal Approximation*

Suppose the probability that a dropped piece of toast lands buttered side down is .8. Over the course of the semester, 50 pieces of toast are dropped on the dining room floor. What is the probability that 42 or more land buttered side down?

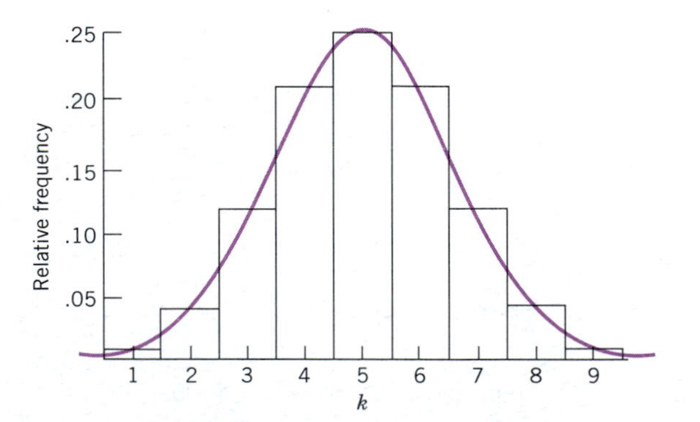

Figure 8.12a Binomial probabilities with 10 trials and p = .5. Matching normal curve is overlaid.

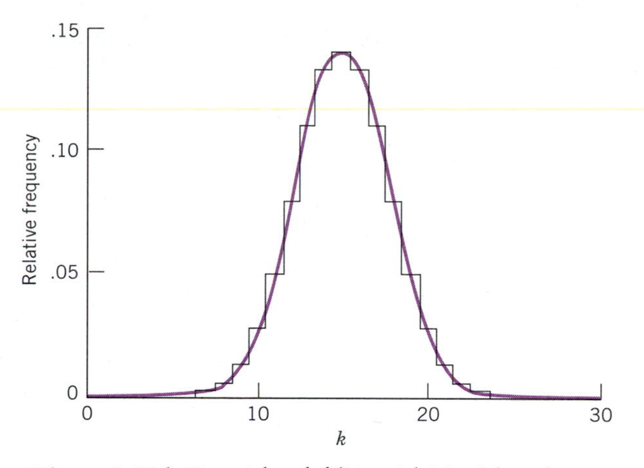

Figure 8.12b *Binomial probabilities with 30 trials and p = .5. Matching normal curve is overlaid.*

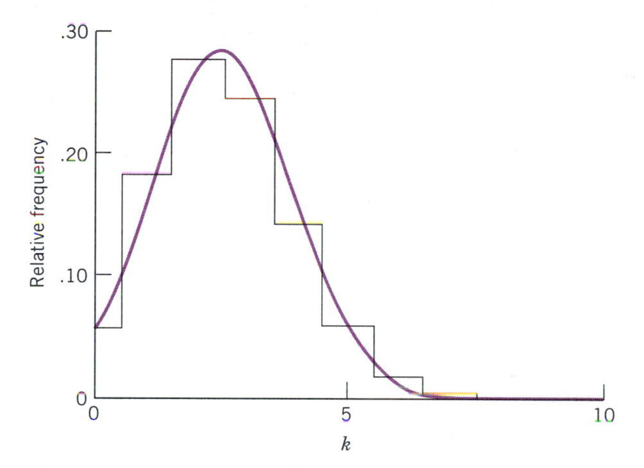

Figure 8.12c *Binomial probabilities with 10 trials and p = .25. Matching normal curve is overlaid. Note that the fit is not very good.*

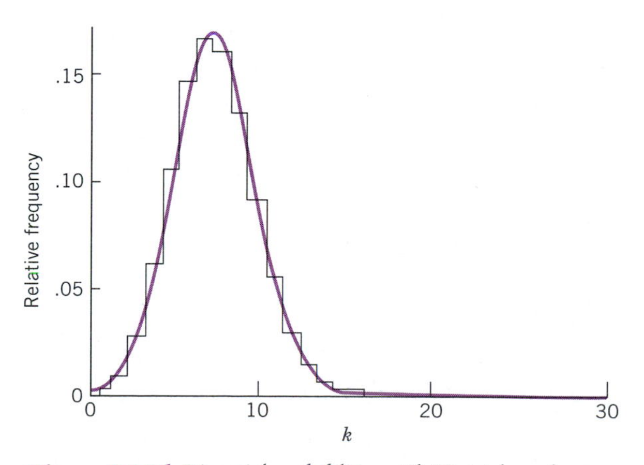

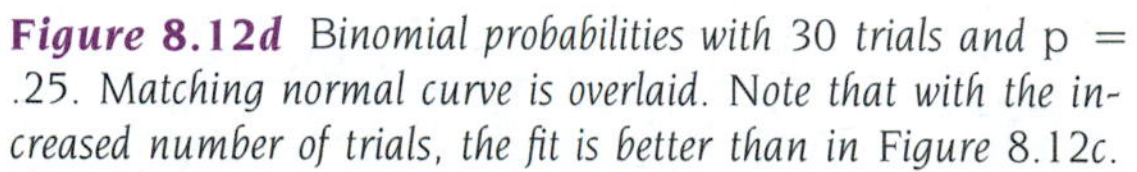

Figure 8.12d *Binomial probabilities with 30 trials and p = .25. Matching normal curve is overlaid. Note that with the increased number of trials, the fit is better than in Figure 8.12c.*

Solution From the statement of the problem, we have

$$n = 50 \text{ and } p = .8$$

so

$$q = 1 - .8 = .2$$

Checking to see whether the normal approximation will be reasonably accurate, we have

$$\text{Expected number with buttered side down: } np = 50 \times .8 = 40$$

and

$$\text{Expected number with buttered side up: } nq = 50 \times .2 = 10$$

Both of these numbers are greater than 5, so the normal approximation will be accurate enough.

Using the formulas for the mean and standard deviation given above, we have

$$\text{Mean} = np = 50 \times .8 = 40$$

and

$$\begin{aligned} \text{Standard deviation} &= \sqrt{npq} \\ &= \sqrt{50 \times .8 \times .2} \\ &= \sqrt{8} = 2.828 \quad \text{(approximately)} \end{aligned}$$

Now the question is: What is the probability of getting a "score" of 42 or more from a normal distribution with mean = 40 and SD = 2.83? We have

$$Z = \frac{42 - 40}{2.828} = \frac{2}{2.828} = .7072$$

So the standard score is, rounding, .71. Looking this up in the normal probability table, we see that about 76% of the time we would be below this score. By subtraction, then, we have

$$\text{Probability of 42 or more} = 1 - .76 = .24$$

This is the answer.

Exercise Set 8.9

33. Suppose that the proportion of the people in the country who favor "harsh penalties for habitual criminals" is .99. You want to figure a probability distribution for a sample size of 100. Can you use the normal approximation to the binomial? Why or why not?

34. Suppose that 50% of the adolescents in a certain community have been arrested at one time or another. You take a random sample of 100. You want to find the probability that fewer than 40 have ever been arrested. Can you use the normal approximation to the binomial? Why or why not?

35. Using the normal approximation, what is the probability there will be 40 or fewer kids with arrest records in a random sample of 100 from a population where half the kids have arrest records?

36. Suppose that 60% of the voters in a population plan to vote for Brown. You randomly select 20 voters from this population. What is the probability that less than half your sample plans to vote for Brown?

8.6 The Central Limit Theorem

Suppose we have a random variable with a known probability distribution. If we make, say, ten observations of the random variable, we expect their average value to be about the same as the average of the original distribution. We also expect the sample standard deviation of these ten values to be about the same as the standard deviation of the original distribution. If we take another ten observations of the same distribution, our expectations for the second sample are the same.

Shifting our focus, how far from the true mean do we expect the sample *averages* to be? This is one of the questions the central limit theorem answers for us.

The central limit theorem also tells us that the averages of repeated samples will—if the samples are large enough—be normally distributed, irrespective of the shape of the original distribution. This is also true of sample sums.

The Central Limit Theorem

A rigorous statement of the central limit theorem is given in Sidebar 8.3. Here is an interpretation:

> *Consider the sum and the average of* n *independent observations from any probability distribution. We repeat the sampling process, so that there are an unlimited number of sample sums and an unlimited number of sample averages. The sample sums and averages have their own distributions (histograms) and each has its own mean and standard deviation. Then:*
>
> > *The larger the sample size* n*, the closer the distributions of the sums and averages will be to a normal distribution.*
> >
> > *The mean of the sample averages will be the same as the mean of the original probability distribution, and the mean of the sample sums will equal the mean of the original probability distribution times* n.
> >
> > *The standard deviation of the sample averages will equal the standard deviation of the original distribution divided by the square root of* n*, and the standard deviation of the sample sums will equal the standard deviation of the original distribution times the square root of* n.

Sidebar 8.3

THE CENTRAL LIMIT THEOREM

Consider the sum and the average of n repeated independent observations of a random variable with any probability distribution. The larger the sample size n, the closer the probability distribution of the sum and the average will be to a normal distribution. The mean of the average will be the same as the mean of a single observation, whereas its standard deviation will be equal to the standard deviation of a single observation divided by the square root of n. The mean of the sum will be equal to the mean of a single observation times n, whereas its standard deviation will be equal to the standard deviation of a single observation times the square root of n.

There is one technical condition needed for the central limit theorem to hold. We must assume that the mean and standard deviation of a single observation both exist and are finite numbers and that the standard deviation is not 0. Unless the distribution is extremely skewed or long-tailed, this assumption will hold. However, there are some very skewed distributions that do occur in real-life situations (such as the size of petroleum deposits or measurements of other things for which a few items can be very large) for which the central limit theorem does not apply.

The formulas described above are set out in Table 8.9.

Table 8.9

	ONE OBSERVATION X	AVERAGE OF SAMPLE OF n OBSERVATIONS $\bar{X}$	SUM OF SAMPLE OF n OBSERVATIONS ΣX
Mean	μ	μ	$n\mu$
Standard deviation	σ	$\frac{\sigma}{\sqrt{n}}$	$\sqrt{n}\,\sigma$

EXAMPLE 8.6 The Central Limit Theorem: Selecting Random Samples of Stocks

Table 8.10 lists 160 stocks that one newsletter recommended for purchase in July 1991.[3] It also gives an approximate return per month for one month. It happens that these 160 stocks average about 1.3% return/month, with a standard deviation of 12.0%. So if we had randomly picked one stock and invested in it at the start of the month, at the end of the month (on average), we could expect to be 1.3% ahead, plus or minus the standard deviation of 12.0%.

What happens if we randomly pick nine stocks, rather than one? According to the central limit theorem, the average of the nine stocks should be close to the population average, and the standard deviation of the averages in the repeated samples of nine should equal the population standard deviation of 12.0% divided by the square root of nine. So the averages of repeated samples of nine should have a standard deviation of around 4%.

We took ten random samples of nine each. The first is shown in Table 8.11 (page 276), with an average return of − 1.3%. The stocks picked for this sample and the other nine random samples are indicated by an "X" in Table 8.10. Note that a stock was not crossed off the list if it was picked once. We left it on the list so that the probabilities did not change as stocks were picked and the list got shorter. Stocks that were picked twice are marked "XX."

The average monthly return for each sample of nine stocks is given in Table 8.12 (page 276). Some of the samples did nicely, with 4% or 5% returns per month, and a couple lost money. On average, the samples averaged 1.3% (rounded to one decimal place in the original data), right on the original population mean of 1.3% (again, rounded to one decimal place). The SD of the sample averages was 4.3%, quite close to the 4.0% predicted by the central limit theorem.

So, from nine randomly picked stocks from the list, we would expect, on average, to get a 1.3% return at the end of month, the same as if we had picked a single stock. But now the "plus or minus" margin of error is 4.0%, rather than 12.0%.

Repeating this process with samples of n = 36, we obtained the results shown in Table 8.13 (page 276). The ten data rows give the results for each individual sample of 36. The first had an average return of 2.63%, so if we had bought those 36 stocks we would have a 2.63% return

Table 8.10

RANK	STOCK	RETURN/ MONTH	1	2	3	4	5	6	7	8	9	10
1	Edmark Corp	65.2										
2	Media Logic Inc	33.9			X							
3	Lsb Industries	33.3										
4	Bull Run Gold M	28.7										X
5	Lennar Corp	27.7		X								
6	Educational Dev	22.2										
7	Homecare Mgmt	21.4										
8	Kcs Energy Inc	21.1									X	
9	Aryt Optronics	19.4										
10	American Enterp	18.7										
11	Johnston Ind-Del	17.6										
12	First Amer Finl	16.4							X			
13	Arctic Alaska F	15.2										
14	Bel Fuse Inc	14.8										
15	Big O Tires Inc	14.3										
16	Micro Systems	13.8				X	X				X	
17	Metalclad Corp	13.8										
18	Biodynamics	12.9						X		X		
19	Gull Laboratorie	12.3										
20	Franklin Elec Pu	12.2						X				
21	Masco Indus	11.7	X									
22	Falcon Products	11.6		X				X				
23	Innovo Group Inc	11.1										
24	Fresh Juice Co	11.0								X		
25	Amer President	10.4			X							
26	Airgas Inc	10.0										
27	Cytrx Corp	9.1		X								
28	Leucadia Natl	8.9			X							
29	Mes Airlines	8.8										
30	Ben & Jerry's H	8.2										
31	Ameriwood Inds	8.1						X				
32	Analystic Surv	8.0										
33	Mylan Labs	7.8						X				
34	Andros Inc	7.1										
35	Arrow Electronics	7.0		X								
36	Biomedical Waste	6.9									X	
37	Comp Tel D Chile	6.7										
38	Great Southern B	6.6					X					
39	Meridian Diagn	6.4										
40	Alza Corp	6.3										
41	Magma Copper	6.3										X
42	Bolar Pharmace	6.2							X			
43	Franklin First F	6.2										
44	Chipcom Corp	5.8						X	X	X		
45	Magna Intl	5.8				X						
46	Gulf State Util	5.7			X							
47	Giddings & Lewis	5.6							X			
48	Medar	5.6								X	X	
49	Durakon Inds Inc	5.6										
50	Gensia Pharmace	5.4										
51	Merry Land Inv	5.1										
52	Merisel Inc	5.1										

(Continued)

RANK	STOCK	RETURN/ MONTH	1	2	3	4	5	6	7	8	9	10
53	Boole & Babbag	5.0										
54	High Plains Corp	4.9										
55	Lancit Media Pr	4.5										
56	Autotote Corp	4.5										
57	Empi Inc	4.5										
58	Jean Phillippe F	4.5										
59	Manor Care Inc	4.1					X					
60	Decorator Indus	3.9						X				
61	Adaptec Inc	3.7										
62	Airtran Corp	3.4										
63	Lunn Industries	3.4	X									X
64	Allied Signal In	3.3										
65	Luxottica Grp Sp	3.3										
66	Chile Fd Inc	2.9	X				X					
67	Mbia	2.7										
68	Mediq Inc	2.7								X		
69	Louisiana-Pacifi	2.7							X		X	
70	Empresa N Elec	2.5										
71	Laser Recording	2.3	X									
72	Eskimo Pie Corp	2.1										
73	Maxtor	2.1		X								
74	Mark IV Industri	2.0	X			X						
75	Banctex Inc	1.9			X							
76	Abs Industries	1.8			X							
77	Michaels Stores	1.7				X						
78	Acc Corp	1.5										
79	Bush Industries	1.5										
80	Adv Marketing S	1.1										
81	Consolidated Rai	1.1										
82	Bombay Co Inc	.7										
83	Mattel Inc	.6										
84	Material Science	.5										
85	Deposit Guaranty	.3										
86	Land's End Inc	.3										
87	Birmingham St	.2							X			
88	Astec Indus	0									XX	
89	Belding Heminw	0										
90	Costar Corp	0										X
91	Flow Intl Corp	0				XX						
92	Gold Reserve C	0		X					X			
93	Kilm Transport	0										
94	Mercury Finance	0										
95	Basic Amer Med	−.7										X
96	California Ampli	−.9										
97	Banyan Strateg	−1.4										
98	Inter-City Prod	−1.8				X						
99	British Airways	−1.9										
100	Heartland Expr	−1.9									X	
101	General Motors	−1.9										
102	Juniper Features	−1.9		X								
103	Exabyte Corp	−2.0										
104	Freept Mcr O&G	−2.1					X					
105	Gander Mountain	−2.1						X				

106	Linear Tech	−2.3										
107	Amer Bus Prod	−2.4										
108	Electric & Gas T	−2.4										
109	Fluke (John) Mfg	−2.6										
110	Drew Industries	−2.6										
111	Gentec Corp	−2.84	X									X
112	Arch Petroleum	−2.9										
113	Applebees Intl	−3.2										
114	Casey's Genl St	−3.3	X									
115	Designs Inc	−3.3										X
116	Falcon Cable Sys	−3.5										
117	Mercer Intl	−3.6			X							
118	Cullen/Frost Ban	−3.7										
119	Henry (Jack)	−3.7										
120	Empire of Carol	−3.9								X		
121	Amer Intl Petrol	−4.3										
122	Ferrofluidic Cor	−4.3			X							
123	Medical Steriliz	−4.3					X					
124	Astro Med	−4.5										
125	Microsoft Corp	−4.9										
126	Idb Comm Group	−5.0										
127	Magnetic Tech	−5.1										
128	Griffin Technol	−5.2										
129	Main St & Main	−5.7										
130	Electro Rent Cor	−6.0										
131	Mgmt Technology	−6.2										
132	Cheyenne Softw	−6.5					X					
133	Astrotech Intl	−6.5		X								
134	Intl Cablecastin	−8.0								X	X	
135	Hathaway orp	−8.3										
136	Audio King Corp	−8.4										
137	Lund Int Hldgs	−8.4										
138	Creative Comp	−8.8				X						X
139	Bailey Corp	−9.1										
140	Condor Services	−9.4										
141	Hadco Corp	−9.7										
142	English China C	−9.9										
143	Input/Output Inc	−10.0										
144	Comptek Res	−12.0							X			
145	Amer Pacific C	−12.0						X				
146	Cam Data Sys	−12.3										
147	Mri Medical Diag	−12.5										
148	Biocontrol Tech	−12.7	X									
149	Mascott Corp	−13.6								X		
150	Ampal Amer Isrea	−14.0			X				X			
151	Hailey Energy	−14.0										
152	Babbages Inc	−15.5	X									
153	Middleby Corp	−15.8		X								
154	Microterra Inc	−17.3										X
155	Lifeway Foods	−19.2										
156	Microage Ince	−19.8										
157	Computer Identic	−25.0										
158	Box Energy	−33.3										
159	Amer Electromed	−33.5					X					
160	Micrografix Inc	−41.4					X					

Mean = 1.349447
SD = 12.00048

Table 8.11 *Sample 1*

RANK	STOCK	RETURN/ MNT
21	Masco Indus	11.7
63	Lunn Industries	3.4
66	Chile Fd Inc	2.9
71	Laser Recording	2.3
74	Mark IV Industri	2.0
111	Gentex Corp	−2.8
114	Casey's Genl St	−3.3
148	Biocontrol Tech	−12.7
152	Babbages Inc	−15.5
	Sample 1 average	−1.3

Table 8.12 *Average monthly return on ten random samples of nine stocks*

SAMPLE NO.	AVERAGE RETURN/MNT
1	−1.33395
2	3.687322
3	4.505556
4	2.147531
5	−9.65
6	5.355556
7	1.207407
8	1.422222
9	4.459259
10	1.190741
Average of averages	1.299164
SD of averages	4.3461

Table 8.13 *Average monthly return on ten random samples of 36 stocks*

SAMPLE	n	MEAN	ST.DEV.	MIN.	MAX.	Q_1	Q_3
1	36	2.63	15.24	−25.00	65.20	−5.53	7.69
2	36	4.49	10.46	−19.80	33.90	−1.24	7.80
3	36	1.50	10.57	−17.30	33.90	−5.03	6.55
4	36	−2.79	11.21	−41.40	14.81	−9.36	4.50
5	36	.60	11.25	−41.40	21.10	−3.53	6.18
6	36	−1.02	11.67	−41.40	33.33	−7.89	5.55
7	36	2.27	11.41	−33.30	28.70	−3.49	6.63
8	36	1.86	13.40	−17.30	65.20	−5.35	7.59
9	36	1.68	8.35	−15.80	16.40	−2.62	8.60
10	36	1.20	11.40	−25.00	33.33	−6.53	8.20
Means		1.242	1.999	−2.790	4.490	.195	2.360

at the end of the month. The second sample of 36 stocks had an average return of 4.49%, so if we had bought that particular group of 36 we would have earned 4.49% in the month, and so on.

Now, recall from the central limit theorem that the sample averages should themselves average about the same as the original distribution and that their standard deviation should be about the same as the original standard deviation divided by the square root of the sample size. Therefore, the average of the samples of $n = 36$ should be about 1.3%, and the standard deviation of the sample averages should be about 12 divided by the square root of 36, which is 2. Statistics on the sample averages are given in the last line of the table. Rounding to one decimal, we see that the sample averages themselves average 1.2% and their standard deviation was (rounding) 2.0%. Things do not always work out this well, but at least in this case the central limit theorem came through nicely.

Sampling from Nonnormal Distributions

The central limit theorem is "central" in the sense that it concerns the mean value, which is a measure of the center of a distribution. It is a "limit" theorem because it says what happens in the limit as the sample size grows larger and larger. It does not say anything definite about any particular sample size *n*, but if the sample size is fairly large, it suggests that the average (and the sum) can be assumed to be approximately normally distributed.

The average of values from a skewed distribution can be normally distributed because when we take the average of many values, only a few (if any) are likely to be far away from the center. Although a single far observation may stand out by itself, when it is averaged with other more central observations, the skewness is reduced.

Consider an extremely skewed distribution, the exponential distribution pictured in Figure 8.13. Whereas most single observations from it are close to 0, occasionally some fairly large values are possible. If we sum pairs of observations from this distribution, the resulting probability distribution (for sums of pairs) shown in Figure 8.14 is already substantially more normal looking. However, although it now has a smooth peak, it is still quite skewed, trailing off much more slowly on the right than the left. The distribution of sums of three observations (Figure 8.15, page 278) is even more normal looking; the skewness, although still there, is further reduced.

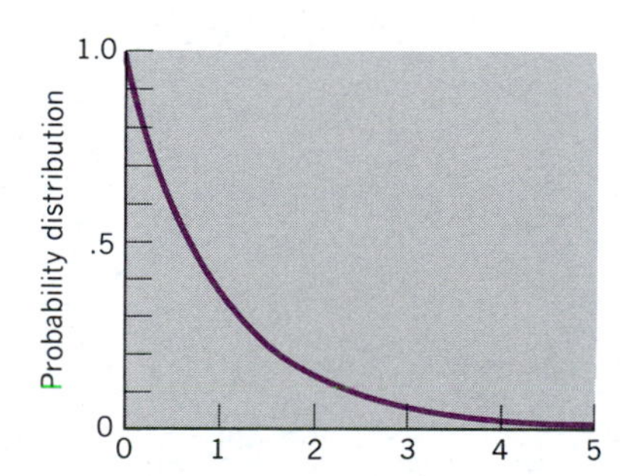

Figure 8.13 *The population: a skewed exponential distribution. Watch the effect of the central limit theorem to see how sums from this population become approximately normally distributed as we add more and more terms.*

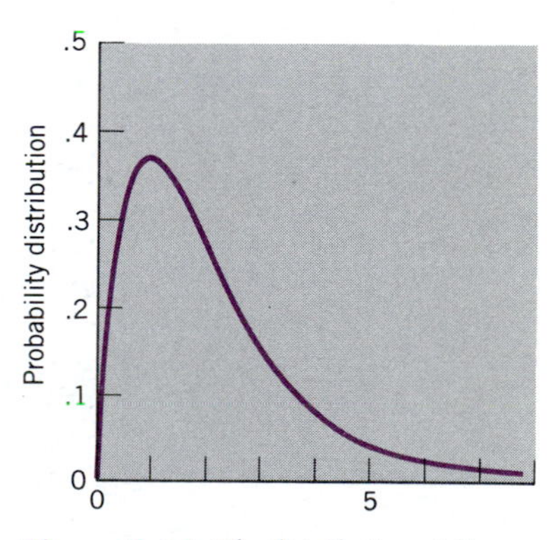

Figure 8.14 *The distribution of the sum of two observations from the population. Still very skewed, but less so than the population itself because of the effect of averaging.*

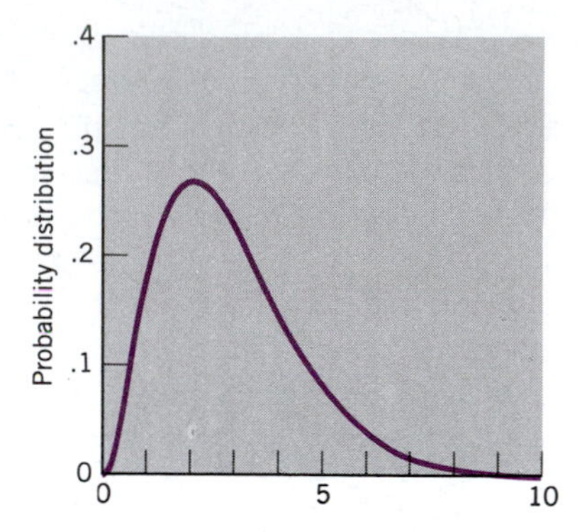

Figure 8.15 *The distribution of the sum of three observations from the population. Less skewed, more symmetric.*

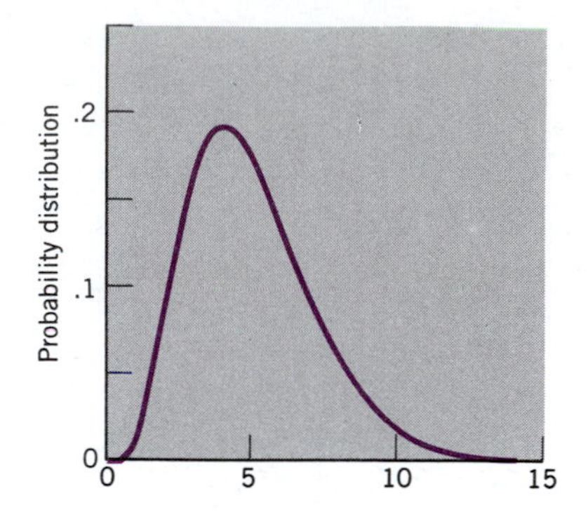

Figure 8.16 *The distribution of the sum of five observations from the population. Fairly symmetric at this point.*

Figures 8.16, 8.17, and 8.18 show the distributions for sums of 5, 10, and 25 observations, respectively, demonstrating the approach to a normal distribution. Careful examination reveals some slight skewness still remaining, even with sums of 25, although the overall impression is that of the normal bell-shaped curve.

Another demonstration of the remarkable fact that the distribution shapes become normal when sums (rather than individual observations) are measured is provided by Figures 8.19, 8.20, 8.21, and 8.22. We begin this time with the symmetric and flat uniform distribution (Figure 8.19) for which observations in the tail are just as likely as observations in the middle. Sums of two observations from this distribution give us the triangular distribution shape in Figure 8.20. We could almost interpret its shape as a simple "stick-figure" approximation to the normal distribution; observations near the center are now much more likely to occur than observations in the tails. Because the distribution of one observation was already symmetric, the approach to the bell-shaped normal curve is quite rapid, indicated by distributions for sums of 3 (Figure 8.21) and sums of 10 (Figure 8.22).

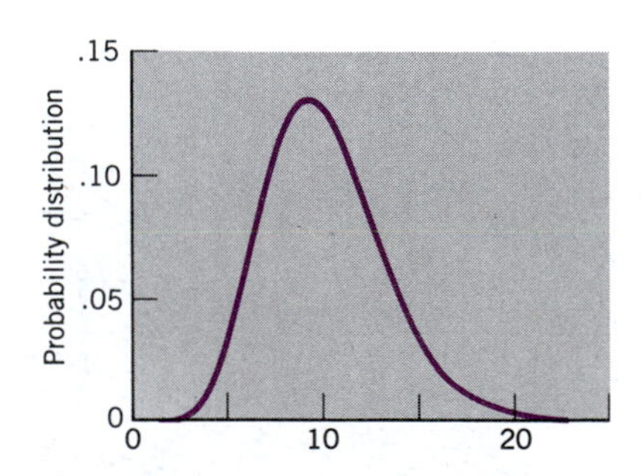

Figure 8.17 *The distribution of the sum of 10 observations from the population. Only slight skewness remains; it looks a lot like a normal distribution.*

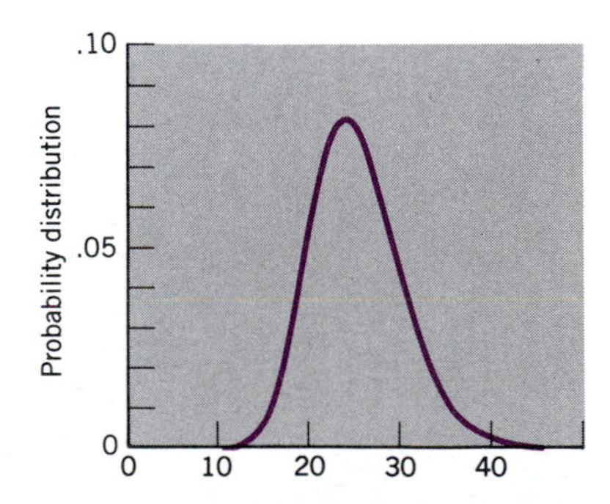

Figure 8.18 *The distribution of the sum of 25 observations from the population. The distribution is nearly normal at this point. A very sharp eye will find very slight skewness remaining.*

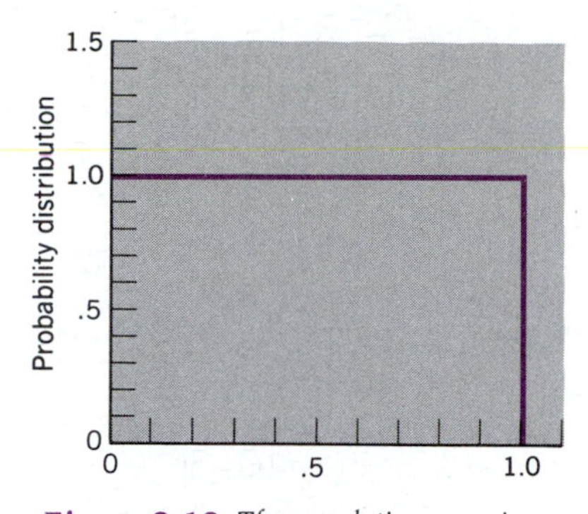

Figure 8.19 The population: a uniform distribution, symmetric from 0 to 1, with much shorter tails than the normal distribution. Watch the effect of the central limit theorem to see how sums from this population become approximately normally distributed as we add more and more terms.

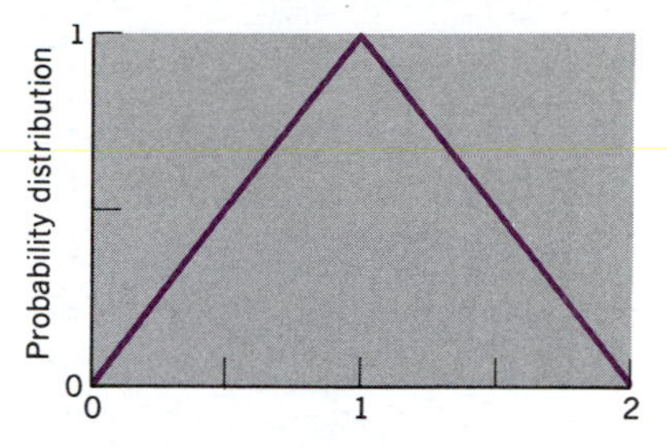

Figure 8.20 The distribution of the sum of two observations from the population. Already, the distribution has developed sloping tails at the edges.

EXAMPLE 8.7 Average from a Skewed Distribution: Colonies of Bacteria

Suppose colonies of bacteria are analyzed in groups of 10 as part of a laboratory experiment. The size of individual colonies has a mean of 350 cells and a standard deviation of 40. Furthermore, the distribution of single-colony sizes has been seen to be somewhat skewed toward larger values, as we often see with counted data.

What can be said about the *averages* of the groups of 10? The central limit theorem applies here, under the assumption that each of the 10 colonies represents an independent replication of the experiment under identical conditions. The conclusions are, first, that the distribution of the average colony size follows a distribution close to the normal distribution. Second, the mean

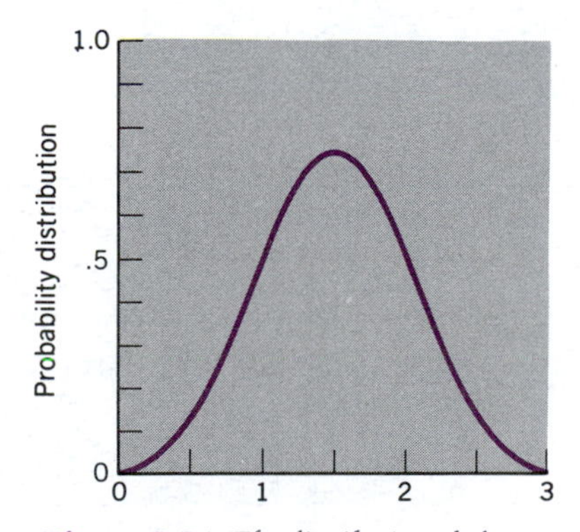

Figure 8.21 The distribution of the sum of three observations from the population. This distribution is already fairly bell-shaped.

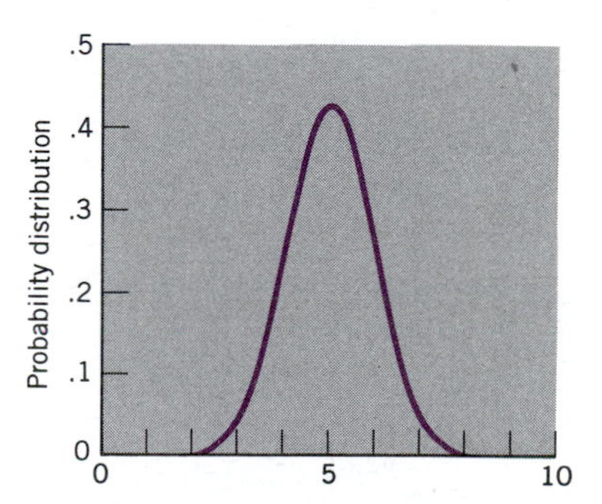

Figure 8.22 The distribution of the sum of ten observations from the population. This is now quite close to a normal distribution.

of the average colony size is still 350 cells. Third, the standard deviation of average colony size is

$$SD = \frac{40}{\sqrt{10}} = \frac{40}{3.16} = 12.6$$

indicating that the variability is reduced from 40 to 12.6 when we look at a batch of 10 instead of only a single colony. The central limit theorem tells us that the extra work of looking at a larger sample size "pays off" in extra precision (reduced variability).

What then can be said about the *sums* of groups of 10? The central limit theorem again gives us three conclusions. First, the sum approximately follows a normal distribution (it will have the same distribution shape as the average because the sum and average differ only by a constant scale term). Second, the mean of the sum of colony sizes will be $10 \times 350 = 3{,}500$ cells. (Naturally, we would expect to have 10 times the amount when examining 10 colonies.) Third, the standard deviation of the sum of colony sizes is

$$40 \times \sqrt{10} = 40 \times 3.16 = 126$$

indicating the increased variability from working with larger numbers. Notice that the variability did not go up tenfold (as one might have expected) but went up less, indicating the extra precision available from the combination of 10 pieces of information.

The distributions of sum and of average are summarized in Figure 8.23. Now that we have

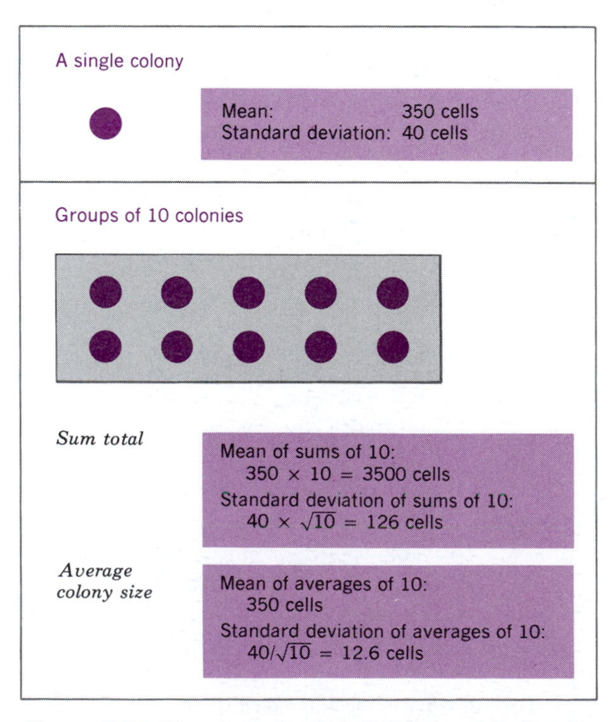

Figure 8.23 *The mean and standard deviation of the number of bacterial cells for three cases: (1) a single colony, (2) the sum of ten colonies, and (3) the average of ten colonies*

the mean and standard deviation for each of these new random variables (the average colony size and the total colony size), we may compute probabilities for them. For example, if we were asked to find the probability that the average colony size is less than 330, we would use the mean (350), the standard deviation (12.6), and the normal probability tables (using the central limit theorem as partial justification). We would see that

$$\begin{aligned}
&\text{Probability that the average colony size is less than 330}\\
&= \text{probability that a standard normal is less than } \frac{330 - 350}{12.6}\\
&= \text{probability that a standard normal is less than } -1.59\\
&= .056 \quad \text{(from the table)}
\end{aligned}$$

Next, to find the probability that the *total* number of cells is at least 3,650, we would proceed similarly but use the mean and standard deviation for the *total* (rather than for the average):

$$\begin{aligned}
&\text{Probability that total number of cells is at least 3,650}\\
&= \text{probability that a standard normal is at least } \frac{3{,}650 - 3{,}500}{126}\\
&= \text{probability that a standard normal is at least 1.19}\\
&= \text{probability that a standard normal is not less than 1.19}\\
&= 1 - \text{probability that a standard normal is less than 1.19}\\
&= 1 - .883 \quad \text{(from the table)}\\
&= .117
\end{aligned}$$

Exercise Set 8.10

37. Suppose you have a random variable with the probability distribution shown in Table 8.14.

Table 8.14

VALUE	PROBABILITY
0	.5
1	.5

a. What is the mean value?

b. What is the standard deviation?

38. Suppose you take many samples of 4 from the distribution in Table 8.14.

a. About what do you expect the sample average to be?

b. About what do you expect the standard deviation of the sample averages to be?

39. For a random variable with mean = 100 and SD = 12, what will be the predicted mean of the sample averages for repeated samples, and the SD of the sample averages, when

a. sample size = 4

b. sample size = 9

c. sample size = 16

40. Suppose groups of 25 students are selected randomly and asked to take an aptitude test that has mean = 500 and SD = 100. The scores in each group are then averaged.

a. About what should the mean of the group averages be?

b. About what should the SD of the group averages be?

8.7 Valuation and Risk

Probability theory can provide some guidance (but no assurances) in some practical decision making. In this section we offer a framework for evaluating economic decisions with a known element of randomness.

Have you ever been offered a bet and wondered whether you should accept it? A guide to the evaluation of bets and other random situations is provided by looking at the mean and the standard deviation of the payoff. The mean represents the "expected return" that you might reasonably achieve, on the average, if the bet were repeated a large number of times. This immediately shows whether the bet is "for you" or "against you." The standard deviation represents the *risk* involved in the situation. A definite (nonrandom) payoff is certain and has 0 standard deviation, representing no risk; but the more random and uncertain the process is, the higher the standard deviation will be, indicating a risky situation.

EXAMPLE 8.8 A Fair Bet

Consider the bet consisting of receiving a dollar if a tossed coin comes up heads and losing a dollar if it comes up tails. Table 8.15 shows the probability distribution of the random variable "winnings for one bet."

MEAN WINNINGS

$$(1)(.5) + (-1)(.5) = 0$$

STANDARD DEVIATION

$$\text{square root of } [(1)(.5) + (1)(.5)] = 1$$

This bet is "fair" to both you and the other person involved in the sense that your "expected" winnings are 0. The risk of 1, as measured by the standard deviation, indicates properly that there is some risk involved: Your winnings may well be higher or lower than the mean value of 0.

What if the stakes were increased to winning or losing $100 instead of $1? The mean return would still be 0 (indicating that the bet is still fair, of course) but the risk would now be 100 times larger. Indeed, it is a much riskier game because there is a real possibility of losing $100 on one coin toss.

Now, consider a slightly more complex bet, which would be harder to evaluate without computing the mean and standard deviation. You and a friend have made some predictions. Your friend offers you $3 if she is wrong; you pay her of $2 if you are wrong. On the surface, it looks like the bet is to your advantage because you would win more (if you win) than you would lose (if you lose). Indeed, it would be to your advantage if the probability was .5 (i.e.,

Table 8.15

WINNINGS	PROBABILITY
1	.5
−1	.5

Table 8.16

WINNINGS	PROBABILITY
3	.3
−2	.7

even) of winning or losing. However, you are realistic about these things and estimate the probability of winning to be only .3. Should you accept the bet? A guide to the answer is provided by examining the probability distribution (Table 8.16) and computing the mean and standard deviation.

$$\begin{aligned} \text{Mean} &= (3)(.3) + (-2)(.7) = -.5 \\ \text{Standard deviation} &= \text{SQRT}[(12.25)\,(.3) + (2.25)\,(.7)] \\ &= \text{SQRT}[5.25] \\ &= 2.29 \end{aligned}$$

Because the mean is negative, you would not want to accept the bet. To accept it would result, on the average, in a loss of 50 cents each time such a bet was played.

The use of the mean and the standard deviation to analyze random payoffs gives us a framework for understanding our intuitive reactions. Consider the following two situations.

A Receive 1 million dollars definitely.

B Receive 2 million dollars, or nothing, based on the toss of a fair coin.

Both of these situations have a mean value of 1 million dollars. However, most of us would choose *A* (certainty) instead of *B*, which is random. This makes sense because, as indicated by the standard deviation for each case, the risk of *A* is 0 (it is, after all, a risk-free offer of 1 million bucks!) whereas the risk of *B* is substantial: There is a real possibility of receiving nothing.

Probability theory cannot tell you exactly when to bet and when not to bet. That is a matter of personal preference. However, it can provide a framework for understanding how favorable or risky a bet is. Generally, if two bets have the same mean but different risks, then the less risky bet would be preferred. Similarly, if two bets have the same risk, then the one that offers a higher expected (mean) return would be preferred. Of course, if one bet offers a higher return with lower risk than another, that one would be preferred on both counts. However, if one bet offers a higher return than another but at the price of a higher risk, then individual preferences for risk-taking (rather than probability theory) must decide which is best.

Exercise Set 8.11

41. Suppose that there is a state lottery. Each ticket costs $1; there is 1 chance in 10 that you will win; and if you do win, you get $5.

- **a.** What are your net payoffs (taking the cost of the ticket into account), if you lose and if you win?
- **b.** What is the predicted average payoff (mean winnings) for people who play this game?
- **c.** What is the predicted standard deviation of their winnings?

42. Suppose that a student faced with a multiple-choice aptitude test runs out of time and answers one block of 20 questions randomly. (You can not really just sit down and think up random answers—

some subconscious processing is going to shape your answers—but for this question suppose that it is possible.) Each question has five possible answers.

a. How many questions do we expect the student to get right, on average?

b. If each correct answer counts 4 points and each incorrect answer counts as 0, what is the expected payoff for guessing on the 20 questions?

c. If each correct answer counts 4 points and each incorrect answer counts as -1 point, what is the expected payoff?

43. Suppose that you are playing American roulette. The wheel has 38 slots; 18 are black, 18 are red, and 2 are green. If you bet a dollar on black and the ball stops on a black slot, you get your dollar back plus another dollar. If you bet a dollar on red and the ball lands in a red slot, you get your dollar back plus another. You are not allowed to bet on green; the house bets green. If the ball lands on green, the croupier reaches out and takes all the money on the table. (And the house does not even put any money on it.)

a. On any given spin, what is the probability the house will win?

b. If the people at the table bet $1,000 total, what are the expected winnings for the house?

c. Of the $1,000 the players bet, how much should they expect, on average, to get back?

44. In the previous exercise, we looked at roulette as a game of all the players against the house. (This is certainly the way the house looks at it.) Now let's look at it from the point of view of an individual playing black, against both red and green. The probability of winning a dollar is $\frac{18}{38}$, and the probability of losing a dollar is $\frac{20}{38}$. (The 20 is the sum of the 18 reds and the two greens.)

a. If you bet a dollar at a time on black, over the course of the evening, about how much should you win or lose on each dollar?

b. If you bet a hundred dollars at a time on black, over the course of the evening, about how much should you win or lose on each hundred dollars?

Summary

A *random variable* represents a numerical outcome of a random experiment. Every random variable has a *probability distribution* that describes the likelihood of occurrence of its possible values. A random variable is said to be *discrete* if we can make a list of its possible outcomes. The probability distribution of a discrete random variable is a list of each possible value with its probability of being observed. A random variable has a *mean* (summarizing the typical central location) and a *standard deviation* (summarizing variability) that can both be computed from the probability distribution. The *variance* of a random variable is the square of the standard deviation. If the random variable represents the gain (or loss) involved in a random situation, then the mean represents the *expected gain,* whereas the standard deviation measures the *risk* of the situation.

The *binomial probability distribution* applies to situations where there are two possibilities (yes/no, heads/tails, male/female, and so on), and the probabilities do not change as we take a random sample. The binomial formula can be used to calculate exact probabilities for these situations. Alternatively, if the sample size and probabilities are such that we expect at least 5 of each result, we can use the *normal approximation to the binomial,* with mean equal to np and standard deviation equal to $\sqrt{npq}$.

The *normal distribution,* the most important probability distribution, is represented by a bell-shaped curve. The probability of a value falling within an interval is represented by the area above that interval and below the curve. For each choice of a mean value and a positive standard deviation, there is a normal distribution. The *standard normal distribution* has mean 0 and standard deviation 1, and its probabilities can be found in a table. To find probabilities for other normal distributions, we first convert to the corresponding values for a standard normal distribution so that we can then use the standard tables. Care must be exercised in using the normal probability tables because although they give exact results for the normal distribution and good approximations for some other distributions, they can yield incorrect and misleading results if the distribution is highly skewed or else is discrete with only a few possible values.

The *central limit theorem* tells us what happens when we look at the distribution of an average (or a sum) of individual independent observations. The distribution tends to be closely approximated by a normal distribution when the number of items summed or averaged is large, even if the distribution of individual observations is not normal. The mean and standard deviation of the average and of the sum can easily be calculated from the summaries for an individual observation and the number of items summed or averaged. Using the mean and standard deviation of the random variable, we can use the tables for the normal distribution to compute approximate probabilities for the sum or the average.

Problems

Basic Reviews

1. Suppose a person chooses a whole number at random from the numbers 1 through 5, and each of the numbers 1, 2, 3, 4, and 5 is equally likely.

a. Find the probability distribution.

b. Find the probability that the number will be 4 or 5.

c. Find the probability that the number will be an odd number.

d. Find the mean value of the probability distribution.

e. Find the standard deviation of the probability distribution.

2. Suppose a person randomly picks whole numbers from 1 to 5, where each is equally likely, and finds their average. Find the expected average, and the standard deviation of the expected average, when the person picks

a. two numbers.

b. three numbers.

b. four numbers.

c. nine numbers.

d. sixteen numbers.

e. twenty-five numbers.

3. Find the following probabilities for a standard normal distribution.

a. The probability of being less than 0.

b. The probability of being less than 1.62.

c. The probability of being greater than 2.14.

d. The probability of being greater than −.68.

e. The probability of being between 1 and 2.

f. The probability of being between −1.5 and .84.

g. The probability of *not* being between −.9 and −.6.

h. The probability of being between −1 and 1.

i. The probability of being between −2 and 2.

j. The probability of being between −3 and 3.

4. Find the value such that the probability a standard normal variable is less than the value is
   a. .001
   b. .01
   c. .05
   d. .5
   e. .75
   f. .90
   g. .95
   h. .99

5. Suppose that a normal random variable has a mean of 80 and a standard deviation of 15. Find the following.
   a. The probability of being less than 75.
   b. The probability of being more than 60.
   c. The probability of being between 90 and 100.
   d. The probability of not being between 60 and 110.

Analysis

6. Consider the random variable "the number of musicians in a randomly selected small recording group." Suppose its probability distribution is given in Table 8.17.

Table 8.17

NUMBER OF MUSICIANS	PROBABILITY
2	.12
3	.27
4	.43
5	.18

   a. Find the probability that a group has three or more players.
   b. Find the probability that a group has four or fewer players.
   c. Find the mean number of players per group.
   d. Find the variance of the number of players per group.
   e. Find the standard deviation of the number of players per group.
   f. Draw a picture representing this probability distribution and indicate the mean and standard deviation on the diagram.

7. The number of defective parts produced per hour in a manufacturing plant has been observed to follow the probability distribution shown in Table 8.18.
   a. Find the probability of having at most one defect during an hour's production.
   b. Find the probability of having either two or three defects.
   c. Find the mean number of defects.
   d. Find the standard deviation of the number of defects.
   e. Draw a picture of the probability distribution and indicate the mean and standard deviation on the diagram.

Table 8.18

DEFECTS IN ONE HOUR	PROBABILITY
0	.61
1	.30
2	.08
3	.01

8. You believe that the home team has a 90% chance of winning the next game. You have made an "even money" bet, so that you win $1 if the team wins, but you lose $1 if it loses.

a. Find the probability distribution.

b. Find your expected return.

c. Find your risk.

d. Briefly explain the meaning behind the numbers you found in parts (b) and (c).

9. Repeat Problem 8, but under the assumption that you believe the following.

a. The home team has a 75% chance of winning.

b. The home team has a 50% chance of winning.

c. The home team has a 10% chance of winning.

d. The home team has only a 5% chance of winning.

10. Find the expected return and the risk of the investment summarized in Table 8.19. State what you are using as measures of expected return and risk.

Table 8.19 *Hypothetical probability distribution of an investment in the stock market*

PROFIT	PROBABILITY
$300	.05
200	.25
100	.35
0	.20
−50	.10
−100	.05

11. Sketch the probability distribution for your profit for the investment summarized in Table 8.19. Indicate the mean and standard deviation.

12. If you had a choice between the investment summarized in Table 8.19 or a straight gift of $100, which should you prefer? Why?

13. Suppose you had to decide between the investment of Table 8.19 and one with an expected return of $150 and a risk of $200. What would your decision depend on?

14. For planning purposes, a company has decided to assume that the number of teddy bears that it may have orders for next year follows a normal distribution. After several meetings, it was decided by the president that they would probably receive orders for 30,000 teddy bears. However, the economy and people's tastes being uncertain, the company decided that the actual number could easily be different by 7,500 in either direction. Taking these numbers as the mean and standard deviation of a normal distribution, find the following.

a. The probability that they receive orders for more than 20,000 teddy bears (this is the event "at least an acceptably good year").

b. The probability that they receive orders for fewer than 15,000 teddy bears (this is the event "a disastrous year").

c. The probability that they receive orders for less than 40,000 teddy bears (this is the event "our current plant capacity can handle the demand").

d. The probability that they receive orders for between 25,000 and 35,000 teddy bears (this is the range the company would like to plan for).

15. An economic forecaster suggests that interest rates will be 9% next year and that the uncertainty in the forecast is indicated by a standard deviation of 2 percentage points. Assume that the actual interest rate is normally distributed with this mean and standard deviation.

a. Find the probability that the forecaster is correct to within 1 percentage point on either side of the 9% forecasted.

b. Find the probability that interest rates are no larger than 12.5%.

c. Find the probability that interest rates will be 8% or smaller.

16. Each Saturday in the summer, a student earns extra money by playing guitar and singing for donations on the sidewalk by the town square. The collections vary from week to week, with a mean of $58 and a standard deviation of $23. The student is trying to plan ahead for the five Saturdays of the month of August and is willing to assume that the money to be received on each of the five Saturdays may be represented as an independent observation of a random variable with these values as its mean and standard deviation.

a. Find the mean value of the total expected to be received for these five Saturdays.

b. Find the standard deviation of this total.

c. We will now assume that the total amount for August approximately follows a normal distribution. What justification is there for this assumption?

d. Find the probability that the student collects a total of at least $250 during the month of August.

e. Find the probability that this total is less than $200.

f. Find the probability that the student makes between $200 and $300.

g. Find the mean value of the average received per Saturday for each of the five days in August.

h. Find the standard deviation of this average.

i. What helps the assumption "the average is approximately normally distributed" to be true?

j. Find the probability that the average earnings are not between $50 and $65.

17. A certain chemical process, still in the laboratory development stage, produces a product that is useful in further experiments. There is random variability in the amount of product produced. The mean is 11 grams and the standard deviation is 3 grams for each run of the process. At this time, the scientists in the lab need 100 grams of the product for further experiments. They decide to run the process 10 times.

a. How much do we expect the total yield to be?

b. What is the standard deviation of this total yield?

18. For the situation described in Problem 17, why is it reasonable to assume that the total yield is approximately normally distributed?

19. For the situation described in Problem 17,

a. what is the probability that the goal is reached (that is, that at least 100 grams are produced)?

b. what is the probability that the goal is not reached?

20. Reconsider the situation in Problem 17. Suppose now that the process is run 12 times, rather than 10. What is the probability now that the required 100 grams of product will be produced?

21. A community has a mean income of $32,000 per family, with a standard deviation of $9,000. As part of a survey, 25 families at random were selected for interview and their average income was computed.

a. Why may we assume that this average for these 25 families is approximately normally distributed?

b. What is the mean (expected value) of the average income for the sample of 25 families?

c. What is the standard deviation of the average income for the sample?

d. What is the probability that the average income for the sample is below $30,000?

e. What is the probability that the average income for the sample is above $35,000?

f. What is the probability that the average income for the sample is between $30,000 and $34,000?

22. Suppose that the weights of the people in a certain office building are close to normally distributed, with a mean of 150 pounds and a standard deviation (SD) of 20 pounds. A randomly selected group of 16 crams into the elevator at the close of business one day.

a. What is the expected *average* weight of those 16 people?

b. What is the expected *sum* of the weights of those 16 people?

c. What is the probable difference between the expected average and the actual average (when "probable difference" is one SD of the sample averages)?

d. What is the probable difference between the expected *sum* and the actual *sum* (where "probable difference" is one SD of the sample sum)?

23. Suppose the elevator described in Problem 22 has a weight-sensitive safety switch and will not move if the load exceeds 2,300 pounds. What is the probability that the combined weight of the 16 randomly selected passengers from this office exceeds 2,300 pounds?

24. Suppose you have a list of 200 stocks you might buy. Based on historical performance of lists compiled the same way, you think the stocks will average a 1.0% return per month, with a standard deviation of 15.0%.

a. If you randomly pick one stock and invest $100,000 in shares, how much do you expect to make in a month? Plus or minus how much?

b. If you randomly pick nine stocks and invest $100,000 in them (shared evenly and ignoring commissions), how much do you expect to make in a month? Plus or minus how much?

c. If you randomly pick 25 stocks and invest $100,000 in them, how much do you expect to make in a month? Plus or minus how much?

25. *(True story)* A small social work agency has an annual fund-raising auction and raffle. At the auction in the summer of 1994 they offered as a prize a gift certificate worth $500 at a local wholesale warehouse. (For purposes of this problem, grant that it was really worth $500.) The agency sold 500 raffle tickets at $2 each. If you bought ten tickets, what is your expected return, judged only in dollars?

chapter 9

Toward Statistical Inference

Now that we understand probabilities and data distributions, we are ready to begin statistical inference. We will, in later chapters, be making probability statements about a population based on the data we have observed. To achieve this, we will first need to learn how to choose a good sample from a population. We will view the entire sampling process as a single random experiment and recognize the sample average as being itself a random variable with its own distribution. The fact that we can assess the variability of the sample average will play a central role in statistical inference, because this will tell us approximately how far away the average of the sample data is from the mean of the population.

9.1 Inference

Beyond Summarizing the Data

Up until now, we have primarily learned about methods for understanding the data itself: how to characterize it, display it, and summarize its various aspects. We have also learned how the rules of probability help us understand what is likely to occur in a random situation. In the coming sections we will put these together and learn how to transcend the data to draw useful and meaningful conclusions about a larger situation or question. We might, for instance, want to predict the winner of an election based on the results of a sample. Much of the power of statistical analysis comes from this ability to draw general conclusions from a limited but representative set of data.

Definition of Inference

The verb *to infer,* from which we obtain the noun *inference,* has as one of its meanings: "to derive by reasoning or implication; to conclude from facts or premises; to derive as a consequence, conclusion, or probability." In this spirit, we will see that statistical inference takes advantage of various methods of reasoning, using all facts and data available to draw likely conclusions about a situation.

Examples of the Need for Inference

Although there are undoubtedly times when we are content with describing the data, in many situations this is not enough. For example, someone running for president may be interested to know that 562 out of 1,000 people who were interviewed said that they would vote for him or her. However, the candidate is probably much more interested in the voting preferences of the many millions of people who will actually vote on election day. Of course, it is impossible to find out the preference of each of these voters and, even if we could, some might change their minds before voting. However, there is certainly some useful information contained in the responses from the sample of 1,000 people. To the extent that the people interviewed are "a lot like" the larger population, it may be possible to infer (at least approximately) what the others think. If only a few people change their minds, then it may even be possible to predict the outcome of the election, although this is a risky business.

The principle of inference about a larger population based on a smaller sample also applies to marketing tests of a new product. A sample of people are tested or interviewed with the hope that their behavior will indicate the extent of success of the new product. Again, to the extent that the sample is representative, analysis of the people in the sample will help with decision making by providing some information about consumers in general.

Discussion

Many of the studies reported in specialized journals rely heavily on the ability to go beyond the original data. For example, reports on cancer incidence in a specific group of laboratory rats are used to indicate the underlying ability of a chemical or food additive to cause cancer in rats in general. Valid conclusions may be drawn when the experiment is designed and analyzed carefully with the aid of statistical reasoning. The next conclusion—that substances causing cancer in rats also cause it in people—is based on reasoning that is not statistical in nature, but relies instead primarily on medical and biological theories.

Understanding statistics is also useful in interpreting conclusions that we read in the newspapers or hear on television. When we hear news about human behavior based on psychological studies or about health and diseases based on medical studies, we must keep in mind that the results are usually based on data from a *sample* of people or animals. Therefore we must be careful in drawing general conclusions. Why do some studies about controversial subjects seem to disagree with other studies on the same issue? Because we will always expect to see randomness in data, differences may be due to chance. In addition, many reports are inferences based on a study with only a small to moderate amount of data. Thus such studies should not be viewed as absolute truth, but only as indications of truth based on inferences from imperfect data. We need to keep things in perspective.

Note that with inference, there will be no guarantee the answer is actually true or correct. Instead, we will be content with inferences that are "probably" correct or "likely" to be true. One of the advances of modern statistical inference is the ability to quantify the degree of correctness of an inference. For example, we will learn about methods that allow us to conclude that we are 95% certain that a specific generalization holds, based on reasoning and calculation from data.

9.2 Populations and Samples

In many statistical studies (and perhaps in much of life in general), a situation divides naturally into two parts: what we would like to know and what we do know. Although these are not difficult concepts, it is important to become familiar with these two components so that they can be properly distinguished. The larger entity, which we would like to understand in all of its details, is called the **population.** The information we have available to us has been obtained from a **sample** chosen as a subset from the population.

This terminology is completely natural when we have, say, selected people for study from a genuine population of people, as in political polling or market research. But the notions of population and sample are much more extensive and flexible and can apply to any situation in which general conclusions are reached based on data from only a part of the situation. What follows are five examples of valid populations that might be studied.

EXAMPLES OF POPULATIONS

1. All companies based in the United States that manufacture durable goods
2. All earthquakes that have ever happened on the west coast of North America
3. All people with diabetes
4. All seventh-grade students in metropolitan areas in industrially developed countries
5. All butterflies in the Rocky Mountains

The list of possible populations goes on and on, but this should give you the idea that a population is simply the well-specified collection of things (often called units) that we would like to understand.

Next are five examples of samples, one for each of the five populations just given.

EXAMPLES OF SAMPLES

1. The 368 companies that returned questionnaires (out of the 500 that were sent out as part of a study)
2. The 53 most serious earthquakes of the past 15 years along the San Andreas Fault in California

3. The 18 people who have recently come to the diabetes clinic of the hospital where the study is being performed
4. Three seventh-grade classes in each of five schools in the metropolitan area of Madison, Wisconsin
5. Observations on 232 butterflies in a field north of Yellowstone National Park

A sample is simply a partial collection of some of the units of the population. We see from these examples that there are many ways in which a sample might be chosen and that some samples are better than others for studying a given situation. We also see that there are practical considerations in choosing a sample. For example, it would usually take too much time or money to study the entire population.

Because information from the sample will be used to draw conclusions about the population, it is extremely important that the sample reflect the population. Ideally, the sample would be exactly like the population in every respect (except that it would be smaller) so that facts discovered about the sample would accurately represent the larger population. When the sample does not reasonably represent the population, mistakes can easily be made. A few examples of particularly bad samples for studying specific problems follow.

EXAMPLES OF BAD SAMPLES

Bad sample A: The basketball team as a sample used to study heights of high school students

Bad sample B: Your 10 best friends as a sample used to predict the winner of the coming election

Bad sample C: The 25 most successful new electronics companies as a sample used to study the financial problems of typical new electronics companies

Bad sample D: The opinions of some of the readers of a particular magazine as a sample used to study the opinions of Americans in general

These bad samples were chosen according to some characteristic that makes them different from the population in a certain very systematic way that tends to exclude a large important segment of the population from study. Thus these samples do not accurately represent the population being studied. The consequences of using each bad sample in a study are discussed briefly below.

CONSEQUENCES OF USING BAD SAMPLES

Consequence of using bad sample A:
The study will conclude that students are taller than they really are because basketball players are taller than most of the rest of us. Note that this sample would be more useful in studying the population of high school basketball players than for students in general.

Consequence of using bad sample B:
Because we tend to choose friends who agree with our views, this sample will tend to reinforce our own preferences rather than provide an accurate indication of the outcome of the election.

Consequence of using bad sample C:
Because only those companies that succeeded were chosen for study, the results can hardly be considered indicative of the problems of "typical" new companies, many of which do not succeed. However, this sample might be very useful for studying the population of new electronics companies that are likely to be successful.

Consequence of using bad sample D:

Different sorts of people read different magazines (compare the readership of *Popular Mechanics* to that of *Working Woman*) and therefore, the opinions of the readership will not be representative of the population at large.

The Problems of Bias

The general term used to describe the problem that exists when the sample is not representative of the population is **bias.** Thus, we use the terms *representative sample* (good) and *biased sample* (bad). Webster's New Collegiate Dictionary gives the relevant definition of *bias* as "a propensity or prepossession; bent; prejudice."[1] So a biased sample has properties of its own that may not match those of the larger population. Naturally, we want the bias (a source of error) to be as small as possible.

There are two sources of bias in survey research. The first is **selection bias,** where there is a systematic difference between the population and the sample, as described in the "bad sample" lists.

The second source of trouble in conducting a survey is **response bias,** which is due to the fact that not everyone being studied will return the survey questionnaire. Response rates of 50% (in which only half of the people contacted returned their forms) are not uncommon. For the sample to then be representative of the population, we often hope or assume that those who did not return the questionnaires are not very different from those who did. Although this hope is sometimes justified, usually there is little we can do about this problem except to accept it and wish for a small amount of response bias.

To see why response bias might be a problem, consider the fact that those who do answer a questionnaire might well represent the more active and vocal members of the population. Those with strong opinions and interests will tend to be overrepresented, whereas those who are quiet and content with the status quo (or who are too busy with other things at the moment) will tend to be underrepresented.

Exercise Set 9.1

1. Suppose the marketing director of a small firm selects a sample by opening the phone book, closing his eyes, and pointing. The next 30 names are selected. All 30 people have the last name Johnson, including 2 Gertrudes, 4 Gilberts, 4 Glens, 4 Gordons, 6 Gregorys, and 4 Gunnars.
   - **a.** Does this sound like a representative sample of the city's population?
   - **b.** Does this sound like a biased sample?
   - **c.** Does it sound like maybe the members of the sample are not independent?
2. Suppose a civic club wants to study day-care needs in the community. One member proposes asking the employees of major corporations how they solved their day-care needs. Would this sample be representative of the larger community? What group(s) (if any) of individuals (wanting day care) might be systematically excluded?
3. Suppose an interviewer, instructed to interview a representative cross section of people, stands on the street corner and interviews every third person she feels comfortable stopping. Will this sample show selection bias? Why or why not?
4. Suppose a marketing firm uses a computer to randomly pick street addresses and mails a questionnaire to all of the picked addresses. Unbeknownst to the firm, the questionnaire is returned by two-thirds of the middle-income families, but by only one-fourth of the low-income families and none of the high-income families. Will this sample show response bias? Explain your answer.

9.3 Choosing a Good Sample

The *random sample* is one of the most powerful tools available for selecting a good sample for statistical purposes. Random sampling helps ensure that the sample represents the entire population, and is not biased or prejudiced toward any particular group or groups within the population. At the very least, it helps eliminate the tendency to select based on a biasing factor (as happened in each of the bad samples discussed earlier).

Random sampling, which we will discuss in some detail, is an important example of an **experimental design** or **study design**. There are other study designs that are also important in statistics, such as *stratified random sampling,* in which a random sample is taken separately from each of several groups of the population.

The Random Sample

Simply put, a **random sample** must satisfy the following two properties.

PROPERTIES OF A SIMPLE RANDOM SAMPLE DRAWN FROM A POPULATION

1. Each unit of the population has an equal chance of being represented in the sample.
2. Units in the sample are chosen independently without regard to one another.

The first property is necessary to ensure that all members of the population are treated equally. This guarantees that there will be no discrimination for or against including any unit of the population. The second property is needed to ensure that the sample contains as much useful information as possible. If some choices had affected others, then some information in the sample might be redundant rather than providing a balanced reflection of the population.

To see why the second property is needed, let's look at a bad sample that satisfies property 1 (so that all units will have the same chance of being chosen) but not property 2 (so that units will not be chosen independently). By definition, we will not have a random sample in this case.

EXAMPLE 9.1 A Sample that Is Not Random

From a group of 100 whites and 100 blacks, choose either a random sample of 15 whites or a random sample of 15 blacks, so that the choices of whites or blacks are equally likely.

Although *individually* each unit of the population of 200 people has the same chance of being in the sample as any other unit, taken *collectively* the sample is neither fair nor random because each sample is either completely whites or completely blacks and therefore cannot accurately represent the population of 200 people, which contains a mixture of both groups.

The lack of independence in this example may be seen from the fact that if the first person of the sample is black, then the second (third, fourth, and so on) will also be black. Because the units chosen for the sample can be predicted from each other, they are dependent on each other. Thus the independence requirement for a random sample fails in this case.

This example was deliberately chosen to show extreme dependence among the members of the sample. However, the basic principle is more general, and dependence can be a serious problem in other, more subtle, situations.

Discussion

Using a random sample in experimental or study design achieves two goals. First, as we have seen, it helps ensure that the sample represents the population by minimizing the sampling bias. Second, it inserts a controlled amount of randomness into the situation. At first glance, having some controlled randomness might not seem to be a great achievement. However, to ultimately reach statistically valid conclusions (such as being 95% certain of something), there must be some well-understood and well-behaved randomness in the situation. Because many populations are very structured (perhaps including randomness that is not understood), it is the randomness from the sampling method that permits us to draw statistical conclusions. Because this randomness is deliberately introduced to the situation from the outside, general statistical methods can be valid even when the internal randomness of the situation is not understood. This is indeed an important achievement.

Although simple random sampling is often a good method of discovery, there are situations that call for more complex study designs. In the example just given, if we know for certain that the population is half whites and half blacks, then we might do better if we put together two half-samples, with one half-sample randomly chosen from each of the groups. This would ensure that the sample has the same proportion of each group that the population does. This is an example of a **stratified random sample,** in which random subsamples are selected from each important group (or *stratum*) of the population.

Another problem with random sampling is that, because it is random, any particular sample may not seem quite representative. This problem often arises when the sample size (the number of units in the sample) is small. Considerations like these make statistics a more lively field than many people realize, with controversies and varied opinions on which is the best procedure to be used in a given circumstance.

Choosing a Simple Random Sample

How is a random sample actually chosen? The best way is to use an external source of randomness, such as a tossed coin, slips of paper in a hat, or (more professionally) a table of random numbers. The use of methods such as drawing slips of paper from a hat can be valid if proper care is used to completely shuffle the slips and if the slips all have the same size and feel to the chooser. But we will now describe a more dependable general procedure for choosing a random sample of a desired number of units from a population.

PROCEDURE: CHOOSING A RANDOM SAMPLE OF ANY DESIRED SIZE

1. Assign a number to each unit of the population (1, 2, 3, . . .).
2. Choose a random place to start (such as the eleventh row, second column) in a table of uniform random numbers, perhaps with the aid of a toss of a die or flip of a coin.
3. Multiply this random number by the population size.
4. Add 1 to the answer in step 3 and throw away any decimal fraction.
5. See if the population unit with that number has already been chosen as part of the sample. If it has not yet been chosen, include it in the sample.
6. Go to the next random number in the table and repeat steps 3 to 6 until the sample has the desired number of units.

We multiply by the population size and then add 1 so that after we throw out decimal fractions our smallest possible answer is 1 and the largest is the population size.

Table 9.1 *Short table of random uniform numbers from 0 to 1*

.4515	.8189	.7099	.7394	.3260	.6457	.0190	.7942
.9487	.4110	.9048	.6239	.5973	.3555	.4236	.8021
.0872	.1237	.4427	.9700	.9431	.4670	.0624	.2899
.0712	.4469	.5378	.3983	.9310	.1380	.7005	.8589
.2070	.8313	.8808	.1156	.9805	.7372	.3293	.9461
.4020	.1886	.1391	.4144	.8624	.3443	.6671	.3245
.6654	.3916						

A short table of random numbers is given in Table 9.1. (A more complete table appears in Appendix B.) Note that these are called "uniform" random numbers. This means that they are evenly spread out (or distributed) between 0 and 1. A stem-and-leaf plot of the 50 numbers in Table 9.1 is shown in Display 9.1.

DISPLAY 9.1

STEM-AND-LEAF PLOT OF THE 50 UNIFORM NUMBERS FROM TABLE 9.1

```
0 | 1 6 7 8
1 | 1 2 3 3 8
2 | 0 8
3 | 2 2 2 4 5 9 9
4 | 0 1 1 2 4 4 5 6
5 | 3 9
6 | 2 4 6 6
7 | 0 0 3 3 9
8 | 0 1 3 5 6 8
9 | 0 3 4 4 4 7 8
```

A perfectly uniform distribution would have rows of even length if it were not for the randomness in sampling. In other words, randomness is why the stem-and-leaf plot in Display 9.1 is not completely rectangular. Nonetheless, it is fairly flat, has very short tails, and shows no particular skewness. How would we decide whether this histogram shows merely a reasonable amount of randomness (on the one hand) or serious deviations from uniformity (on the other)? This is a "hypothesis testing" question that can be answered by the chi-square test, discussed in Chapter 14. In fact, the randomness we see here is well within the usual amount we expect when ten categories are equally likely to occur.

EXAMPLE 9.2 *Choosing a Random Sample*

To choose a random sample of 10 people from a group of 83 that are eligible to participate in a study, we can proceed as follows.

1. First, we label each person with a number from 1 to 83. (It does not matter how this is done so long as it is done first; the randomness in the next stages will remove any systematic

effects of the particular numbering scheme even if, for example, all of the men are numbered first and receive the lowest numbers.)

2. Second, we choose a random place to start in the table. For example we might blindly open the phone book and use the last digits of the first numbers on the page to get a "2" and an "8." We would then take the random number in the second row and eighth column, which is .8021.
3. We multiply this random number by the population size.

$$.8021 \times 83 = 66.57$$

4. We add 1 and throw away the decimal fraction to get

$$(66.57 + 1) = 67.57 \rightarrow 67$$

5. Person number 67 is therefore included in the sample.
6. Repeating steps 3 to 6 until we have 10 different people selected for the sample, we calculate

$$.0872 \times 83; \quad (7.24 + 1) = 8.24, \text{ which truncates to } 8$$

Person 8 is included in the sample.

$$.1237 \times 83 = 10.27; \quad (10.27 + 1) = 11.27, \text{ which truncates to } 11$$

Person 11 is included in the sample.

$$.4427 \times 83 = 36.74; \quad (36.74 + 1) = 37.74, \text{ which truncates to } 37$$

Person 37 is included in the sample.

$$.9700 \times 83 = 80.51; \quad (80.51 + 1) = 81.51, \text{ which truncates to } 81$$

Person 81 is included in the sample.

$$.9431 \times 83 = 78.28; \quad (78.28 + 1) = 79.28, \text{ which truncates to } 79$$

Person 79 is included in the sample.

$$.4670 \times 83 = 38.76; \quad (38.76 + 1) = 39.76, \text{ which truncates to } 39$$

Person 39 is included in the sample.

$$.0624 \times 83 = 5.18; \quad (5.18 + 1) = 6.18, \text{ which truncates to } 6$$

Person 6 is included in the sample.

$$.2899 \times 83 = 24.06; \quad (24.06 + 1) = 25.06, \text{ which truncates to } 25$$

Person 25 is included in the sample.

$$.0712 \times 83 = 5.91; \quad (5.91 + 1) = 6.91, \text{ which truncates to } 6$$

(*Note:* Person 6 had been previously selected, so this choice is ignored.)

$$.4469 \times 83 = 37.09; \quad (37.09 + 1) = 38.09, \text{ which truncates to } 38$$

Person 38 is included in the sample.

The final sample of people is 67, 8, 11, 37, 81, 79, 39, 6, 25, and 38. If we place these in order, which is both convenient and one way to be sure that there are no duplicates, the random sample consists of those people who were numbered earlier as

6 8 11 25 37 38 39 67 79 81

Having selected the subjects randomly in this way for study, we have some confidence that problems that can arise from selection bias will not trouble us. Although there can be no guarantee that this particular sample represents the larger population of 83 people exactly, this sample will be representative enough for many purposes. Furthermore, we will be able to consider using many of the powerful standard statistical procedures that require a random sample for validity.

Exercise Set 9.2

5. Suppose you have a population of 100 people, and you have numbered them from 1 to 100. You are drawing a random sample, and the next random number between 0 and 1.000 is .490. Which person is selected for the sample?

6. Suppose that you are drawing a random sample from the population given in Table 9.2. You already have the numbered list, and you have gotten the following five random numbers between 0 and 1.0.

   .9749 .9811 .8771 .0997 .7841

Table 9.2 *Hypothetical information about 24 people taking a class together*

NO.	NAME	SEX	NSIBS	NO.	NAME	SEX	NSIBS
1	Aaron	M	1	2	Barclay	M	0
3	Clinton	M	1	4	Duncan	M	1
5	Elbert	M	2	6	Fred	M	2
7	Glenn	M	0	8	Hugh	M	1
9	Ivar	M	1	10	Justin	M	1
11	Kirby	M	1	12	Lars	M	3
13	Mike	M	2	14	Nan	F	0
15	Opal	F	0	16	Pamela	F	1
17	Quintin	M	0	18	Rachel	F	0
19	Sandy	F	1	20	Terry	F	1
21	Ursula	F	5	22	Valerie	F	2
23	Wanda	F	0	24	Xavier	M	4

Nsibs = number of siblings
Population N = 24
Mean number of siblings = 1.25
Standard deviation, number of siblings = 1.29
Proportion male = 15/24 = .625

   a. Convert these to integers (whole numbers) from 1 to 24.
   b. What members of the class are included in the sample?
   c. What is the average number of siblings for these people?

7. Draw a random sample of four from the population in Table 9.2. Use the following sequence of random numbers.

   .4180 .2387 .9392 .9459 .7313

   a. What members of the class are included in the sample?
   b. What is the average number of siblings for these people?

8. Use the random number table in Appendix B to pick a random sample of four from the population in Table 9.2.
   a. What members of the class are included in the sample?
   b. What is the average number of siblings for these people?

Computer Applications: Random Sampling

There are two ways computer programs can simplify drawing a random sample.

1. If the population list has been entered into a computer file, we can ask the computer to select a random sample and identify the cases for us.
2. If the population list has not been entered into the computer, we can use the computer to generate a list of random integers, much as we did by hand, that we can then use to identify the sample.

EXAMPLE 9.3 *Sampling from a Computer File*

The class list from Table 9.2 has been saved as a data file called "CLASSLST.MTW," with the names in column 1, the ID number in column 2, sex coded as 1 = male and 2 = female in column 3, and Nsibs in column 4.

The following Minitab commands retrieve the worksheet, draw a random sample of four from the class, and store the sample in columns c6 to c8.

```
MTB > Retrieve 'CLASSLST.MTW'.
MTB > Sample 4 'ID'-'NSibs' C6-C8.
```

In this particular case, we happened to get the sample shown in Table 9.3.

Table 9.3

ID	SEX	NSIBS
23	1	4
8	1	1
1	1	1
24	2	0

EXAMPLE 9.4 *Generating a List of Random Numbers*

If we do not already have the list entered on the computer, we can use the computer to generate a list of random numbers over the appropriate range. In Minitab, starting with an empty worksheet (restart Minitab to get an empty worksheet), we use the following commands to generate four random integers in the interval 1 to 24 and store them in column 1.

```
MTB > Random 4 c1;
SUBC > Integer 1 24.
```

In this particular case, we get these following numbers.

19 12 16 7

9.4 Population Parameters, Sample Statistics, and the Sampling Variability of the Sample Average

Suppose we are trying to estimate some population value by using a sample from that population. In general, we call the population value a **parameter,** and the sample value used to estimate it a **statistic.** Because this is an important conceptual distinction, we have ways of identifying whether a quantity has been computed from a sample or for the larger population.

In standard notation, population parameter values are indicated with Greek letters. Thus, a lowercase Greek mu, μ, indicates the mean value of a variable for a population. The average or mean for a sample is indicated by $\overline{X}$. The standard deviation of a population mean is indicated by a lowercase Greek sigma, σ, while the sample standard deviation is indicated by S.

Generally, the distinction can be intuitively thought of as follows. The summary value computed from the sample is the sample average, but the unknown quantity representing the entire population is the population mean. We like to think of the sample average (which we do know) as giving us a good indication of the value of the population mean (which we do not know). This expresses the statistical fact that the sample average is an estimate of the population mean. If we knew the value of the population mean, we would know how good that estimate is. But because we do not know the population mean, we usually do not know how good the estimate is.

How, then, can we assess how well the sample average estimates the population mean in a practical situation? Statistics provides several methods that help us do this in general terms. These methods rely on the properties of the average as computed for a random sample and provide a guide to the amount of uncertainty in the estimate.

Examples of Sample Average and Population Mean

Here are two examples to illustrate the difference between the sample average and the population mean.

1. To study heights of high school students, a random sample of 589 students was used. The sample average of these 589 data values was computed to be 5 ft, 5.8 in. The mean for the entire population of all high school students is not known. However, the mean can be thought of as follows. If the height of every high school student could be measured, then the population mean would be the average of all those (approximately) 14 million data values.
2. A candidate for political office has commissioned a poll. The results, computed for a random sample of 1,508 voters, show that 58.2% of these people intend to vote for the candidate. This percentage is, in effect, the sample average of the responses (as was discussed in the section on yes/no data, Section 4.4). The mean for the entire population of registered voters is not known by anyone. Although its value is unknown, the mean does nonetheless exist in the sense that if we could interview all of the registered voters and they told us whom they intend to vote for, then the resulting percentage would be the population mean.

Based on these examples, you should now have a good idea of the important conceptual difference between sample average and population mean. Do not confuse these with each other!

Randomness of the Average

When the average is computed from a random sample, we know that its value is not exactly the same as the population mean, but we expect it to give a good estimate of the population mean. The average "inherits" randomness from the random sample from which it was calculated, and the amount of randomness in the average can be measured, at least approximately. This assessed variability of the average is called its **standard error.** Before we define this precisely, we must first consider the concept of randomness of the average.

To conceive of the "variability of the average," we need to have some justification for considering the existence of many potential averages that could have been computed, even though there is usually only one actual average value for the study. One way to think about this variability of the average is to consider the other samples that could have been chosen. In the candidate's poll, for example, if a different random sample of 1,508 people had been chosen, then a different average value would almost certainly have been computed. Although the average value from a particular study seems to be a single exact number, it is in fact only the value for that particular random sample. It is the randomness in the choice of one particular sample, from the many potential samples, that gives the average a component of randomness.

A different, but related, way to think about the variability of the average is to consider the consequences of actually repeating the entire experiment or survey several times, much as we did in discussing the central limit theorem. Instead of only one study interviewing, for example, 1,508 people, imagine repeating this entire study over and over many times. Each time the study is repeated, a new and different sample of 1,508 people is chosen and a new and different average value computed. We would then have a list of average values and would see the variability in those values. That variability represents the differences between an average from one study and another average from a different (repeated) study.

To see what actually happens in a study, let's look first at Figure 9.1. From the population, we take one sample of n observations and then compute the average. To put the computed average value into perspective, we see in Figure 9.2 (page 304) what might have happened if we were to repeat the process some number k of times. In each case, we obtain a sample of n observations and compute an average. This gives us a new kind of sample, consisting of the k sample averages (each one different because of randomness). The histogram of these sample averages then would show us the distribution and variability of the average of a sample of size n. Visualizing the process in this way and using lots of sample averages help us interpret the sampling distribution of a single sample average in the same way that a histogram of data shows the distributional pattern for a single observation.

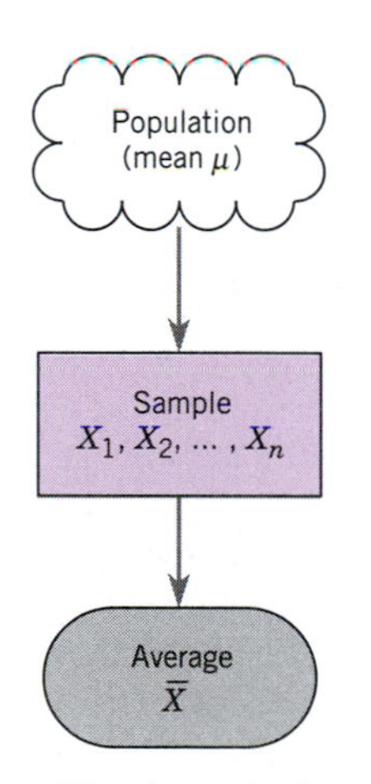

Figure 9.1 *The actual sampling process. We select a single sample of data and compute the average.*

Discussion

Of course, in practice, we would not usually repeat an expensive study in its entirety. Nonetheless, the notion that we could repeat it (if only we had enough resources) provides an important basis for assessing and interpreting the variability of the average. The ability to conceive of the possibility that different data might potentially have been observed is an important step in

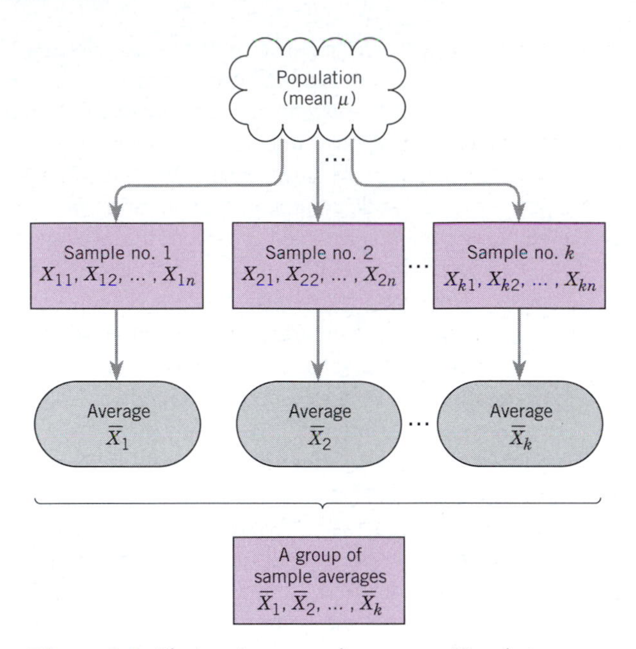

Figure 9.2 *The imaginary sampling process. We select many samples, find the average of each sample, and then view these sample averages as a single group of numbers with its own distribution and variability. This captures the essence of the variability of the average.*

statistical reasoning. Many historical data sets represent studies that could never actually be repeated exactly. Nevertheless, we can conceive of the randomness that existed at the time and can consider the other possible data values (and resulting averages) that might have been recorded instead.

Exercise Set 9.3

Recall the ages of the students in the hypothetical statistics class introduced in Exercise Set 2.3:

16 17 17 17 18 18 18 18 19 19 19 19 19 20 20 20 20 20
21 21 21 21 22 22 22 23 23 23 24 24 25 29 30

The mean age of this class is 20.758 years, with a standard deviation of 3.192 years.

9. Draw a (real) random sample of four from the list of ages above. (*Hint:* This calls for a random number table, a statistical calculator with a random number generator, or a computer. If you have a computer handy, you can use the Minitab "Sample" command to draw your sample.) Calculate the mean for this sample of four.

10. Draw another (real) random sample of four from the list of ages above. Calculate the mean for this sample of four.

11. Draw another (real) random sample of four from the list of ages above. Calculate the mean for this sample of four.

12. Write a very brief paragraph comparing the three sample averages with the true population average (20.758). (You might want to consider using the SD of the population as the basis for the comparisons.)

9.5 Standard Error of the Average

The variability of the average value computed from a random sample is measured by the quantity called the *standard error of the average*. As we have seen, the average itself is a random quantity, reflecting the underlying randomness in the selection of the random sample. It is therefore reasonable for this variability to be computed using the standard deviation of the sample, which we already know how to compute.

The (usual) standard deviation of a sample, as we have defined it, reflects the typical variability of *each number* in the sample, rather than the variability of the average itself. Because the average combines all of the data values, we naturally would expect it to be less variable (and more precise) than any individual data value. It is from combining many individual pieces of information that statistical methods gain their strength and power. Granted, then, that the average is less variable than individual observations (as measured by the standard deviation), how much less variable is it? The answer is provided, in principle, by the central limit theorem (Chapter 8).

Since we do not know the true population values used in the central limit theorem, we substitute their sample estimates. The **standard error of the average** is then computed by taking the standard deviation and dividing it by the square root of the number of data values.

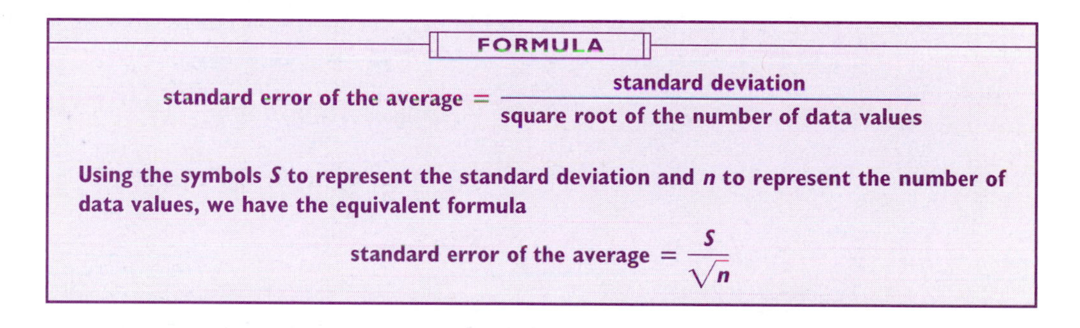

FORMULA

$$\textbf{standard error of the average} = \frac{\textbf{standard deviation}}{\textbf{square root of the number of data values}}$$

Using the symbols *S* to represent the standard deviation and *n* to represent the number of data values, we have the equivalent formula

$$\textbf{standard error of the average} = \frac{S}{\sqrt{n}}$$

This formula shows us how the sample size *n* affects the precision of the average. To be twice as precise, we need four times as large a sample (not just twice as large a sample). Thus, larger samples give more precise estimates than smaller samples, but not as precise as we might want them to be.

Sidebar 9.1

WHAT IF THE SAMPLE IS A BIG PART OF THE POPULATION?

Let's start by noticing that sometimes the population is fairly small. For example, the hypothetical class in Table 9.2 only has 24 members. Surely, if we "sample" all 24 of them, we will know exactly what their average number of siblings is, and the standard error should be 0. But that is not what we get when we use the formula.

There are two general ways out of this problem. The first is to ask what the formula does give us. The other is to use a formula that gives us the correct answer.

First, when we use the formula

$$\text{standard error} = \frac{S}{\sqrt{24}}$$

we are calculating the standard error for a random *sample* of 24 from a *much* larger population. This tells us how far this class's average is likely to be from the average of this much larger population (if we are willing to assume that the class is a random sample from that much larger population).

If the class of 24 *really is* the population we are interested in, then we have to use a correction factor. The correction factor is

$$\sqrt{1 - \frac{n}{N}}$$

where n is the sample size and N is the population size. To get the corrected standard error, we multiply the result from the regular formula by this correction factor.

$$\text{corrected standard error} = \frac{S}{\sqrt{n}} \times \sqrt{1 - \frac{n}{N}}$$

If the sample is actually the entire population, then the correction factor works out to 0:

$$\sqrt{1 - \frac{n}{N}} = \sqrt{1 - 1} = \sqrt{0} = 0$$

When we multiply the regular standard error by this, we get 0 for the corrected standard error. This is what we want for the situation when we really do know the population value.

When the sample is a *very* small part of the population, n/N is about 0, and we have

$$\sqrt{1 - \frac{n}{N}} \cong \sqrt{1 - 0} = \sqrt{1} = 1$$

When we multiply the regular standard error by this, we get the same value we started with.

EXAMPLE 9.5 Standard Error: Cost of Shopping Trips

Suppose we computed an average cost of \$35.61 and a standard deviation of \$2.52 from a sample of 36 shopping trips. The standard error of the average would then be computed as

$$\begin{aligned} \text{standard error of the average} &= \frac{2.52}{\sqrt{36}} \\ &= \frac{2.52}{6} \\ &= .42 \end{aligned}$$

giving a standard error of \$.42 or 42 cents. This number expresses the degree of variability or uncertainty in the average value of \$35.61. Note, however, that the average value itself does not appear in the standard error calculation. Note also that the average, which combines all 36 data values, has much less variability (standard error of .42) than the individual data observations (standard deviation of 2.52). Thus, although it is difficult to predict what will happen on any individual shopping trip (due to its larger variability), the average of 36 trips is much easier to predict (due to its smaller variability).

Discussion: Interpretation of the Standard Error

The standard error of the average has an interpretation that is similar to that of the standard deviation, except that it describes the variability of the average, instead of the variability of individual observations. This indicates that the average could reasonably be expected to vary by about one standard error or so in either direction (either slightly larger or slightly smaller) due to the randomness of sampling.

To understand the role of the population mean here, remember that the sample average estimates the population mean. Once we have the standard error of the average, we have an assessment of *how well the sample average estimates the population mean*. Even though we (usually) do not know the exact value of the population mean or exactly how well the sample average approximates it, the standard error provides a good guess (an estimate) of how far off the sample average is from the true, unknown population mean. But do not forget that it is only an approximate guide to the error of estimation.

In terms of repeating the entire study many times (as we discussed in Section 9.4), an interpretation can be made as follows. Imagine that in addition to the one real sample we actually have, we could (hypothetically) repeat the entire study many times. Then the standard error (which was calculated from just the one real sample) would indicate about how different the averages computed for the hypothetical samples would be from one another.

If these hypothetical averages were normally distributed, then we could say more. We would expect that about two-thirds of the averages would be within one standard error of the middle of this normal distribution. This idea is crucial to the development of statistical inference for the mean of the population (which is the center of the distribution of the averages from repeated sampling), as we will see in the next few sections.

For now, it will be enough to think about the standard error as indicating the amount of variability we should expect from the average. This indicates the precision of the average as a measure of the center of the population (the mean) and reminds us of the true uncertainty in the situation. Sometimes we would rather not be reminded and prefer to think of the average as an exact number, but this would be unrealistic. It is much better and safer to compute the standard error and recognize that, because of sampling variability, the average is not quite as exact a measure as it might have first appeared.

EXAMPLE 9.6 *Repeated Samples and the Standard Error: Pain Relief*

The comfort of a patient just after a medical operation helps ensure a safe and proper recovery. Because of variability from one person to another, the amount of pain medication required is a random quantity. The average value over many patients provides a useful description of the typical requirements and might also help a hospital be sure sufficient stock is at hand. In a study by Nunn and Slavin (1980),[2] the number of doses of pain medication required by 12 people just after an operation was reported as:

1 2 2 3 2 3 6 1 4 4 3 1

The average, standard deviation, and sample size are easily found:

average = 2.67 doses
standard deviation = 1.50 doses
sample size = 12 people

From these, the standard error of the average is computed as

$$\text{standard error of the average} = \frac{1.50}{\sqrt{12}} = \frac{1.50}{3.464} = .43 \text{ dose}$$

To properly interpret this standard error, we must identify the appropriate population from which these data were sampled. In this case (based on the title of the article from which the data were obtained), a good choice for the population might be "all patients from a certain geographical area who might need a cholecystectomy operation." A bad choice for the population might be "all patients who have any operation at all"; this is too large a group and an unrealistic choice since less serious operations would require less medication than would be represented by this sample. One might hope that conclusions could be drawn for a population of patients from a larger geographical area than the data represent, who are hospitalized for the same operation. This might be justified by arguing that there is very little difference between people from the different regions.

Once the population has been identified as "all patients from a certain geographical region who might need a cholecystectomy operation," the population mean is identifiable as the unknown mean (or average value) for the entire population. The average (from the sample) of 2.67 doses provides the estimate of the unknown population mean, whereas the standard error of the average (.43 dose) indicates about how far the average of 2.67 is from the true, unknown population mean. Of course, we do not know exactly how far the population mean is from 2.67 and we do not know in which direction (that is, whether it is too small or too large). Nonetheless, the .43 is a very useful indication of the approximate goodness of the average as an estimate of the mean: We know that we are roughly within about a half of a dose or so of the true value.

Now, let us perform a hypothetical experiment for purposes of illustration. (Such an experiment would not ordinarily be done when reporting statistical results, but it will help explain what is happening here.) Suppose we could repeat the entire study four more times, under the exact same conditions.[3] Suppose the results of these four additional studies were as follows.

First additional study
data: 6 2 1 3 4 2 1 2 2 4 3 1
average = 2.58 standard deviation = 1.51

Second additional study
data: 3 1 3 1 1 3 3 4 2 4 2 2
average = 2.42 standard deviation = 1.08

Third additional study
data: 3 1 2 4 1 6 4 3 2 2 3 2
average = 2.75 standard deviation = 1.42

Fourth additional study
data: 4 4 6 3 4 3 1 2 3 2 2 4
average = 3.17 standard deviation = 1.34

Note that the data values are randomly different from one additional study to another. Therefore, the averages and standard deviations computed for each study are also somewhat different from each other. Remember that each additional study represents a set of results that we might easily have seen in place of the actual data.

The average values from the original study and the four additional studies provide us with a new data set consisting of five average values, each computed from a sample of 12 people:

2.67 2.58 2.42 2.75 3.17

We are now at a higher level because this data set consists of averages instead of direct observations. Nonetheless, we may compute the standard deviation of this new data set in the usual way, thinking of it as a data set of five numbers. The result is

$$\text{standard deviation for the five averages} = .28$$

This number is an estimate that directly and clearly represents the variability from one average to another in the hypothetical repeated sampling. This is the number that the standard error of the average (.43, which was computed from just the first actual sample of 12 people) is also trying to capture. We are not surprised or upset that .43 and .28 are different; in fact, as an approximate indication of variability, they are fairly close. Because of randomness in sampling we would not have expected them to be exactly equal and are satisfied that they are close.

The purpose of this example has been to show how the standard error of the average, which is easily computed from just one sample, provides a reliable indication of the variability in the averages that would have been observed if we had been able to obtain extra entire samples from the population.

Exercise Set 9.4

13. Use these data to calculate a standard error for the sample average.

a. Sample average = 250, Sample SD = 36, Sample n = 16

b. Sample average = 250, Sample SD = 36, Sample n = 81

c. Sample average = 250, Sample SD = 36, Sample n = 144

14. Suppose you take a random sample of four people in a class and ask them how many siblings they have. You get

Kirby	1
Lars	3
Wanda	0
Hugh	1

a. What is the sample average?

b. What do you estimate the mean number of siblings for the whole class to be?

c. What is the standard deviation of the number of siblings in this sample?

d. What is the standard error for your estimate of the population mean?

15. Suppose you take a random sample of four people in a class and ask them how many siblings they have. You get

Ursula	5
Rachel	0
Valerie	2
Hugh	1

a. What is the sample average?

b. What do you estimate the mean number of siblings for the whole class to be?

c. What is the standard deviation of the number of siblings in this sample?

d. What is the standard error for your estimate of the population mean?

9.6 Standard Error of a Percentage

In this section, we discuss the standard error of a sample percentage. What, for instance, is the standard error in a survey showing that 48% of 700 probable voters plan to vote for candidate Smith?

In the section on binary (yes/no) data (Section 4.4), we saw how the proportion or percentage of times that something happens can be thought of as the average of a data set consisting only of 1's and 0's. This correspondence makes it easy for us to now find the standard error of a proportion (which can be turned into a percentage) because the previous results for the standard error of an average can be applied immediately to the yes/no data situation. Because we know how to compute the proportion and the standard deviation for this kind of data set, the standard error of the proportion is again just the standard deviation divided by the square root of the total number of data values (when we count the numbers of both the 0's and the 1's in the total).

We can also produce the same number by using a method based directly on the proportion of Yes responses (the average) and the total number of responses (the sample size).

PROCEDURE: COMPUTING THE STANDARD ERROR OF A PROPORTION USING THE DIRECT METHOD

1. Subtract the proportion from 1.
2. Multiply this by the proportion itself.
3. Divide by 1 less than the sample size.
4. Take the square root.

The formula is:[4]

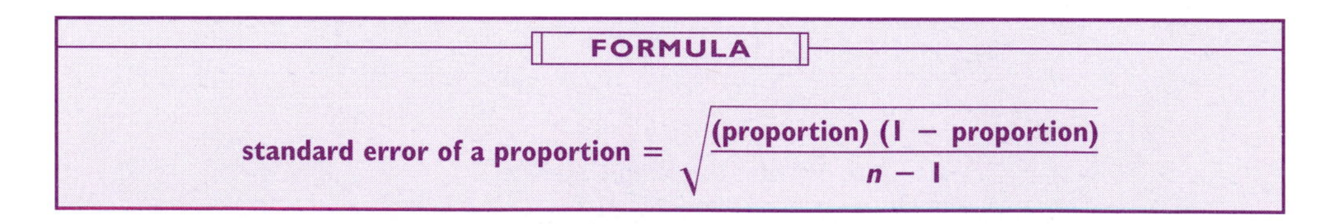

FORMULA

$$\text{standard error of a proportion} = \sqrt{\frac{(\text{proportion})\,(1 - \text{proportion})}{n - 1}}$$

The proportion can then be converted to a percent by multiplying by 100.

This direct method is especially useful for situations in which only a percentage and a sample size are given because it proceeds directly from these values.

EXAMPLE 9.7 *Standard Error of a Percent: Chicken Preferences*

The advertising column of the business section of *The New York Times*[5] once discussed an advertising campaign based on a taste test of two "premium brand names" of chicken. Out of

371 respondents who regularly served chicken at home, 197 indicated that they preferred the "Cookin' Good" brand after they had compared this brand to a different (and very popular) brand. The proportion, 197/371 = .531, indicates that of this sample, .531, or 53.1%, preferred the "Cookin' Good" brand. The precision indicated by 53.1% is probably excessive. The last decimal place, .1, is probably not reliable as an indication of what the larger population of home cooks who serve chicken regularly would have said (although 53.1% is perfectly accurate as an indication of what brand these particular 371 people preferred).

The standard error is easily computed. Using the full procedure, we first find the standard deviation:

$$\text{standard deviation} = \sqrt{\frac{197 \times (371 - 197)}{371 \times 370}}$$
$$= \sqrt{.24971}$$
$$= .49971$$

Then, based on this, we can find the standard error of the proportion:

$$\text{standard error} = \frac{\text{standard deviation}}{\sqrt{371}}$$
$$= \frac{.49971}{19.261}$$
$$= .02594 \quad \text{or approximately .026}$$

The direct computational method, based just on the proportion of .531 and the sample size of 371, proceeds as follows:

$$\text{standard error} = \sqrt{\frac{.531 \times (1 - .531)}{371 - 1}}$$
$$= \sqrt{\frac{.531 \times .469)}{370}}$$
$$= \sqrt{.00067308}$$
$$= .02594 \quad \text{or approximately .026 (as before)}$$

Ordinarily, of course, we would complete only one of these two standard error calculations, choosing the method based on which data values are handy. We computed the standard error both ways in this example only to demonstrate that they do indeed give the same answer.

The result says that the sample proportion of .531 is precise as an estimate for the population only to within about .026 to either side (that is, it could reasonably be this much larger or this much smaller than .531). To convert to percentages, both of these numbers must be multiplied by 100. The conclusion would be that 53.1%, with a standard error of 2.6%, prefer the "Cookin' Good" brand.

As we suspected, 53.1% appears much more precise than it really is. Based on the standard error of 2.6%, any values from 53.1 − 2.6 = 50.5% to 53.1 + 2.6 = 55.7% or slightly beyond could easily have been expected to occur in place of 53.1. Thus, the last decimal digit (.1) is not very reliable, and even the 3 in 53.1 is open to some question. Nevertheless, 53.1% is still the best estimate available based on the data. To acquire more precision, a larger sample would have been necessary.

The next example does use a larger sample size. Note how much more precise the average is, as indicated by its smaller standard error.

EXAMPLE 9.8 Standard Error of a Percent with a Large Sample: An Opinion Poll

Public opinion of government aid to the poor was studied by *The New York Times* and CBS in a 1982 poll.[6] Of the 1,545 adults interviewed, it was reported that 56% believe most of the poor people in the United States really do need government aid. To find the precision of this reported percentage, we convert it to the proportion .56 and find the standard error using only the proportion and the sample size (the direct method):

$$\text{standard deviation} = \sqrt{\frac{.56 \times (1 - .56)}{1{,}545 - 1}}$$

$$= \sqrt{\frac{.56 \times .44}{1{,}544}}$$

$$= \sqrt{.000015985}$$

$$= .012633 \quad \text{or appoximately } .013$$

Thus, the uncertainty caused by the random selection of 1,545 people from the population of all adults in the United States shows that the sample average value of 56% is precise to within approximately 1.3% on either side. The reported value is therefore correct to within about a percentage point or so.

Discussion

Note that Example 9.8 has greater precision (standard error of 1.3) than does (Example 9.7) (standard error of 2.6), primarily because the sample size is so much larger (1,545 compared to 371 people). Other factors being equal, because of the presence of the square root of the sample size in the computation of the standard error, to reduce the standard error by a factor of 2, it should be necessary to increase the sample size by a factor of 4. This is because the square root of 4 is 2. Unfortunately, the net effect seems to be against us: We have to make four times the effort to obtain one-half (not one-fourth) of the error.

Indeed, in these two examples the ratio of sample sizes is 1,545/371 = 4.16, which is approximately 4 and did result in half the standard error (from 2.6 to 1.3). This comparison is valid partly because the percentages were not very different from each other (56% compared to 53.1% for the two examples). The fact that one example is measuring something about poverty and the other is measuring something about chickens does not matter to the computations; the underlying statistical structure of the problem is the same in each example.

The standard error directly indicates the reliability of the computed percentage by indicating about how far it is likely to be (on either side) from the true or actual value for the entire population. The interpretation of the standard error for the percentage is the same as the interpretation of any standard error of an average.

To claim that "91% of doctors recommend . . ." is usually pretending to have precise information when, in fact, the figures are only an estimate. With the standard error, one might be able to say, for example, that "91% (with a standard error of 2.8%) of doctors recommend something." The standard error adds the information that the 91% is not completely precise, but is known only to within approximately 2.8% on either side (in the sense that the standard deviation of something indicates the variability of that something). From now on, whenever you see percentages being reported remember that they are not as exact as they might seem.

Also, remember that these formulas apply to simple random samples, where each member of the population has an equal chance of being selected. However, many studies use stratified or cluster samples. In a **stratified sample,** a population is classified on a variable (say, income), and then a certain number are sampled from each stratum (so many upper, so many middle, and so many lower income). In a **cluster sample,** some population unit other than the individual is used for a first round of sampling (say, a random sample of classrooms in a school district), and then individuals are randomly selected within these clusters (say, ten students from each selected class). All of these designs require corrections in the calculation of the standard error.

Stratified and cluster samples are likely to be used when either (a) some information is already available about the population (like the incomes of the individuals) and stratification can be used to improve the statistical efficiency of the study, or (b) interviewers are conducting face-to-face interviews and there are substantial time or travel costs involved.

Something close to a simple random sample can be conducted on the telephone using random-number dialing, and many opinion surveys are conducted this way. In such cases, the formulas in this chapter are appropriate.

Exercise Set 9.5

16. What is the standard error of the proportion, if the sample proportion is .45 and the sample size is 25?

17. What is the standard error of the proportion, if the sample proportion is .45 and the sample size is 100?

18. Suppose you take a random sample of four from a (very large) class and ascertain their genders. You get these results:

Kirby	M
Justin	M
Wanda	F
Hugh	M

a. What is your sample proportion of males?

b. What do you estimate the class proportion of males to be?

c. What is the standard error of your estimate of the class proportion?

19. Suppose you take a random sample of four from a class and ascertain their genders. You get these results:

Valerie	F
Mike	M
Pat	F
Barclay	M

a. What is your sample proportion of males?

b. What do you estimate the class proportion of males to be?

c. What is the standard error of your estimate of the class proportion?

20. Suppose you conduct a random telephone survey and find that 45% of 400 respondents think the local mayor is doing a good job. What is the standard error of the percent of "favorable" responses?

21. Exercises 16, 17, and 20 all dealt with samples where the sample proportion was .45, or 45%.

a. What is the relationship of the standard errors in these problems?

b. How do you explain this relationship?

Summary

Statistical *inference* takes us beyond a summarization of the data itself and, using probability, allows us to make generalizations about the larger situation from which the data were selectively obtained. Usually, the population is too large to be completely studied, so a smaller sample is used instead. Statistical methods help us make general statements about the population based on this smaller sample. Although these inferences are not always correct, with statistical reasoning we are able to produce statements that are "probably true" or "likely to be true," together with some indication of the likelihood of correctness. Some background is needed in populations, samples, and the sampling behavior of the average to set the stage for these inferences.

A *population* can be any collection of identifiable things (often called *units*) that could conceivably be measured and are of interest. A *sample* is any collection of units from the population for which information is available. The sample is what we know about; the population is what we would like to know about. The number of units in the sample is called the *sample size*. A sample can be chosen in many different ways. A good sample will be representative of the characteristics of the population, whereas a bad sample will not. A bad sample is said to be *biased*.

One way to choose a good sample is to use a *random sample*. This is an experimental design in which (1) each unit of the population has an equal chance of being selected and (2) units in the sample are chosen independently of each other. Random samples tend to be representative of the population, and they introduce a controlled amount of randomness into the situation that can be exploited later by the methods of statistical inference. A random sample can be chosen by any of several methods, including the use of a *table of random numbers*.

In estimating a population value by using a sample from that population, we call the population value a *parameter* and the sample value used to estimate it a *statistic*. The *population mean* is the arithmetic average of the entire population (a parameter that is usually unknown), whereas the *sample average* is the usual arithmetic average computed from the sample (a statistic that is known). Thus, the (known) sample average provides us with an estimate of the (unknown) population mean. This distinction between these two words (average and mean) is helpful, but is not universally accepted. When the average is computed for a random sample, it may be regarded as a random quantity in itself because different random samples result in different average values. This *sampling variability of the average* forms the basis for many methods of statistical inference about the population mean.

The *standard error of the average* is found by dividing the (usual) standard deviation by the square root of the sample size. It gives an indication of the variability of the sample average. In the same way that the standard deviation summarizes variability for individual values in a data set, the standard error of the average summarizes the variability of the different average values that would have been obtained from entirely different samples. The standard error therefore indicates approximately how far the sample average is from the population mean. The standard error is never larger than the standard deviation, reflecting the fact that the sample average (which combines information from the entire sample) is a more precise estimate of the population mean than the observations themselves taken individually.

The *standard error of a percentage*, based on binary (yes/no) data that follow a binomial distribution, is computed as follows:

$$\sqrt{\frac{(\text{proportion})\ (1 - \text{proportion})}{n - 1}}$$

Problems

Basic Reviews

1. Identify the population and the sample in each of the following situations.

a. To test the safety and effectiveness of a new treatment for migraine headaches, a study is performed on some headache patients at a clinic in Boston.

b. To study the problem of noise near airports, monitoring equipment is set up at 20 selected airports in the United States.

c. To assess compliance with the tax laws, the Internal Revenue Service randomly selects some households. For each household, the IRS performs a detailed audit of the tax return.

2. Suppose you have a random sample of 4 lines of text taken from a term paper. The lines average 12 words with a standard deviation of 4 words. What is the standard error for the average number of words per line?

3. Suppose you have a random sample of 16 lines of text taken from a term paper. The lines average 12 words with a standard deviation of 4 words. What is the standard error for the average number of words per line?

4. Suppose a shipment of nine bicycles has just arrived at your store, and you would like to open the boxes before you sign the shipping form just to make sure that everything is all right. However, there is not enough time to open all of them before the delivery people would become exasperated. You have decided to choose three at random and open only those. The shipping labels on the boxes are listed as follows:

A71 B35 B49 C82 D11 D13 D35 F85 G19

a. Select three of these at random, starting with the first entry of the 11th row of the table of random numbers in the appendix.

OR

b. Select three of these at random, using a computer program such as Minitab.

c. (*Bonus question*) How exasperated do the delivery people get while you do (a) or (b)?

5. Using the alphabetical list of abbreviations for the 50 states of the United States shown at the top of page 316, select 8 states at random. You may use the table of random numbers, or you may use a computer. If you use the table of random numbers, say where you started. If you use a computer, include the computer instructions with your answer.

AK	AL	AR	AZ	CA	CO	CT	DE	FL	GA
HI	IA	ID	IL	IN	KS	KY	LA	MA	MD
ME	MI	MN	MO	MS	MT	NC	ND	NE	NH
NJ	NM	NV	NY	OH	OK	OR	PA	RI	SC
SD	TN	TX	UT	VA	VT	WA	WI	WV	WY

6. Display 9.2[7] shows a computer analysis of the tree data originally presented in Chapter 6, where we looked at stem-and-leaf plots and compared the means, medians, standard deviations, and interquartile ranges. For this question, assume that the trees in the data base are a random sample from a particular forest plot.

DISPLAY 9.2

ANALYSIS OF TREE DATA

MTB > Describe 'DIAMETER'-'VOLUME'.

	N	MEAN	MEDIAN	TRMEAN	STDEV	SEMEAN
DIAMETER	31	13.248	12.900	13.156	3.138	0.564
HEIGHT	31	76.00	76.00	76.15	6.37	1.14
VOLUME	31	30.17	24.20	28.87	16.44	2.95

	MIN	MAX	Q1	Q3
DIAMETER	8.300	20.600	11.000	16.000
HEIGHT	63.00	87.00	72.00	80.00
VOLUME	10.20	77.00	19.10	38.30

a. What is the best prediction of the mean diameter of trees in that plot?

b. About how far is your prediction of the mean diameter likely to be from the true mean diameter (if you ever establish it)?

c. What is the best prediction of the mean height of trees in that plot?

d. About how far is your prediction of the mean height likely to be from the true mean height (if you ever establish it)?

e. What is the best prediction of the mean volume of trees in that plot?

f. About how far is your prediction of the mean volume likely to be from the true mean volume (if you ever establish it)?

Analysis

For problems 7–10, use Table 9.4[8] as though it listed an entire population, even though it is much smaller than most of the populations we deal with. It is, in fact, the population of the 10 largest cable TV operators in 1983. The purpose of this exercise is to understand what happens when we can see the population, so we may gain intuition about what happens in the more usual situation when we cannot see the whole population.

7. Find the population mean, which is simply the arithmetic average of these population values.

8. Draw a random sample of four cable operators. Use the table of random numbers and say where you started, or use a computer and include the instructions.

Table 9.4 Number of franchises of the ten largest cable system operators in 1983

ID NUM	NAME	NUMBER OF FRANCHISES
1	TeleCommunications	428
2	Time Inc.	119
3	Westinghouse	140
4	Cox Cable	58
5	Warner Amex	146
6	Storer Communications	122
7	Times Mirror	67
8	Rogers/Rogers UA	22
9	Newhouse	64
10	Continental	66

a. Compute the average for the sample.

b. Compute the standard deviation for the sample. State what this measures the variability of.

c. Compute the standard error of the average for the sample. Explain what this measures the variability of.

d. Compute the difference between the sample average and the population mean. (Remember that we usually would not have access to the mean value, but for this small population, we do.) Compare this difference to the standard error of the average.

9. Draw a second random sample from the companies listed in Table 9.4. List them and their numbers of franchises.

a. Compute the average for the sample.

b. Compute the standard deviation for the sample. State what this measures the variability of.

c. Compute the standard error of the average for the sample. Explain what this measures the variability of.

d. Compute the difference between the sample average and the population mean. Compare this difference to the standard error of the average.

10. Briefly compare the samples drawn in Problems 8 and 9 with each other and with the population.

11. Suppose a (hypothetical) study has found that 256 of 895 students randomly selected to be interviewed prefer morning to afternoon classes.

a. Express this data set as a percentage of people who prefer morning classes.

b. Find the standard error of the *proportion*.

c. Interpret the standard error in terms of the *percentage* computed in part (a).

12. Suppose a (hypothetical) study based on 92 randomly selected adults has found that 14% of them prefer their toast very dark.

a. Compute the standard error using the direct method.

b. Interpret its meaning with respect to the number 14%.

c. Based on this same situation, compute the standard error of 86%, the percent who do not prefer their toast very dark. Compare this standard error to your answer to part (a).

13. To see the effect of a change in sample size, suppose a larger study of 828 people (nine times as many as in Problem 12) found that 15.1% prefer their toast very dark. (Note that, from sample to sample, results such as averages usually change somewhat.) Find the standard error of this 15.1%. Compare it to the standard error from Problem 12a. In particular, how much more precision (ratio of standard errors) is obtained from a ninefold increase in sample size?

14. Following are some measurements of the wing length (in millimeters) of killer bees from French Guiana.[9]

8.56 8.51 8.51 8.56 8.96 8.82 8.39 8.54 8.62

a. Calculate their average and standard deviation.

b. Assuming this is a random sample, what is your best estimate of the mean wing length for all killer bees?

c. What is the standard error of this estimate?

15. In some earlier exercises, we used data on the first 15 summits listed in the *AMC White Mountain Guide*, and the high points of the first 15 hikes in *Best Hikes with Children in Western Washington.* Does just choosing the first 15 items sound like a good sample design? What kinds of bias could it introduce? (Be ruthless.) If our intention was to get an estimate of the altitudes mentioned in these two books, what should we have done?

16. We took a simple random sample of nine hikes from *Best Hikes with Children.* Here are their high points:

848 2,900 900 2,981 2,300 5,180 3,982 3,161 5,699

a. Calculate their average and standard deviation.

b. What is the best estimate (based on this sample) of the mean high points of the hikes in this book?

c. What is the standard error of this estimated mean high point?

17. We took another simple random sample of nine hikes from *Best Hikes with Children.* Here are their high points:

3,982 300 4,508 6,000 2,061 3,982 3,800 1,120 4,900

a. Use this sample to estimate the mean high point of all the hikes in this book.

b. State your conclusion and provide an estimate of sampling error in a sentence or two.

18. We took yet another simple random sample of nine hikes from *Best Hikes with Children.* Here are their high points:

5,400 4,500 5,092 3,700 700 5,400 3,800 4,461 5,180

a. Use this sample to estimate the mean high point of all the hikes in this book.

b. State your conclusion and provide an estimate of sampling error in a sentence or two.

19. *(True story)* The board of directors at a small social work agency decided to do a survey of the agency donors, to help them better focus their fund-raising efforts. Some board members drew up a questionnaire and had an administrative assistant at the agency mail it to 300 people on the list of donors. The agency received 54 questionnaires back and asked the administrative assistant to compile the results.

a. In principle, does this survey appear to be vulnerable to any bias in its results?

b. It happens that over half the people who responded were over 70 years old. What does this suggest about the representativeness of the results?

20. Do the sailboat data in Table 2.19 appear to be a true random sample from a larger data set? Why or why not?

21. Suppose you wanted to get more information on the sailboats in the sailboat data set from Table 2.14, but you do not have the time or patience to collect it for all of them. Describe how you would go about picking a group of sailboats for further examination, such that you would have a known relationship between these sailboats and the population of sailboats.

22. This problem is based on the work of Deming.[10] Suppose that a company grows crystal beads in a vat. Due to impurities in the raw ingredients, 20% of the beads are defective and will have to be rejected. With each production run, each worker in the bead plant stirs the bead vat, reads a saying or two from the collected works of Adam Smith, then randomly selects 100 beads from the vat. Without looking at the beads, the workers bring them to the inspector, who identifies defective beads. Workers with more than 25 defective beads are placed on probation.

a. About what percent of the workers will be placed on probation, per run?

b. Would the rate of defective beads decrease if the workers chanted the sayings of Adam Smith, rather than simply reading them?

c. Would the rate of defective beads decrease if management decided to "get serious" and fire workers with more than 25 defective beads?

d. Based on the information in the problem, what should the company do to cut the rate of defective beads?

23. Suppose that the probability that a newborn baby is male is 52%. In one hospital in the month of July there are 30 live births; in another hospital in the same month there are 120 live births. In which hospital is there a better chance that half or more of the newborns are female? Explain.

24. A *quota sample* is one in which the investigators assign themselves a quota for the number of cases meeting selected demographic criteria. The basic idea is that if the sample matches the population on these criteria—the same percent female, the same percent college educated, whatever—the chances are improved that the sample will match the population on the characteristics being studied. Unfortunately, quota samples are wide open to response bias. *The Janus Report* (1993),[11] for instance, is based on 2,795 people who volunteered to be interviewed about their sex lives. The investigators made sure that the age distribution in their sample matched the age distribution in the United States population; to make sure they had enough older people in the sample, they interviewed older clients at sex therapy clinics. The study says that 72% of males and 68% of females age 18–26 report having sex at least once a week, and that 69% of males and 75% of females over 65 report having sex at least once a week. On the other hand, results from a carefully constructed random survey[12] report much lower figures: 57% of the males and 58% of the females age 18–26, and 17% of the males and 6% of the females over age 65, report having sex once or more a week. Explain this discrepancy. Which figures are more likely to be right?

chapter 10

Confidence Intervals

Now that we have developed the background ideas such as populations, samples, and standard errors that are necessary for doing statistical inference, it is time to put these ideas to work. The *confidence interval* provides a definite, exact statement about the unknown population mean based on the average and variability computed from a sample. Rather than the indefinite statement "the average is about a standard error from the population mean," with the confidence interval we can be more precise and say that "we are 95% certain that the population mean is within a certain distance from the average."

Because the confidence interval statement is more definite, it will often require some extra assumptions. Be sure to pay attention to these assumptions when they come up. Also be sure to keep in mind the basic framework of population and sample, in which the population mean is an unknown (but fixed) number and the sample average is a known (but random due to random sampling) number.

This implies that the confidence interval itself will be a *random interval* (even though for any particular random sample, it will not look random). Such random intervals will contain the fixed, nonrandom, but unknown population mean 95% of the time.

10.1 Confidence Interval for the Mean of a Normal Distribution

Consider an experimental treatment for high blood pressure that achieved a 12.3-point reduction, on average, for 25 randomly selected patients from a larger population of patients. While this average blood pressure reduction, 12.3, is an important summary number, it is not really what we want (although it is the best estimate we have). Instead of how well the treatment worked for these particular 25 patients, we would like to know how it would have worked (on average) had we been able to apply it to the entire population of patients. It is somewhat remarkable that we can obtain information about this unobserved population number using only information from the sample.

Once the confidence interval is calculated, we can make statements like this: "We are 95% sure that, had we applied the treatment to the population from which the 25 patients were randomly sampled, the population mean blood pressure reduction would have been somewhere between 7.7 and 16.9 points." Thus the confidence interval gives an indication of how accurate the sample average (12.3) is as an estimate of the population mean by making an exact confidence statement about this unknown quantity.

Sidebar 10.1

THE *t*-TABLE

The *t*-table expresses two facts about random averages. First, because they are approximately normally distributed (remember the central limit theorem), they will be about 2 of their own personal standard deviations (that is, the standard deviation of the sample average) from their mean value about 95% of the time. This is why most values in the *t*-table are approximately 2, more or less. Second, because we use the standard error (a sample statistic, the estimated variability) in place of the unknown standard deviation of the sample average (a population parameter, the true variability), 2 is not large enough to achieve the full 95%. This is why values in the *t*-table are larger when the sample size n is smaller. This larger *t*-table value expands the confidence interval to reflect the lack of precision when the standard deviation is estimated with less information from a smaller sample. Figure 10.1 shows how *t*-curves compare to the normal curve.

We begin with the *t*-table approach because it covers the general case of sampling from a population with unknown variability and, at the same time, covers the particular case (treated in Section 10.3) where the population variability is known exactly.

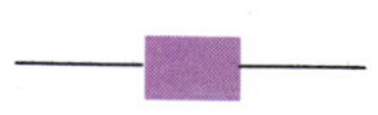

Finding a 95% Confidence Interval

Suppose we have a group of numbers that we can interpret as a random sample from a population with a normal distribution (this is the only assumption necessary here, but it assumes quite a lot). Suppose the average value is 9.7. We know that this is the best single estimate of the population mean; we also know that it is probably not exactly equal to the population mean. We will use the sample information on variability to construct an interval that has a known probability of including the population mean. This **95% confidence interval** (CI) for the mean

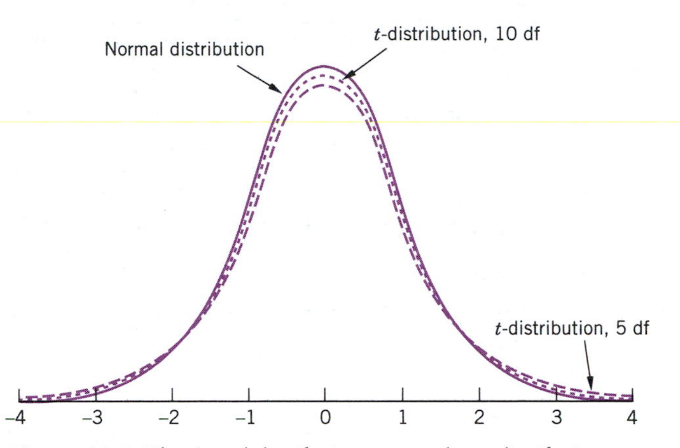

Figure 10.1 *The normal distribution compared to t-distributions with 5 and 10 degrees of freedom (df). Note that t-distributions have longer tails than the normal distribution. This is why, to achieve the same probability, we find t-table values that are larger than we would use for a normal distribution. Note also that when the degrees of freedom number is large, t-distributions are close to the normal distribution.*

extends on each side of the average by an amount reflecting the variability of the average. The procedure is as follows.

PROCEDURE: FINDING A 95% CONFIDENCE INTERVAL

1. From the data, find the average, the standard deviation, and the standard error of the average.
2. Subtract 1 from the sample size to find the **degrees of freedom** (df). Look in the *t*-table (Table 10.1) to find the corresponding entry, called the **t-value** or the **critical value.**

Table 10.1 *t-table: Two-sided 5% critical values for 95% confidence intervals (also one-sided 2.5% critical values for 97.5% one-sided confidence intervals)*

DEGREES OF FREEDOM	CRITICAL VALUE	DEGREES OF FREEDOM	CRITICAL VALUE	DEGREES OF FREEDOM	CRITICAL VALUE
1	12.706	11	2.201	21	2.080
2	4.302	12	2.179	22	2.074
3	3.182	13	2.160	23	2.069
4	2.776	14	2.145	24	2.064
5	2.571	15	2.131	25	2.060
6	2.447	16	2.120	26	2.056
7	2.365	17	2.110	27	2.052
8	2.306	18	2.101	28	2.048
9	2.262	19	2.093	29	2.045
10	2.228	20	2.086	30	2.042
				40	2.021
				50	2.009
				60	2.000
				120	1.980
				Infinity	1.960

3. Multiply this *t*-value (from the table) by the standard error to get a margin of error.
4. Subtract this margin of error from the average value to find the lower confidence limit. Add the margin of error to the average value to find the upper confidence limit. These two limits specify the 95% confidence interval.

Formula for the Confidence Interval

This procedure can also be written as a mathematical formula. If we let t_{n-1} represent the *t*-value from the *t*-table corresponding to $n - 1$ degrees of freedom, then the 95% confidence interval is from

$$\text{average} - [t_{n-1} \times \text{standard error}]$$

to

$$\text{average} + [t_{n-1} \times \text{standard error}]$$

In terms of mathematical symbols for the average, the standard deviation, the sample size, and the *t*-value, a formula for the 95% confidence interval may be written as follows (note that the square root of the sample size is required to convert the standard deviation to the standard error).

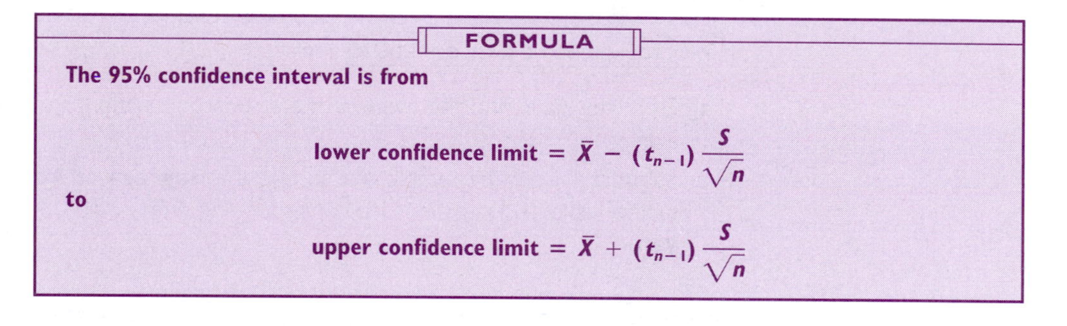

FORMULA

The 95% confidence interval is from

$$\textbf{lower confidence limit} = \bar{X} - (t_{n-1})\frac{S}{\sqrt{n}}$$

to

$$\textbf{upper confidence limit} = \bar{X} + (t_{n-1})\frac{S}{\sqrt{n}}$$

The t-Table

A statistical table contains the results of long and extensive computations and saves us a good deal of trouble when we perform a routine statistical calculation (such as a confidence interval for the mean).

As noted in step 2 of the procedure, we find the *t*-value for a 95% confidence interval in a table of critical values such as Table 10.1. For example, the *t*-value corresponding to 6 degrees of freedom is 2.447. The *t*-value is a *two-sided* value because the confidence interval extends on both sides of the average. The 5% refers to the proportion of time we make the error of not including the population mean within the confidence interval. Being 95% sure means being wrong 5% of the time.

Not every value of the degrees of freedom is listed in the table. If the value we need is not listed, then an approximation will be close enough. For example, the *t*-value corresponding to 35 degrees of freedom is somewhere between 2.042 and 2.021, which are the *t*-values listed for 30 and 40 degrees of freedom, respectively. So a value of 2.03 might be appropriate to use for 35 degrees of freedom.

A safe, conservative way to proceed when the exact number of degrees of freedom is not listed is to use the t-value for the next smaller number of degrees of freedom that is listed. For example, with 35 degrees of freedom, it would be safe to use 2.042 (the value for 30 degrees of freedom). This approach is conservative in the sense that the resulting confidence interval will cover at least 95% of the time and possibly more often.

EXAMPLE 10.1 *Calculating a 95% Confidence Interval: Killer Bees*

Killer bees do not appear very different from ordinary honeybees. Without observing their behavior, we need careful statistical analysis of many different measurements to distinguish one kind from the other. Identifying dead killer bees can help track their spread to new territories.

From a study to characterize a population of killer bees from French Guiana, a sample of bees had wing widths as shown in Table 10.2.[1]

From this data set, we can compute the following basic relevant quantities:

sample size = 20 bees
average = 2.858 mm
standard deviation = .0516 mm
standard error = .0115 mm

From step 2 of the confidence interval procedure, we subtract 1 from the sample size to find the degrees of freedom. We have

$$20 - 1 = 19 \text{ degrees of freedom}$$

Looking up the value in the t-table for 19 degrees of freedom, we find 2.093. For step 3, we multiply this by the standard error.

$$2.093 \times .0115 = .024$$

In step 4, we add and subtract this from the average:

$$\text{lower 95\% confidence limit} = 2.858 - .024 = 2.834$$
$$\text{upper 95\% confidence limit} = 2.858 + .024 = 2.882$$

Finally, we are ready to report these two values as a confidence interval:

The 95% confidence interval is from 2.834 *to* 2.882.

We are 95% sure that the mean wing width for this population is between 2.834 and 2.882 mm.

Table 10.2 *Wing widths, in mm, for 20 killer bees from a population in French Guiana*

2.88	2.94	2.83	2.91	2.86
2.83	2.80	2.88	2.91	2.88
2.97	2.80	2.77	2.80	2.88
2.83	2.80	2.86	2.86	2.88

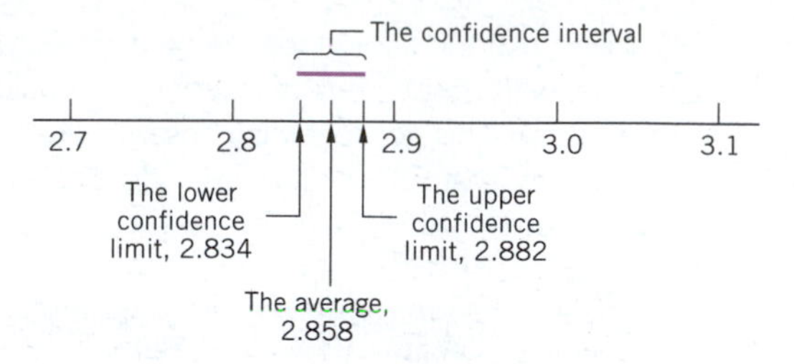

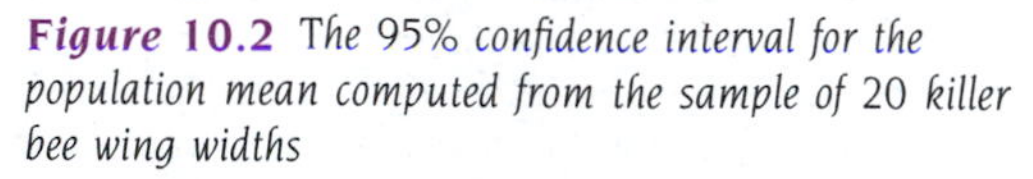

Figure 10.2 *The 95% confidence interval for the population mean computed from the sample of 20 killer bee wing widths*

This confidence interval is displayed in Figure 10.2 to show how it represents an interval of possible values in which we hope the unknown population mean will be. For reference, Figure 10.3 provides a histogram of the data itself. Notice that most of the data values are not inside the confidence interval. They do not have to be; it is the population mean that we intend to capture within this interval. Remember that the confidence interval is built up from all of the data in the sample and reflects the combined values, not the individual data values. For a larger sample size, the average and the standard deviation should be about the same (except for randomness), but the standard error and the confidence interval should be smaller (reflecting the increased amount of information contained in the larger sample).

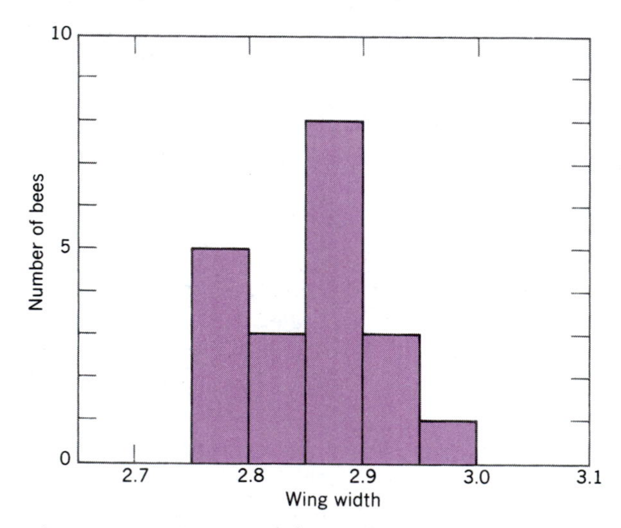

Figure 10.3 *Histogram of the sample data of 20 killer bee wing widths*

Exercise Set 10.1

1. If you have a sample of $n = 3$, how many degrees of freedom (df) do you have?
2. If $n = 4$, what does df equal?
3. If $n = 4$, what t-value do you use when calculating a 95% confidence interval?
4. Suppose you have a sample of $n = 4$, with a sample SD of 12. What is the standard error for the sample average?

5. Suppose the sample of 4 with an SD of 12 has a sample average of 25.

a. What is the lower bound for a 95% confidence interval?

b. What is the upper bound for a 95% confidence interval?

6. a. Find a 95% confidence interval (CI) when you have a sample of size 9, a sample average of 150, and a sample standard deviation of 12.

b. Find a 95% CI when you have a sample of size 9, a sample average of 150, and a sample standard deviation of 21.

c. What is different and what is the same in the answers to parts (a) and (b)? Why is there a difference?

7. a. Find a 95% CI when you have a sample of size 36, a sample average of 150, and a sample standard deviation of 12.

b. What is different and what is the same when comparing this interval to the interval calculated in Problem 6a? Why is there a difference?

8. Suppose you take a random sample of 36 athletes who turn out for men's basketball at all major universities. Your sample averages 75 inches tall with a standard deviation of 3 inches. Find a 95% confidence interval for the population mean. (*Note*: There is no entry for df = 35 in the table. You can use the entry for df = 30, which gives a *t*-value bigger than you really need. This is conservative, in the sense that your error band will be a little wide. You are not claiming more precision than you really have.)

Discussion: Interpretation of the Confidence Interval

The simplest statement that can be made about the confidence interval we computed for the killer bees is the following.

> *We are 95% sure that the mean wing width for the entire population of killer bees in French Guiana falls within the computed confidence interval.*

Notice that this is a statement about an unknown quantity, namely, the mean wing width for a very large and unmeasured population of bees in French Guiana. But if these relatively few bees are a random sample, then we are able to make this very definite statement about our estimate and its uncertainty.

But how definite is the confidence statement really? With randomness we cannot expect to be completely sure of anything, but the 95% confidence allows us to make a definite statement in an uncertain situation. The general facts about a confidence interval may be stated as follows.

1. We do not know for certain whether the true population mean is located inside or outside the computed confidence interval.
2. The confidence interval is random in the sense that if we repeated the entire experiment, we would have a different random sample and hence a different confidence interval.
3. The chances are 95% that the true population mean will be located within the computed confidence interval.
4. Conversely, the chances are 5% that the true population mean will not be located within the computed interval.
5. Unfortunately, we will never know (for this particular confidence interval) whether we were right or wrong in claiming that the population mean was within the interval.
6. These statements can break down if the assumptions are not satisfied. In particular, if the sample were not a random sample, then our conclusions might be incorrect because the sample might not be representative of the population. In addition, if it were not a random

sample, the 95% claim might be incorrect because it relies on the "controlled randomness" introduced when a random sample is drawn.

Discussion: What Does 95% Confidence Really Mean?

The desired confidence level is our choice; it is not the result of a calculation, but a subjective decision on our part. The usual choice of 95% as a level of confidence used to construct confidence intervals is a statistical convention based partly on tradition. We certainly want to have a high degree of confidence in our statements, and ideally, we would like to be 100% certain of what we claim. Unfortunately, due to randomness it is impossible to be completely certain, so we settle for some level of confidence less than but reasonably close to 100%.

The particular choice of 95% sets a standard that says we will be right .95 of the time, or 19 times out of 20. Many people believe that to be wrong more than 1 time out of 20 is too often (although occasionally 90% is used for some difficult problems with large amounts of randomness). Often, the data with their randomness prevent us from routinely demanding a higher level than 95%, although for some particularly nice situations with small amounts of randomness, you might see 99% or even 99.9% confidence levels used. However, the 95% level is fairly universally accepted and used as a standard value.

Now that we know where 95% came from as a statistical tradition, what does it really mean? In particular, it seems that whenever we compute a confidence interval, it either does contain the unknown mean or else it does not, and we do not know which is the case. How then do we interpret this 95%?

One interpretation proceeds by imagining that the entire experiment is repeated many times. For each repetition, a different random sample would be drawn and analyzed, and a different confidence interval would be computed. The mathematical theory behind the confidence interval (and partly represented in the *t*-table that we use) says that out of many repetitions of the experiment, the true population mean would be contained inside the (random) confidence intervals 95% of the time. Figure 10.4 shows the result of an actual experiment, whereas Figure 10.5 illustrates the hypothetical results of 50 repetitions of that experiment.

Note in Figure 10.4 that the confidence interval just happened to contain the population mean (although we usually would not know whether it did or not). However, Figure 10.5 shows

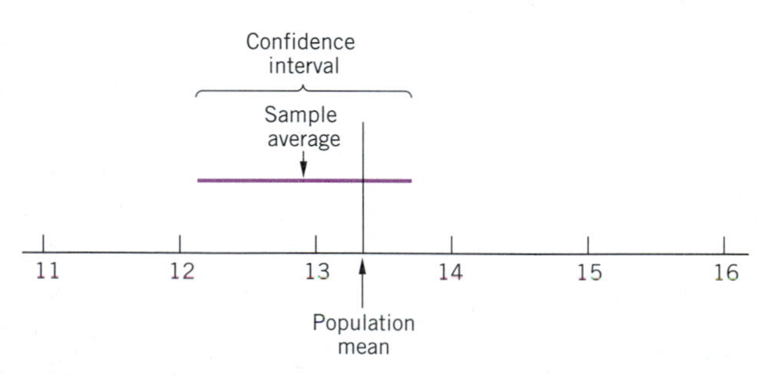

Figure 10.4 *An example of a 95% confidence interval. The interval is centered at the sample average. In this particular case, we have been lucky: The population mean just happens to be inside the confidence interval. (Note: The populaton mean is indicated for illustration only; in practical situations, the population mean is unknown and we do not know whether it is within the confidence interval or not.)*

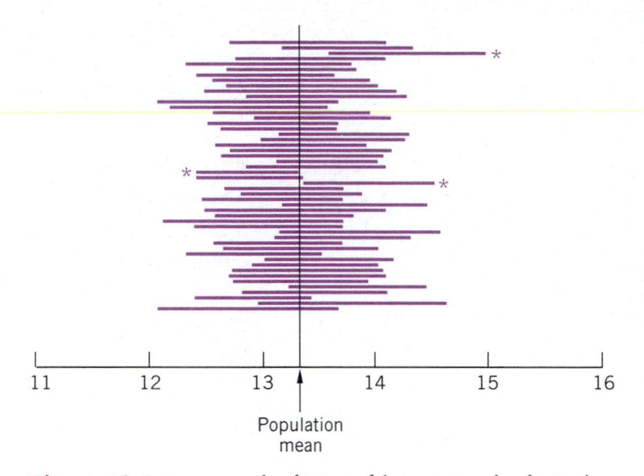

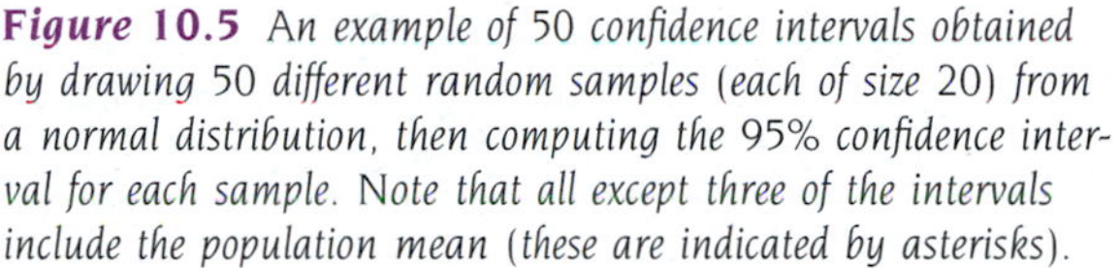

Figure 10.5 *An example of 50 confidence intervals obtained by drawing 50 different random samples (each of size 20) from a normal distribution, then computing the 95% confidence interval for each sample. Note that all except three of the intervals include the population mean (these are indicated by asterisks).*

50 confidence intervals that might have been computed instead of the one shown in Figure 10.4. Note that 3 out of these 50 intervals, or about 6% of them, did not contain the population mean. When we compute a confidence interval, it is in this sense that we have a 95% chance of containing the population mean: Out of the many samples that might have been chosen, 95% of them contain and 5% of them do not contain the population mean.

A second interpretation of the 95% level is to consider your "track record" as a statistician. Over your lifetime, you may well compute many different confidence intervals for many different situations. Although you will never know whether you were right (in the sense of having captured the population mean within the interval) in any given case, you will have the satisfaction of knowing that your track record consists of 95% successes and only 5% failures. The uncertainty about which particular ones were successes does not trouble statisticians. We simply do the best we can in a random world where exactness is rarely possible.

Computer Calculation of 95% Confidence Intervals

Using Minitab as an example, we first enter the killer bee wing width data into a file. Before proceeding with the calculation of a confidence interval, we should check our assumptions.

The first assumption is that the data are a random sample from the larger population. In most cases we cannot check this assumption, so we either rely on the assurance of the original investigators or we decide that for the purposes of carrying out the calculation we simply will accept it.

The second assumption is that the data are from a distribution that is fairly normal. More particularly, we assume the absence of erratic extreme values or outliers. The issue is not really normality; a rectangular or a pyramid-shaped distribution would often be fine. We can check this assumption; for example, the stem-and-leaf plot of the data in Display 10.1 (page 330) shows a fairly normal distribution of wing widths, so we can proceed.

The program for calculating a confidence interval assumes that we want a 95% CI. (You may specify other confidence levels.) The output for a 95% CI is given in Display 10.2 (page 330).

DISPLAY 10.1

COMPUTER PLOT OF KILLER BEE WING WIDTHS

```
MTB > STEM-
AND-LEAF
'WIDTH'.
STEM-AND-
LEAF OF
WIDTH N =
20
 1      27      7
 1      27
 5      28      0 0 0 0
 8      28      3 3 3
 8      28
(3)     28      6 6 6
 9      28      8 8 8 8 8
 4      29      1 1
 2      29
 2      29      4
 1      29      7
```

DISPLAY 10.2

COMPUTER-GENERATED 95% CI FOR WING WIDTHS

```
MTB > TInterval 95.0 'Width'.
              N     MEAN      STDEV     SEMEAN    95.0 PERCENT C.I.
WIDTH        20    2.8585    0.0516    0.0115    (2.843, 2.8827)
```

Looking at this display, we see several familiar elements. We have the name of the variable (from the top of the data column) at the very left. The number of data points, 20, is under N. The sample average, called the MEAN by this program, is next, followed by the standard deviation of the individual data values in the sample. Then we have the standard error of the mean—the amount by which we expect the averages of samples of 20 to differ from the true population mean.

Finally, without printing the intermediate *t*-value, the program produces the 95% CI. The results agree with our results, within rounding error. The program computes the *t*-value for a 95% CI for a sample with $n - 1$ degrees of freedom and uses it, but does not print it.

10.2 The Confidence Interval When the Standard Error Is Known Exactly

When the standard error of some estimated value is known exactly, there is no need to estimate it from the sample. This might happen if we know the variability of the estimated value from a large amount of past experience. For example, we might give an aptitude test with a known standard deviation to a small sample of students. Because we already know the standard deviation, we do not have to estimate it from the small sample. The estimated value might be an average (as we considered earlier), but might also be a single observation or even the result of a complex calculation.

To form the 95% confidence interval in this case, we proceed much as we did before from

$$\text{lower confidence limit} = \text{estimated value} - [1.96 \times \text{standard error}]$$

to

$$\text{upper confidence limit} = \text{estimated value} + [1.96 \times \text{standard error}]$$

where we use the t-value in the table for "infinity," that is, for a standard error that was estimated for a large sample. This is because by assuming that the standard error is known exactly, we are assuming that it represents the perfect information available from a very large amount of data.

EXAMPLE 10.2 *Known Variability: Measuring Water Quality*

Suppose that occasionally when you turn on the water for the first time in the morning, you notice that it is slightly discolored. Although the discoloration is subtle, you seem to notice it each morning and especially when you return from a trip. You decide to have the water tested to see whether it is a serious problem needing correction. From consulting with a local laboratory, you find that it may just be a small amount of iron from the pipes. (You hope that this is the explanation because it would not pose a health hazard.) To find out for sure, you request that the laboratory perform a test.

The analysis of the sample of slightly discolored water reveals an iron concentration of 8 parts per million.

However, being a statistically aware person, you realize that analytical instruments are not perfectly accurate. In the fine print of the report, you find a statement to the effect that the instruments that produced the measurements were carefully calibrated and adjusted so that

the standard error of measurement is .9 part per million.

This is the standard error associated with the concentration measurement. You consider the standard error of .9 part per million to be fixed and known because it is based on extensive past experience on the part of the laboratory (not just from a few data values). You can now easily compute the 95% confidence interval. The standard error times the t-value for infinity is

$$.9 \times 1.96 = 1.8$$

so the 95% confidence interval goes from

$$8 - 1.8 = 6.2 \text{ parts per million of iron}$$

to

$$8 + 1.8 = 9.8 \text{ parts per million of iron}$$

Now you are 95% sure that the iron concentration in your water is between 6.2 and 9.8 parts per million. Because this represents a high concentration of iron, you decide that the discoloration problem is probably just iron from the pipes. You also decide to have another test done on the clear water later in the day, so that you can be more certain that your understanding of the situation is correct. This would also make good statistical sense: to design a careful comparative experiment to see whether iron is or is not responsible for the slight discoloration of the water in the morning.

Exercise Set 10.2

9. What value do you use for t when you know the standard error and want to calculate a 95% confidence interval?

10. Suppose you *know* that a certain aptitude test has a standard deviation of 15. You are going to administer this test to a small random sample of students in a local school and calculate a 95% confidence interval for the average of the whole school, if all students took the test. You give the test to 9 students.

a. What is the standard error of the sample average?

b. What t-values should you use in calculating a 95% CI?

11. Suppose that a college admission test has a known standard deviation of 100 points. The admissions office at one university pulls a random sample of 400 applications and calculates the sample average (505) and the sample standard deviation (95) on the admissions test. What is the value of the standard error for the mean score of applicants to this university?

12. Suppose that a simple optical distance measuring device has a known standard deviation of plus or minus 1 foot at a distance of 100 feet. You take nine measurements of the lot line at a friend's parent's house with this device, and get an average measurement of 100 feet. What is the standard error of this average?

10.3 Other Confidence Levels

Although 95% is the most common level of confidence used in reporting confidence intervals, it is possible to construct confidence intervals with other levels of confidence. This is done by using a different t-value; everything else about the confidence interval computation stays the same.

Appendix C includes t-tables giving values for 80%, 90%, 95%, 98%, 99%, 99.8%, and 99.9% confidence intervals to supplement the 95% table given earlier.

EXAMPLE 10.3 *Calculating Alternative Confidence Intervals*

The data on killer bee wing widths (Table 10.2) had a sample average of 2.858 mm, a standard deviation of .0516 mm, and a standard error of .0115 mm, with $n = 20$ and df $= 20 - 1 = 19$. We calculated that the 95% confidence interval for the population's mean wing width was from 2.834 mm to 2.882 mm. We were 95% sure that this interval included the mean.

Now suppose that we want an interval that is even more likely to include the population mean. For example, say we want to be 99% sure that the confidence interval includes the population mean. The calculations are the same as for a 95% confidence interval, except that we use a *t*-table for 99% confidence intervals to find the appropriate *t*-value. For 19 degrees of freedom, this value is 2.861, so we have

$$\text{margin of error} = 2.861 \times .0115 = .0329$$

and

$$\text{lower 99\% confidence limit} = 2.858 - .0329 = 2.825$$

to

$$\text{upper 99\% confidence limit} = 2.858 + .0329 = 2.891$$

Thus we can be 99% sure that the population's mean wing width is included in the interval 2.825 mm to 2.891 mm.

However, it might be that the 95% confidence interval (not to mention the 99% CI) is too wide for our purposes and that we need a narrower interval. Trading certainty for precision, we calculate a 90% confidence interval. The procedure is the same as for the other confidence intervals, except that we use the *t*-table for 90% confidence intervals. For 19 degrees of freedom, we find a *t*-value of 1.729, so we have

$$\text{margin of error} = 1.729 \times .0115 = .0199$$

and

$$\text{lower 99\% confidence limit} = 2.858 - .0199 = 2.838$$

to

$$\text{upper 99\% confidence limit} = 2.858 + .0199 = 2.878$$

Thus, we are 90% sure that the population's mean wing width is included in the interval 2.838 mm to 2.878 mm.

The three confidence intervals we have calculated from this data are compared in Figure 10.6. This figure clearly shows that larger intervals are required to be more certain that the population mean is covered.

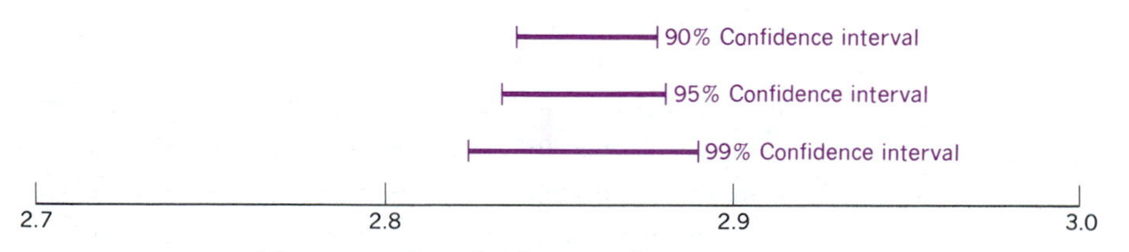

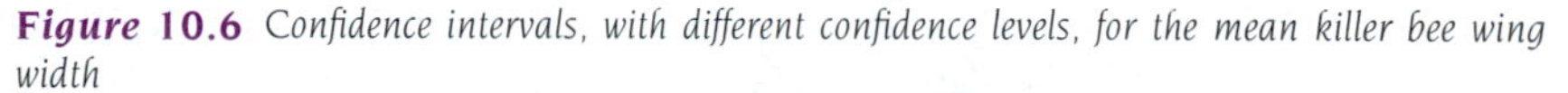

Figure 10.6 *Confidence intervals, with different confidence levels, for the mean killer bee wing width*

Discussion

Nothing is free. If we want confidence, we pay for it by having a larger confidence interval. The larger interval will be more likely to contain the unknown population mean, but the apparent precision of the estimate will be reduced. This is reflected by the larger *t*-values in the tables for 98% and 99% confidence; for example, with 20 degrees of freedom, the *t*-values are 2.086 (for 95% confidence), 2.528 (for 98%), and 2.845 (for 99%). Similarly, if we are willing to accept a lower level of confidence, then we are entitled to a smaller interval with a smaller probability of containing the unknown population mean. For example, with 20 degrees of freedom, the *t*-value is 1.725 for 90% confidence.

To determine which confidence level should be used in a given situation we rely partly on tradition: The most standard choice is 95%. But we also consider the amount of randomness in the data. In situations with large quantities of data with only a little randomness, we can demand higher confidence levels (99% or even more) while still obtaining intervals of a manageable size. Example situations might include cell-counting experiments, where sample sizes are in the thousands. On the other hand, in situations with small samples and appreciable randomness, lower confidence levels (90% or even less) might be used so that the confidence interval is not too large. This approach might be used in the interpretation of human behavior such as in advertising testing in which large sample sizes are prohibitively expensive and peoples' reactions vary widely.

■ Exercise Set 10.3

13. Suppose you have a sample of size 25, a sample average of 40, and a sample SD of 10. Find:
  a. the 90% CI for the population mean.
  b. the 95% CI for the population mean.
  c. the 98% CI for the population mean.
  d. the 99% CI for the population mean.

14. Suppose you have a sample of size 36, a sample average of 99, and a sample SD of 12. Find:
  a. the 90% CI for the population mean.
  b. the 95% CI for the population mean.
  c. the 98% CI for the population mean.
  d. the 99% CI for the population mean.

15. Suppose you have a sample of size 36, a sample average of 100, and a sample SD of 24. Find:
  a. the 90% CI for the population mean.
  b. the 95% CI for the population mean.
  c. the 98% CI for the population mean.
  d. the 99% CI for the population mean.

16. Explain (in one or two sentences) what we are trying to do when we calculate a confidence interval. Why do we bother?

17. Suppose you ask four randomly selected classmates their ages, and are told 18, 19, 19, and 22.
  a. What is the 80% confidence interval for the mean class age?
  b. What is the 98% confidence interval for the mean class age?

18. From the data on killer bees in Table 10.2 (page 325), we have a sample size of 20, a sample average of 2.858 mm, a standard deviation of .0516 mm, and a standard error of .0115 mm. We have already calculated the 95% CI, from 2.834 mm to 2.882 mm.
  a. Find the 90% CI for this situation.
  b. Find the 98% CI for this situation.

19. Using one of the problems in this exercise set as an example, write a brief paragraph about the relationship between the size of our confidence interval and our level of confidence.

Computer Calculation of Other Confidence Intervals

To use the computer to calculate other confidence levels, we simply specify in the instructions the level we want. For example, to find a 90% confidence interval for the killer bee wing width data, we use 90.0, as in Display 10.3. Similarly, we calculate 98% and 99% confidence intervals by using 98.0 and 99.0 in the command, as in Display 10.4.

DISPLAY 10.3

CALCULATING A 90% CONFIDENCE INTERVAL

```
MTB > TInterval 90.0 'Width'.
            N      MEAN      STDEV     SEMEAN     90.0 PERCENT C.I.
WIDTH      20     2.8585    0.0516     0.0115     (2.8385, 2.8785)
```

DISPLAY 10.4

CALCULATING 98% AND 99% CONFIDENCE INTERVALS

```
MTB > TInterval 98.0 'Width'.
            N      MEAN      STDEV     SEMEAN     98.0 PERCENT C.I.
WIDTH      20     2.8585    0.0516     0.0115     (2.8292, 2.8878)

MTB > TInterval 99.0 'Width'.
            N      MEAN      STDEV     SEMEAN     99.0 PERCENT C.I.
WIDTH      20     2.8585    0.0516     0.0115     (2.8255, 2.8915)
```

10.4 One-Sided Confidence Intervals

So far, our confidence intervals have been two-sided in the sense that they extend an equal distance to each side of the average. A one-sided confidence interval is limited on one side but not the other. This allows for statements about the population mean being "at least as large as" some value or "no larger than" some value. These are the two kinds of one-sided confidence intervals, depending on whether the interval is limited on the smaller or larger side.

Consider an ordinary two-sided 95% confidence interval. There is a 95% chance that the population mean is inside the interval, a 2.5% chance that (by mistake) the mean is smaller than the entire interval, and a 2.5% chance that the mean is larger than the interval. If we extended the interval to the right indefinitely, as illustrated in Figure 10.7, then we would no longer make that upper 2.5% mistake. In fact, we would have a one-sided 97.5% confidence interval and could make the statement that we are 97.5% sure that the mean is at least as large as the lower confidence limit.

If we do not want a 97.5% one-sided confidence interval, preferring to keep 95% confidence, we start with a 90% two-sided confidence interval and then extend it indefinitely on one side. The result is also shown in Figure 10.7 (page 336). To keep 95% confidence, we have shortened

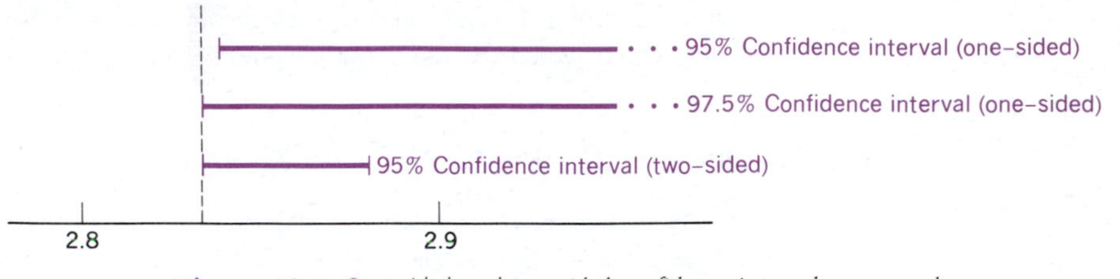

Figure 10.7 *One-sided and two-sided confidence intervals compared*

the interval slightly on one side and extended it greatly on the other. These balance out because it is more likely that the mean will be close to the center of the interval than farther away. Therefore, by extending indefinitely far, we are not much more likely to include the population mean.

To find a one-sided confidence interval, we use the *t*-tables as follows:

- For a one-sided 95% confidence interval, use the *t*-table for a two-sided 90% confidence interval.
- For a one-sided 99% confidence interval, use the *t*-table for a two-sided 98% confidence interval.

Basically, we double the distance between 100% and the desired one-sided confidence level and then use the two-sided *t*-table with that confidence level instead. Then the interval is restricted on one side of the average only and extends indefinitely on the other.

EXAMPLE 10.4 *Calculating an Upper One-Sided Confidence Interval: Mileage*

Suppose it has been decided for the purposes of a study that we need a statement that each car's mileage is "at least" a certain amount. For one of these cars, the following measurements have been obtained.

average = 26.1 mi/gal
standard error = 2.9 mi/gal
sample size = 15 observations
degrees of freedom (= 15 − 1) = 14

To construct an upper one-sided 95% confidence interval, we will use the critical value from the two-sided 90% *t*-table for 14 degrees of freedom, which is 1.761. Then, the confidence interval extends from

$$26.1 - 1.761 \times 2.9 = 21.0$$

to

infinity (no upper bound)

Thus, we could say that

we are 95% confident that this car's mileage is at least 21.0 *mi/gal.*

EXAMPLE 10.5 ***Calculating a Lower One-Sided Confidence Interval: Calorie Content***

Suppose an experimental "new and improved" food is of interest only if it has fewer calories than other foods of its kind. Therefore, we need a confidence statement that it has "no more than" a certain number of calories. For one of these experimental foods, it was found that

$$\begin{aligned} \text{average} &= 92 \text{ calories} \\ \text{standard error} &= 8 \text{ calories} \\ \text{sample size} &= 10 \text{ observations} \\ \text{degrees of freedom } (&= 10 - 1) = 9 \end{aligned}$$

To construct a lower one-sided 99% confidence interval, we will use the critical value from the two-sided 98% t-table for 9 degrees of freedom, which is 2.821. Then, the confidence interval extends from

0 (because no food can have a negative amount of calories)

to

$$92 + 2.821 \times 8 = 115$$

Thus, we could say that

we are 99% confident that this food has no more than 115 calories.

Discussion

One-sided confidence intervals should be used carefully and only if there is a reason to do so. Otherwise, two-sided confidence intervals should be used. One-sided confidence intervals are useful in situations for which we are interested only in the one side and are not the least bit interested in the other. In such cases, we would compute only one (either an upper or a lower) of the one-sided confidence intervals but not the other. To compute both one-sided confidence intervals is misleading; it is better to compute the two-sided confidence interval if both an upper and a lower limit are needed.

Exercise Set 10.4

20. Suppose you are running a bakery, and as part of your quality control program you weigh 20 loaves of bread. Their average weight is 1.05 pounds with an SD of .03 pound. You want to assure the public that, on average, your loaves weigh at least a pound.

a. What is the 95% lower one-sided confidence limit for the mean weight?

b. What is the 99% lower one-sided confidence limit for the mean weight?

21. Suppose you work for a water pollution control agency. You take 16 samples from a local factory's discharge pipe and find that they average 20 parts per million (ppm) on a certain pollutant, with a sample SD of 8 ppm. You are concerned that the factory may be discharging, on average, too much of the pollutant. You are not at all worried if they discharge too little.

a. Find the one-sided 95% CI for the most the factory is discharging, on average.

b. Find the one-sided 99% CI for the most the factory is discharging, on average.

10.5 A Confidence Interval for a Nonnormal Distribution

If the population has a distribution that does not look symmetric and bell-shaped like the normal distribution, there are two ways to approach the problem of finding a confidence interval for the population mean. The first approach is to decide that the normal theory approach will be "close enough" and proceed exactly as if the population were normal—that is, to compute as we did in the previous sections, using the t-table to compute the confidence interval. This approach may be justified in some cases, as we will argue in this section. The second approach is the *nonparametric* approach, which is quite different and appropriate even with extremely nonnormal distributions; it will be discussed in Section 10.7.

Justifications for Using the Normal Theory Approach

If we decide that the normal theory approach is "close enough," there are two main justifications for arguing that the confidence interval is robust, which means that it still works well even if the distribution of the data is not exactly normal.

The first justification argues intuitively by considering a population that is basically normal, except for a few outliers every now and then. If the sample does not contain an outlier, then there is no problem with proceeding as if we had a normal distribution. But if there is an outlier in the data, then it causes two effects that cancel out each other. The first effect is to distort the average, making it far away from the population mean. The second effect is to inflate the standard deviation (as we saw when we showed that the standard deviation is not robust), making the confidence interval much larger than it would ordinarily be. The net result is that occasionally a big confidence interval will have its center far away from the mean in such a way that it can still contain the mean a good deal of the time.

Statistical research has shown that this intuitive concept—that combining these two nonrobust procedures (the average and the standard deviation) in this way results in a robust method (the confidence interval based on the t-table)—actually works in many instances for nonnormal distributions. Therefore, we are safe when we determine a confidence interval pretending that the distribution was normal.

However, if the distribution is extremely nonnormal, this safe method may not be efficient. A big confidence interval may contain the population mean, but it may be so wide that it conveys very little information. (For example, it may be correct to state that, on the average, families have somewhere between .8 and 6.2 children, but this interval is not very useful.) A better confidence interval might be found using the nonparametric approach (see Section 10.7).

The second justification for proceeding as if the distribution is normal is to rely on the central limit theorem (see Section 8.6) which says that even if the population is not normally distributed and the sample does not look normally distributed, the sample average itself will often nonetheless be approximately normally distributed. This is possible because the average combines many values from the sample and so can have a distribution that is more smooth than that of each individual data value.

Recall that the central limit theorem is a mathematical fact about the averages of *large* samples drawn randomly from a population. When sampling repeatedly from a population and computing the average over and over, using a larger sample size has the following two consequences:[2]

1. The average has less variability and tends to be closer to the population mean (as is reflected by a smaller standard error).
2. The distribution of these average values (one per sample) looks more and more like a normal distribution.

The consequences for finding confidence intervals are as follows. Strictly speaking, to use the t-tables we have to assume that the population is normally distributed. However, the confidence interval is based on the average value that (by the central limit theorem) is likely to be nearly normally distributed. This suggests that we need not require that the population be normally distributed; research has shown that this is often true (Benjamini, 1983).[3]

The conclusion is that we should feel free to compute the confidence interval using the t-tables even if the distribution is somewhat skewed or otherwise not normally distributed. However, if the distribution is extremely nonnormal or has one or more very large outliers, then the t-distribution may not apply and the nonparametric approach may be preferable.

10.6 The Confidence Interval for a Percentage

Remember that with binary (yes/no) data, in which each response is represented as either a 1 or a 0, the data are quite definitely not normally distributed. However, the arguments just presented (especially the central limit theorem) apply in this case, and we can compute the 95% confidence interval in the way we have learned, using the t-table. This is very fortunate because this kind of data is very common and there is often a need for the corresponding confidence interval.

EXAMPLE 10.6 Popularity of President Clinton's Budget

A survey conducted by Yankelovich Partners, Inc.,[4] for *Time*/CNN found that out of 800 people, 46% thought that President Clinton's first approved budget represented "a major change in the direction of the country." Another 45% thought it did *not* represent a major change. (Note that 9% had no opinion or refused to answer.) If we focus on the positive responses, the data set looks like a sequence of 1's (representing people who agreed it represented a major change) and 0's (representing those who did not agree it represented a major change), in which case 46% of the data consists of 1's. Note that we are looking here only at "positive" as opposed to "not positive," with the not-positive category containing those who had no opinion in addition to those who had a negative response.

To compute the 95% confidence interval for the percent of people who had a positive response, we proceed in the usual way by first computing the basic quantities:

$$\text{sample size} = 800 \text{ people}$$
$$\text{proportion} = .46$$

$$\text{standard error} = \sqrt{\frac{.46 \times (1 - .46)}{800 - 1}} = .017$$

Next, we need a t-value from Table 10.1 (page 323). The degrees of freedom are $800 - 1 = 799$, which is larger than the last finite value in the table, so we will use the value for infinity, yielding a t-value of 1.96. Multiplying this t-value by the standard error, we have

$$\text{margin of error} = (t\text{-value}) \times (\text{standard error}) = 1.96 \times .017 = .035$$

Subtracting and adding this to the average, we find a 95% confidence interval from

$$\text{lower confidence limit} = .46 - .035 = .425$$

to

$$\text{upper confidence limit} = .46 + .035 = .495$$

The final confidence interval, found by multiplying these proportions by 100 and rounding to units of percent, goes from 42% to 50%. The corresponding confidence statement would be as follows:

We are 95% sure that 42% to 50% of adults thought Clinton's first budget represented a major change in the direction of the country at the time the poll was conducted.

We included the qualification "when the poll was conducted" because people's opinions change over time, so a given survey can be only a "snapshot" of the population.

Discussion: Reporting of Confidence Intervals for Survey Results

Ideally, the results of a poll would always be reported with the confidence limits in addition to the percentage, so one could see how exact the values are. Now that you know how, you can compute the confidence interval yourself when you need to, provided you can find out how big the sample size is and are willing to assume it was a simple random sample.

Estimates of sampling error are often reported with survey results, but the clarity with which they are explained varies. One group to be commended for including an assessment of the error of estimation is *The New York Times*/CBS News Poll.[5] In a brief column describing the method used in conducting their poll, it was reported that:

In theory, it can be said that in 95 cases out of 100 the results based on the entire sample differ by no more than 3 percentage points in either direction from what would have been obtained by interviewing all adult Americans.

This quotation conveys in ordinary language the meaning of the confidence interval. It says that the 95% confidence interval for the poll results extends at most three percentage points to either side of the reported percentage, a degree of accuracy that is typical for a survey involving approximately 1,500 people. The population has even been identified so that it is clear that an inference is being drawn: From data on the 1,545 people who were interviewed, inferences are being drawn about the much larger population of all adult Americans.

Somewhat less clearly, the *Time*/CNN poll on Clinton's budget states, in full, "Sampling error is ±3.5%." It turns out (since we did the arithmetic) that this is the margin of error for a 95% confidence level, rather than the standard error. But at least they did include some indication that there is sampling error.

Exercise Set 10.5

22. Suppose you have a sample of 30 couples from a particular neighborhood, 15 of which have dual incomes. Find the 95% CI for the percent of couples with dual incomes in this neighborhood.

23. Suppose a random sample of 1,600 voters indicates that 54% think the incumbent president is doing a satisfactory job or better. What is the 95% confidence interval for the percent of all voters who think the incumbent is doing a satisfactory job or better?

24. Suppose a survey of 1,600 voters finds that 48% plan to vote for Brown.

a. What is the standard error for this percent?

b. What is the 95% confidence interval for the population percent?

25. Suppose a survey of 400 voters finds that 48% plan to vote for Brown.

a. What is the standard error for this percent?

b. What is the 95% confidence interval for the population percent?

26. Suppose a survey of 100 voters finds that 48% plan to vote for Brown.

a. What is the standard error for this percent?

b. What is the 95% confidence interval for the population percent?

27. Using Exercises 24–26 as an example, write a brief paragraph about the relationship between sample size and the standard error.

10.7 A Nonparametric Confidence Interval for the Median

Sometimes the data are from a distribution that is probably not normal (like house prices, salary levels, or pollutant measurements), and the sample is too small to take comfort (or refuge, if you will) in the central limit theorem. In such cases, there is a **nonparametric approach** to finding a confidence interval for the center of the population distribution. The word *nonparametric* here refers to a collection of methods for statistical inference (primarily confidence intervals and hypothesis tests) that do not assume the population distribution to be normal. By relaxing this restrictive assumption, these procedures are applicable in more general situations than the classical procedures that are based on the normal distribution. In particular, they are robust to outliers.

The confidence interval for the median is very easy to find. It is based on the notion of **ranks,** which we have used before, and is closely related to the so-called *sign test* of nonparametric statistics. Because many robust statistical methods (such as the median and the quartiles) are based on ranks, this seems to be a sensible approach.

The Assumptions

Assumptions are always needed for statistical methods to be valid. In this case, we require only one:

The data must be a random sample from a population.

This assumption is very general, but keep in mind that it still requires two conditions: The data values must be independent of each other, and the sample must properly represent the population in the sense that each population value has an equal chance of being included in the sample.

PROCEDURE: FINDING A CONFIDENCE INTERVAL FOR THE MEDIAN

1. Put the data in order from smallest to largest.
2. Look up the rank of the confidence limits in a table of ranks such as Table 10.3 (page 342).

Table 10.3 *Ranks for the nonparametric 95% confidence interval for the median*

SAMPLE SIZE	RANK OF 95% CONFIDENCE INTERVAL LIMITS	SAMPLE SIZE	RANK OF 95% CONFIDENCE INTERVAL LIMITS
8	1	30	10
9	2	31	10
10	2	32	10
11	2	33	11
12	3	34	11
13	3	35	12
14	3	36	12
15	4	37	13
16	4	38	13
17	5	39	13
18	5	40	14
19	5	41	14
20	6	42	15
21	6	43	15
22	6	44	16
23	7	45	16
24	7	46	16
25	8	47	17
26	8	48	17
27	8	49	18
28	9	50	18
29	9	[a]	[a]

[a] For sample sizes larger than 50, a good approximation to this rank may be found by taking the square root of the sample size, multiplying by 1.96, subtracting this from the sample size, adding 1, dividing by 2, and ignoring any decimal places (truncating) so as to obtain a whole number.

3. Count in that many data values starting from the smallest. This is the lower confidence limit for the population median.
4. Count in that many data values starting from the largest. This is the upper confidence limit for the population median.

Table of the Ranks of the 95% Confidence Limits

In Table 10.3 the ranks for the upper and lower confidence limits have been computed based on a simple statistical principle. Terms have been defined so that half of the units of the population are larger than the median and half are smaller. It follows that each unit sampled has an equal chance of being greater than or less than the population median. Because these sampled units are independent of each other, the chances of, say, at least two being above and at least two being below the median can be computed (even though the population distribution is not known); the data values two in from each end would then contain the median between them a known proportion of the time. (In place of two, this can be computed for any number.) The smallest rank for which the median falls within those values at least 95% of the time is the number reported in the table.

EXAMPLE 10.7 *Mercury in Swordfish*

Mercury, a metal that is liquid at room temperature, is a very unhealthy substance to eat. In an effort to identify dietary sources of mercury, the amount of mercury, in parts per million (ppm), was measured for a sample of swordfish. The results were as follows:[6]

1.2 2.1 1.6 1.5 .9 1.1 1.0 1.3 .3 1.2 .6 .8

To find the nonparametric confidence interval for the population median, we first put these numbers in order. In Table 10.4, the ranks (from the top as well as from the bottom) are also shown.

Table 10.4 *Mercury in swordfish*

	RANKS	
MERCURY LEVEL	FROM THE BOTTOM	FROM THE TOP
.3	1	12
.6	2	11
.8	3	10
.9	4	9
1.0	5	8
1.1	6	7
1.2	7	6
1.2	8	5
1.3	9	4
1.5	10	3
1.6	11	2
2.1	12	1

Next, we consult Table 10.3. For a sample size of 12, as we have here, we find that the rank of the confidence limits is 3. (Note that we use sample size for this table, not degrees of freedom). The two data values at rank 3 are .8 (third from the bottom) and 1.5 (third from the top). Our 95% confidence interval for the median is therefore from .8 to 1.5 parts per million. We may make the confidence statement that

We are 95% certain that the median mercury level for the population of swordfish (from which these were sampled) is somewhere between .8 and 1.5 parts per million.

The interpretation for this confidence interval is very much the same as the interpretation we learned earlier. The (unknown) population median (for all swordfish that might have been caught in place of these) would be contained in the (random) confidence intervals such as this one about 95% of the time if we were to draw repeated samples from the population.

Discussion: Comparing the Median CI with the Mean CI

For symmetric population distributions, the population median is the same as the population mean. Thus a confidence statement for the median would also be a confidence statement for the mean. However, the median is usually different from the mean (especially for skewed distributions or distributions with outliers) so we have to be careful to clearly state that this rank-based procedure generates a confidence interval for the median (not for the mean).

For the sake of comparison, let us also compute the confidence interval for the mean for the swordfish data set. After computing the standard error, .137, and using the value for 12 − 1 = 11 degrees of freedom from the *t*-table, 2.201, together with the average, 1.13, we find that

> *We are 95% certain that the population mean mercury level falls somewhere between .83 and 1.43 parts per million.*

(This is so because .83 = 1.13 − 2.201 × .137, and 1.43 = 1.13 + 2.201 × .137.) It is reassuring that in this case, in which the population is fairly normally distributed to begin with, the classical and the nonparametric methods give approximately the same confidence limits for the center of the population.

Discussion: The Effects of an Outlier on the Confidence Intervals

When the distribution is highly nonnormal, we would expect these two approaches to setting confidence limits to be different from each other. In particular, if an outlier is present, we would expect the classical (*t*-based) confidence interval for the mean to change much more than the nonparametric (rank-based) confidence interval for the median.

Let us now introduce an outlier into the swordfish data set. Consider what happens if we replace the largest mercury value, 2.1, in the example with the outlier 5.0. The ranks now look like those shown in Table 10.5.

Despite the presence of an outlier, the data values at rank 3 have not changed at all. Therefore, the nonparametric 95% confidence interval for the median still goes from .8 to 1.5. However, the outlier does have an effect on the average, which is now 1.38, and its standard error, which is now .346. Therefore, the 95% confidence interval for the mean now goes from .62 to 2.14, which is very different from what it did before.

This tendency for the classical confidence interval for the mean to be much more sensitive to outliers than the nonparametric interval for the median is illustrated in Figure 10.8. Note that the outlier has two primary effects on the confidence interval for the mean: The confidence interval is wider (reflecting the increased variability as measured by the standard error) and its center is pulled toward the outlier (reflecting the increase in the average value for the sample).

Table 10.5 *Mercury in swordfish (with outlier)*

	RANKS	
MERCURY LEVEL	FROM THE BOTTOM	FROM THE TOP
.3	1	12
.6	2	11
.8	3	10
.9	4	9
1.0	5	8
1.1	6	7
1.2	7	6
1.2	8	5
1.3	9	4
1.5	10	3
1.6	11	2
(Outlier) 5.0	12	1

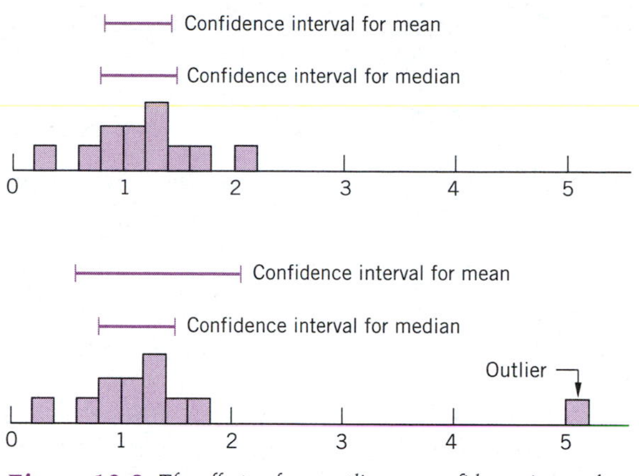

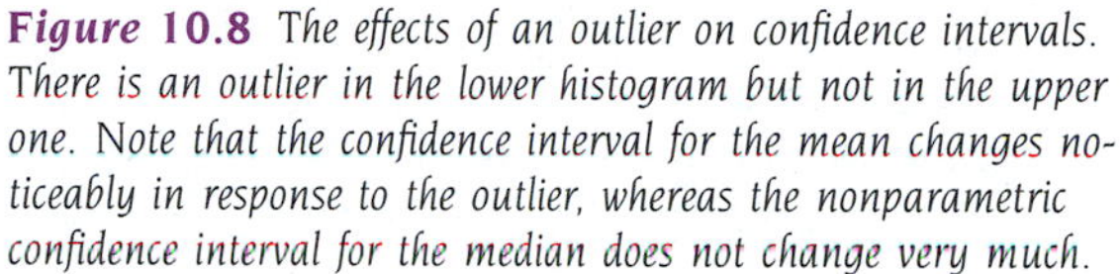

Figure 10.8 *The effects of an outlier on confidence intervals. There is an outlier in the lower histogram but not in the upper one. Note that the confidence interval for the mean changes noticeably in response to the outlier, whereas the nonparametric confidence interval for the median does not change very much.*

We will not "get something for nothing" by using a nonparametric confidence interval. We will pay for its greater generality, ease of use, and relaxed assumptions by being somewhat less efficient when the population truly is normal, a case in which either method could legitimately be used. This might result in a wider confidence interval for the nonparametric procedure, reflecting less precision. Remember that nonparametric procedures are mainly for problems in which the data are very nonnormally distributed. Further, nonparametric procedures can be used with data measured as ordered categories.

We have learned that the median is much more stable than the average as a measure of center when skewness or outliers are a problem. It is for this reason that the nonparametric procedures produce a confidence interval for the population median, not the population mean.

■ Exercise Set 10.6

28. Suppose you have a sample of 24. How many do you count in from each end of the ordered data to find the limiting value for the confidence interval for the median?

29. Construct a 95% CI for the population median using the following sample data.

0	0	0	1	1	1	1	4	4	4	8	8
8	8	10	19	24	34	44	55	66	77	88	99

30. Here, in thousands of dollars, are the prices of all the homes on the east side of Lake Washington advertised by one particular broker in *The Seattle Times* on Sunday, 30 July 1995. (The last house is a waterfront home in a *very* nice area.)

140	145	150	170	184	185	190	240	250	252	259	279	295	299	300	315
320	349	349	349	382	394	399	425	450	475	515	540	565	639	899	2,975

Assuming that these prices are a random sample of this broker's listings, find a 95% CI for the median price of this broker's listed homes.

31. Suppose you have a random sample of runaway kids, and you ask each of them how many different foster homes they have lived in. The answers are:

0 0 0 0 0 0 1 1 1 1 2 2 2 3 13

Construct a 95% CI for the median number of foster homes in this population of runaway kids.

Summary

The average of the data values in a random sample is a good estimator of the mean of the population that was sampled, but by itself the average value does not show us how accurate that estimate is. We can get an indication of the accuracy of estimation from the *confidence interval,* which consists of a lower and an upper limit within which we are fairly certain that the unknown population mean is. Keep in mind that the *population mean is fixed* but the *confidence interval is random* (due to the random sample on which it is based). To construct a confidence interval, we use the *average,* the *standard error of the average,* the *degrees of freedom,* and the *t-tables.* This procedure works very well if the random sample of data was drawn from a normal distribution.

The central limit theorem tells us what happens when the sample size is increased: (1) The average has less variability and tends to be closer to the population mean and (2) the distribution of the average becomes more like a normal distribution (even if the distribution of the individual values in the population is not normal). One consequence is that a larger sample will generally result in a smaller confidence interval, reflecting the greater accuracy of estimation that is possible when more information is available. Another consequence is that the confidence interval constructed from the *t*-tables will be (approximately) valid for percentage (yes/no) data and other nonnormal distributions, provided they are not too long-tailed and the sample size is not too small.

The most commonly used level of confidence is 95%, which says that the population mean will be contained in about 95% of the confidence intervals computed for it if the entire procedure of drawing a random sample from the population and computing the confidence interval were to be repeated many times. Because we usually compute only one confidence interval for a given situation, we can also think of the 95% as indicating our track record over many different problems: Only 5% (1 in 20) of our confidence intervals will be wrong (that is, will fail to contain the population mean). Unfortunately, we will usually never know which ones did and which did not; this is inevitable under uncertainty. The marvel is that we can assess the error rate at all with such accuracy.

When the standard error is known exactly, we use the *t*-table values for a very large number (infinity) of degrees of freedom, representing the perfect information that would be available from a very large sample. Additional *t*-tables are available for confidence levels other than 95%. To make one-sided confidence interval statements like "We are 95% sure that the population mean is larger than. . . ," or "We are 95% sure that the population mean is smaller than. . . ," we use the *t*-tables as follows: For a 95% one-sided confidence interval, use the 90% two-sided *t*-table; for a 99% one-sided confidence interval, use the 98% two-sided *t*-table, and so on.

If the distribution is not normal, we can use one of several approaches. One approach is to rely on the central limit theorem, calculating as though the population were normal. This approach works fairly well if the sample size is large enough and the distribution is not too skewed. For example, the confidence interval for a percentage may be computed in this way because the normal distribution provides an approximation to the binomial distribution. Another approach is to use a *nonparametric* procedure, which does not require that the population be normally distributed.

One easy-to-use procedure is based on a table of ranks of the sample and provides a *nonparametric confidence interval for the median* of any population. The only assumption required for its validity is that we have a random sample from the population (note that normality is not required). Even when the population is very long-tailed, very skewed, or has extreme outliers, this nonparametric confidence interval will be valid. In such cases, it is often more efficient in providing limits for the center of the population than is the confidence interval based on the t-tables. On the other hand, when the population is normal, the nonparametric confidence interval will be somewhat less efficient and less precise.

Problems

Basic Reviews

1. How many degrees of freedom are there for calculating a confidence interval for the mean from a sample of 35 historical museums?

2. What t-value would be used in calculating a 95% confidence interval for the mean from a sample of 11 ice cream factories?

3. What t-value would be used in calculating a 95% confidence interval for the mean based on a very large sample of defective calculators?

4. Consider the data presented in Display 10.5[7].

DISPLAY 10.5

ANALYSIS OF TREE DATA

Selected statistical results

	N	MEAN	MEDIAN	TRMEAN	STDEV	SEMEAN
DIAMETER	31	13.248	12.900	13.156	3.138	0.564
HEIGHT	31	76.00	76.00	76.15	6.37	1.14
VOLUME	31	30.17	24.20	28.87	16.44	2.95

a. Construct a 99% confidence interval for the mean diameter of the trees in this forest plot.

b. Construct a 95% confidence interval for the mean height of the trees in this forest plot.

c. Construct a 99% confidence interval for the mean height of the trees in this forest plot.

d. Construct a 95% confidence interval for the mean volume of the trees in this forest plot.

e. Construct a 99% confidence interval for the mean volume of the trees in this forest plot.

f. Construct a 95% confidence interval for the mean diameter of the trees in this forest plot.

Analysis

5. The following are measurements in centimeters of the petal length for a sample of iris flowers.[8]

1.4	1.4	1.3	1.5	1.4	1.7	1.4
1.5	1.4	1.5	1.5	1.6	1.4	

a. Compute the average, sample size, standard deviation, and standard error.

b. Find the degrees of freedom, then find the appropriate *t*-value from the table for a 95% confidence interval.

c. Compute the 95% confidence interval for the mean petal length of the population from which this was sampled.

6. Given a random sample of 23 rabbits with an average length of 43.6 cm and a standard deviation of 11.3 cm, find the 95% confidence interval for the mean length for the entire population of rabbits from which these were sampled.[9]

7. Suppose a sample of 257 rabbits is chosen at random from the same population as in Problem 6. This time, the average length is found to be 45.3 cm with a standard deviation of 9.2 cm.

a. Find the 95% confidence interval for the mean rabbit length.

b. Compare this to your answer to Problem 6.

c. Why isn't the standard deviation a lot smaller, given that the sample size is so much larger in this problem than in Problem 6?

8. A survey of 30 people in a dormitory shows that 18 of them are satisfied with the food. Assuming that these people represent a random sample of all dormitory residents, make a 95% confidence statement about the percentage of residents who are satisfied with the food.

9. Suppose a random sample of 1,545 adults found that 37% were willing to postpone cost-of-living increases in Social Security to help reduce the federal deficit. Place this percentage in perspective by computing the 95% confidence interval and briefly interpreting it.

10. Find the critical value from the *t*-table for each of the following:

a. A 90% confidence interval for the proportion of households that are likely to buy a refrigerator this year, based on a sample of 28 households

b. A 99% confidence interval for the concentration of chlorine in a swimming pool, based on a single measurement with a known standard error

c. A one-sided 95% confidence interval based on 17 interviews

d. A one-sided 99% confidence interval based on 10 measurements

11. Find the 90% and 99% confidence intervals for the situation described in Problem 8.

12. Find the 90% and 99% confidence intervals for the situation described in Problem 9.

13. Find the one-sided upper 95% confidence interval for the situation described in Problem 9 and complete the statement that "We are 95% sure that at least ______ were willing to postpone cost-of-living increases. . . ."

14. Which sample ranks would be used to find a nonparametric 95% confidence interval for the median based on 39 measurements of immunoglobulin concentration?

15. A survey of a random sample of tourists at state parks resulted in the following data set of length of stay at the campsite (in days).

2 1 8 2 1 1 4 1 2 16 3 1 42 1 5

a. Complete a histogram of this data set and describe the distribution.

b. Are the necessary assumptions satisfied so that we may proceed with a confidence interval for the mean length of stay (for all tourists at these parks) using the *t*-tables? If not, and if we did one anyway, would it be valid?

c. Find the 95% confidence interval for the mean.

d. Find the 95% confidence interval for the median.

e. Briefly compare your answers to parts (c) and (d).

16. Here are some measurements of the wing lengths of killer bees from French Guiana.[10]

8.56 8.51 8.51 8.56 8.96 8.82 8.39 8.54 8.62

They average 8.61 with a sample SD of .175.

a. How many degrees of freedom are there in this data set?

b. What value of *t* would we use to calculate a 95% confidence interval for the population mean?

c. Calculate the 95% confidence interval for the mean length of killer bee wings.

17. We took a simple random sample of nine hikes from *Best Hikes with Children*. Here are their high points:

848 2,900 900 2,981 2,300 5,180 3,982 3,161 5,699

Calculate a 95% confidence interval for the mean high point of the hikes in this book, based on these sample data.

18. We took another simple random sample of nine hikes from *Best Kikes with Children*. Here are their high points.

3,982 300 4,508 6,000 2,061 3,982 3,800 1,120 4,900

Calculate a 95% confidence interval for the mean highpoint of the hikes in this book, based on these sample data.

19. We took yet another simple random sample of nine hikes from *Best Hikes with Children*, Here are their high points:

5,400 4,500 5,092 3,700 700 5,400 3,800 4,461 5,180

Calculate a 95% confidence interval for the mean high point of the hikes in this book, based on these sample data.

20. (*Big Problem*) Consider the following data on the number of deaths caused by 15 major earthquakes.

8	62	81	115	146	250	1,000	1,000
1,200	1,300	1,621	4,000	4,200	40,000	55,000	

Assume that these are a random sample from some hypothetical population of potential earthquake fatalities.

a. Construct a 95% confidence interval for the (hypothetical) mean number of fatalities in the (hypothetical) population of potential earthquake fatalities, using the untransformed numbers.

b. Do a log transformation of the numbers, construct a 95% confidence interval for the mean using the transformed values, then back-transform the limits.

c. Comment on any obvious differences between the confidence intervals calculated in parts (a) and (b).

d. Calculate a 95% confidence interval for the *median* number of deaths caused by the 15 major earthquakes.

e. Comment on this confidence interval, compared to the ones calculated in parts (a) and (b).

21. (*Real problem*) Wagner and Morgan (1993)[11] studied the incidence of learning disabilities in adolescents staying at emergency shelters. They were able to get complete evaluations on a small but carefully drawn random sample (the exact size is not critical) and calculated an 80% confidence interval running from 51% to 87%, for the percent with diagnosable learning disabilities. How sure are they, based on these numbers, that 51% or more of the population had diagnosable learning disabilities? Explain your answer.

chapter 11

Testing a Hypothesis About the Mean

It is easier to show that something is false than to prove that it is true. To show that a claim is false, you merely have to come up with evidence that could not have occurred if the claim were true: "No, Mortimer, I could not possibly be the one who stole your gold watch. In fact, I was on vacation in India at the time the theft occurred." If you really were in India at that time, then you could not have stolen the watch. Certainly it is possible to accumulate supporting evidence in favor of a claim: "But we found the watch in your pocket and we have a copy of your letter to Terry, which suggests that jealousy was the motive." However, proof of a claim is difficult because supporting evidence is almost never enough (maybe the butler committed the crime and put the watch in your pocket).

Methods of *hypothesis testing* have been carefully developed to reflect these considerations, which apply in the real world to situations from courtroom dramas to scientific revolutions. First we make assertions (called *hypotheses*) and then we look at the data as evidence either for or against those

assertions. If an assertion could not reasonably have produced such data, then we reject it. Otherwise we have the weaker conclusion that the assertion cannot be rejected.

There is a close connection between confidence intervals and hypothesis testing. Your hard work with the last chapter's material will pay off by making basic hypothesis testing almost easy. Just be sure to pay attention to the different nature of the various hypotheses and the different interpretations of the possible results of a test.

11.1 Hypotheses About the Population Mean

Suppose that a new, experimental medical treatment has lowered the cholesterol levels of 50 randomly selected patients with heart disease by an average of 15.3 points. Would we be justified in claiming that this treatment really lowers cholesterol?

Let's first determine what is being asked here. While it is clear that the cholesterol levels of these *particular* patients were lowered (on average), we are really concerned with a much larger group of patients that we would like to help. So the real question, which is about the larger population of patients and not just about our sample, is "If we gave the treatment to all patients in the larger population, would cholesterol levels, on average, be lowered?" This is now a question about a population mean.

The answer depends on how much randomness there is in the sample data. If the sample estimate of the difference (15.3 points) is close to the population mean difference, then we do have strong evidence of lowered cholesterol for all patients. But if the standard error is large, then we may not be able to reject the claim that cholesterol levels would be unchanged in the population. After all, if there is substantial randomness, perhaps some samples of 50 patients would show an *increase* in cholesterol of about the same size. We would not want to make medical policy decisions based on random results that indicated the "luck of the draw" more than the actual effectiveness of the treatment.

To answer questions like this, we use **hypothesis testing.** First, we turn the question into a pair of statements (hypotheses) to be tested, such as: "If we gave the treatment to all patients in the larger population, then cholesterol levels, on average, would be *unchanged*" and "If we gave the treatment to all patients in the larger population, then cholesterol levels, on average, would be *lowered*."

In this chapter, we will be concerned with only certain kinds of hypotheses, namely, those concerning population means. We will restrict ourselves for now to situations in which there are two hypotheses. One hypothesis, called the **null hypothesis,** denoted H_0, is very specific and claims that the mean is equal to a definite, specific value. We call this the **hypothesized value.** The other hypothesis is called the **alternative hypothesis,** denoted H_1, and claims that the mean is some other value—in fact, any other value—than the hypothesized value. The result of testing these hypotheses based on a sample of data is that one of them will be accepted and the other rejected. This is a fancy way of saying that we will decide to believe one but not the other.

EXAMPLE 11.1 *Null and Alternative Hypotheses for Measured Values: Candy Bar Weights*

Suppose we suspect that a certain brand of candy bar does not, on the average, weigh the 1 ounce claimed on the wrapper. In fact, we have weighed a few bars and found that, although they weigh close to an ounce, some weigh slightly more and some weigh slightly less. The *null hypothesis* will be that the mean weight of the population of all candy bars of this brand is exactly 1 oz. The *hypothesized value* is 1 oz. The *alternative hypothesis* is that the candy bars do not, on average, weigh 1 oz, but that the mean weight is either less than or greater than 1 oz. A sample of candy bars would be randomly selected, each weighed, and the sample average compared to the claimed value of 1 oz by the methods we will learn in this chapter.

EXAMPLE 11.2 *Null and Alternative Hypotheses for Sample Percents: Voting*

Suppose we would like to predict whether a certain candidate will win or lose an election. We design a survey and ask registered voters whom they plan to vote for. The *null hypothesis* is that the candidate will receive 50% (or half) of the vote (if we assume a two-person race) and that the election will be an undecided tie. Note that the null hypothesis had to specify a definite value and that, for this situation, the natural value to specify is the break-even point that divides winning from losing the election. We do not really believe that the election will result in the candidate's receiving exactly 50% of the vote; this hypothesis merely says that we cannot tell who will really win. The *alternative hypothesis* is that the candidate will not receive 50% of the vote, but will either win or lose.

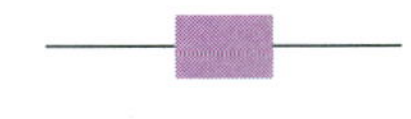

Exercise Set 11.1

1. Suppose you know that American 12-year-old boys weigh, on average, 83 pounds. You see a group of such boys turn out for a middle school football program and decide that their weights probably do not average 83 pounds. You decide to plan a formal study of this issue. You are going to take a random sample of the kids who turned out for football and see if there is a "statistically significant" difference between their average weight and the average weight of American 12-year-old boys, which is 83 pounds.

  a. State your null hypothesis.

  b. State your alternative hypothesis.

2. Suppose you have the results of a random sample of pollution discharges from a group of factories. They average 10 units of pollutants, which would be legally acceptable. However, some observers assert that the real pollutant discharge level is different. You propose a statistical analysis of the sample data to resolve the issue.

  a. State the null hypothesis.

  b. State the alternative hypothesis.

3. Suppose the student guide at your college states that students are expected to spend two hours on outside assignments for every hour in class. Halfway through your introductory statistics course, you are beginning to think that this may not be accurate. You decide to poll a random sample of introductory statistics students, then do a test of statistical significance.

  a. State your null hypothesis.

  b. State your alternative hypothesis.

11.2 Statistical Significance

Many **tests of statistical significance** can be thought of as variations on one idea: Construct a confidence interval around an observed value. Any value outside the confidence interval is said to be *significantly different* from the observed value; any value inside the confidence interval is said to be *not significantly different.*

To test whether the population mean is equal (null hypothesis) or is not equal (alternative hypothesis) to a hypothesized value, a sample is selected and measured, the average calculated, and a decision made. This decision will be based on how far the sample average is from the hypothesized value, taking into account the variability of the sample average (as measured by its standard error). Naturally, if the average is very different from the hypothesized value, we will decide that the population mean is significantly different from the hypothesized value. Thus we will reject the null hypothesis and accept the alternative hypothesis. If we decide that the population mean could reasonably be equal to the hypothesized value, we will accept the null hypothesis.

Another way to look at the result of the test is to directly compare the sample average value (which, after all, is what we have to work with) to the hypothesized value we are testing against. Naturally, they will (nearly always) be different, but what we will want to test is whether that difference is statistically *significant.* The word significant has a specialized meaning in statistics.

Rejecting the null hypothesis says that we have decided the hypothesized value is probably not the population mean. In this case, we will declare that the sample average is significantly different from that hypothesized value.

Accepting the null hypothesis says that we have decided the population mean could reasonably be equal to the hypothesized value. In this case, we will declare that the sample average is not significantly different from that hypothesized value.

Curiously, a difference does not have to be very big in absolute terms to be significant (although it often is). Statistical significance refers instead primarily to whether the difference is or is not larger than the underlying sampling variability of the situation.

Relationship Between Testing and Confidence Intervals

We should have few technical difficulties in learning how to make a decision for the hypothesis test of a population mean against a hypothesized value; the technical details have already been overcome by the work we have done in constructing confidence intervals. An important, fundamental relationship exists between testing a hypothesis about a population mean value (on one hand) and constructing a confidence interval for that population value (on the other). This fundamental relation is as follows (see also Figure 11.1):

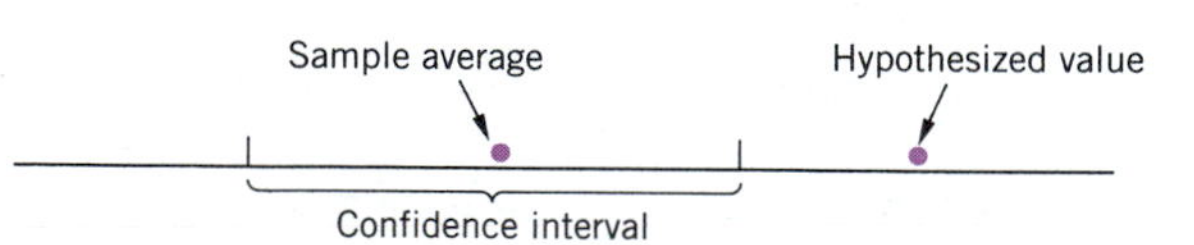

Figure 11.1a *Situation where the hypothesized value is outside the confidence interval. There is a statistically significant difference.*

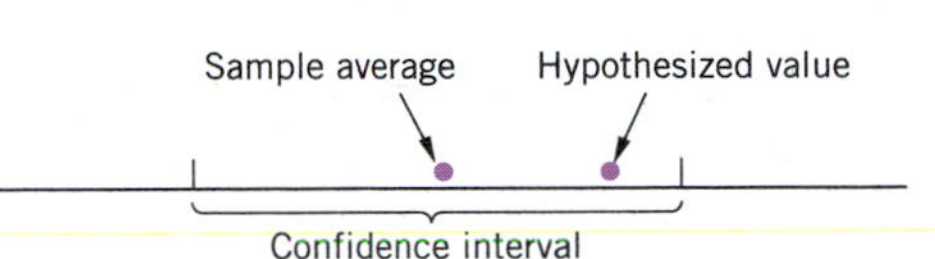

Figure 11.1b *Situation where the hypothesized value is inside the confidence interval. The difference between the sample average and the hypothesized value is not statistically significant.*

1. If the hypothesized value we are testing is *outside* the range of the confidence interval, then we conclude that the population mean is significantly different from the hypothesized value. In this case, the null hypothesis is rejected and the alternative hypothesis accepted.
2. If the hypothesized value we are testing against is *inside* the confidence interval, then we conclude that the population mean does not differ significantly from the hypothesized value. In this case, the null hypothesis is accepted.

Exercise Set 11.2

4. Suppose you weigh a random sample of nine 12-year-old boys who turn out for middle school football (see Exercise 1, p. 353). Doing the arithmetic, you find that their average weight is 100 pounds, with a standard deviation of 12 pounds. The 95% confidence interval runs from 91 pounds to 109 pounds.

a. Can you reject the null hypothesis that the average weight of 12-year-old boys turning out for football equals 83 pounds?

b. Is the observed value of 100 pounds significantly different from the hypothesized value of 83 pounds? [*Hint:* The answers to parts (a) and (b) should be the same; these are just different ways of saying the same thing.]

5. Refer to Exercise 2, page 353. Suppose you do the arithmetic on the random sample of factory pollutants, which averaged 10 units of pollutant, and find that the 95% confidence interval runs from 8 units to 12 units.

a. Can you reject the null hypothesis that true mean value is 20 units?

b. Is the observed average value of 10 units significantly different from the hypothesized value of 20 units?

6. Suppose that a random sample of introductory statistics students shows that, on average, they spend 2.5 hours on outside assignments per hour of class time, and that the 95% confidence interval for the population mean runs from 2.1 to 2.9 hours. Can we reject the null hypothesis that the true mean for introductory statistics students is 2.0 hours? State your conclusion in a sentence. If you decide to reject the null hypothesis, include a statement of what you now believe.

7. Suppose that on a neighboring campus a friend takes a random sample of introductory statistics students which shows that, on average, they spend 2.3 hours on outside assignments per hour of class time, and that the 95% confidence interval for the population mean runs from 1.9 to 2.7 hours. Can your friend reject the null hypothesis that the true mean for introductory statistics students is 2.0 hours? State your friend's conclusion in a sentence. If your friend decides to reject the null hypothesis, include a statement of what your friend now believes.

11.3 Student's t-Test for the Mean of a Normal Distribution

A commonly used test of a hypothesis is *Student's t-test,* named after the pen name "Student" used by W. S. Gosset in his scientific publications. (See Sidebar 11.1, page 356). Because there are other tests with this name, this one is sometimes referred to more specifically as Student's one-sample t-test because it involves only one sample of data. Gosset was the first person to discover the correct values to use for small samples when testing or constructing confidence intervals for the mean of a normal distribution. These t-values, summarized in Table 10.1 (page 323) become smaller as the sample size grows larger, reflecting increased precision in the sample standard deviation as an estimate of the population standard deviation. (This is a different, but related, effect of the increased precision in the average, which is reflected by the square root of the sample size used in finding the standard error of the average.)

Sidebar 11.1

THE GUINNESS BREWERY, GOSSET, AND STATISTICS

W. S. Gosset, originator of the t-test, lived from 1876 to 1937. He worked as Chief Brewer at the Guinness Brewery in London, where some of his pioneering statistical work involved refining methods for studying the relationship between the raw materials (barley, hops, and production methods) and the finished product (beer). With small sample sizes, it was important to be able to detect a real difference between two brewing methods and to tell when only randomness was present.

The opening paragraphs of Gosset's 1908 paper[1] provide a clear statement of some of the problems, assumptions, and methods of statistics.

> Any experiment may be regarded as forming an individual of a "population" of experiments which might be performed under the same conditions. A series of experiments is a sample drawn from this population.
>
> Now any series of experiments is only of value in so far as it enables us to form a judgment as to the statistical constants of the population to which the experiments belong. In a great number of cases the question finally turns on the value of a mean, either directly, or as the mean difference between the two quantities.
>
> If the number of experiments be very large, we may have precise information as to the value of the mean, but if our sample be small, we have two sources of uncertainty: (1) owing to the "error of random sampling" the mean of our series of experiments deviates more or less widely from the mean of the population, and (2) the sample is not sufficiently large to determine what is the law of distribution of individuals. It is usual, however, to assume a normal distribution, because, in a very large number of cases, this gives an approximation so close that a small sample will give no real information as to the manner in which the population deviates from normality: since some law of distribution must be assumed it is better to work with a curve whose area and ordinates are tabled, and whose properties are well known. This assumption is accordingly made in the present paper, so that its conclusions are not strictly applicable to populations known not to be normally distributed, yet it appears probable that the deviation from normality must be very extreme to lead to serious error. We are concerned here solely with the first of these two sources of uncertainty.

Why Gosset published as "Student" rather than use his real name is not known, although various explanations have been offered. Perhaps his employer did not want any competitors to realize that his apparently abstruse publications had economically significant real-world applications.

Let us now reconsider some of the general examples described in the previous section and see how actual results can be computed for these situations once data are available.

EXAMPLE 11.3 *Significance Test with Measured Values: Candy Bar Weights*

Suppose the sample candy bar weights from Example 11.1 are (in ounces)

1.03 1.01 .97 .95 1.01 .98

and we wish to test whether the population mean differs from 1 oz. The average value is .992 oz, which is less than 1. To determine whether it is significantly less, we first construct the confidence interval based on the average value of .992, the standard deviation of .0299, and the sample size of 6, giving a standard error of .0122. With a sample of size 6, we then look up the t-value for $6 - 1 = 5$ degrees of freedom, which is 2.571. We compute the confidence interval, which goes from $.992 - .0122 \times 2.571$ to $.992 + .0122 \times 2.571$. After calculation, this reduces to the following confidence statement.

The 95% confidence interval goes from .961 *to* 1.023.

That is, we are 95% certain that the true population mean weight falls within these limits. Note that the hypothesized value, 1 oz, does fall within the interval because 1 is indeed between .961 and 1.023. (See Figure 11.2.)

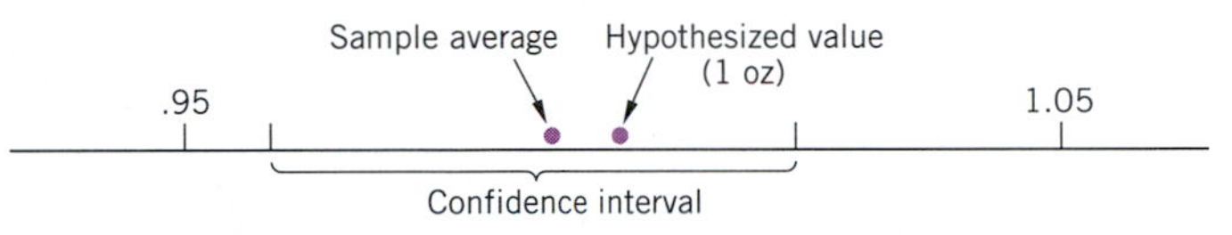

Figure 11.2 *Test of significance for candy bar example*

Based on this data set, we would therefore

accept the null hypothesis

that the mean weight (for the population of lots and lots of candy bars) is equal to 1 oz. Another way to state the outcome is that we would find that the average weight for the sample, .992, is

not significantly different

from the hypothesized value of 1 oz. Note that the sample average (.992) is indeed different from 1, but that this difference is within the range of variability we would expect from an average of a sample of this size (only 6 candy bars). In this case, we could not accuse the candy bar company of cheating the customers because the evidence is not conclusive.

EXAMPLE 11.4 Testing Significance with Percents: Voter Preference

Suppose we find, based on interviews of 1,251 registered voters, that 779 of them intend to vote for our preferred candidate. This yields a percentage of 62.3%, which appears to be a strong majority. Is this percentage significant? Can we conclude that the entire population of registered voters includes a majority in our favor?

To answer these questions, we first construct the confidence interval. Working with the proportion .623 (instead of percents), we determine that the standard error is .0137. The t-value for a sample this large is about 1.96, and the 95% confidence interval goes from $.623 - 1.96 \times .0137$ to $.623 + 1.96 \times .0137$. That is, it goes from .596 to .650. Converting back to percentages, we obtain the following confidence interval.

The 95% confidence interval goes from 59.6% *to* 65.0%.

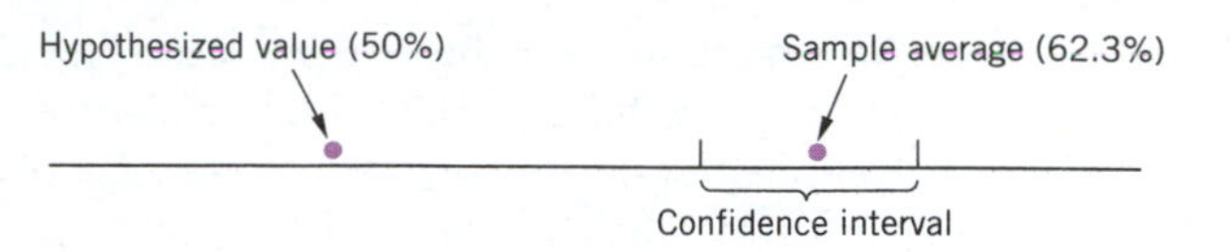

Figure 11.3 *Test of significance for a percent*

We are 95% certain that the population preference is within this range. (See Figure 11.3.) Fortunately, these computations are still correct even though the population is not normally distributed, because the central limit theorem applies to a percentage from a large yes/no data set.

The hypothesized value being tested (which is 50% because 50% divides winning from losing) is outside the confidence limits (it is not between 59.6% and 65.0%). We therefore

reject the null hypothesis

that the population is evenly divided and

accept the alternative hypothesis

that the population does indeed have a preference. In more practical language, we conclude that the apparent majority, 62.3%, is indeed

statistically significant

in the sense that it differs from the hypothesized value of 50% well beyond the usual extent of sampling error we would expect from a survey of this sort.

Discussion

Note that deciding in favor of one or the other hypothesis does not make that hypothesis true. We still do not know which one is actually true; we are merely making the best decision we possibly can based on the data that are available. In the candy bar example, the decision might be correct. On the other hand, the company may well be shorting its customers by an amount so small that we would need to measure a sample much larger than 6 to detect the problem. A larger sample gives more information, which makes it easier to detect small differences.

Statistical Significance Versus Substantive Importance

In a way, the use of the term *statistically significant,* especially when shortened to *significant,* is very unfortunate. It seems to say that the finding is important. That is, after all, one of the ordinary English meanings of the word *significant.* But in the context of statistical testing we are using the term in a narrow, specialized way. When we say that a finding is statistically significant, we are saying only that it is unlikely to be just on the basis of chance. With a large enough sample any real difference can be shown to be statistically significant.

Whether the difference has any substantive importance is a separate question. Substantive importance is a matter of the size of the difference, its theoretical implications, and (among other things) any economic or ethical considerations that are engaged by its practical implications.

Exercise Set 11.3

8. Suppose we have a sample of size 9 with a sample average of 150 and a standard deviation of 12. There is a theory that the mean is really 100. We want to do a statistical test of the theory, on the assumption that the data are a random sample from the population the theory applies to.

a. What is the null hypothesis?

b. What is the 95% confidence interval for the population mean, based on the sample data?

c. Is the hypothesized mean inside the confidence interval?

d. State your conclusion.

9. Suppose we have a sample of size 9 with a sample average of 150 and a standard deviation of 12. There is a theory that the mean is really 152. We want to do a statistical test of the theory, on the assumption that the data are a random sample from the population the theory applies to.

a. What is the null hypothesis?

b. What is the 95% confidence interval for the population mean, based on the sample data?

c. Is the hypothesized mean inside the confidence interval?

d. State your conclusion.

10. Suppose you buy a new car and want to check its gas mileage. You keep track of the miles between fill-ups, and the amount of each fill-up. Your first four mileage estimates (which you are willing to treat as a random sample) are 22.4, 23.1, 21.6, and 23.8 mpg. The city gas mileage estimate posted for this model is 23.0 mpg. Is there a statistically significant difference between your sample average and the posted rating? (Show your work, and state your conclusion.)

11. Suppose that the situation is exactly the same as in the last problem, but now the posted city mileage rating is 25 mpg. Is there a statistically significant difference between your sample average and the posted rating? (Show your work, and state your conclusion.)

Computer Calculations

To illustrate the use of the computer for confidence interval calculations and tests of significance, we entered the following numbers in column C1 of a Minitab worksheet.

10 11 12 13 14 15 16

Assuming that these numbers are a random sample from some population, we now instruct the computer to calculate a 95% confidence interval for the mean value of the population from which they were drawn. The output is shown in Display 11.1.

DISPLAY 11.1

```
MTB > TInterval 95.0 C1.

        N      MEAN      STDEV     SEMEAN     95.0 PERCENT C.I.
C1      7     13.000     2.160      0.816     ( 11.002, 14.998 )
```

The first line, following the MTB > prompt, tells the program to calculate a 95% confidence interval, using t-values, based on the data in column C1. The next two lines report that there are 7 values in column C1, their mean is 13.000, their standard deviation is 2.160, and the standard error for estimating the population mean is .816. The last column reports that the 95% confidence interval runs from 11.002 to 14.998.

If we had hypothesized that the true population mean was something outside the interval

11.002 to 14.998—say, 9—we would reject that hypothesis. There is a statistically significant difference between the sample data and a hypothesized value of 9.

The t-Statistic

There is another way to perform a *t*-test. It is more trouble, involves computations with less direct meaning, and always gives the same answer as the method based on the confidence interval. However, it does parallel the way many other statistics tests are conducted.

To use the alternative method we first compute the *t*-statistic

$$t\text{-statistic} = \frac{\text{sample average} - \text{hypothesized value}}{\text{standard error}}$$

and then decide the test based on whether the *t*-statistic (in absolute value) is larger or smaller than the value in the *t*-table.

If the absolute value of the t-statistic is larger than the value in the t-table, then we reject the null hypothesis that the population mean equals the hypothesized value.

If the absolute value of the t-statistic is smaller than the value in the t-table, then we accept the null hypothesis that the population mean equals the hypothesized value.

This procedure is algebraically equivalent to constructing the confidence interval first and then seeing whether the hypothesized value is in the interval, as we have done. The disadvantage of the *t*-statistic is that the calculation of the *t*-statistic does not have immediate, direct relevance to the problem: It is an intermediate step that requires us to compare it to a value in a table. A rough rule of thumb is that *t*-statistics larger than 2 in absolute value are significant (that is, we can reject the null hypothesis), but the *t*-table must be checked anyway because with smaller sample sizes, larger values are required for significance.

EXAMPLE 11.5 A Classic *t*-Test of Significance

Suppose we have a random sample of nine 12-year-old boys who turned out for middle school football. Their sample average weight was 100 pounds with a standard deviation of 12 pounds. We want to know whether this is significantly different from a hypothesized mean value of 83 pounds.

PROCEDURE: PERFORMING A CLASSIC *t*-TEST

1. Find the standard error:

$$SE = \frac{SD}{\sqrt{n}} = \frac{12}{3} = 4$$

2. Calculate the *t*-statistic:

$$t = \frac{\text{sample average} - \text{hypothesized value}}{\text{standard error}}$$

$$= \frac{100 - 83}{4} = \frac{17}{4} = 4.25$$

3. Find the critical value of the t-statistic.

   For df = 9 − 1 = 8, at a 95% confidence level, t-critical = 2.306.

4. Note whether the calculated value of t is greater than the critical t-value.

$$4.25 > 2.306$$

5. Make a decision:

   If t-statistic is greater than t-critical, the difference is statistically significant.

   If t-statistic is *not* greater than t-critical, the difference is not statistically significant.

   In this case, the difference is statistically significant.

p-Values and Significance

Much of statistical theory and practice was developed before computers simplified the calculations. We have presented tests of significance, so far, within the traditional framework. We do a series of calculations and then decide whether we can or cannot reject the null hypothesis at a predetermined level (usually 5%).

Computer software typically includes an additional calculation, the *p-value,* which is the probability that the test statistic would be as far or farther from 0 if the null hypothesis is true. Small p-values indicate that these results are unlikely just on the basis of chance, if there are no real differences. It is becoming common practice to report the p-value as the "observed level of significance." We will do this as we examine computer output in the rest of the book.

Computer Calculations

The classic t-test is also available on Minitab. The program makes two assumptions: (1) We are testing a null hypothesis that the population mean value is 0, and (2) the alternative hypothesis is that the population mean is something—anything—other than 0.

Using the same data in column C1 as before—10, 11, 12, 13, 14, 15, 16—the output is as shown in Display 11.2. The first line tells the program to do a t-test of the null hypothesis that the population mean value is 0, using the data (which are of course assumed to be a random sample from the population) in column C1. The subcommand on the next line tells the program to use alternative hypothesis number "0," which is that the mean value is different from the hypothesized value (either higher or lower).

DISPLAY 11.2

```
MTB > TTest 0.0 C1;
SUBC> Alternative 0.

TEST OF MU = 0.000 VS MU N.E. 0.000

          N      MEAN       STDEV      SEMEAN       T        P VALUE
C1        7      13.000     2.160      0.816        15.92    0.0000
```

The first line of the results restates the test being conducted: We are conducting a test of the null hypothesis that the population mean MU = 0.000 VS the alternative hypothesis that MU is N.E. (not equal) to 0.000.

The last two lines give the results. The summary values for the sample mean, standard deviation, and standard error are reported first. Then, rather than a confidence interval, the t-value is reported. The 15.92 reported for this example says that the sample average of 13 is 15.92 standard errors from the hypothesized population mean of 0.0. The p-value, the probability of this happening (given the assumption of a random sample), works out to 0.0000. We can safely reject the null hypothesis.

We may want to test against a hypothesized value other than zero for the population mean. For example, if we test the hypothesis that the population mean was really "10.0," we would get the output shown in Display 11.3. The command lines are the same, except that we use " 10.0" rather than "0.0" in the first line.

DISPLAY 11.3

```
MTB > TTest 10.0 C1;
SUBC> Alternative 0.

TEST OF MU = 10.000 VS MU N.E. 10.000

          N      MEAN      STDEV     SEMEAN      T      P VALUE
C1        7      13.000    2.160     0.816       3.67   0.010
```

Our summary results are the same, of course, except that the t-value is now 3.67. Thus the sample average of 13.0 is only 3.67 standard errors from the hypothesized population mean. Still, under the assumptions we are basing the calculations on, this has a probability of only .01. This is less than the standard significance level of .05, so we still reject the null hypothesis.

Exercise Set 11.4

12. For the data 10, 11, 12, 13, 14, 15, 16, the output of a t-test run on Minitab is shown in Display 11.4.

DISPLAY 11.4

```
MTB > TTest 9.0 C1;
SUBC> Alternative 0.

TEST OF MU = 9.000 VS MU N.E. 9.000

          N      MEAN      STDEV     SEMEAN      T      P VALUE
C1        7      13.000    2.160     0.816       4.90   0.0027
```

a. State the null hypothesis being tested.

b. What is the value of t?

c. Can the null hypothesis be rejected at the .05 significance level?

13. For the same data (10, 11, 12, 13, 14, 15, 16), the output of a t-test run on Minitab appears in Display 11.5.

a. State the null hypothesis being tested.

b. What is the value of t?

c. Can the null hypothesis be rejected at the .05 significance level?

DISPLAY 11.5

```
MTB > TTest 11.50 C1;
SUBC> Alternative 0.

TEST OF MU = 11.500 VS MU N.E. 11.500

          N      MEAN      STDEV     SEMEAN      T      P VALUE
   C1     7    13.000      2.160      0.816   1.84       0.12
```

11.4 Interpreting a Hypothesis Test

The result of a hypothesis test, also known as a test of significance, can be interpreted on several levels. We will first consider the practical interpretation, and then we will try to understand the theory of testing by considering all of the various underlying possibilities. In this section, we are still considering the test of whether the population mean is or is not equal to a hypothesized value, based on a random sample of data. However, much of what we will discuss also applies to more general hypothesis testing situations.

The Practical Interpretation

There are two possible outcomes, so let's consider them one at a time. First, consider the outcome of accepting the null hypothesis that the mean value equals the hypothesized value, concluding that any observed difference between the sample average and the hypothesized value is not statistically significant. In this situation, we are not saying that we believe the mean *equals* the hypothesized value. We are actually saying that there is no compelling evidence against this null hypothesis.

This suggests that the null hypothesis has a very favored position: It is given the benefit of the doubt. We will accept it as true until and unless the data state very strongly that it is not true. A standard analogy may be found in the U.S. legal system, in which a person is innocent until proven guilty. For hypothesis testing, the null hypothesis is accepted until the data strongly suggest otherwise.

Now, let's consider the second possible outcome. Suppose that the null hypothesis is rejected and the difference between sample average and hypothesized value is declared to be statistically significant. This says that we declare that the population mean is *not equal* to the hypothesized value because the sample average is so different from the hypothesized value that the difference *is unlikely on the basis of sampling randomness alone*. The difference is so great that, even giving it the benefit of the doubt, the null hypothesis has to be abandoned as an explanation of the situation based on the strong evidence against it from the data.

In rejecting the null hypothesis, we make a much stronger statement than if we had accepted it. Accepting it makes the somewhat weak statement that we really cannot be very sure about our decision. Rejecting it makes the strong statement that the data are so inconsistent with a mean equal to the hypothesized value that we are quite sure the hypothesized value could not be the population mean.

Because of this, researchers are usually much happier when the null hypothesis can be rejected. In such a case, their findings are declared significant because the data could not be explained by sampling variability alone. It is as though they now have a license to explain the (nonstatistical) reasons for the difference. Had the difference not been found to be significant, they would not have this freedom to explain because they might well be trying to explain randomness.

Requiring that findings be statistically significant helps assure that the scientific (and other) journals are not filled with detailed explanations of random, nonrepeatable quirks of an individual sample. In statistical terms, it is general inference about the population that is important, rather than the special properties of a particular sample that might not reflect the population.

The Deeper Theoretical Interpretation

If we take a wider view of the situation, we see that there are actually four possible complete situations for each test. True, there are two possible decisions that might be made, but there are also two possibilities for the true situation itself, which is unknown to us both before and after the test is done (although we try to do our best to find out). In other words, because there are two possible decisions and our decision may be either correct or not, there are four possibilities: two kinds of correct decisions and two kinds of errors. We can use a two-by-two diagram as shown in Table 11.1 to illustrate these four possibilities.

Naturally, we hope to make the right decision in each situation. This can happen in two ways: Either we correctly accept the null hypothesis that the mean equals the hypothesized value, or else we correctly reject this hypothesis. If it were not for randomness, we could make correct decisions all of the time because the data would contain complete information about the true situation. Unfortunately, randomness inevitably introduces the possibility of error in our decision. We next examine those errors and how they are controlled.

Type I Error. The **Type I error** occurs when the mean, in fact, really does equal the hypothesized value but we just happened to draw a random sample that strongly suggested otherwise. Thus, we wrongly decide that the mean does not equal the hypothesized value. Of course, we do not know that we have made an error because all we have to go on is the sample of data. Nonetheless, this error can be controlled and should be controlled because it wrongly gives us license to explain a difference that does not really exist in the population.

The Type I error (wrongly declaring a difference to be significant) is conventionally controlled so that it happens at the 5% rate. This is the result of basing our test on the 95% confidence interval: The confidence interval contains the population mean 95% of the time, which says that it does not contain it during the remaining 5% of the time.

Thus, we will wrongly declare significance (and make a Type I error) 5% of the time when

Table 11.1

		OUR DECISION	
		MEAN EQUALS HYPOTHESIZED VALUE	MEAN DOES NOT EQUAL HYPOTHESIZED VALUE
THE TRUE SITUATION	MEAN EQUALS HYPOTHESIZED VALUE	Correct decision, congratulations!	*Wrong!* Type I error
	MEAN DOES NOT EQUAL HYPOTHESIZED VALUE	*Wrong!* Type II error	Correct decision, congratulations!

the null hypothesis is true. This probability is called the **level** or the **significance level** of the test.

DEFINITION

The significance level of a test is the probability of a Type I error—that is, the probability of wrongly rejecting the null hypothesis.

The Greek letter alpha, α, is often used as notation for the significance level. Thus the statement "$\alpha = .05$" means the test is at the 5% significance level.

Suppose we completed many repeated tests, each based on a new random sample. If the mean really was equal to the hypothesized value, then about 5% of the repeated tests would wrongly find a significant difference (because of sampling variability). Another way to think about this 5% error rate is that our lifetime track record as a statistician will include mistakes: 5% of the time we will mistakenly reject the null hypothesis, even though it is true.

Unfortunately, we do not know, for a given test, whether we are right or wrong. But in the presence of randomness, this is the best analysis that can be done.

Type II Error. The **Type II error** occurs when the mean is not equal to the hypothesized value, but we decide that it is. We "fail to reject" the null hypothesis when, in fact, we should reject it. This error is difficult to control. If the mean is close to the hypothesized value, sampling variability will overwhelm this small difference and render it undetectable. In effect, a decision that the mean is equal to the hypothesized value is really a statement that there is no conclusive evidence against their being equal.

The lack of control of Type II errors can be a source of frustration to the researcher who believes that there truly is a difference, but that the statistical methods have simply been unable to detect and demonstrate it. This might happen because a sample size is too small. Sometimes, a small experiment hints at a difference that is not significant, and so a larger experiment will then be run to see whether the situation can be "pinned down" by using the extra precision (and lower sampling variability) of the average that occurs when the sample size is larger. However, if the difference is extremely small, then even with a larger sample size a significant result still might not be found.

EXAMPLE 11.6 *Type I Error: Mistakenly Rejecting the Null Hypothesis*

Suppose a friend claims the ability to psychically influence the results of a coin toss. (He cannot really do this.) You conduct a little experiment by tossing a coin while your friend tries to psychically influence the results so that the coin shows heads. Suppose that in three flips, you get heads twice. "See," your friend says, "it landed heads most of the time because of my psychic power." You decide that you would like to conduct an experiment and test the statistical significance of the results.

You do some arithmetic and decide that there is about a 5% chance of more than seven heads in ten flips of an honest coin. (In fact, by the binomial theorem, there is a 5.46% chance.) You flip the coin ten times while your friend thinks hard about making it come up heads. You get eight heads in the ten flips. "See," your friend says, "that did not happen just by chance!" He has mistakenly rejected the null hypothesis. This is a Type I error.

EXAMPLE 11.7 *Type II Error: Not Rejecting a False Null Hypothesis*

Suppose that in the population at large, most (well over half) voters are planning to vote for Smith. A polling organization phones 30 likely voters and finds that exactly 15 are planning to vote for Smith. They conclude that they cannot reject the null hypothesis that exactly 50% are planning to vote for Smith (and the election will be a tie). They have failed to reject a null hypothesis that is false. This is a Type II error.

Exercise Set 11.5

14. A friend of yours with time on his hands flips a coin 200 times and writes down the number of heads. He repeats this again and again until he has done it 100 times. Then he goes down the list of numbers of heads and does a test of significance of the hypothesis that heads will come up half the time. In 5 cases, or 5% of the time, he finds that the actual number of heads is "significantly different from the hypothesized number." Which of the following statements is most likely true?

a. This is an example of a Type I error, where we reject a null hypothesis even though it is true.

b. This is an example of the well-known ability of coins to "change their spots" and show more or fewer heads than expected based on (1) what they did last time and (2) the fervor of the wishes of the person tossing them.

15. Suppose a new drug really is better than a standard drug. In a limited pretest, four volunteers are randomly assigned to conditions where they get either the new (and actually better) drug or the standard drug. Two get the new drug, two get the standard drug. In each condition, one gets better and the other does not. Even though the sample size is small, the drug company—which is in a hurry to get the new drug on the market—goes ahead and does a test of significance. According to this test, there is no significant difference between the tests. (The director of product development hands in his resignation, and the marketing director advocates the use of color graphics in discussing this with shareholders.) Keeping in mind that the new drug really is better than the standard drug, decide which of the following these results illustrate.

a. A Type II error, where we fail to reject a null hypothesis that is not true.

b. Another example of the futility of using random assignment procedures when we are sure of the truth before we do the experiment.

c. A Type I error, where we accidentally reject a true null hypothesis.

16. The U.S. Department of Health and Human Services and the Department of Agriculture (Nov. 5, 1990) recommended that people get no more than 30% of their calories from eating fat (animal fat, butter, margarine, cooking oils, etc.). Suppose that for all the aerobics instructors in your community, the mean percent of calories from fat really is 30%, just like the recommendation, but that you do not know this. You want to find out what the percent of calories from fat is in a highly motivated, trained group of individuals, so you take a random sample of four aerobics instructors and determine the percent of calories from fat for each of them. You find that they average 40% of their calories from fat, and that the 95% CI for the population mean runs from 32% to 48%. You reject the null hypothesis that the population of aerobics instructors is averaging 30% calories from fat. Have you made a mistake? If so, what kind?

17. Suppose that the aerobics instructors in your city really average 34% of their calories from fat, rather than the 30% the government recommends, but you do not know this. You take a random sample of 16 aerobics instructors and find that they average 32% of their calories from fat, and that the confidence interval for the population mean runs from 30% to 34%. You conclude that you cannot reject the null hypothesis that the aerobics instructors are averaging 30% calories from fat. Have you made a mistake? If so, what kind?

11.5 When the Standard Error Is Known Exactly: The z-Test

When the standard error is known exactly, we already know how to find the confidence interval: We use the t-table entry for an infinite number of degrees of freedom (representing our assumed perfect knowledge about the variability). As before, we accept the null hypothesis if the hypothesized value is within the confidence interval and reject the null hypothesis if the hypothesized value is outside the interval.

EXAMPLE 11.8 *One-Sample z-Test: Water Quality*

Consider again the water quality example, in which a single measurement of iron concentration in drinking water was found to be

$$\text{concentration} = 8 \text{ parts per million}$$

and for which it is known that a single measurement of this type has a standard error of

$$\text{standard error} = .9 \text{ part per million}$$

Let's test whether this concentration differs significantly from the amount of iron in sea water, which is about .02 part per million. That is,

$$\text{hypothesized value} = .02 \text{ part per million}$$

The test will rely on the 95% confidence interval, which was computed using the t-value for an infinite number of degrees of freedom (1.96) to go from

6.2 *parts per million*

to

9.8 *parts per million*

Because the hypothesized value, .02, is outside the interval, we reject the null hypothesis that the drinking water sample contains the same amount of iron (.02 part per million) as sea water and we conclude that

The sample has an iron concentration that is significantly different *from that of sea water.*

Exercise Set 11.6

18. Suppose that a psychological questionnaire has a series of questions about what makes you angry and how often you get angry. The number of Yes answers is added up to get an "anger score." The test has been given to many, many teenagers, and it is known that for American teenagers as a group the average score is 10 with a standard deviation of 4. The test is going to be given to a random sample of 16 clients of the Division of Juvenile Rehabilitation to see whether they show more anger than teenagers in general, while assuming that the standard deviation is unchanged from that of the larger group.

a. What SD should you use when calculating the standard error?

b. If you take a random sample of 16, what is the standard error of the mean?

c. What value of t do you use to calculate the 95% confidence interval?

19. Suppose the problem is the same as in Exercise 18, but now you are going to test 36 clients of the Division of Juvenile Rehabilitation. Is this going to change the t value you use in calculating the 95% confidence interval?

11.6 Other Significance Levels

Although 5% is the most common significance level used in hypothesis testing, we can also use other levels. We do this by using the t-tables for other levels, or equivalently, by using a confidence interval constructed at a different level of confidence. For example,

- To test at the 10% significance level, we see whether the 90% confidence interval contains the hypothesized value.
- To test at the 5% significance level, we see whether the 95% confidence interval contains the hypothesized value.
- To test at the 1% significance level, we see whether the 99% confidence interval contains the hypothesized value.

In general, the significance level is 100% minus the confidence level.

EXAMPLE 11.9 Significance Test at the 10% Level: Advertising Effectiveness

An advertising testing agency is interested in how a particular ad compares to a standard benchmark, a score of 3 on a scale from 0 to 5. This particular ad, based on a test involving a group of people, had the following characteristics.

average score = 3.57
standard error = .31
sample size = 25 people
degrees of freedom = 25 − 1 = 24

The agency typically tests at the 10% level because it would prefer to occasionally include an ad of typical performance, rather than incorrectly classify a really good (or really bad) advertisement. The t-value for the 90% confidence level is 1.711, so the 90% confidence interval goes from

$$3.57 - 1.711 \times .31 = 3.04$$

to

$$3.57 + 1.711 \times .31 = 4.10$$

Because the hypothesized value 3 is outside the confidence interval, we conclude that

> *The ad differs* significantly *from the benchmark value and is* significantly more effective *at the 10% level.*

Curiously, the 95% confidence interval, from 2.93 to 4.21, does contain the hypothesized value of 3. Therefore,

> *The ad* is not significantly different *from the benchmark value at the 5% level.*

How can these both be true (significant at the 10% level, but not significant at the 5% level)? Remember that a hypothesis test is a decision, not necessarily a fact. We still do not know whether the ad is truly different from the hypothesized value, and performing a hypothesis test does not change this. But by setting the error probabilities in different ways, we control how easy or hard it is to make errors in the decision-making process. Unfortunately, because of the randomness in the measurements, errors cannot be eliminated—only controlled—and the results will depend on how these errors are set.

Discussion

With the different choices of significance level we are controlling the probability of a Type I error—the probability of wrongly deciding against the null hypothesis. The smaller this probability is, the less chance we have of falsely rejecting the null hypothesis. With a test at the 1% level, we would make this mistake only about once in 100 tests.

Why then do we not always make this error probability as small as possible? Why not 0? The reason is that the Type I error must be considered against the Type II error, and the smaller one of them is, the larger the other must be. So we typically set the significance level (Type I error probability) as high as we can tolerate, so that the Type II error will be as small as possible under the circumstances.

The smaller the significance level is, the more evidence we need before we can reject the null hypothesis. Thus, with a small significance level, it is impressive when the null hypothesis is rejected, and we can feel secure in our decision to reject it. However, if we do not reject the null hypothesis, then the conclusion is much weaker: We do not know whether the null hypothesis is actually true or we merely lack sufficiently strong evidence against it.

Exercise Set 11.7

20. You want to conduct a test at the 2% significance level. What confidence level should you use?

21. You find that a hypothesized mean is outside the 90% confidence interval. This finding is significant at what level?

22. You find that a hypothesized mean is outside the 99% confidence interval. This finding is significant at what level?

23. The killer bee wing width data (Table 10.2, page 325) was analyzed by computer. We found a 90% CI for the population mean wing width of 2.8385 mm to 2.8787 mm. Suppose we hypothesize that the true population mean is 2.8900. Can we reject this null hypothesis on the basis of these calculations? (State your conclusion in a sentence. Include the significance level of the test.)

24. Starting with the killer bee wing width data (Table 10.2), and doing the calculations by computer, we find a 99% CI for the population mean wing width of 2.8255 mm to 2.8915 mm. Suppose we hypothesize that the true population mean is 2.8900. Can we reject this null hypothesis on the basis of these calculations? (State your conclusion in a sentence. Include the significance level of the test.)

11.7 One-Sided Testing

There are situations in which we are interested specifically in whether a quantity is *significantly larger than* or *significantly smaller than* a hypothesized value, rather than simply whether it is significantly different from the value. If we are careful, we can test either of these hypotheses using a one-sided test.

Care is required because a one-sided test can be used only when we are interested solely in that one side and *would not have been interested had the data come out on the other side.* By performing a one-sided test, we give up the right to have any interest in the other side. If both sides of the hypothesized value are of interest to us, we should perform a two-sided test. The two-sided test is recommended for most situations.

EXAMPLE 11.10 One-Sided Significance Test: Fishing Line Strength

A certain society of people who enjoy fishing is very particular. Specifically, this society only uses fishing lines that are even stronger than their claimed rating. The society therefore includes a group that performs statistical testing of equipment. They test each brand of fishing line to see whether it is at least as strong as it claims to be. Their testing procedure is to perform a one-sided test with the following hypotheses.

NULL HYPOTHESIS

The fishing line is exactly as strong as (or perhaps weaker than) claimed.

ALTERNATIVE HYPOTHESIS

The fishing line is stronger than claimed.

A one-sided test is appropriate here because the group is interested in only one issue: fishing lines with extra strength. The group is relinquishing the right to test the other one-sided hypothesis and cannot decide that a line is significantly weaker than claimed.

For one particular brand of fishing line, the claimed strength is 20 pounds of force. The group's measurements of the force required to break the line are

$$\begin{aligned} \text{average breaking force} &= 28.1 \text{ pounds} \\ \text{standard error} &= 2.3 \text{ pounds} \\ \text{sample size} &= 60 \text{ repeated tests} \\ \text{degrees of freedom} &= 60 - 1 = 59 \end{aligned}$$

Using the t-table for a one-sided 95% confidence interval (the same as for a two-sided 90% confidence interval), they found 1.671 for 60 degrees of freedom, which is close enough for the 59 degrees of freedom here. The 95% one-sided confidence interval therefore goes from

$$28.1 - 1.671 \times 2.3 = 24.3 \text{ pounds of force}$$

to

infinity (no upper bound) pounds of force

Thus, the group is 95% sure that the mean breaking force for the fishing line in general is at least 24.3 pounds. (We subtracted here because we know that the confidence interval should contain the average value, 28.1, which occurs if we subtract rather than add.) Notice that the hypothesized value, 20 pounds, is outside the interval. The group therefore concluded that

This fishing line significantly exceeds *the claimed rating of* 20 *pounds of force (at the* 5% *level).*

For a different test involving another brand of fishing line, the claimed rating of 35 pounds fell within the confidence interval (from 32.8 pounds to infinity). The group therefore concluded that

This other brand of fishing line did not significantly exceed *its claimed rating.*

Note that this is the only conclusion available in this second case. We cannot decide whether it was equal to or is actually less than the claimed strength. To do this would have required a standard two-sided test in the first place.

Discussion

The decision to use a one-sided test (rather than a two-sided test) should be based on the situation and not the data. Unfair "peeking" occurs if we first notice that the sample average is larger than the hypothesized value and then decide to use the "larger than" one-sided test. If we are unsure whether to perform a one- or a two-sided test before looking at the data, it is best to do a two-sided test. If the two-sided test is significant, then you may conclude that the value is significantly larger or significantly smaller than the hypothesized value according to whether the average is greater or less than the hypothesized value.

Once the decision is made to perform a one-sided test, we simply use the appropriate one-sided confidence interval. Then we proceed as usual, determining that the result is significant if the hypothesized value is not in the confidence interval and is not significant if the hypothesized value is in the interval. (See Figure 11.4, page 372.)

Even though we give up the right to be interested in the other side when we complete a one-sided test, we gain a more powerful test. A one-sided test makes correct decisions more often when the one-sided alternative hypothesis we are interested in is really true. Technically, we make fewer Type II errors.

Exercise Set 11.8

25. Suppose you are running a bakery, and as part of your quality control program you weigh 20 loaves of bread. Their average weight is 1.05 pounds with an SD of .03 pound. You want to reject the hypothesis that the mean weight is really 1 pound or less, at the 5% significance level.

a. State the null hypothesis.

b. Find the one-sided 95% confidence interval.

c. Is the hypothesized mean below the lower limit?

d. State your conclusion.

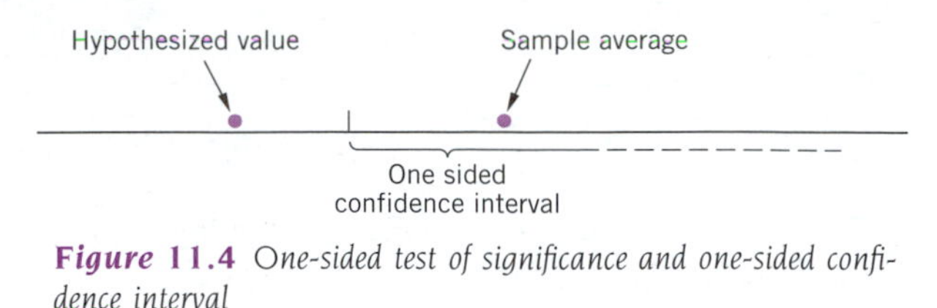

Figure 11.4 *One-sided test of significance and one-sided confidence interval*

26. What *t*-value would you use to do a one-sided test of significance at the 1% significance level if you had $n = 20$?

27. Rework Problem 25, using the 1% significance level.

Computer Calculations: One-Sided Tests of Significance

We can specify an alternative hypothesis with a direction. Using the data set from p. 359 (10, 11, 12, 13, 14, 15, 16), let us ask whether the results are significantly more than a hypothesized mean value of, say, 11. The output is shown in Display 11.6.

DISPLAY 11.6

```
MTB > TTest 11.0 C1;
SUBC> Alternative 1.

TEST OF MU = 11.000 VS MU G.T. 11.000

        N      MEAN      STDEV     SEMEAN      T      P VALUE
C1      7      13.000    2.160     0.816       2.45   0.025
```

Notice that we specified the hypothesized value of 11 in the first line. In the second line we specified alternative hypothesis "1," which is that the results are significantly "greater than"—not just different from—the hypothesized value. The output gives us a t of 2.45, with a probability of .025. This is statistically significant at the .05 level.

For comparison, let's run the same test but with the alternative hypothesis stated as "not equal" 11. The results are shown in Display 11.7. Notice that there are only two differences: the statement of the alternative hypothesis and the p-value. The p-value is twice as large, because we have twice as many possibilities of rejecting the null hypothesis.

DISPLAY 11.7

```
MTB > TTest 11.0 C1;
SUBC> Alternative 0.

TEST OF MU = 11.000 VS MU N.E. 11.000

        N      MEAN      STDEV     SEMEAN      T      P VALUE
C1      7      13.000    2.160     0.816       2.45   0.050
```

11.8 A Nonparametric Test for the Median

When the population distribution is extremely nonnormal, the t-test for testing the center of the population may be neither valid nor efficient in its ability to detect a significant difference.

Instead of testing the mean value, which is very sensitive to subtle changes in a long-tailed distribution, it may be preferable instead to test the median, which is more stable. A nonparametric test based on ranks (which are robust quantities) instead of averages and standard deviations (which are not robust) would be useful in these situations.

The Sign Test

We will learn a technique that is sometimes known as the **sign test** because it is often thought of as being based on the number of data values above the hypothesized value (which might be marked as "plus" signs) and the number of data values below the hypothesized value (which might be marked as "minus" signs). The probability theory is based on the binomial distribution.

Just as we can compute a confidence interval for the population median based on the data values from the sample at the correct ranks (chosen from a table of ranks), so we can use this confidence interval to test the hypothesis that the population median takes on a given hypothesized value. If the hypothesized value falls within the confidence interval, we accept the null hypothesis of no significant difference. On the other hand, if the hypothesized value falls outside the interval, then the difference is significant and the null hypothesis is rejected and the alternative hypothesis is accepted.

Even though the population might be extremely nonnormally distributed, the close relationship between confidence intervals and hypothesis testing still applies. The only assumptions needed for the validity of this method are that the data consist of a random sample from a population. This requires only that the data values be obtained independently from each other and that each population unit have an equal chance of being chosen.

EXAMPLE 11.11 *Mutual Savings Banks*

In the time from 1981 to 1982, the mutual savings banks of the state of Washington did not appear to do very well. Their changes in net worth over that year were mostly negative, indicating losses. Placed in order, the data set looks like Table 11.2.[2]

Based on this data set, let's test the hypothesis that the median change was 0, a number that indicates the break-even point between profits and losses. The null hypothesis would be that the median is 0, whereas the alternative hypothesis would be that it is not 0. The use of robust and nonparametric methods seems appropriate here because of the apparent outliers, one positive and one negative.

Table 11.2 *Changes in net worth of Washington state mutual savings banks*

Savings Bank of Puget Sound	35,045,000
Horizon Mutual	−245,000
State Mutual	−869,000
Mount Baker Mutual	−1,012,000
First Mutual	−1,449,000
United Mutual	−1,726,000
Prudential Mutual	−4,197,000
Lincoln Mutual	−6,335,000
Washington Mutual	−21,293,000

First, we find the nonparametric 95% confidence interval for the median. From Table 10.3 on page 342, we see that the data values at rank 2 provide this for a sample of size 9. Therefore,

The 95% confidence interval for the median goes from −6,335,000 *to* −245,000 *dollars.*

We are 95% certain that the (imaginary, in this case) population median falls within these limits.

Next, to test whether 0 is an acceptable hypothesized value for the population median, we look to see whether the confidence interval contains this value. It does not. Therefore,

Because 0 *is not between* −6,335,000 *and* −245,000, *we will* reject the null hypothesis *that the median is* 0 *and* accept the alternative hypothesis *that the median is not* 0.

Practically speaking, the banks had a significantly bad year as a group. Even though one did very well, and even allowing for sampling variability, we conclude that it is highly unlikely that these banks all had an even chance of turning a profit.

Discussion

When we compute a confidence interval or do a hypothesis test, we should consider the assumptions we are implicitly making. Can we reasonably assume that the list of banks in Example 11.11 represents data gathered on a random sample from a population? The answer is perhaps no and perhaps yes, depending on our point of view.

If we follow a strict interpretation of the rules, we see that this list is not a random sample of anything because it is a complete list of all of the mutual savings banks of the state of Washington. No randomization was done in selecting this sample to be representative of any population. Based on this interpretation, it would not be valid to do a confidence interval or a hypothesis test; the necessary assumptions would not be satisfied.

However, guided by a broader interpretation of the rules, we can use our imagination in applying statistical inference. We could say that these banks were chosen out of the imaginary population of all conceivable mutual savings banks that could have existed at this time (see Sidebar 11.1). Although the process might not have been perfectly random, we are free to assume that it was approximately random and to evaluate the consequences of this assumption. Based on this interpretation, a confidence interval or hypothesis test would measure something about the financial health of "mutual savings banks of Washington in the abstract" by considering the data set to be a sample of these.

In Example 11.11 we use the broader interpretation. In particular, the null hypothesis that the population median is 0 says that half of the banks made a profit, whereas half suffered losses. Choosing randomly from such a population, is it conceivable that we could find a data set like the one we have, which is so dominated by losses? The hypothesis test answers this question.

Keep in mind that the conclusions reached in this particular example are based on questionable assumptions about the population and sample. Sometimes, however, we must make uncertain assumptions to proceed with any analysis at all. When this is the case, we are permitted to make the necessary assumptions, provided we do it openly and report the results with an appropriate warning to the reader of the report.

Exercise Set 11.9

28. Suppose you have the following random sample data.

0 0 2 3 5 6 8 9 10 12 14

Someone suggests that the population median is really 10. You want to test this hypothesis on the basis of these sample data.

a. State the null hypothesis.

b. What are the limiting ranks for a 95% confidence interval?

c. Is the hypothesized median inside the confidence interval?

d. State your conclusion.

29. Suppose you have the following random sample data.

.2 1.4 2.2 .3 0 1.8 1.3 .5 .6 1.0 .9 .7

Someone suggests that the population median is really 1.5. Can you reject this hypothesis on the basis of the sample data? Show your work.

30. *The Seattle Times* for 14 August 1995 lists these asking prices for used Toyota 4Runners:

18,000	15,950	15,899	15,500	13,900	11,888
11,500	9,900	9,384	8,000	6,200	5,999

Assuming these are a random sample of the available used 4Runners, can we reject the hypothesis that the median price of a used 4Runner is $13,000? (Show your work; state your conclusion.)

31. Using the same data and assumptions as in Exercise 30, can we reject the hypothesis that the median price of a used 4Runner is $7,000? (Show your work; state your conclusion.)

Summary

Statistical hypothesis testing is making decisions based on data. In this chapter, we make a decision between two possibilities about the population mean based on a sample of data: "Is the mean equal to a certain fixed and hypothesized value, or is it not equal to that value?" The *null hypothesis* claims that the mean and the hypothesized value are equal, whereas the *alternative hypothesis* claims that they are different. The decision will result in one of these hypotheses being accepted as true and the other being rejected as false. Because the test is based largely on the difference between the hypothesized value and the sample average (which will ordinarily be somewhat different because of sampling randomness), we are also testing whether this apparent difference is *statistically significant,* or whether this observed difference could be explained by randomness alone.

There is a simple close connection between the confidence interval for a population quantity and a hypothesis test of this quantity against a hypothesized value. If the hypothesized value is within the confidence interval, then we accept the null hypothesis that the mean equals the hypothesized value. On the other hand, if the hypothesized value falls outside the confidence interval, then we reject the null hypothesis and accept the alternative hypothesis that they are different. We also declare the difference between the sample average and the hypothesized value to be statistically significant.

This connection is fortunate because no new work is involved in doing a hypothesis test once the confidence interval has been computed. If the population is normally distributed, then the confidence interval for the mean (as computed using the *t*-tables) provides an accurate test of hypothesis. This test is called the *Student's t-test* and is named for Student, the pen name of W. S. Gosset. If the population is approximately normal (including the case of percentages arising from yes/no data situations), then this procedure is also accurate enough for practical applications.

The practical interpretation of the result of a hypothesis is as follows. If we accept the null hypothesis that the population mean is equal to the hypothesized value, then we conclude that a population with a mean equal to the hypothesized value could reasonably have generated the sample that we have and that the difference between the sample average and the hypothesized value could simply be due to random sampling variability. On the other hand, if we reject the null hypothesis and accept the alternative hypothesis that the population mean and the hypothesized value are different, then we declare that the difference between the sample average and the hypothesized value is significant and probably not due to randomness alone. Finding a significant difference is often interpreted as granting a "license" to explain that difference.

A deeper theoretical look at the situation shows that there are four possibilities in a testing situation: two possible true situations (mean is or is not equal to the hypothesized value) and two possible decisions (mean is or is not equal to the hypothesized value). Naturally, we hope the decision is the same as the true situation, but it will not always be. There are two kinds of errors that may occur. The *Type I error* refers to incorrectly deciding that the mean does not equal the hypothesized value (incorrectly declaring a significant difference) and is controlled, conventionally, at the 5% level. The *significance level* of a test is the (controlled) probability of a Type I error. The *Type II error* refers to incorrectly deciding that the mean is equal to the hypothesized value (incorrectly failing to declare a significant difference). The type II error cannot easily be controlled.

When the standard error is known exactly, we use the critical value from the t-table with infinity as its degrees of freedom, as we learned before for constructing the confidence interval in this case. For significance levels other than 5%, we use the appropriate t-table for the desired significance level, basing our test on a confidence interval whose confidence level is 100% minus the desired significance level.

One-sided tests should be used only with caution. They are appropriate if only values on one particular side of the hypothesized value are of interest. The decision to use a particular one-sided test must be based only on the situation and not the particular data values. A one-sided test is easily performed based on the appropriate one-sided confidence interval. Users of one-sided tests gain power in detecting significant results on that one side of the hypothesized value, but give up the right to make significance statements about results that come out on the other side. *When in doubt, use a two-sided test.*

If the population is very nonnormally distributed due to skewness, long tails, or extreme outliers, then a nonparametric test for the median called the *sign test* might be more appropriate. This test is based on the nonparametric confidence interval for this kind of situation.

Problems

Basic Reviews

1. Suppose a random sample produces an average value of 100 on some measure, and the 95% confidence interval for the population value goes from 90 to 110. Is a hypothesized population mean value of 120 significantly different from these results?

2. Suppose a random sample produces an average value of 100 on some measure, and the 95% confidence interval for the population value goes from 90 to 110. Is a hypothesized population mean value of 105 significantly different from these results?

3. Suppose that someone proposes that a certain species of iris flower has petals that average 1.9 cm in length. You take a random sample of iris flowers of this species. On doing the calculations, you find that your sample has an average petal length of 1.4615 cm with a 95% confidence interval of 1.398 cm to 1.525 cm. Is the hypothesized population value of 1.9 cm significantly different from the sample results?

4. Suppose that someone proposes that the mean length of the rabbits on a certain island is 40 cm. You take a random sample of 23 rabbits. On doing the calculations, you find the sample averages 43.6 cm in length and the 95% confidence interval for the population mean is from 38.72 cm to 48.49 cm. Are these results significantly different from the hypothesized value?

5. Suppose that someone proposes that the mean length of the rabbits on a certain island is 40 cm. You take a random sample of 257 rabbits. On doing the calculations, you find the sample averages 45.3 cm in length and the 95% confidence interval for the population mean is from 44.175 to 46.425 cm. Are these results significantly different from the hypothesized value?

6. Suppose that someone proposes that killer bee wings average 8.6 mm in length. You take a random sample of killer bees and find that the sample average length is 8.61 mm and the confidence interval for the population value runs from 8.47 mm to 8.74 mm. Is there a statistically significant difference between the hypothesized population value and the sample results? State your conclusion as a single sentence.

7. Suppose that someone proposes that killer bee wings average 9.0 mm in length. You take a random sample of killer bees and find that the sample average length is 8.61 mm and the confidence interval for the population value runs from 8.47 mm to 8.74 mm. Is there a statistically significant difference between the hypothesized population value and the sample results? State your conclusion as a single sentence.

8. Does the sample average of 3.48 defects per 1,000 differ significantly from a hypothesized value of 5 per 1,000, given that the standard deviation is 6.2 and the sample size is 36?

9. Does the sample average of 122.3 tweets per hour differ significantly from a hypothesized value of 109.1 tweets per hour (established by extensive previous studies), given that the standard deviation is 3.10 and the sample size is 8?

Analysis

10. From a particular day's chocolate ice cream production, the average weight of a random sample of cartons was found to be 1.21 pounds with a 95% confidence interval from 1.09 to 1.33 pounds. The cartons are supposed to weigh exactly 1 pound, but this cannot be controlled perfectly. Given this information, should we accept or reject the hypothesis that the mean weight of the entire day's production is 1 pound?

11. A company plans to market an additive to gasoline by claiming that it increases the gas mileage and will therefore save you money. In fact, the company completed a study of 20 cars that showed an average improvement of 2.6 mi/gal. However, in the fine print, they also report a standard error of the mean of 1.7 mi/gal. They claim that because the standard error is less than the average improvement, they have proven that their product really works. What do you think about their claim? Base your answer on a test of the hypothesis at the 5% level that the mean improvement is 0 (that is, that their additive makes no difference).

12. A random poll of 355 people shows that 54.2% of them plan to vote in favor of building a new bridge into town. May we safely conclude that this initiative would pass if the election were held right away? Why or why not?

In Student's original 1908 paper, he gives some examples of hypothesis tests. The data represent the additional sleep gained by using experimental soporific (sleeping pill) medications. The results from 10 patients are shown in Table 11.3 (page 378). Use these results to answer Problems 13–16.

Table 11.3

PATIENT	MEDICATION 1: (DEXTRO HYOSCYAMINE HYDROBROMIDE)	MEDICATION 2: (LAEVO HYOSCYAMINE HYDROBROMIDE)
1	.7	1.9
2	−1.6	.8
3	−.2	1.1
4	−1.2	.1
5	−1.0	−.1
6	3.4	4.4
7	3.7	5.5
8	.8	1.6
9	.0	4.6
10	2.0	3.4

13. a. Compute the average, standard deviation, and standard error for the extra sleep obtained under medication 1.

b. Find the 95% confidence interval for the extra sleep obtained under medication 1.

c. Using a *t*-test, decide whether medication 1 is effective by testing whether the extra sleep obtained is 0 or not.

14. a. Compute the average and standard error for the extra sleep obtained under medication 2.

b. Find the 95% confidence interval for the extra sleep obtained under medication 2.

c. Using a *t*-test, decide whether medication 2 is effective.

15. a. For medication 1, find the nonparametric 95% confidence interval for the extra amount of sleep.

b. Using the sign test, test whether the median extra amount of sleep from medication 1 is significantly different from 0.

16. a. For medication 2, find the nonparametric 95% confidence interval for the extra amount of sleep.

b. Using the sign test, test whether the median extra amount of sleep from medication 2 is significantly different from 0.

17. A researcher honestly believes that a new medication reduces blood pressure. However, a test on five people did not show a significant average reduction in blood pressure. Is it still possible that the medication is effective? What might be done to find out?

18. The amount of color in an oil-based paint has been measured as 8.3 g/kg of paint with a known standard error of .17 g, based on extensive past experience.

a. Test, at the 5% level, whether this amount differs significantly from the amount, exactly 8 g, that the manufacturer is supposed to insert into each can of paint.

b. Perform the same test, but at the 10% level.

c. Perform the same test, but at the 1% level.

d. Compare and explain any differences between your answers to parts (a), (b), and (c).

For Problems 19–21, suppose the quality control department of a factory will make adjustments to the compact disc pressing machinery only if the day's production of 5,000 discs shows *significantly more* than 3% defects.

19. a. Why would a one-sided test be appropriate here?

b. Today's production showed 173 defects. What is the defect rate?

c. Do a one-sided test at the 5% significance level. Would the quality control department make adjustments to the machinery in this case?

d. If the machinery is actually producing defects at a rate of exactly 3% over the very long term, about how often (that is, what proportion of days) would adjustments be called for if the one-sided test as in part (c) is used?

20. Suppose the factory's production had resulted in 245 defects per 5,000 discs.

a. What is the defect rate?

b. Do a one-sided test at the 5% significance level against a hypothesized value of 3% defects. Would the quality control department make adjustments to the machinery in this case?

21. Suppose the factory's production had resulted in 96 defects per 5,000 discs.

a. What is the defect rate?

b. Do a one-sided test at the 5% significance level. Would the quality control department make adjustments to the machinery in this case?

22. Suppose that following an earthquake, an insurance company does a statistical sampling survey to assess the extent of damages and expected claims. For the 200 policyholders surveyed, the average damage was $4,800 with a standard deviation of $1,300.

a. Find the standard error. What does it measure?

b. Find the 95% confidence interval for the mean damage for all policyholders in the earthquake area.

c. The insurance company, based on past experience, had hypothesized that for an earthquake of this size in this area, the mean damage would be $5,000. Test this hypothesis using the information about the sample just given.

23. Suppose your company has been charged with discrimination. The fact is more men than women have executive positions: There are 19 men and only 17 women.

a. What is the sample size?

b. What is the percentage of women?

c. What is the standard error of this percentage?

d. What is the 95% confidence interval for this percentage?

e. Test the hypothesis of discrimination.

f. Is there conclusive evidence of discrimination?

24. A market survey has shown that people will spend an average of $2.34 each for your product next year, based on a sample survey of 400 people. The standard deviation of the sample was $.72.

a. Indicate how accurate the $2.34 figure is as an indication of the mean of the entire population by providing a 95% confidence interval.

b. You had thought earlier that the mean would be exactly $2.00. Is this still a reasonable possibility? Perform a hypothesis test at level .05 to find out whether the $2.34 observed differs significantly from this preconceived value of $2.00.

25. Suppose a survey of 200 randomly selected consumers in the market for a new car showed that they were willing to pay an average of $15,862 with a standard deviation of $2,815.

a. Compute (and give the name for) the quantity that indicates the uncertainty of the average value.

b. Compute the 95% confidence interval for the average amount a consumer in the market would pay for a new car.

c. The retail price of a new kind of car, the Buttersworth, has been set at $20,448. Does this price significantly exceed the average price a consumer shopping for a new car would pay? (In answering this question, clearly state what you mean by "significant.")

PART THREE

More Than One Group of Numbers

chapter 12

Comparing Two Groups of Numbers

In this chapter, we discuss new kinds of data sets. When we have *two* related single groups of numbers presented as a single data set, we would like to compare them if at all possible. In doing this, we will be answering the questions "How different are the groups, and in what ways?" (an *exploration* and *estimation* consideration) and "Are they really different at all?" (a *hypothesis testing* consideration).

We have already done some comparisons of simple groups of numbers—for instance, using box plots. Now we will also be testing to see whether apparent differences are reasonably explained as accidents of random sampling or whether the differences are *statistically significant.*

12.1 Onward to Richer Data Structures

Data sets come to us in many different forms and structures. So far, we have developed the standard statistical procedures on a single group of numbers. Through this simplest kind of data set, we were able to illustrate the basic kinds of statistical activities, including exploration, summarization of center and variability, estimation, assessment of the error due to sampling randomness, and hypothesis testing.

As we proceed to study richer data structures, we will see that many of the techniques and concepts that we already know can be extended in a straightforward way to fit the new situation. There will be relatively few completely new ideas. However, creativity is involved in adapting familiar methods to these more complex situations. But the payoffs will be great because the structure contained in a more complicated data set is usually harder to see without the use of these methods.

Always keep in mind that your job as a statistician includes thinking up creative ways of looking at the data. There are always simpler data structures contained within larger, more complex data sets. Sometimes a data set should be broken into pieces, and each piece analyzed separately. In particular, most data sets contain individual groups of numbers, and using the methods we know can help us understand the individual groups as part of the larger process of building a more complete understanding of all of the connections and interactions.

In this chapter we will be concerned with ideas and methods for the comparison of two single groups of numbers. That is, we will be looking for similarities and differences between two situations. Exploratory methods—directly comparing the histograms or box plots by eye—naturally come first, because these are the only way to see and characterize the entire set of data with all of its individual patterns, quirks, exceptions, and problems. Next will come the more formal ideas of statistical comparison: estimating how different the population mean values are, constructing a confidence interval for this mean difference, and testing hypotheses about whether we can really detect any difference at all beyond mere sampling variability.

12.2 Exploring Two Groups of Numbers

Recall (from our initial discussion of single groups of numbers) that it is impossible to directly compare just any two groups of numbers. To do a proper comparison, at a minimum we will have to require that the two groups be measured in the same units. Note that in each of the examples below, the same kind of measurement is being performed in each of the two situations.

EXAMPLES OF TWO GROUPS THAT CAN BE COMPARED

1. The heights of tall buildings in each of two cities (measured in the same units)
2. The reductions in blood-sugar level in each of two groups of people who received different medications (measured in the same units)
3. The heights of the children in two different classes (measured in the same units)
4. The weights of the children in two different classes (measured in the same units)

It is important to note that we are making two restrictions. The first is that we focus on a single variable, such as the heights of children. We are not talking about the *heights* and *weights* of a single group of children. We are talking about the heights of *two groups* of children. (Of course, the relationship between two variables, like height and weight, is also of interest and is discussed in Chapter 15, on bivariate data.) The second restriction is, in principle, obvious: The units used to measure the variable have to be the same for both groups. If we have the heights of the students in a European class measured in centimeters and those in an American class measured in inches, we have a problem that we have to take care of before we try to compare them.

Visually Comparing Two Groups

To compare two groups of numbers, we can do two histograms, one for each group, being careful to display them on the same scale and near enough to each other so that a comparison can be made properly. We could use box plots (also on the same scale) instead for a somewhat less detailed comparison. Recall that we used multiple box plots in Chapter 3 to compare several groups of numbers.

By "comparison" we mean the search for similarities and differences in central values, variabilities, and distribution shapes. The simplest case, and the one we often hope we will see, occurs when the centers are different but the variabilities and distribution shapes are the same. Indeed, many statistical methods tacitly assume this simple situation and will not work properly otherwise. However, because the real situation is often more complicated, it is important to explore the data a bit before turning it over to calculations that require restrictive assumptions that might not be satisfied by a particular data set.

What follows is a series of examples showing the kinds of results that might be obtained just by comparing histograms.

Two Identically Normally Distributed Groups

The very simplest possible case to consider is when the distributions are essentially identical (except for some ever-present randomness) in all respects: the center, variability, and distribution shape in each of the groups. If the common distribution is normal, so much the better for applying statistical inference later on. Such a case might occur when the groups are chosen without any regard to (that is, independently of) the variable being measured or other variables like it. This case is also the one that will represent the null hypothesis of "no difference between groups" when we move along from exploration to inference.

EXAMPLE 12.1 *Comparing Normal Distributions: Gasoline Mileage*

Suppose we are planning a study of automobile gasoline mileage. As part of our preliminary studies, we take a collection of 70 cars and randomly divide them into two groups with 35 cars in each group. If we measure the mileage for each car by the same technique the EPA uses to estimate highway mileage, then the (hypothetical) results could be displayed as two histograms, one for each group of cars, as shown in Display 12.1 and Figure 12.1 on page 386.

Because the groups were chosen randomly, there should not be any systematic differences in the measurements of mileage. (We would, however, expect to see some differences if the groups were not selected at random, for example, if we divided them into groups according to how heavy or how old they were.) The only differences that should show up here are random ones, arising from several sources: the sampling variability arising from the random selection of groups, the variation from one car to another in the original group of cars before it was divided up, and all of the usual natural sources of experimental variation including measurement error.

Indeed, the histograms just presented do appear very similar. They have about the same central values, approximately the same variability, and their distribution shapes look fairly normally

DISPLAY 12.1a

GASOLINE USED (IN TENTHS OF GALLONS) FOR THE FIRST RANDOM GROUP OF 35 CARS

Stem	Leaves
19	7
20	
21	3
22	1 4 5
23	2 5 7 7 8
24	0 0 0 2 5 6 7
25	0 0 1 2 2 3 6 6 6 8 9
26	3 7 8 9
27	3 3
28	
29	4

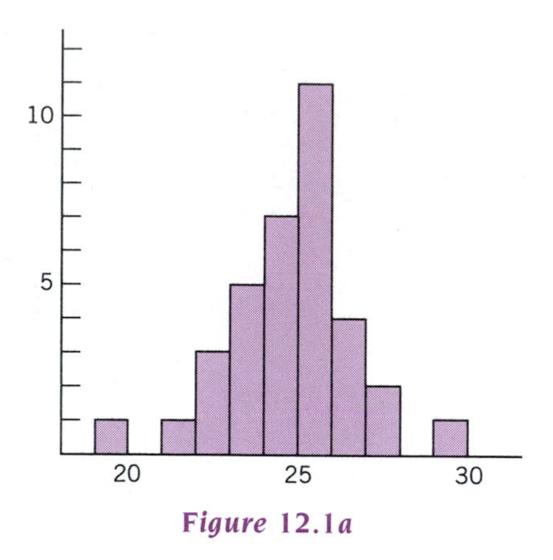

Figure 12.1a

DISPLAY 12.1b

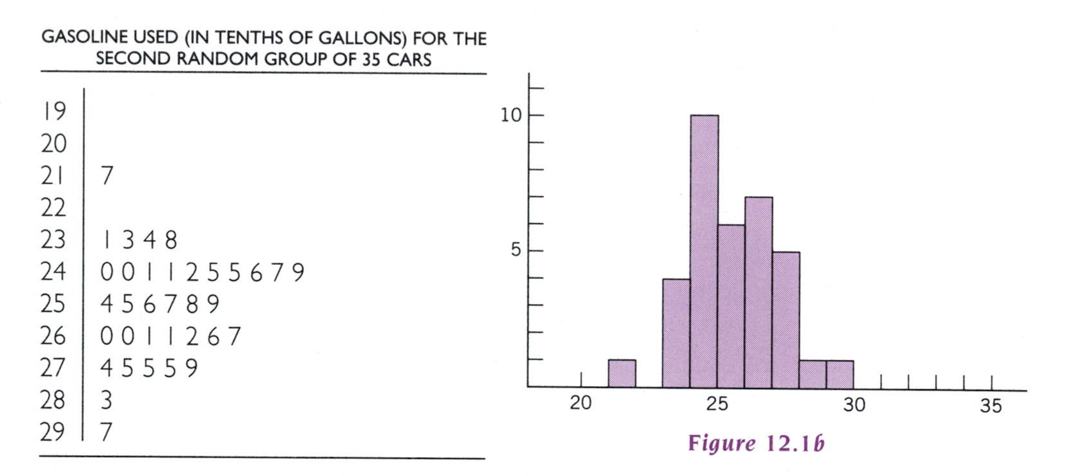

GASOLINE USED (IN TENTHS OF GALLONS) FOR THE SECOND RANDOM GROUP OF 35 CARS

Stem	Leaves
19	
20	
21	7
22	
23	1 3 4 8
24	0 0 1 1 2 5 5 6 7 9
25	4 5 6 7 8 9
26	0 0 1 1 2 6 7
27	4 5 5 5 9
28	3
29	7

Figure 12.1b

distributed—at least as an approximation. Of course, the averages and standard deviations will not be exactly equal because of the sampling variability. To assess some of these differences, we will use the confidence intervals and hypothesis tests discussed later in this chapter.

Two Identically Nonnormally Distributed Groups

Sometimes the data values will not follow a normal distribution, yet will still represent the same population. Again, this would be represented by two histograms with similar centers, variabilities, and distribution shapes even though the (common) shape does not follow the normal distribution curve.

EXAMPLE 12.2 *Comparing Nonnormal Distributions: Savings Accounts*

Suppose that once a month a bank randomly selected 75 of its $1,000 minimum-balance deposit certificate accounts and examined the sample in detail to make sure everything was all right and to keep track of what was going on with the accounts in general. The worker who usually ran the sampling program was out sick on Monday, so someone else had the computer select a sample for analysis. On Tuesday, the regular worker returned and, without realizing it had already been done, ordered the computer to select a sample, which of course was a different random sample of 75 accounts than the one selected Monday.

The result was that the research department received the results of two different random samples of accounts. By noticing the amount of the differences in the results, they could see the degree of the effects of random sampling. (Partly as a result of this, they decided to increase the sample size to achieve a more accurate estimated picture of the population of savings accounts.)

The hypothetical histograms of the two independent random samples are shown here in Display 12.2 and Figure 12.2. We see in these histograms a clear indication that the groups are the same except for random variations. Because the histograms are on the same scale, we can directly compare them, and we see that the distributions are both very skewed toward high values and concentrated in about the same places. The fact that the outliers to the right in each picture look a little different in the two samples should be no surprise. Because these observations in the right-hand tail each reflect just one data value, they are expected to be more variable than, say, the center of the distribution.

DISPLAY 12.2a

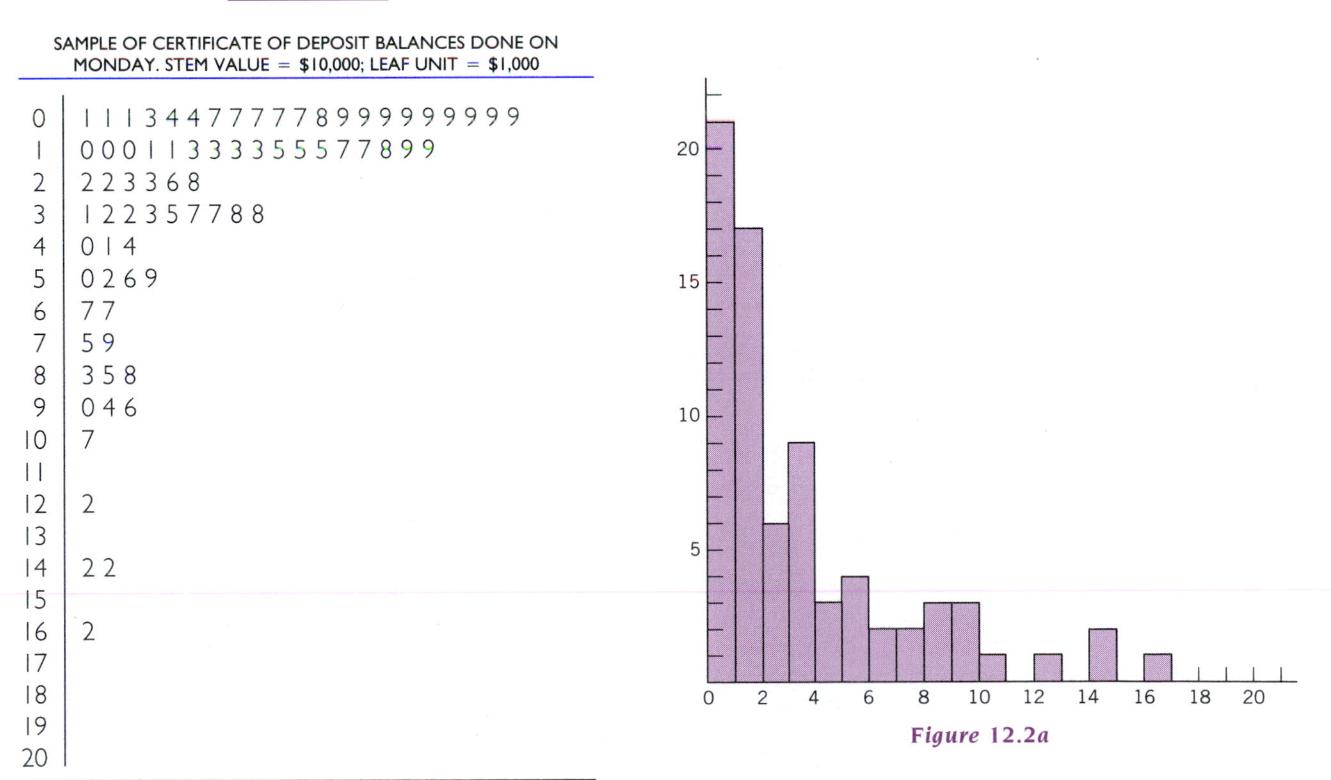

SAMPLE OF CERTIFICATE OF DEPOSIT BALANCES DONE ON MONDAY. STEM VALUE = $10,000; LEAF UNIT = $1,000

Stem	Leaves
0	1 1 1 3 4 4 7 7 7 7 7 8 9 9 9 9 9 9 9 9
1	0 0 0 1 1 3 3 3 3 5 5 5 7 7 8 9 9
2	2 2 3 3 6 8
3	1 2 2 3 5 7 7 8 8
4	0 1 4
5	0 2 6 9
6	7 7
7	5 9
8	3 5 8
9	0 4 6
10	7
11	
12	2
13	
14	2 2
15	
16	2
17	
18	
19	
20	

Figure 12.2a

DISPLAY 12.2b

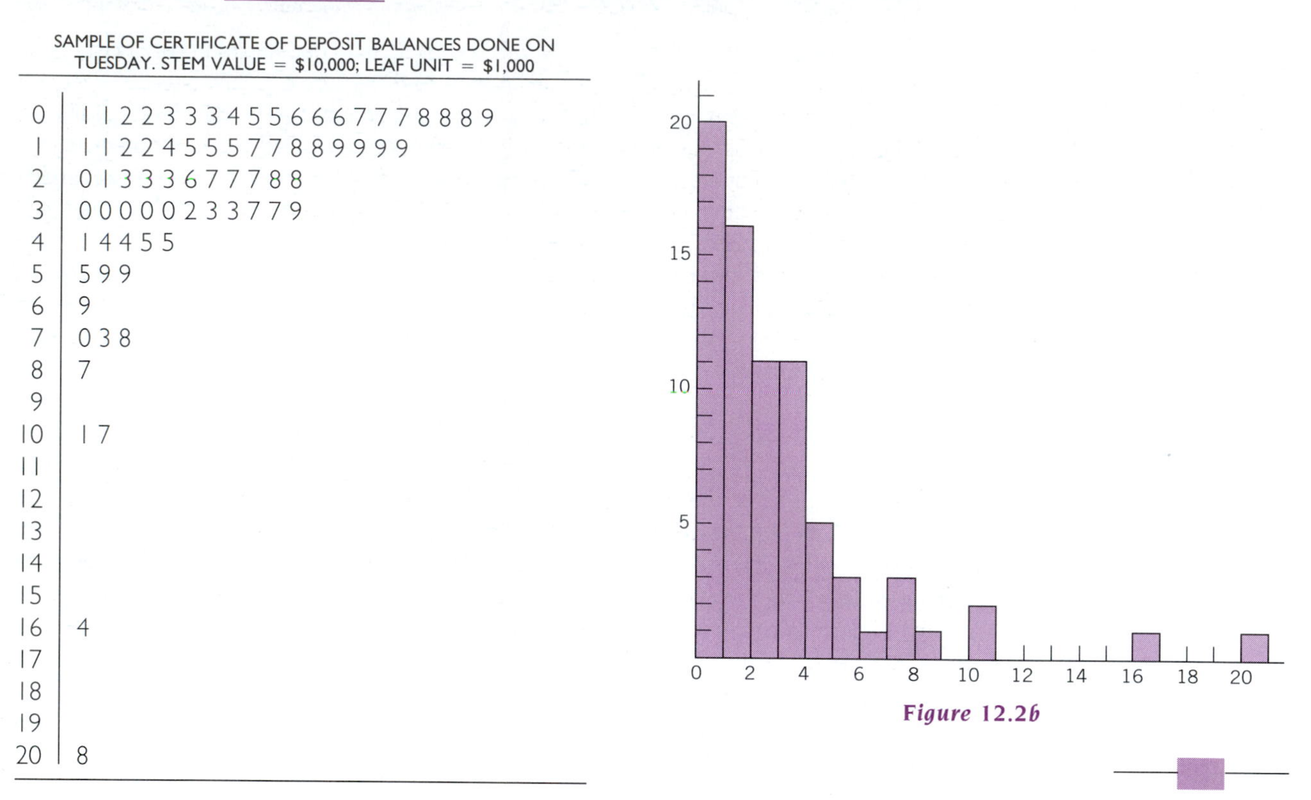

SAMPLE OF CERTIFICATE OF DEPOSIT BALANCES DONE ON TUESDAY. STEM VALUE = $10,000; LEAF UNIT = $1,000

Stem	Leaves
0	1 1 2 2 3 3 3 4 5 5 6 6 6 7 7 7 8 8 8 9
1	1 1 2 2 4 5 5 5 7 7 8 8 9 9 9 9
2	0 1 3 3 3 6 7 7 7 8 8
3	0 0 0 0 0 2 3 3 7 7 9
4	1 4 4 5 5
5	5 9 9
6	9
7	0 3 8
8	7
9	
10	1 7
11	
12	
13	
14	
15	
16	4
17	
18	
19	
20	8

Figure 12.2b

Two Normally Distributed Groups with the Same Variability

Because it is the easiest situation to deal with in a statistical way, we are happiest in comparing two groups of numbers when it happens that (1) the variabilities are the same in the two groups (approximately, of course), (2) the data seem to follow a normal distribution, and (3) only the centers are different. In cases like this, we would legitimately be able to say that one group is just like the other one, except for generally being a certain amount larger or smaller.

EXAMPLE 12.3 Normal Distributions with the Same Variability: Iris Flowers

The Fisher–Anderson iris data,[1] a famous data set in statistics, include several kinds of measurements on flowers of several species. (See Figure 12.3.) In particular, consider the lengths of the petals of a sample of 50 flowers from each of two species: *Iris virginica* and *Iris versicolor.* The histograms, one for each species, are shown in Figure 12.4.

Examining these histograms of petal length, several facts become clear. First, the distributions have similar shapes that seem fairly close to that of a normal distribution. Second, the widths of the distributions are similar, indicating similar variabilities in each of the two species. Third, the centers of the distributions are different. *Iris virginica* generally has larger petals than *Iris versicolor.* Note that the distributions look as though one is a "shifted" version of the other, shifted approximately 10 mm, expressing the fact that the petals of one species tend to be about 10 mm longer than the petals of the other species.

We see that, although the distributions show that *Iris virginica* flowers are generally larger

Figure 12.3

than *Iris versicolor,* there is nevertheless some overlap. Some individuals seem to go against the trend. For example, the smallest *Iris virginica* has a petal length that would be about typical for *Iris versicolor.* Later on, when we do statistical inference in a situation like this, we will concentrate on the more general differences between populations as reflected in the samples and not so much on the particular individual data values that may or may not follow the trend.

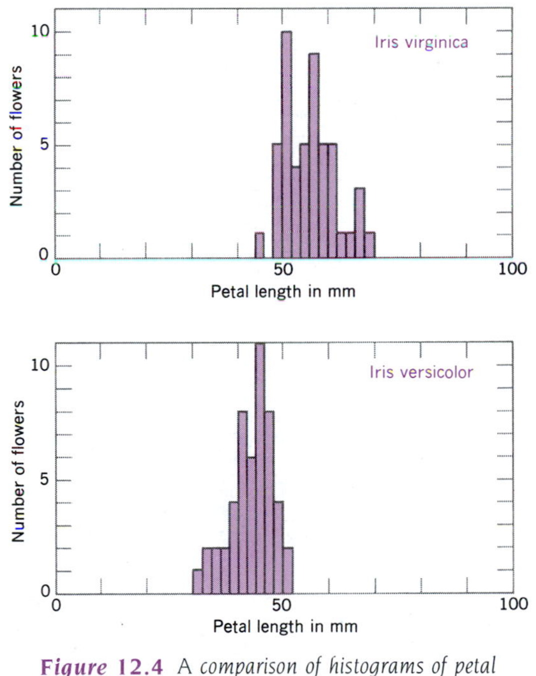

Figure 12.4 *A comparison of histograms of petal lengths for two species of iris flowers*

Two Groups with Different Centers and Different Variabilities

Life is not always so simple that only the central values of the two groups are different. It is fairly common to see data sets in which the variabilities are also different and perhaps skewed (making them not normally distributed) as well. It is important to recognize this situation because some statistical procedures we will learn about would be wrong or improper to use in such a case. In some of these situations we can use a transformation to set things right, so that the transformed groups of data are normally distributed with similar variabilities. Then the only difference would be their central values.

EXAMPLE 12.4 *Different Centers and Different Variabilities: Incomes*

Different neighborhoods may have different income distributions. Sometimes a landmark such as the railroad track will divide a town into two parts with different socioeconomic levels. Suppose that a random sample of households is examined on each side of the tracks to look for such a difference. The (hypothetical) results are shown in Display 12.3 and Figure 12.5 in the form of a histogram for each group.

These two histograms differ in almost every way. In terms of centers, it appears that the west side of the tracks is more affluent, with a higher income level in general than the east side. But the variabilities are also different: The west side appears to have more dispersion than the east side, perhaps because its larger center gives the distribution more room to spread out.

Both distributions appear to be slightly skewed toward high values, although not to an extreme extent. In a situation like this, where the variabilities are different and there is some skewness, transformation of the data could work very well in simplifying the analysis. The desired results that a transformation might bring about would include (1) making the distributions more symmetric (and therefore more normal) and (2) making the variabilities more similar from one group to the other. We should, of course, be careful to use the same transformation for both data sets so that a comparison of the transformed values will be a valid comparison of the underlying situations.

DISPLAY 12.3a

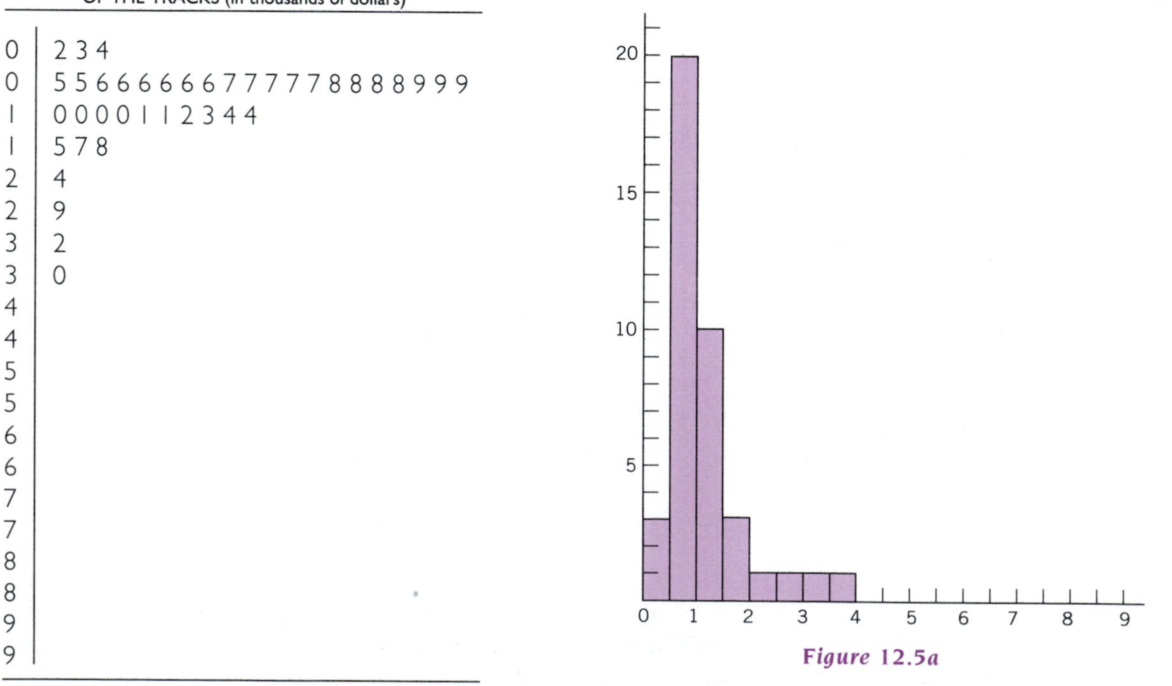

MONTHLY INCOME FOR 40 HOUSEHOLDS TO THE EAST OF THE TRACKS (in thousands of dollars)

Stem	Leaves
0	2 3 4
0	5 5 6 6 6 6 6 6 7 7 7 7 7 8 8 8 8 9 9 9
1	0 0 0 0 1 1 2 3 4 4
1	5 7 8
2	4
2	9
3	2
3	0
4	
4	
5	
5	
6	
6	
7	
7	
8	
8	
9	
9	

Figure 12.5a

DISPLAY 12.3b

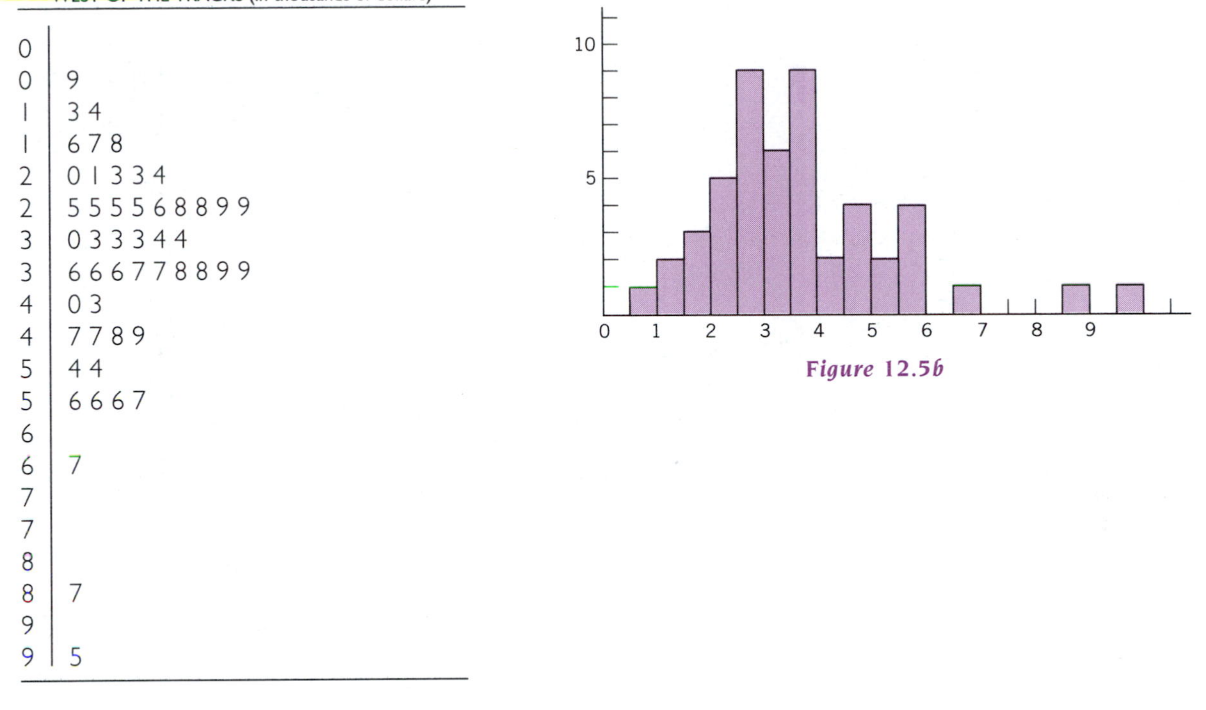

MONTHLY INCOME OF 50 HOUSEHOLDS TO THE WEST OF THE TRACKS (in thousands of dollars)

0	
0	9
1	3 4
1	6 7 8
2	0 1 3 3 4
2	5 5 5 5 6 8 8 9 9
3	0 3 3 3 4 4
3	6 6 6 7 7 8 8 9 9
4	0 3
4	7 7 8 9
5	4 4
5	6 6 6 7
6	
6	7
7	
7	
8	
8	7
9	
9	5

Figure 12.5b

DISPLAY 12.4a

LOGARITHM OF MONTHLY INCOME FOR 40 HOUSEHOLDS TO THE EAST OF THE TRACKS

7	
7	7
8	2 4 6 6
8	7 7 7 8 8 8 8 8 9 9 9
9	0 0 0 0 1 1 1 2 2 2 2 3 3 4 4
9	5 6 6 7 8
10	1 2 3
10	5
11	
11	

Figure 12.6a

DISPLAY 12.4b

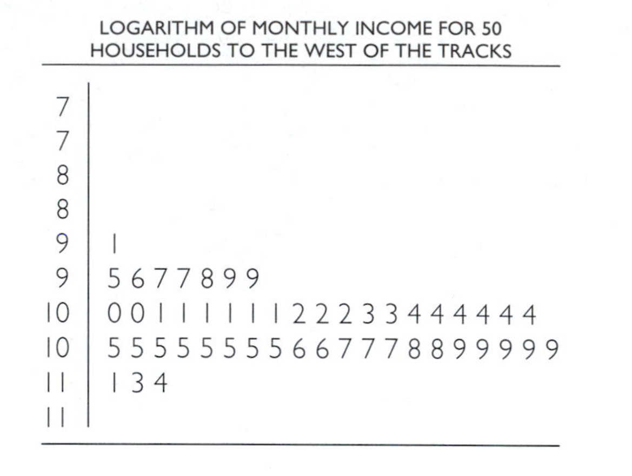

LOGARITHM OF MONTHLY INCOME FOR 50 HOUSEHOLDS TO THE WEST OF THE TRACKS

Stem	Leaves
7	
7	
8	
8	
9	1
9	5 6 7 7 8 9 9
10	0 0 1 1 1 1 1 1 1 2 2 2 3 3 4 4 4 4 4 4
10	5 5 5 5 5 5 5 5 6 6 7 7 7 8 8 9 9 9 9 9
11	1 3 4
11	

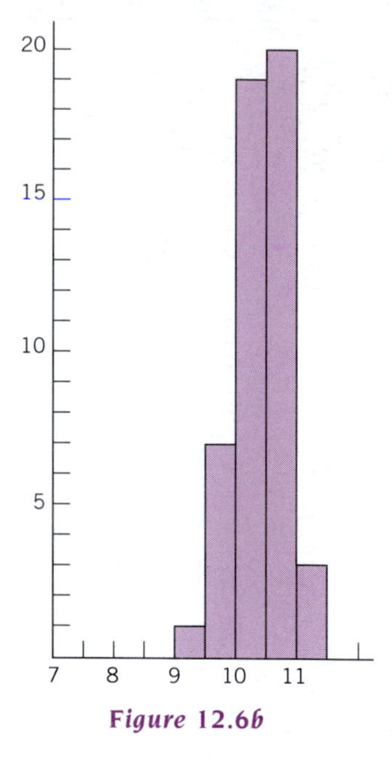

Figure 12.6b

If we take the logarithms of the monthly income values in each group, histograms of the transformed data values look like what appears in Display 12.4 and Figure 12.6. Now, even though the two original data sets of incomes were different in center and variability, the transformed groups differ only in the center. Their distribution shapes are nearly alike, and their variabilities are much more similar to each other than they were on the original untransformed scale.

Discussion

A transformed configuration with roughly equal variability is both easier to think about and more amenable to further statistical treatment than the original untransformed values. Note also that the distribution is more symmetric and normal looking after the transformation. It is as though the data had been measured in the wrong kind of units to begin with, and transformation was needed to set things right for several aspects (normality and equal variabilities) of the situation.

How would we "transform back again" to report the results of an analysis? Remember that logarithms are easy to interpret because they convert multiplication to addition. In this case, a constant additive shift in the distribution on the log scale would convert back to a constant multiple in the original scale of incomes. We could summarize the situation by saying that incomes on the west side of the tracks are just like incomes on the east side, except that they are larger by some factor (which we might estimate to be somewhere between 2 and 5). That is, people on the west side make about 2 to 5 times as much as those on the east side. This certainly does seem to be an uneven distribution of wealth, but to say for sure, we would have to test that hypothesis (which we will soon learn to do).

Although transformations may not always help, there are many situations in which we come closer to the ideal—normal distributions with equal variability differing only in central location—by using a suitable transformation of the data values.

12.3 *Inference About the Mean Difference*

Now that we understand how to look at and compare data sets consisting of two single groups of numbers, let's consider inferences about the populations that generated the data. It is not *always* appropriate to go beyond summarization of the data and into more general inferences; there are restrictions. For example, there must be a meaningful population more general than the data set itself, at least in our imagination, and certain assumptions about the distributions must be satisfied. But when there is a valid population, there is no substitute for the power of statistical inference and its ability to make definite statements in the face of uncertainty.

It is usually the population that we are really interested in. Many comparisons boil down to the question "How different is this from that?" This question might be answered by giving an estimate of the mean difference (perhaps with a confidence interval for the difference and maybe even a significance test of it). The following example will help solidify these ideas.

EXAMPLE 12.5 *Testing a Difference Between Sample Averages: Dirty Versus Clean Air Filters*

To understand the effect of a dirty air filter on gasoline mileage, suppose we take 20 cars and divide them randomly into two groups of 10 each. Then, one group has a clean air filter installed in each car, whereas the other group uses old dirty filters (of the same type as the clean filters). The gasoline mileage of each car is measured, forming the data set of two groups of numbers.

This data set would be interesting to look at for general features, such as shape, centers, and variabilities. However, our particular interest is in the difference between the sample averages in the two groups, which would give information about any mileage difference due to the condition of the air filter. Further, we want to infer something definite about dirty air filters in general rather than just about the cars in these two samples. This involves making a statement about a larger population—cars in general—based on a particular sample—these 20 cars. Suppose we get the mileage figures shown in Table 12.1.

Table 12.1 *Gas mileage in cars with clean and dirty air filters*

CLEAN	DIRTY
19.0	16.0
22.0	20.0
24.0	21.0
24.5	21.5
25.0	23.0
25.0	21.0
25.5	22.5
26.0	25.0
28.0	25.0
31.0	27.0

The question "How much effect does a dirty air filter have on gasoline mileage?" corresponds to the following question about the statistical populations: "What is the difference between the (population) mean gasoline mileage when a clean air filter is used as compared to when a dirty air filter is used?" The general question is

How different are the population means?

An estimate of the answer to this question, which is the best that we can obtain from data, is given by

How different are the sample averages?

because the sample averages are known to us but the population means are not.

The hypothetical results are summarized in Table 12.2. The difference between the sample averages is easily calculated:

$$\text{average difference} = 25.00 - 22.20$$
$$= 2.80 \text{ mpg}$$

This suggests that there is a penalty of 2.80 mpg for using a dirty air filter. But is this apparent penalty real? For all that we know, it could be that the dirty air filters just happened to be assigned a few cars with higher fuel consumption than average. After all, there is some variability, so it is unlikely that the two groups would have identical averages. How can we find out how accurate the 2.80 mpg penalty is? How do we test whether it means anything at all?

Table 12.2 *Summary of gas mileage data*

	CLEAN FILTER GROUP	DIRTY FILTER GROUP
Number of cars (sample size)	10	10
Average fuel use (mpg)	25.00	22.20
Standard deviation mileage	3.21	3.09

Assumptions Are Necessary. In general, the assumptions needed to proceed with inference about a difference between the population means are as follows.

1. Each of the two groups of data follows an approximately *normal distribution.*
2. The two distributions have nearly *identical variabilities.*
3. Each group is a random sample from its population.
4. All observations are *independent* of one another.

Be careful! If these assumptions are not satisfied, the classical inferences based on the standard error are not necessarily valid. Fortunately, the first two assumptions (normality and equal variances) need be only approximately satisfied because these inferences are robust and will still remain valid, to a good approximation.

It is also fortunate that when these first two assumptions are not satisfied, a transformation can often help the distribution be symmetric (and therefore more normal) and at the same time help the variabilities be closer to each other on the transformed scale. When we use the same transformation on both groups of numbers and then test based on the transformed data, we are still testing for a difference between the two groups.

Inference Is Based on the Standard Error

If the variabilities of the two populations are not equal, we can perform a slightly different version of the two-sample *t*-test. We can use a separate standard error estimate for each sample and use a different formula for the degrees of freedom. We examine this procedure in the computer analysis section (page 401).

If the population variances are equal, statistical theory provides us with a standard error to accompany the average difference, and the *t*-ratio has a *t*-distribution. As usual, this standard error indicates how much the sample estimate is likely to vary from the real population mean difference. Just as we computed confidence intervals and performed hypothesis tests before, based on estimates and standard errors, we will use the same procedures here. The only difference is in some of the methods of computation. Fortunately, the ideas involved are remarkably similar to those we have already studied.

From the standard error for the average difference, we construct a confidence interval that indicates the precision of the estimate of mean difference and construct a test of hypothesis about whether the difference is or is not equal to a given standard value. Of particular importance is the test of a standard value equal to 0, which tests the hypothesis of there being "no difference" (at least in population mean values) between the two populations that were sampled.

We will proceed exactly as we did for constructing confidence intervals and for testing hypotheses about a mean value based on an estimated mean value (the average) together with its standard error and degrees of freedom.

PROCEDURE: FINDING THE STANDARD ERROR OF AVERAGE DIFFERENCE

1. Find the average, the standard deviation, and the sample size for each of the two groups.
2. To find the *pooled variance estimate,* square each of the standard deviations and then multiply each square by 1 less than the sample size for its group. Add these two results and divide the sum by 2 less than the sum of the sample sizes. This is basically an averaging procedure that combines information from both samples to estimate their common variabilities.
3. Find the reciprocals of the two sample sizes and add them together.
4. Multiply the results of steps 2 and 3.
5. Find the square root of this number. This is the *standard error* for the *difference* between the average values of the two samples.

FORMULA

Standard Error of Average Difference

To find the pooled variance estimate, or estimated population variance, S^2, we have

$$S^2 = \frac{(n_1 - 1)S_1^2 + (n_2 - 1)S_2^2}{n_1 + n_2 - 2}$$

where n_1 and n_2 are the sample sizes, and S_1 and S_2 are the standard deviations in each of the two groups.

We use the averaged estimate of the population variance to calculate a standard error for the difference between two sample averages:

$$\text{SE dif} = S_{\bar{X}_1 - \bar{X}_2} = \sqrt{S^2\left(\frac{1}{n_1} + \frac{1}{n_2}\right)}$$

EXAMPLE 12.6 Standard Error of the Average Difference: Fuel Use

For the fuel use problem of Example 12.5, we can compute the standard error of the average difference, which was 2.80 mpg, using these five steps, as follows.

1. Averages are 25.00 and 22.20, standard deviations are 3.21 and 3.09, and sample sizes are 10 and 10, respectively, for the two groups (clean air filter and dirty air filter groups).
2. The pooled variance estimate is

$$S^2 = \frac{(10 - 1) \times 3.21^2 + (10 - 1) \times 3.09^2}{10 + 10 - 2}$$
$$= \frac{9 \times 10.30 + 9 \times 9.548}{18}$$
$$= \frac{92.70 + 85.93}{18}$$
$$= 9.924$$

3. The sum of the reciprocal sample sizes is

$$\frac{1}{10} + \frac{1}{10} = .1 + .1$$
$$= .2$$

4. Multiplying these two results, we find

$$9.924 \times .2 = 1.9848$$

5. Finally, taking the square root, we find that

$$\text{Standard error} = \sqrt{1.9848}$$
$$= 1.4088 \quad \text{or about } 1.409$$

This number, 1.409, estimates the variability of the sample average difference that is due to sampling randomness. It is like the standard deviation of a normal data distribution, or the standard error of a sample average, in the sense that about two-thirds of the time the observed value will be within this much of the true population mean difference.

For the two-sample case we are considering here (with equal population variability), the degrees of freedom that we use will be found by adding together the degrees of freedom contributed by each group individually. Because each group contributes 1 fewer than its sample size, we have

$$\text{Degrees of freedom} = (\text{sum of sample sizes}) - 2$$
$$= (\text{size of first sample} - 1) + (\text{size of second sample} - 1)$$

As before, the number of degrees of freedom reflects how good an estimate of the underlying variability we have. We have "pooled" (combined together) the information from both samples about variability. This is reflected in the degrees of freedom of the standard error.

In this example, there are 10 cars in the clean air filter group and 10 cars in the dirty air filter group. The number of degrees of freedom is therefore

$$\text{Degrees of freedom} = 10 + 10 - 2 = 18$$

To construct the confidence interval for the mean population difference between two groups, we will use the *t*-tables. The 95% confidence interval for the difference in population mean values therefore goes from

$$(\text{difference in sample averages}) - (t_{\text{degrees of freedom}} \times \text{standard error})$$

to

$$(\text{difference in sample averages}) + (t_{\text{degrees of freedom}} \times \text{standard error})$$

To test whether the difference is equal to a hypothesized fixed value, we simply see whether that hypothesized value is within the confidence limits. If it is inside the limits, we accept the hypothesis that the mean difference is (or could be) equal to the hypothesized value. If the standard value is outside the limits of the confidence interval, we conclude that the mean difference is not equal to the hypothesized value and declare that the sample average difference is significantly different from the hypothesized value.

With 18 degrees of freedom (because we have sample sizes of 10 and 10 in the two groups), the *t*-table value is 2.101. Recall the average difference is 2.80 mpg with a standard error of 1.409 mpg. The 95% confidence interval therefore goes from

$$2.80 - 2.101 \times 1.409 = -.16$$

to

$$2.80 + 2.101 \times 1.409 = 5.76$$

Now we can formally test the hypothesis to see whether there is any mileage difference between clean and dirty air filters. The null hypothesis is that the true mean population difference is equal to the fixed hypothesized value of 0, which expresses the condition of "no difference." The alternative hypothesis is that this difference is anything other than 0.

To perform the test, we look at the confidence interval to see whether it contains 0. It does. We therefore

accept the null hypothesis of no difference

and conclude that there is no difference in gas mileage between clean and dirty air filters. In particular,

the difference of 2.80 mpg is not statistically significant

and could easily have been due to sampling variability alone.

Discussion

This confidence interval and test have the same interpretation as before. Confidence intervals calculated this way will contain the true mean population difference about 95% of the time, in the long run (but we do not know in any particular case whether it does or not). The test will incorrectly declare that a significant difference exists about 5% of the time when, in fact, there is no difference (and we again do not know for any particular test).

This test for whether the mean difference is equal to a fixed hypothesized value is known as the **two-sample t-test.** Except for the details in computing the standard error and the degrees of freedom, it is just like the one-sample *t*-test that we used for testing a single population mean on the basis of a single sample.

The computed confidence interval, from −.16 to 5.76, puts a whole new light on the situation in the examples. Our best estimate for the difference in fuel consumption due to a dirty air filter was 2.80 mpg (the average difference), but the confidence interval indicates that this estimate was not very precise. We are not even confident that the mean population difference has the same first digit (2) as the average suggested. Having computed the confidence interval, we have a much clearer picture of the quality of the estimate, 2.80. We find that we cannot have great faith in its accuracy, because the sampling randomness shows that the population mean difference could be anywhere from −.16 to 5.76 mpg.

We can interpret this finding of no difference one of two ways. First, it could be that there truly is no difference between fuel consumption with clean and dirty air filters, and that would be the end of that. Second, it could be that there is a difference, but that it was masked and obscured by randomness. It may be that we could refine the analysis to show the difference. For example, it is possible that with another experiment with larger sample sizes (which lead to more precise estimates of the mean difference), significance might be found. However, with the present experiment and the analysis so far, we can conclude only that there is no evidence that the state of the filter (clean or dirty) made any difference in gas mileage.

Exercise Set 12.1

1. In 1991, the 36 tournaments on the men's professional golf tour paid an average first prize of $186,110, with a standard deviation of $31,740. The 26 events on the women's professional golf tour paid an average of $74,230, with a standard deviation of $29,550.

a. What is the observed difference between men's and women's average first-place prizes?

b. Assuming that these data are from independent random samples from hypothetical populations, what is the 95% confidence interval for this difference?

c. Is the observed difference statistically significant?

2. Back in 1979, the average of the men's first-place prizes in 35 events was $51,310 with a standard deviation of $8,720. The women's tour averaged $16,967, with an SD of $5,034, for 30 events.

a. What is the observed difference between men's and women's average first-place prizes?

b. Assuming that these data are from independent random samples from hypothetical populations, what is the 95% confidence interval for this difference?

c. Is the observed difference statistically significant?

3. Looking at the data in Exercises 1 and 2, we notice that both the men's and the women's prizes are substantially greater in 1991 than they were in 1979. Part of this no doubt results from the effects of inflation, but have the prizes changed more than can be accounted for just by inflation? Checking the *Statistical Abstract of the United States, 1994,* Table 747, we find that the Consumer Price Index in 1979 was at 72.6, and in 1991 it was at 136.2. Doing the arithmetic, we find that a 1979 dollar turns into $1.876 in 1991, just on the basis of inflation. Applying this inflation correction factor to all the women's prizes in 1979 and recalculating our summaries, we find that the women's average 1979 prize was worth $31,830 in 1991 dollars, with an SD of $9,440. Is the difference between this and the actual 1991 women's average prize statistically significant?

Computer Analysis: Two-Sample *t*-Test

The hypothetical gas mileage figures for cars with clean and dirty air filters are stored in two columns of a Minitab worksheet, and named "Clean" and "Dirty." We can use Minitab to do a two-sample *t*-test of the null hypothesis that the two populations really have the same mean, as shown in Display 12.5.

DISPLAY 12.5

```
MTB > TwoSample 95.0 'Clean' 'Dirty';
SUBC> Alternative 0;
SUBC> Pooled.

TWOSAMPLE T FOR Clean VS Dirty

          N      MEAN      STDEV     SEMEAN
Clean    10      25.00     3.21      1.0
Dirty    10      22.20     3.09      0.98

95 PCT CI FOR MU Clean - MU Dirty: (-0.2, 5.76)

TTEST MU Clean = MU Dirty (VS NE): T = 1.99 P = 0.062 DF = 18

POOLED STDEV = 3.15

MTB >
```

The first command line tells the program to conduct a two-sample test of the difference between "Clean" and "Dirty." The second command line specifies the alternative hypothesis, number 0, which is that there is no difference. The third command line tells the computer to use a pooled estimate of the variance. The output from the program starts with a listing of the summary information for each group of numbers (N, MEAN, and so forth).

The results of the test calculations are presented in two equivalent ways. First, there is the calculation of a 95% confidence interval for the difference between the population mean values. It extends from $-.2$ to 5.76, which, except for rounding error in the second decimal place, is the same as the interval we calculated. Since the hypothesized value of 0 is included in the interval, we cannot reject the hypothesis that there is no difference.

The next-to-last line of the output presents the results as a classic *t*-test. The observed difference is divided by the standard error for the difference and $t = 1.99$ is obtained. With 18 degrees of freedom this has a probability of .062, which is not statistically significant.

A probability of .062 is not far from statistical significance. Not many people will abandon an idea about how the world works when the evidence testing the theory comes this close to supporting it. And it happens that in this case we have a legitimate refinement of the analysis available that will, under some circumstances, allow us to claim statistical significance.

We really do have a theory about the direction of the difference. We think that cars with clean air filters should have better gas mileage than cars with dirty air filters. Because we are willing to specify the direction of the difference in advance, we can use a one-sided test of significance, provided that we are willing to give up any right to claim a significant difference in the other direction, no matter how strongly it is supported by the data.

DISPLAY 12.6

```
MTB > TwoSample 95.0 'Clean' 'Dirty';
SUBC> Alternative 1;
SUBC> Pooled.

TWOSAMPLE T FOR Clean VS Dirty

          N      MEAN      STDEV     SEMEAN
Clean     10     25.00     3.21      1.0
Dirty     10     22.20     3.09      0.98

95 PCT CI FOR MU Clean - MU Dirty: (-0.2, 5.76)

TTEST MU Clean = MU Dirty (VS GT): T = 1.99 P = 0.031 DF = 18

POOLED STDEV = 3.15

MTB >
```

The computer output for the one-sided test is shown in Display 12.6. We have specified alternative hypothesis 1, which is that the difference is greater than (not just different from) 0. (You do have to keep careful track of which variable is being subtracted from which to make sure you get the direction right.)

The confidence interval calculation in Minitab remains the same as for a two-sided test. (The program calculates a two-sided CI, rather than a one-sided CI, as we do.) On the t-test itself, the null hypothesis is stated as MU Clean = MU Dirty, versus an alternative hypothesis that the difference is greater than 0. (Notice that since "Dirty" is being subtracted from "Clean," the difference will be greater than 0 if "Clean" gets better mileage than "Dirty.") The value of t is the same as when we calculated a two-sided test (1.99), but now the probability, $p = .031$, is half as much. Again, this is because we are only looking at half of the possibilities of being that far from 0.

Since $p = .031$ is less than the classic threshold of $p = .05$ for declaring statistical significance, we can declare that clean air filters lead to better mileage than dirty air filters and that the difference is statistically significant. We can reject the null hypothesis of no difference in favor of the alternative hypothesis that clean air filters lead to better mileage.

Let us repeat an ethical caution here. One-sided tests are legitimate only in cases where you have specified the expected direction of the difference *in advance of* looking at the data and are willing to give up the right to claim significance in the other direction. The problem is that some fields are so complicated that reasonable explanations can be proposed for almost anything *after* you have looked at the data. Looking at the data and then deciding on the direction of the test is the equivalent of giving yourself a second flip of the coin.

Populations with Different Variability

So far, we have been assuming that the two populations we have samples from have the same variability. We may not want to make this assumption, either because we have definite reason to think that it is wrong or because we are trying to be conservative. We may think the assumption is wrong because we have a theory that says the populations should have different variability or because our sample variabilities are very different. We may want to be conservative in the sense that we will make it harder for the differences to be statistically significant. If we do not assume that the populations have equal variability, it is often harder to show a statistically significant difference, so we are usually being conservative if we do not assume equal variability.

EXAMPLE 12.7 *Using Separate Estimates of Variability: Car Speeds*

Consider a field experiment to determine the effect of various methods (signs, parked police cruisers, active patrolling, etc.) to reduce car speed on a particular section of road. The raw data from the data set CARSPEED.MTW are shown in Display 12.7. For now, we will focus on the results for methods 1 and 4.

DISPLAY 12.7

```
MTB > Retrieve 'C:\MTBWIN\STUDENT1\CARSPEED.MTW'.

MTB > Print 'METHOD 1' 'METHOD 4'.

ROW     METHOD 1     METHOD 4

  1        72           55
  2        55           52
  3        63           62
  4        61           54
  5        74           55
  6        52           50
  7        56           58
  8        67           55
  9        68           48
 10        68           62
 11        71           58
 12        66
```

Of course, it is difficult to say anything just looking at the raw numbers, so we prepare a dot plot, shown in Display 12.8. By eye, it appears that speeds are generally lower with method 4, that both distributions are roughly normal, and that the results of method 1 may be more variable than the results of method 4. A two-sample *t*-test of whether the differences are statistically significant is given in Display 12.9 (page 402). Since we do not tell the program to assume common variability (the "pooled" subcommand), it uses the formulas for different population variabilities.

DISPLAY 12.8

```
MTB > DotPlot 'METHOD 1' 'METHOD 4';
SUBC>  Same.
      .     . .      .  .   . . :   . .   .
-----+---------+---------+---------+---------+---------+-METHOD 1


           .
 .  .  .  .:    :      :
-----+---------+---------+---------+---------+---------+-METHOD 4
  50.0      55.0      60.0      65.0      70.0      75.0
```

DISPLAY 12.9

```
MTB > TwoSample 95.0 'METHOD 1' 'METHOD 4';
SUBC> Alternative 0.

TWOSAMPLE T FOR METHOD 1 VS METHOD 4

              N      MEAN      STDEV      SEMEAN
METHOD 1     12      64.42     7.10       2.1
METHOD 4     11      55.36     4.46       1.3

95 PCT CI FOR MU METHOD 1 - MU METHOD 4: (3.9, 14.2)

TTEST MU METHOD 1 = MU METHOD 4 (VS NE):
     T = 3.69 P = 0.0017 DF = 18
```

It does indeed appear that method 4 reduces speed substantially more than does method 1 and that the difference is statistically significant. The observed reduction is a little over 9 miles per hour, which is statistically significant at $p = .0017$.

Notice the degrees of freedom, which are reported as df = 18. By the formula we used when we assumed equal variabilities, we would have had df $= n_1 + n_2 - 2 = 21$. The smaller degrees of freedom are required because we are estimating the two variabilities separately. The actual calculation usually results in a fractional part for the degrees of freedom (for example, df = 18.63), but Minitab discards the fractional part.

Exercise Set 12.2

4. We have data on the arcs of bird claws provided by Alan Feduccia.[2] An arc of 180 means that the claw curves around in a half circle; an arc of 90 means that it curves around in a quarter circle. It appears, in general, that trunk-climbing birds have claws that are more curved than those of birds that live on the ground. The data for two species, one a trunk climber and the other a ground dweller, are shown below. Clearly, these samples are not random with respect to the population of all trunk climbers and all ground dwellers. But we are willing to assume that the samples are random with respect to species. We want to see whether, in comparing these two species, we can conclude that their claws indeed have different arcs.

Dendrocopus pileatus, a trunk climber, species code 31
Claw arcs: 152 152 154 154 160 163 165 167 170 174

Neomorphus geoffroyi, a ground dweller, species code 19
Claw arcs: 64 68 68 73 81 81 81 82 85 93

We enter the data in a Minitab file and conduct a two-sample *t*-test with the results shown in Display 12.10. Using the output in the display, answer the following questions.

a. What is the average claw arc of species 31's claws?

b. What is the average claw arc of species 19's claws?

c. From part (a) and (b), what is the difference in the average claw arcs of the two species?

d. What is the 95% confidence interval for the population value of the difference in claw arcs?

e. Is a value of 0 (no difference) included in the confidence interval?

f. Can we reject the null hypothesis of no difference?

g. Is the difference in claw arcs statistically significant?

DISPLAY 12.10

```
MTB > TwoT 95.0 'ClawArc' 'Species';
SUBC> Alternative 0;
SUBC> Pooled.

TWOSAMPLE T FOR ClawArc

        N      MEAN        STDEV     SEMEAN
31      10     161.10      7.94      2.5
19      10     77.60       9.05      2.9

95 PCT CI FOR MU 31 - MU 19: (75.5, 91.5)

TTEST MU 31 = MU 19 (VS NE): T = 21.94 P = 0.0000 DF = 18

POOLED STDEV = 8.51
```

h. What is the value of the t-statistic for the difference in average claw arcs?

i. What is the probability of getting a t-statistic this large (or larger), if the null hypothesis is true?

j. What is the value of the pooled estimate of the standard deviation?

12.4 Difference Between Sample Medians

The t-test makes assumptions that the data do not always meet. In that case, we can use the **Mann–Whitney *U* test** for two independent samples. In broad conceptual terms, we can think of the Mann–Whitney test as the nonparametric alternative to a two-sample t-test. It tests for the statistical significance of the difference between the sample medians, but does so in a way that relies on the ranks of the data values rather than their numerical values. Thus, the data need be measured only at the level of ordered categories. However, for the probability calculations to be correct, the data must still meet two assumptions: The two samples must be independent random samples, and the two populations from which they are drawn must have the same distribution shape (whatever it may be).

The Mann–Whitney U Test

For the sake of simplicity, we will assume that we have a computer with a statistics package to perform the calculations. The basic procedure for the Mann–Whitney U test is this: The values from both samples are arranged in order and ranked together. The smallest observation in either sample will be ranked 1, the next smallest 2, and so on. Tied observations are all assigned their average rank. In essence, the test is conducted by summing the ranks of the data in one sample and comparing this sum to the sum of the ranks in the other sample. If the two sums are about the same, the difference may be due to chance variation. If one sum is significantly larger than the other, we can reject the hypothesis that the difference is due to chance.

EXAMPLE 12.8 *The Mann–Whitney U Test: Feelings of Crowdedness*

The data for this example are from a psychological experiment in which separate samples of men and women were asked to rate how crowded they felt after sitting in a green room for a period of time. (Data are from data set CROWDS.MTW, distributed by Minitab.) The possible ratings are from 1 to 30. Notice that these ratings are basically ordinal measures, comparable to checking off a series of very finely graded categories (for example, "I feel very, very, very crowded" vs "I feel very, very, very, very crowded"). Since they are measured at the ordinal level, we cannot treat them as numerical measurements comparable to weighing people on a scale, and the *t*-test may not be appropriate. Thus we will use the Mann–Whitney test for a difference of medians.

Dot plots of the data are shown in Display 12.11. By eye, we see that the lowest rating was about 3 and the highest was 21. In general, the two sets of ratings appear very comparable, with the exception of the one very low rating.

DISPLAY 12.11

```
MTB > Retrieve 'C:\MTBWIN\STUDENT8\DATA\CROWDS.MTW'.

MTB > DotPlot 'FEM RATE' 'MALERATE';
SUBC> Same.

                          .  .        .
                          :  :     .  :       :        .
 .                      : :  :  :  :  :  :  : :  :     :
-----+---------+---------+---------+---------+---------+-FEM RATE

                     .
                     :
                     :    .     .  .     :  : .     .
                     :  : :  :  :  :  .  :  : :  .  :
-----+---------+---------+---------+---------+---------+-MALERATE
    3.5       7.0      10.5      14.0      17.5      21.0
```

Display 12.12 gives the results of the Mann–Whitney test. The first few lines of the output are familiar and easy to interpret. But the line giving the "point estimate for ETA1 − ETA2 is 1.000" is new material; it is the estimated average difference in population medians. This value is not equal to the observed difference between the sample medians (14.5 − 14.0 = .5, not

DISPLAY 12.12

```
MTB > Mann-Whitney 95.0 'FEM RATE' 'MALERATE';
SUBC>   Alternative 0.

Mann-Whitney Confidence Interval and Test

FEM RATE N = 36   Median =   14.500
MALERATE N = 36   Median =   14.000
Point estimate for ETA1-ETA2 is   1.000
95.1 Percent C.I. for ETA1-ETA2 is (-0.999,2.000)
W = 1383.5
Test of ETA1 = ETA2 vs. ETA1 ~= ETA2 is significant at 0.4371
The test is significant at 0.4355 (adjusted for ties)

Cannot reject at alpha = 0.05
```

1.0) because it is computed on the basis of the difference between all of the ranked pairs. The 95% confidence interval is also based on all the pairwise differences and makes the additional assumption that the data are measured numerical values. (The actual interval reported is the 95.1% CI because with this data set, that was the CI closest to 95% that could be calculated precisely. With other data sets, different CIs near 95% will be reported.) The reported value for "W" is the sum of the ranks of the first sample and is used in the program's internal probability calculations.

Finally, the output gives three statements of the results of the significance test. The first two are statements in terms of p-values, with the first declaring that the test is "significant at 0.4371," with the term "$p =$" left implicit, and the second reporting the significance level after correcting for ties. Clearly, .4371 is (a lot) bigger than .05, so this difference is not statistically significant at any of the standard significance levels. The third statement makes this explicit, stating "Cannot reject at alpha = 0.05."

The test confirms our impression from the dot plots: The difference in the sample medians is small (14.5 versus 14.0) and not statistically significant.

12.5 Testing for a Difference in Percentages

In many cases we are interested in counted or yes/no data, such as the number of people planning to vote for a candidate. Recall that some of the techniques we have studied can be used with yes/no data by treating the sample proportion of Yes responses as the sample average and then calculating a standard deviation using the number of Yes responses and the number of No responses. Once we have calculated the sample proportions and standard deviations (using the formulas given in Chapter 4), we can go on to do a test of the difference of two-sample proportions using the procedures developed in this chapter.

EXAMPLE 12.9 Testing for a Difference in Percentages: Voter Preferences

Suppose you are running for the office of social director of a club. You have had an exciting and controversial campaign so far, and last week your assistants conducted a survey that found that 61.4% of the 83 people sampled who expect to vote are planning to vote for you. You were pleased to hear that and naturally attributed the good news to your ability to relate to people in meaningful ways.

Now, with the election only 2 days away, an assistant says that the latest poll, with a sample size of 91 people, finds that only 57.1% plan to vote for you, down from last week's 61.4%. The trend seems clear: Your support is eroding and some last-minute campaigning is needed to win the election.

Considering this development, you remember something from your statistics course: It is possible to test whether this apparent loss of support is real or whether it could merely be an artifact of sampling variation and not representative of voters in general. You also recall that sample sizes of 83 and 91 are not terribly large and that more than a little sampling variability is always to be expected. In fact, many professional polling operations try to interview 1,500 people to obtain accuracy of around 1%. So perhaps you should worry less about the polls' results.

However, you decide to formally test the null hypothesis that your supporters have not

Table 12.3 *Summary of voter surveys*

	GROUP 1 (LAST WEEK'S SURVEY)	GROUP 2 (TODAY'S SURVEY)
Average (proportion)	.614	.571
Standard deviation	.490	.498
Sample size	83	91

wavered (that only the particular random samples are different and the population has remained unchanged) against the alternative hypothesis that a change has occurred. You call us in as consultants, and we compile the summaries of the situation as shown in Table 12.3.

To compute the standard deviation, we used the formula from page 127, where we found X (the number from group 1 who said they would vote for you) by multiplying $.614 \times 83 = 50.962$, which rounds to 51 people. The standard deviation of .490 for group 1 is therefore the square root of $[51 \times (83 - 51)/(83 \times 82)]$. Similar computations lead to the standard deviation .498 for group 2.

Now, to test for erosion of support, we will test for the existence of a difference in the mean population values. We will proceed as we have before, with a two-sample *t*-test, after we check the assumptions listed next.

1. *Normal distributions.* These are not normal distributions, but we know that averages of percentage-type data usually have a distribution that is close to the normal distribution because of the central limit theorem.
2. *Identical variabilities.* The standard deviations certainly do seem very close to each other (.490 and .498), so this should not be a problem. In fact, when testing a difference between two percentages or proportions, this assumption is not crucial to the validity of the test.
3. *Independent random samples.* Your assistant assures you that each group was randomly selected using a special table of random numbers and that each group was selected independently, without regard to the other group.

Next, we need to compute the change in your support, which is the average difference.

$$\begin{aligned}\text{apparent loss of support} &= .614 - .571 \\ &= .043, \quad \text{or } 4.3\%\end{aligned}$$

We also need the standard error for this loss, which is computed with the same formula we have been using for the two-sample problem:

$$\begin{aligned}&\text{standard error (of the loss of support)} \\ &= \sqrt{\frac{82 \times .490 \times .490 + 90 \times .498 \times .498}{172} \times \left(\frac{1}{83} + \frac{1}{91}\right)} \\ &= \sqrt{.00563} \\ &= .075, \quad \text{or 7.5 percentage points}\end{aligned}$$

Now we are in a position to construct the confidence interval. Even before we do any calculations, we see that the loss of support (4.3%) is of the same order of magnitude as its standard error (7.5%), which is a clear indication that it could be due to random sampling fluctuations. To confirm this, we look up the *t*-value for 172 degrees of freedom and are content to use the last value, 1.960, for very large samples. The confidence interval for the population loss of support therefore goes from

$$.043 - 1.960 \times .075 = -.104 \quad (\text{or } -10.4\%)$$

to

$$.043 + 1.960 \times .075 = .190 \quad (\text{or } 19.0\%)$$

We come upon a sobering thought. We are 95% confident that the change (if any) in voter preference is somewhere in the wide range between a 10.4% gain (a gain because the "loss" was negative) and a 19.0% loss. It seems that those numbers (61.4% and 57.1%) were not as accurate as they first looked and, in particular, that the last decimal digit in 57.1% is hardly meaningful at all as an indication of what voters plan to do.

The hypothesis test itself will accept the null hypothesis that there was no change (0 change) in population preference because 0 is contained within this confidence interval. The net results are twofold. First, the direct result of the test says there is no evidence that your support has eroded. Any apparent (sample) loss of support is not significant and is within the margin of error that we expect from random samples of these sizes. The second result is that you now have a much better idea of the accuracy of these polls.

In particular, you begin to wonder about that 57.1% voter preference you have now. Is it significantly different from 50%? Or could it be that your opponent is equally matched to you after all? After quickly testing this hypothesis using methods from Chapter 11, you make a note to require that standard errors be reported for each percentage value in the future (this would be the standard error for an individual percentage, not the standard error for a difference between two percentages that we just used).

You then leave your room to deliver some more campaign speeches.

Exercise Set 12.3

5. Suppose that you have random samples of two brands of jelly beans, each of size 100. In one group, 46% are green. In the other group, 65% are green.

a. What is the standard deviation for the percent green in the first group?

b. What is the standard deviation for the percent green in the second group?

c. What is the difference in the percent green between the two groups?

d. What is the 95% confidence interval for the difference in the percent green?

e. Is the difference in the percent green statistically significant?

6. Suppose you take a stratified random sample of the Ph.D.'s holding research faculty appointments at a major university. Of the eight males, 6 have part-time appointments and 2 have full-time appointments. Of the eight females, 2 have part-time appointments and 6 have full-time appointments. Is the difference in the proportions of full-time vs. part-time appointments statistically significant?

12.6 Procedures for Paired Data

So far, we have been considering the comparison of two groups that are completely separate from each other, in the sense that the data values must be independently obtained for the confidence intervals and t-test to be valid. However, **paired data,** an important kind of data not contained within this framework, can also be handled very easily by taking the differences of the data values at the start of the analysis. We thus obtain one single group of numbers—the **paired differences**—on which we may base our inferences using the methods we learned for the one-sample case.

Paired Differences and the Paired *t*-Test

This technique—working with the differences between the paired members of the groups—may be used whenever the numbers in the two groups correspond in a natural way so that subtraction makes sense. For example, the numbers in one group might be the heights of "husbands" and those in the second group the heights of "wives." Then our paired difference is the difference in height for each "couple." In general, the first number in group 1 must correspond to the first number in group 2, the second numbers must correspond, and so on. In particular, the two groups must have the same number of data values.

Such a situation often arises when each unit of the population produces two values, one for each group. For example, we might have sets of "before" and "after" observations on the same individuals. If these two values are in the same units, subtracting them produces a meaningful measure of change or difference for each unit of the population, and a simple analysis may proceed based only on the differences. A *t*-test based on the single group of differences is known as a **paired two-sample t-test** or simply a **paired t-test** and is the recommended way to test for a difference between population means when the two groups are paired in this way.

Assumptions Needed. The assumptions necessary for the paired *t*-test to be valid are that (1) the paired differences are normally distributed and (2) the paired differences represent a random sample from the population.

EXAMPLE 12.10 *Paired Differences in Testing: Chocolate Desserts*

Suppose people in a sample are presented with two kinds of chocolate desserts, in randomized order, one made with nuts and one without. The people are then asked to rate each dessert. (Ratings are, of course, ordinal level measures. But for this example we will treat them as if they were measured numerical values.) The resulting data set consists of two groups of numbers, one group for "with nuts" and one for "without nuts." The data might look like Table 12.4, in which a scale from 1 to 5 was used, with larger numbers indicating a more favorable rating.

Table 12.4 *Ratings on chocolate desserts*

PERSON	RATING FOR "WITH NUTS"	RATING FOR "WITHOUT NUTS"
Kevin	3	2
Carolyn	5	4
Quentin	3	1
Andy	5	5
Ann	2	1
Suzanne	4	3
Average	3.667	2.667
Standard deviation	1.211	1.633
Standard error of the average	.494	.667

Note that the data values here are not all independent of each other because each person rated both desserts. Thus, it would not be altogether proper to compute the standard error for the average difference or to complete a confidence interval or hypothesis test based on that standard error because the required assumption of independence is not satisfied.

If we explore the data in the usual way, by doing a histogram for each group, the comparison is not very clear-cut (see Display 12.13 and Figure 12.7). Although there may be a slight hint of a preference for "with nuts" in these histograms, the main impression is that there is quite a lot of variation in scores from one person to another. This high level of random variation in scores would ordinarily make it difficult to reach any definite conclusion about people's preferences.

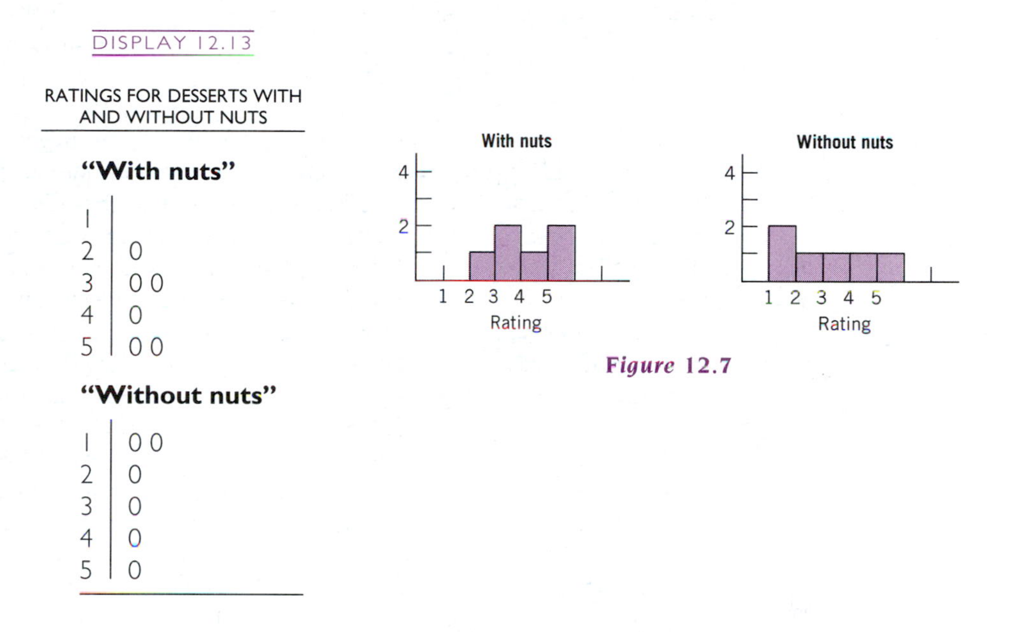

Figure 12.7

These histograms, done separately for each group, do not make use of the fact that the two groups are related. The information about paired observations for each person has been lost. However, by subtracting the two ratings contributed by each person, we obtain a single group of numbers that does reflect the pairing of the data (see Table 12.5).

Table 12.5 *Paired differences for dessert ratings*

PERSON	DIFFERENCE (WITH NUTS − WITHOUT NUTS)
Kevin	1
Carolyn	1
Quentin	2
Andy	0
Ann	1
Suzanne	1
Average	1.0
Standard deviation	.632
Standard error of the average	.258

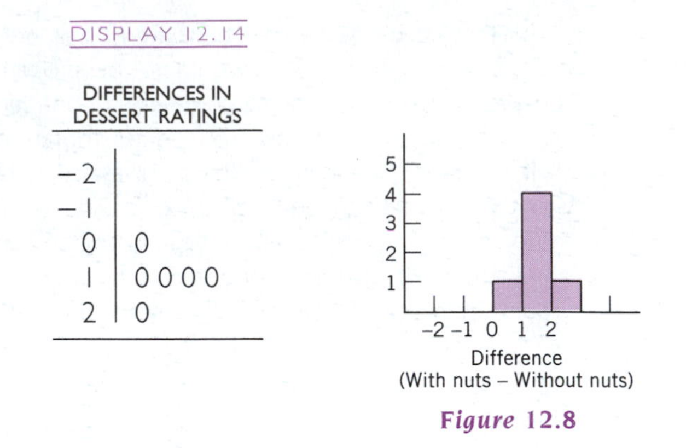

DISPLAY 12.14

DIFFERENCES IN DESSERT RATINGS	
−2	
−1	
0	0
1	0000
2	0

Figure 12.8

On this difference scale, positive values indicate a preference for "with nuts," 0 indicates no preference, and negative values indicate a preference for "without nuts." A histogram of these differences looks more definitive in its suggestion that there is a preference for "with nuts" because of the dominance of positive-valued differences and a lower level of variability (see Display 12.14 and Figure 12.8). This stronger indication of a difference is also seen in the smaller standard error of the differences (.258) as compared to the standard errors of the averages for the separate groups (.494 and .667, respectively). This indicates that estimation of the mean difference is more precise when the paired difference approach is used. Incidentally, it will always be true that the average of the differences (1.000 here) is equal to the difference between the averages for the two groups (3.667 − 2.667).

Before proceeding with our inferences, we should check the assumptions. First, these differences should be normally distributed; this is only approximately true, and we hope this will not cause trouble. Second, the differences should represent a random sample; they do here because each paired difference was computed for a different person from the sample. Even though the original two groups of ratings were not all independent of each other, within the single group of differences, the numbers are independent of each other.

Now, to complete our inferences about the population preference, we note that the sample average difference of 1.000 unit is an estimate of the population mean preference for "with nuts" over "without nuts." Using the standard error for the average difference of .258 and the *t*-value of 2.571 with $6 - 1 = 5$ degrees of freedom, we add and subtract $2.571 \times .258$ from 1.000 to find that

the 95% confidence interval goes from .337 *to* 1.663.

To test whether the apparent preference for "with nuts" is significantly different from 0 (which would indicate no preference), we see that 0 is not within the confidence interval. Therefore, the null hypothesis of "no preference" is rejected, and we conclude that there is a slight but significant preference for chocolate desserts "with nuts" over those "without nuts." With significance established, we have a right to proceed to explain the difference (not a statistical activity in itself!) by, for example, pointing out how well roasted nuts complement the basic flavor of chocolate.

Discussion

Pairing works so well because it tends to eliminate the variability from one sample unit to another and concentrates instead on the differences, which are often much less variable. In this example, there are strong variable differences from one person to another in their basic liking for chocolate. However, there was a consistent trend in that people tended to favor the "with nuts" dessert by a small but consistent margin. A small but consistent trend such as this is masked by the person-to-person variability when the groups are considered separately.

Computer Applications: Paired *t*-Test

Let's look again at the gas mileage data, used in developing Example 12.5. Suppose that there is a natural pairing in the numbers: Each pair of numbers is from one car, tested once with a dirty air filter and once with a clean air filter (with the order of testing randomized). We can do a paired *t*-test by first calculating a third variable (we will call it "dif") that is the difference between the two. Note that once the basic data have been entered on a computer worksheet, the computer can be used to find the values of "dif." The results of this calculation are shown in Table 12.6.

Table 12.6 *Paired differences for gas mileage data*

CLEAN	DIRTY	DIF
19.0	16.0	3
22.0	20.0	2
24.0	21.0	3
24.5	21.5	3
25.0	23.0	2
25.0	21.0	4
25.5	22.5	3
26.0	25.0	1
28.0	25.0	3
31.0	27.0	4

Now we use the one-sample *t*-test to see whether the average of the paired differences is significantly different from 0. Looking at the output (Display 12.15), we see that it is. Doing the test first as a confidence interval calculation, we see that the confidence interval runs from 2.142 to 3.458 mpg. This interval does not include 0, so we can reject the null hypothesis. Repeating the analysis as a classic *t*-test, we get a *t*-value of 9.64 with a probability (under the null hypothesis) of 0.000. Again, we can reject the null hypothesis.

DISPLAY 12.15

```
MTB > TInterval 95.0 'dif';

          N      MEAN     STDEV    SEMEAN     95.0 PERCENT C.I.
DIF      10     2.800     0.919     0.291     ( 2.142, 3.458 )

MTB > TTest 0.0 'dif';
SUBC > Alternative 0;

TEST OF MU = 0.000 VS MU N.E. 0.000

          N      MEAN     STDEV    SEMEAN        T       P VALUE     DF
DIF      10     2.800     0.919     0.291     9.64       0.000       9
```

Exercise Set 12.4

7. Suppose that you take a random sample of 100 likely voters on Tuesday. You weigh them. On Friday, one of the candidates becomes embroiled in a financial scandal. On Saturday, you go out and *take another random sample of 100 voters* and weigh them. You intend to look at any difference in their average weights to see whether the scandal has caused binge eating and weight gain. Should you use a two-sample *t*-test or a paired *t*-test?
8. Suppose that you take a random sample of 100 voters on Tuesday. You weigh them. On Friday, one of the candidates becomes embroiled in a financial scandal. On Saturday, you go out and *again weigh the same 100 voters you weighed on Tuesday.* You intend to look at any difference in their weight to see whether the scandal has caused binge eating and weight gain. Should you use a two-sample *t*-test or a paired *t*-test?
9. Suppose you have the measurements shown in Table 12.7 on six (randomly selected) people.

Table 12.7

BEFORE	AFTER	CHANGE (AFTER − BEFORE)
3	5	2
4	4	0
5	6	1
5	4	−1
6	8	2
8	9	1

   a. On average, what is the difference between the "after" and the "before" measurements?
   b. Is it appropriate to do a paired *t*-test, or should you do a two-sample *t*-test?
   c. Is the average change statistically significant?

12.7 A Nonparametric Test for a Sample Difference with Paired Data

We originally learned the sign test as a way to test for the significance of the difference between a sample median and a hypothesized population median of 0. We can also use the sign test for the difference between two groups when we have paired data. As with the paired *t*-test, we create a single group of numbers by finding the difference between each pair. We then do calculations on these differences. If we can reject the hypothesis that the population median may be 0, we can conclude that there is a difference between the groups.

EXAMPLE 12.11 *The Sign Test for Paired Differences: Chocolate Desserts*

Let's return to the example of rating of chocolate desserts (Example 12.10). Recall that the data consist of a series of ratings of chocolate desserts with and without nuts. Each individual rates each dessert (in randomized order), and we find which they prefer by finding the differences in the ratings (Table 12.5). A positive difference means they prefer the dessert with nuts, a negative difference means they prefer the dessert without nuts, and a difference of 0 indicates a tied score. Since the ratings are ordinal level measures, the sign test is arguably preferable to the paired *t*-test.

DISPLAY 12.16

```
MTB > STest 0.0 'dif';
SUBC>   Alternative 0.

SIGN TEST OF MEDIAN = 0.00000 VERSUS N.E. 0.00000

          N       BELOW       EQUAL       ABOVE       P-VALUE       MEDIAN
dif       6       0           1           5           0.0625        1.000
```

While we can do the calculations by hand, as we did earlier, we can also use a computer program to do the arithmetic. The results for a Minitab analysis, where the differences are stored as the variable "dif," are given in Display 12.16. The basic score is that there are no differences below 0, one equal to 0, and five above 0. The probability of this if the population value is really 0—if there is really no preference one way or the other—is .0625, which does not quite meet the typical significance level of .05. On the basis of this sample, at a significance level of .05, we cannot reject the possibility that people have no preference between chocolate desserts with and without nuts.

Exercise Set 12.5

10. Suppose you have these income data on several dual-income couples:

MALE$	FEMALE$	DIFF (F$ − M$)
198,000	65,000	−133,000
45,000	55,000	10,000
30,000	35,000	5,000
25,000	30,000	5,000
25,000	27,000	2,000
25,000	26,000	1,000
25,000	25,500	500
25,000	25,250	250
25,000	25,125	125

a. For these data, do the males, overall, earn more than the females?

b. Look at the data couple by couple. Which member of the couple typically has the higher salary?

c. Is the tendency you note in part (b) statistically significant?

d. What is it about this data set that allows the females typically to earn more than their male partners, while at the same time the males overall earn more than the females?

Summary

We have seen in this chapter that all we have learned for a single group of numbers applies separately when we have two groups, but we have the added possibility of performing comparisons from one group to the other. To compare one group to another, we ordinarily require that the measurements all be of the same type and in the same units so that a direct comparison of the numerical values makes sense.

Exploring two groups of numbers consists of doing a histogram for each, aligned near each other and on the same scale so that we may observe similarities and differences for central values, variabilities, and distribution shapes. The easiest kind of comparison occurs when the two groups have normal distributions with the same variability, so that only the mean values are different. However, if this simple situation does not take place, we can apply a transformation to both data sets to help eliminate skewness and equalize the variabilities.

Inference about the difference between the population mean values using the pooled two-sample test is based on the *difference between the sample averages* and requires that three *assumptions* be satisfied: (1) each group population follows a normal distribution, (2) the variabilities of the groups are equal, and (3) each group is a random sample from its population and all observations are independent of one another. The first two assumptions need only be very approximately satisfied. After we check the validity of these assumptions, a *standard error for the average difference* can be computed based on the standard deviations and sample sizes for each group. The number of degrees of freedom for this situation is equal to the sum of the sample sizes minus 2. An unpooled test may be used when variances are unequal.

We compute a *95% confidence interval for the difference in population means* from one group to the other in the usual way, based on the average difference (obtained by subtracting the average values for the two groups), its standard error, the degrees of freedom, and the *t*-table. We also test the hypothesis that the population mean difference is equal to a hypothesized value in the usual way: We see whether the hypothesized value being tested is in the confidence interval or not. If it is not in the interval, we decide that the population mean difference is not equal to the hypothesized value; otherwise, we decide that it is. This test is called the *two-sample t-test.* The particular case of testing whether the two populations are different corresponds to the particular choice of a hypothesized value of 0.

When the data do not satisfy the normal distribution assumption of the two-sample *t*-test, we can use a nonparametric alternative, such as the *Mann–Whitney U test.* We use the Mann–Whitney test to test whether the medians of two independent random samples are significantly different. This test is valid even if the population distributions are very nonnormal, provided they have the same shape.

These methods apply equally well to percentage data. With them we can find a confidence interval for the difference between two percentage values (provided we know the sample sizes) and test whether they are significantly different from each other.

When the two groups of numbers come as *paired data* with one number in each group (for example, "before" and "after" measurements done on a single person), then special considerations apply. Because of this pairing, the assumption of independence of all data values cannot be made so the usual two-sample standard error, confidence interval, and hypothesis test do not apply. Instead, we reduce the two groups to a single group of numbers by subtracting the two numbers in each pair. Naturally, this only applies when subtraction from one group to the other makes sense; it can often be interpreted as the change, or difference, in each case being studied.

Once the subtraction is done, the usual methods of statistical inference (confidence intervals and tests of hypothesis) can be applied to this single sample, provided the assumptions of random sampling from a normal population apply to the *paired differences* (they need not apply to the two original groups). The results tell us how the population means of the two original groups differ from one another. One definite advantage of the paired approach is that there may be much less variability in the paired differences than there was within each group of data. Naturally, when there is less variability, it is easier to reach a useful conclusion about a situation. If the differences do not have a normal distribution, the one-sample sign test for the median may be used as a nonparametric one-sample test for paired differences.

Problems

Basic Reviews

1. Suppose that you have two random samples, one with an average value of 100 and the other with an average value of 105, on some characteristic. You note that the difference is 5 units. You wonder whether this difference is real or possibly the result of random error in drawing the two samples. On doing the arithmetic, you find that the confidence interval for the difference runs from 2 to 8 units.

a. Is a difference of 0 included in the confidence interval?

b. Is the observed difference of 5 units statistically significant?

c. Can you reject the null hypothesis that the observed difference is the result of random sampling error?

2. Suppose that you have two random samples, one with an average value of 100 and the other with an average value of 115, on some characteristic. You note that the difference is 15 units. You wonder whether this difference is real or possibly the result of random error in drawing the two samples. On doing the arithmetic, you find that the confidence interval for the difference runs from −1 to 31 units.

a. Is a difference of 0 included in the confidence interval?

b. Is the observed difference of 15 units statistically significant?

c. Can you reject the null hypothesis that the observed difference is the result of random sampling error?

Analysis

3. Consider the heights in feet of tall buildings (over 600 ft) in Chicago and New York City, in 1993 (Table 12.8).[3]

a. Explore these two groups of numbers by doing histograms or dot plots for each on the same scale, aligned so that a comparison is easy to do by eye.

b. Describe the distributions. In particular, are they normally distributed?

c. Chicago has the world's tallest building, the Sears Tower, at 1,454 ft. Based on that fact alone, we might expect Chicago to dominate New York City more generally in the number of tall buildings. Is this the case?

Table 12.8 *Height of buildings over 600 ft in Chicago and New York City, in 1993 (in feet)*

Chicago									
1,454	1,136	1,127	970	901	891	871	859	852	775
727	700	697	695	673	662	648	645	605	601
New York City									
1,368	1,362	1,250	1,414	1,046	950	927	914	850	813
808	802	792	778	764	756	750	750	745	743
741	739	730	725	724	716	707	705	700	697
687	687	685	680	679	674	673	673	670	664
653	650	650	648	648	645	640	640	634	630
630	630	630	628	628	628	625	625	620	620
618	615	615	610	610	609	603	600		

4. Voter turnout statistics are important as an indicator of the extent to which the population is actively participating in the election process. Many fewer people actually vote than are entitled to (by being old enough); even those who register to vote do not all follow through on election day. Consider the voter turnout statistics for the 1980 presidential election, as measured by the percentage of registered voters who voted and also by the percentage of the voting-age population who voted (Table 12.9).[4]

Table 12.9 *Percent of registered voters voting and percent of voting-age population voting in 1980 presidential election, by state*

STATE	REGISTERED	VOTING AGE	STATE	REGISTERED	VOTING AGE
Alabama	62.9%	49.7%	Missouri	73.9%	58.8%
Alaska	60.7	61.3	Montana	73.3	65.0
Arizona	78.0	49.1	Nebraska	74.7	56.2
Arkansas	70.6	53.6	Nevada	81.7	45.7
California	75.6	50.6	New Hampshire	73.5	58.4
Colorado	82.6	57.8	New Jersey	79.0	55.1
Connecticut	82.4	60.6	New Mexico	69.9	52.5
Delaware	78.4	56.1	New York	n/a	48.1
Wash., D.C.	60.2	36.6	North Carolina	66.9	45.8
Florida	76.7	53.6	North Dakota	n/a	64.3
Georgia	n/a[a]	43.6	Ohio	72.8	55.6
Hawaii	75.3	46.2	Oklahoma	78.2	54.0
Idaho	75.3	69.0	Oregon	75.3	61.9
Illinois	76.2	59.0	Pennsylvania	79.3	52.7
Indiana	n/a	58.2	Rhode Island	78.4	60.5
Iowa	76.7	63.0	South Carolina	71.9	42.9
Kansas	75.9	55.8	South Dakota	73.2	67.6
Kentucky	71.0	51.2	Tennessee	75.3	50.5
Louisiana	76.8	55.7	Texas	68.4	47.1
Maine	68.8	66.2	Utah	77.3	67.1
Maryland	74.6	50.7	Vermont	68.4	59.4
Massachusetts	80.1	58.7	Virginia	81.0	48.9
Michigan	68.3	59.6	Washington	79.8	62.3
Minnesota	86.9	69.2	West Virginia	71.0	54.4
Mississippi	60.2	54.1	Wisconsin	n/a	66.0
			Wyoming	80.5	52.8

[a] n/a stands for "not available."

a. Explore these two groups of numbers by doing histograms for "registered voter percent voting" and for "voting age percent voting."

b. Describe the distribution shapes. Are they similar? Do they appear to be normally distributed?

c. Compare the centers and the variabilities. Are they the same or different from one group to the other? If different, in what way?

d. Write a brief paragraph about what you have learned about voter turnout statistics.

e. Do a box plot for each group of numbers, complete with the adjacent values and outliers labeled by the state name. Be sure each plot is drawn to the same scale and aligned so that visual comparison is easy. Describe the resulting display.

5. Suppose a utilization study was conducted to see how often two rooms of a sports facility were being used during the noon hour. The number of people in each room was counted at 12:30 P.M. each Monday for 30 weeks. The hypothetical results are shown in Table 12.10.

***Table* 12.10** *Sports facility utilization*

Number of people in weight room									
11	13	6	8	12	9	13	7	8	10
3	10	8	12	10	7	4	6	6	5
6	10	9	13	11	7	13	7	8	7
Number of people in gymnasium/dance studio									
85	78	75	92	86	110	85	100	82	103
73	80	89	99	87	93	91	101	94	80
83	66	84	89	96	94	97	103	99	80

a. Explore these two groups of numbers using histograms on the same scale, aligned for visual comparison.

b. Describe the distribution shapes.

c. Compare the centers and variabilities for the two groups.

d. Take the square roots of the data and repeat parts (a) through (c).

e. Take the logarithms of the data and repeat parts (a) through (c).

f. Which transformation, if any, of the data do you prefer? Why?

6. Suppose that in one sample the average number of sea lions sighted near Big Sur, California, was 5.6 per day with a standard deviation of 3.1, based on 10 days of observation. In a second sample, taken farther north, the average number sighted was 10.2 per day with a standard deviation of 3.7 based on 15 days of observation.

a. Find the average difference, the standard error of the average difference, the degrees of freedom, and the 95% confidence interval for the mean population difference, based on these two samples.

b. Use a two-sample *t*-test to decide whether the two groups are different on the average or not.

7. Suppose that in one day of stock market trading within the high technology group, a sample of 32 stocks lost an average of $1.16 per share with a standard deviation of $.62. On the same day within the heavy industry group, a sample of 83 stocks lost an average of $.86 per share with a standard deviation of $.49.

a. Find the average difference, the standard error of the average difference, the degrees of freedom, and the 95% confidence interval for the mean population difference.

b. Use a two-sample *t*-test to decide whether the two groups are different on the average or not.

8. Suppose a sample of 350 downtown shoppers found that 16.6% of them had purchased more than one item. A sample of 260 suburban shoppers found that 24.6% of them had purchased more than one item.

a. Find the average difference, the standard error of the average difference, the degrees of freedom, and the 95% confidence interval for the mean population difference.

b. Use a two-sample *t*-test to decide whether the two groups are different on the average or not.

9. Suppose the A team, which has 11 players, finished with an average gain of 3.6 points per player and a standard error of .85. The B team, also with 11 players, gained only 2.3 points on the average per player with a standard error of .67.

a. Find the average difference, the standard error of the average difference, the degrees of freedom, and the 95% confidence interval for the mean population difference.

b. Use a two-sample *t*-test to decide whether the two groups are different on the average or not.

Question 10 is based on the data from Student's 1908 paper representing additional sleep gained by using experimental soporific (sleeping pill) medications. The results from ten patients are listed in Table 12.11.

Table 12.11 *Extra sleep, in hours, for ten patients under two different medications*

PATIENT	MEDICATION 1: (DEXTRO HYOSCYAMINE HYDROBROMIDE)	MEDICATION 2: (LAEVO HYOSCYAMINE HYDROBROMIDE)
1	.7	1.9
2	−1.6	.8
3	−.2	1.1
4	−1.2	.1
5	−1.0	−.1
6	3.4	4.4
7	3.7	5.5
8	.8	1.6
9	0.0	4.6
10	2.0	3.4

10. a. Are these paired data? Why or why not?

b. Would it be appropriate to do a two-sample *t*-test on the two groups (one group for each medication)? Why or why not?

c. Take differences to compute the single group of numbers representing, for each patient, the additional sleep gained under medication 2 as compared to medication 1.

d. Compute the basic statistics (average, standard deviation, and standard error) for the differences from part (c).

e. Compute the 95% confidence interval for the general (population) difference between the two medications based on your answers to part (d).

f. Test, using the paired *t*-test, whether the two medications are different or not.

11. A small hypothetical company is being sued for discrimination. The ten men employed there have an average salary of $37,493 per year, whereas the six women have an average salary of only $28,854 per year.

a. Compute the average difference in salary between men and women.

b. Can you compute a 95% confidence interval for this difference or do a statistical test for the existence of discrimination, based only on the information given here? Why or why not?

c. Suppose the standard deviation of the men's salaries is $6,152 and the standard deviation of the women's salaries is $4,550. Do a formal statistical test to see whether there is evidence of gender-based salary differences.

d. Could a discrepancy this large in salary levels be explained by the hypothesis that men and women are treated equally, except for random variations?

12. With respect to Problem 11, suppose that the standard deviation of the men's salaries had been larger, say $10,134, and that the standard deviation for the women's salaries had been $7,493, but that the sample sizes and averages are the same as previously reported. Can you conclude, in this situation with larger variabilities, that there is significant evidence of gender-based salary differences?

13. In Exercises 16 and 17 from Chapter 10, we took two random samples of hikes from *Best Hikes with Children* and noted their high points. The samples are:

Sample 1:
848 2,900 900 2,981 2,300 5,180 3,982 3,161 5,699

Sample 2:
3,982 300 4,508 6,000 2,061 3,982 3,800 1,120 4,900

(We hope you saved your work.)

a. What is the average difference in the high points of the two samples?

b. What is the standard error for the differences in average high points?

c. Construct a 95% confidence interval for the population mean differences.

d. Note whether 0 is in the confidence interval, and state your conclusion as a test of significance.

14. A random sample of altitudes mentioned in the Appalachian Mountain Club's *White Mountain Guide*—which is organized differently than *Best Hikes with Children,* so the reporting of high points on individual hikes is not consistent—produced these results:

3,590 3,056 4,397 2,403 2,743 3,630 3,625 2,071 2,253

a. Calculate the average height and the standard deviation of heights for this sample.

b. What is the difference in average heights between this sample and sample 1 in Problem 13?

c. Is the differences in heights between this sample and sample 1 from Problem 13 statistically significant?

15. Display 12.17 (page 420) shows dot plots of the petal lengths for three species of iris flowers from Fisher's data set.

a. Are the three distributions reasonably close to normal?

b. Do the variabilities of the three distributions appear to be about the same? If not, what differences do you see?

c. Is the assumption of equal variability really critical, or can we get results that are approximately correct when the variabilities are somewhat different?

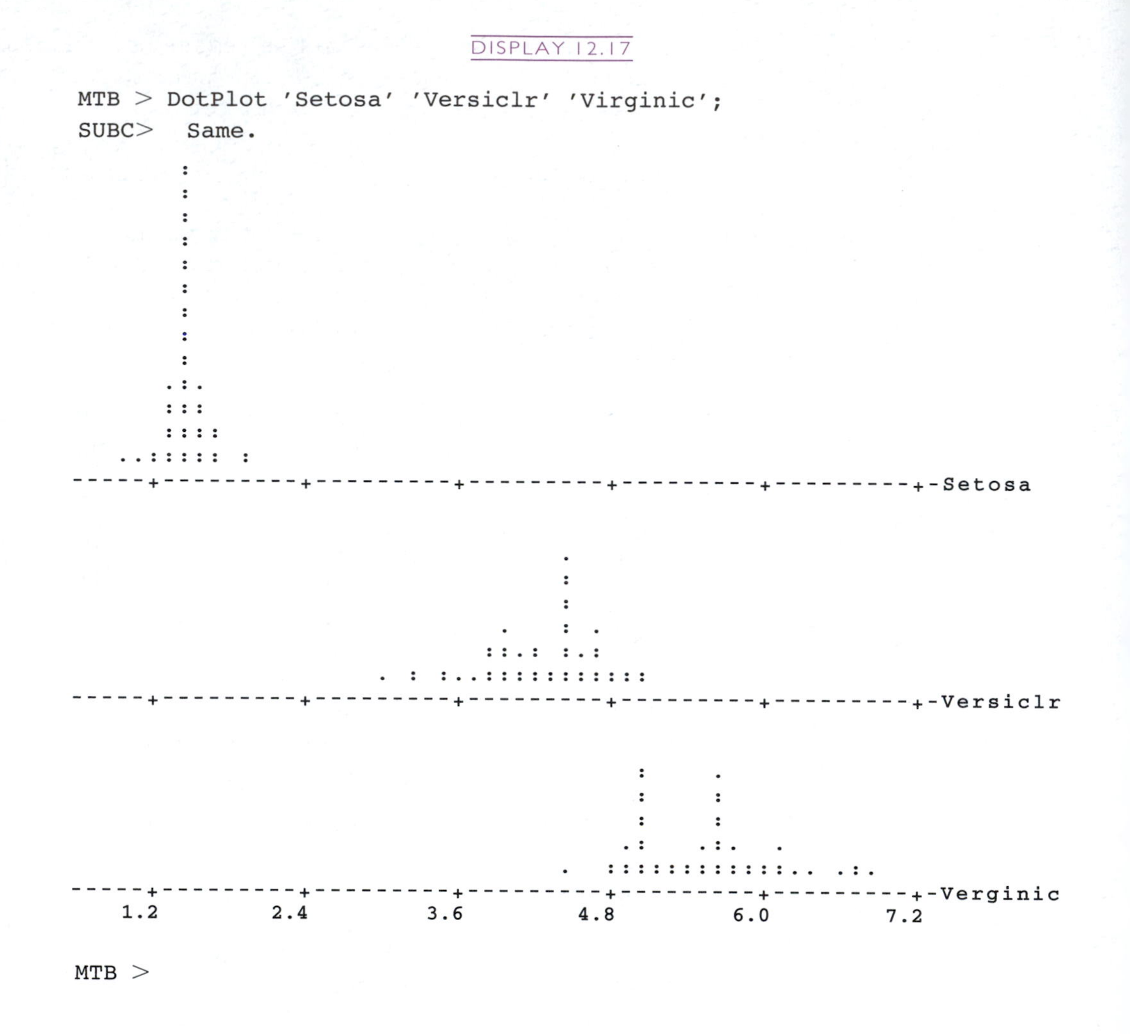

DISPLAY 12.17

16. Display 12.18 is Minitab output for an analysis of petal length of *Iris versicolor* ("Versiclr") versus *Iris virginica* ("Virginic").

DISPLAY 12.18

```
MTB > TwoSample 95.0 'Versiclr' 'Virginic';
SUBC>  Alternative 0;
SUBC>  Pooled.

TWOSAMPLE T FOR Versiclr VS Virginic

             N      MEAN      STDEV     SE MEAN
Versiclr    50     4.260      0.470      0.066
Virginic    50     5.552      0.552      0.078

95 PCT CI FOR MU Versiclr - MU Virginic: (-1.495, -1.089)

TTEST MU Versiclr = MU Virginic (VS NE): T = -12.60
                                P = 0.0000 DF = 98

POOLED STDEV =   0.513
```

a. What is the mean length for *I. versicolor* petals?

b. What is the mean length for *I. virginica* petals?

c. What is the average difference in their length?

d. What is the 95% confidence interval for the population mean difference in petal length for these two species?

e. Can we reject the null hypothesis that there is really no difference in their average length?

f. State your conclusion as a test of statistical significance.

17. Display 12.19 is Minitab output for an analysis of petal length of *Iris setosa* ("Setosa") versus *Iris virginica* ("Virginic").

DISPLAY 12.19

```
MTB > TwoSample 95.0 'Setosa' 'Virginic';
SUBC>  Alternative 0;
SUBC>  Pooled.

TWOSAMPLE T FOR Setosa VS Virginic

                N      MEAN      STDEV     SE MEAN
Setosa         50     1.462      0.174     0.025
Virginic       50     5.552      0.552     0.078

95 PCT CI FOR MU Setosa - MU Virginic: (-4.252, -3.928)

TTEST MU Setosa = MU Virginic (VS NE): T = -49.99
             P = 0.0000 DF = 98

POOLED STDEV =   0.409
```

a. What is the mean length for *I. setosa* petals?

b. What is the mean length for *I. virginica* petals?

c. What is the average difference in their length?

d. What is the 95% confidence interval for the population mean difference in petal length for these two species?

e. Can we reject the null hypothesis that there is really no difference in their average length?

f. State your conclusion as a test of statistical significance.

18. Display 12.20 (page 422) is Minitab output for an analysis of petal length of *Iris setosa* ("Setosa") versus *Iris versicolor* ("Versiclr").

a. What is the mean length for *I. setosa* petals?

b. What is the mean length for *I. versicolor* petals?

c. What is the average difference in their length?

d. What is the 95% confidence interval for the population mean difference in petal length for these two species?

e. Can we reject the null hypothesis that there is really no difference in their average length?

f. State your conclusion as a test of statistical significance.

DISPLAY 12.20

```
MTB > TwoSample 95.0 'Setosa' 'Versiclr';
SUBC>  Alternative 0;
SUBC>  Pooled.

TWOSAMPLE T FOR Setosa VS Versiclr

             N      MEAN      STDEV     SE MEAN
Setosa      50     1.462      0.174      0.025
Versiclr    50     4.260      0.470      0.066

95 PCT CI FOR MU Setosa - MU Versiclr: (-2.939, -2.657)

TTEST MU Setosa = MU Versiclr (VS NE): T = -39.49
         P = 0.0000 DF = 98

POOLED STDEV =  0.354

MTB >
```

19. Table 12.12 shows the EPA estimates of city and highway gas mileage for two-seat car models available in the United States in 1992.

a. Would a paired *t*-test or a two-sample *t*-test be more appropriate for this data?

b. What is the average difference between highway gas mileage and city gas mileage?

Table 12.12 EPA *mileage estimates for two-seat car models* (1992)

CAR	CITY MPG	HIGHWAY MPG
Acura NSX	18	24
Alfa Romeo Spider	21	28
Cadillac Allante	15	22
Chevrolet Corvette	17	25
Dodge Viper	13	18
Ferrari 348	13	19
Ferrari F40	13	19
Ferrari Testarossa	10	15
Honda CRX	32	35
Lamborghini Diablo	9	14
Lotus Elan SE	24	31
Lotus Esprit Turbo SE	17	28
Maserati Spyder	15	18
Mazda MX-5 Miata	25	30
Mazda RX-7	17	25
Mercedes-Benz 300SL	15	21
Mercedes-Benz 500SL	15	21
Nissan NX	28	38
Porsche 911	16	24
Porsche 968	16	25
Toyota MR2	22	28
Vector W8 Twinturbo	8	13

c. Do a statistical test of the null hypothesis that there is no difference in average mileage, against the alternative hypothesis that the difference is greater than 0 (that is, that highway mileage is greater than city mileage). State your conclusion. (You may want to use a computer for the arithmetic.)

20. Recall that Feduccia[5] has suggested that the arc of a bird's claws depends in part on their ecological niche. Display 12.21 gives dot plots of claw arc measurements for *Sitta carolinensis* ('Sitta c.') (white-breasted nuthatch), a trunk climber, and *Momotus momota* ('Momotus') (blue-crowned motmot), a percher.

DISPLAY 12.21

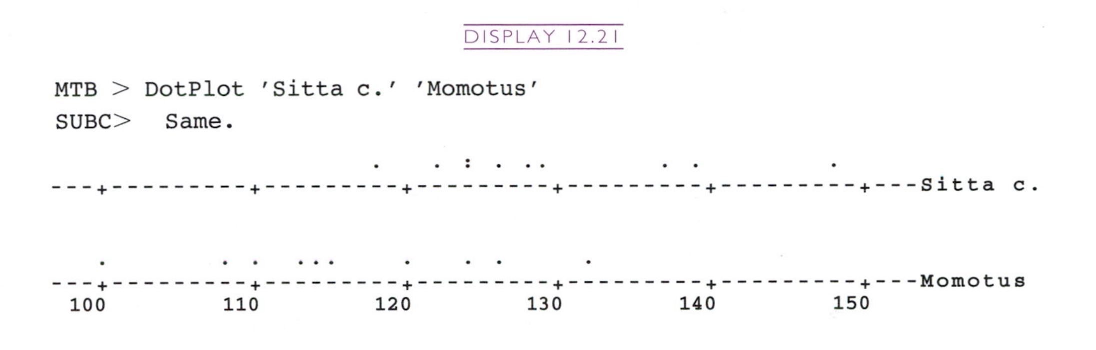

a. Do the claw arcs appear to be (approximately) normally distributed?

b. Do the variabilities appear to be about the same?

21. Display 12.22 is Minitab output for a two-sample test for a difference of means, for the *Sitta* and *Momotus* claw arc data.

DISPLAY 12.22

```
MTB > TwoSample 95.0 'Sitta c.' 'Momotus';
SUBC>   Alternative 0;
SUBC>   Pooled.

TWOSAMPLE T FOR Sitta c. VS Momotus

               N      MEAN       STDEV     SE MEAN
Sitta c.      10     129.50      9.14       2.9
Momotus       10     116.20      9.46       3.0

95 PCT CI FOR MU Sitta c. - MU Momotus: (4.6, 22.0)

TTEST MU Sitta c. = MU Momotus (VS NE): T = 3.20
               P = 0.0050 DF = 18

POOLED STDEV =   9.30
```

a. What is the mean claw arc for *Sitta?*

b. What is the mean claw arc for *Momotus?*

c. What is the average difference in their arc?

d. What is the 95% confidence interval for the population mean difference in claw arc for these two species?

e. Can we reject the null hypothesis that there is really no difference in their average claw arc?

f. State your conclusion as a test of statistical significance.

22. Display 12.23 gives dot plots of claw arc measurements for *Pteroglossus frantzi* (collared aracari), a percher, and *Momotus momota*, another percher.

DISPLAY 12.23

```
MTB > DotPlot 'Pteroglo' 'Momotus';
SUBC>  Same.

          ..           .       .     :.    .      .        .
-----+---------+---------+---------+---------+---------+-Pteroglo

 .            .  .  . ..       .     .  .      .
-----+---------+---------+---------+---------+---------+-Momotus
  104.0     112.0     120.0     128.0     136.0     144.0
```

a. Do the claw arcs appear to be (approximately) normally distributed?

b. Do the variabilities appear to be about the same?

23. Display 12.24 is Minitab output for a two-sample test for a difference of mean for the *Pteroglossus* and *Momotus* claw arc data.

DISPLAY 12.24

```
MTB > TwoSample 95.0 'Pteroglo' 'Momotus';
SUBC>  Alternative 0;
SUBC>  Pooled.

TWOSAMPLE T FOR Pteroglo VS Momotus

             N      MEAN      STDEV     SE MEAN
Pteroglo     10     122.9     11.8      3.7
Momotus      10     116.20    9.46      3.0

95 PCT CI FOR MU Pteroglo - MU Momotus: (-3.3, 16.7)

TTEST MU Pteroglo = MU Momotus (VS NE): T = 1.40
          P = 0.18 DF = 18

POOLED STDEV =  10.7
```

a. What is the mean claw arc for *Pteroglossus*?

b. What is the mean claw arc for *Momotus*?

c. What is the average difference in their arcs?

d. What is the 95% confidence interval for the population mean difference in claw arcs for these two species?

e. Can we reject the null hypothesis that there is really no difference in their average claw arcs?

f. State your conclusion as a test of statistical significance.

24. Using the data in Table 2.22 (data file HEART.MTW), do younger and older men have different heart fill rates and oxygen utilization capacities? Are any apparent differences statistically significant?

25. Is there a statistically significant difference in the total body length of male versus female bowhead whales (see Table 2.21, data file WHALES.MTW)?

26. Bowhead whales make their living by straining plankton out of the water, so they have very large mouths equipped with baleen straining plates. It is at least possible that males and females might have evolved slightly different feeding requirements and that these might be reflected in slightly different relative mouth sizes. Use the data in file WHALES.MTW to test this hypothesis.

27. John Gottman (1994) has done 20 years of research on why marriages succeed or fail. His basic finding is that there must be a ratio of at least five positive acts to each negative act for the marriage to last. Suppose you decide to replicate this study. You recruit 13 couples and make videotapes of each couple discussing an important issue in their relationship. You then count the number of positive acts (compliments, smiles, agreements, and so forth) in each interaction, as well as the number of negative acts (scowls, insults, and so on). From these numbers you then calculate the ratio of positive to negative acts, as well as the total number of acts. Three years later you relocate the original couples and see whether they have broken up (1 = broke up, 0 = did not break up). These data are in Table 12.13[6] and data file MARRIAGE.MTW.

Table 12.13 *Positive and negative actions among couples, and whether they break up within 3 years*

POSITIVE	NEGATIVE	RATIO	TOTAL	BREAK UP
25	6	4.1667	31	1
25	5	5.0000	30	0
25	10	2.5000	35	1
25	15	1.6667	40	1
10	2	5.0000	12	0
10	1	10.0000	11	0
10	1	10.0000	11	0
10	5	2.0000	15	1
50	10	5.0000	60	0
50	5	10.0000	55	0
50	15	3.3333	65	1
50	25	2.0000	75	1
75	15	5.0000	90	0

a. Is there a statistically significant difference between the couples that broke up and those that stayed together in their ratios of positive to negative acts?

b. Formulate and test at least two more hypotheses about what determines whether a couple breaks up. State your hypotheses explicitly, and support your conclusions with appropriate computations and/or graphics.

28. A *Newsweek* article on adults who had been diagnosed hyperactive as children, compared to adults who had not, reports that by early adulthood the hyperactive group was "nine times as likely to have served time in prison."[7] Suppose there were 100 men in the hyperactive group and 100 in the not hyperactive group. Make up a data set where "nine times" as many hyperactives have spent time in prison, and test this difference for statistical significance.

29. The *Newsweek* article described in Problem 28 also reports that by early adulthood the hyperactive group was "twice as likely to have arrest records." Suppose there were 100 men in the hyperactive and 100 in the not hyperactive group. Make up a data set where "twice" as many hyperactives have arrest records, and test this difference for statistical significance.

30. Consider the data shown in Table 12.14[8] on adult status of males who had been diagnosed as being hyperactive or not as children.

Table 12.14 *Adult status of males diagnosed as hyperactive ($n = 91$) or not hyperactive ($n = 90$) as children. Numbers are percents of samples.*

	HYPERACTIVE	NOT HYPERACTIVE
High school dropout	25	2
Finished college	12	50
Professional job	4	12

a. Is the difference in high school dropout rate statistically significant?

b. Is the difference in the percentages finishing college statistically significant?

c. Is the difference in the percentages holding professional jobs statistically significant?

31. Consider the data shown in Table 12.15[9] on adolescent status of males who had been diagnosed as being hyperactive or not as children.

Table 12.15 *Adolescent status of males diagnosed as hyperactive ($n = 103$) or not hyperactive ($n = 100$) as children. Numbers are percents of samples.*

	HYPERACTIVE	NOT HYPERACTIVE
Drug abuse problem	16	4
Antisocial personality	27	8

a. Is the difference in the percentages with drug abuse problems statistically significant?

b. Is the difference in the percentages diagnosed as antisocial personality statistically significant?

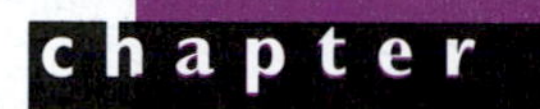

chapter 13

Analysis of Variance: Several Groups of Numbers

In Chapter 3 we used box plots to look at several groups of numbers simultaneously. By inspecting the box plots, we were able to compare their median values, their variability, and their extremes. But are any of the apparent differences we see real? Or might they be the result of random sampling? This is a question of statistical significance.

In Chapter 12 we used *t*-tests to compare two groups, but *t*-tests are not appropriate when we have more than two groups. With more than two groups we use the analysis of variance (ANOVA), which allows us to look at all of the groups simultaneously.

The step from two groups of numbers to more than two groups is a big one. Although the basic goal of comparing the groups to estimate and test for differences remains the same, the methods must be different because there is no longer a single number (such as the difference between the two

sample averages) that adequately summarizes the possible differences among several groups. We will learn how to explore this kind of data, how to test whether the groups are different at all, and (if there are any differences) how to decide which particular groups are indeed different from others.

13.1 Exploring Several Groups of Numbers

To examine and compare more than two groups of numbers (that are all expressed in the same units of measurement), we will proceed in basically the same way as we did to compare just two groups. We will do histograms or box plots, side by side on the same scale, to help us see comparisons of center, variability, distribution shape, and outliers. Box plots can be especially useful because when we look at several groups of numbers simultaneously, the extra detail of a histogram can be distracting. Box plots help us quickly judge the basic features of the distributions, making it easy to compare many situations at once.

Transforming to Stabilize the Variability

Comparison of distributions is simplest if they are alike in shape and variability and differ only in central value. This would imply that each distribution is like the others except for an additive shifting to a higher or lower value. Because it is a simpler situation, analysis is easier to do and understand when the variabilities are the same. It is also easier to apply formal statistical procedures, like hypothesis tests to see whether the groups are significantly different from each other, when the variabilities are equal across the groups.

Our first task, then, is to check for roughly equal variability. If the groups do not show equal variability, we will look for a transformation that, when applied to all numbers in all the groups of data, makes the variabilities of the transformed data sets as alike as possible. In particular, after a good transformation is applied, there should be no relationship between central values and variabilities. Further analysis using the transformed scale will usually be more valid and also more effective than blindly turning untransformed data over to the formal methods of statistical inference.

Naturally, we will be careful to transform all of the groups using the same transformation, so that the largest group remains largest, the smallest stays smallest, and so on. The analysis is made easier, but is not changed in a fundamental way by careful transformation. After the analysis is complete, we should transform back to the original units of measurement so we may report in understandable language what has been learned about the data set.

Our primary goal is to find a transformation that makes the variability about the same in each group. Square roots or logarithms are often the best choice when the data represent counts of events, such as repeated determinations of radioactive counts in several different environments, the number of defects in the day's production on various assembly lines, or the number of sea lions spotted each day in different locations.

The two goals of transformation—symmetry of the distribution within each group of numbers and stable variability across groups—relate to each other in two ways. First, it often happens that the same transformation will achieve both goals, avoiding the need to choose between them. Second, the need for stable, equal variabilities from one group to another is probably the more important goal of transformation in the context of a comparison of several groups of numbers.

EXAMPLE 13.1 Comparing Several Groups: Populations of Counties

We all know that the states of the United States are divided into regions (usually called counties) with varying populations. Some states are more populous than others, and within a state there is variability in the size of its counties due to the locations of large cities and other factors. Let us try to understand the patterns of variation in the sizes of the counties of seven states: Massachusetts, Maine, New Hampshire, New Jersey, South Carolina, Vermont, and Wyoming.

The data[1] we will use represent results of the 1990 census. The five-number summaries of the data sets are shown in Table 13.1 (recall that a five-number summary actually reports six numbers: the number of data values, the smallest, the lower quartile, the median, the upper quartile, and the largest).

Table 13.1 *Five-number summary of county populations*

STATE	N	MIN	Q1	MEDIAN	Q3	MAX
MA	14	6,012	122,037	445,793	665,450	1,398,468
ME	16	18,653	33,147	48,358	113,243	243,135
NH	10	34,828	37,796	72,525	151,465	336,073
NJ	21	65,294	134,498	395,066	527,962	825,380
SC	46	8,868	23,470	44,053	105,564	320,167
VT	14	5,318	22,973	34,399	54,273	131,761
WY	23	2,499	6,518	12,373	29,370	73,142

These summaries provide enough information for drawing quick box plots, which can be useful, but to draw more detailed box plots we need the actual data values so that outliers and adjacent values can be identified. Thus, the first diagram, Figure 13.1, shows the box plots of the data values for the group of county populations for each of these seven states.

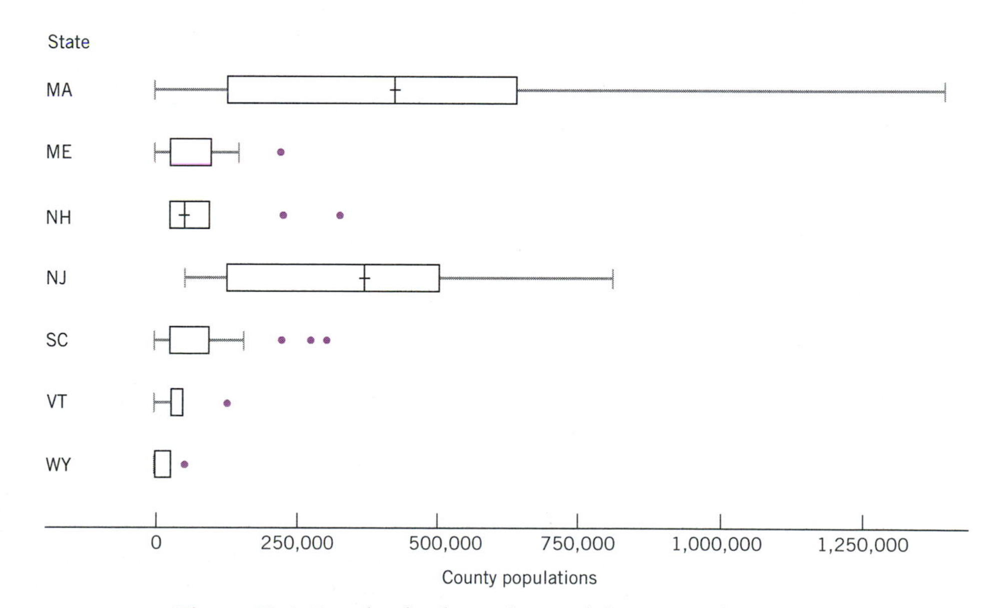

Figure 13.1 *Box plots for the populations of the counties of seven states*

When looking for structure in data, it is important to make use of the ordering of the objects. With one or two groups, ordering of the groups makes little or no difference, but with more than two it can make a crucial difference. The states have been ordered so far by alphabetical ordering of their two-letter abbreviations, not a meaningful ordering for most purposes relating to population. Instead, it seems natural to order them by size, and the natural measure of size for each group is its center. For exploring, we will use the median as a measure of center. (Feel free in other situations to order the categories based on their average values instead.)

Placing these same box plots in order by median value, we have the situation displayed in Figure 13.2. From these reordered groups, it is clear that center and variability are related. As we progress from top to bottom, the typical (median) counties become larger and so do the variabilities of the counties (expressed here by the width of the box that measures the interquartile range). In fact, it is difficult to see any detail about the sparser states—Wyoming and Vermont—because they have been "squashed" in order to fit on the same scale with more populous states such as New Jersey and Massachusetts, which have several times the variability of the sparser states.

We can see that we need a transformation to stabilize these unequal variabilities. Because the variabilities increase as the central values increase, we should try smaller powers of the data such as square roots or logarithms to compress the scale at higher values. Thus, taking the square roots of all of the *original* population values, recomputing the outlier thresholds and redrawing the box plots, we arrive at Figure 13.3.

This diagram does look a bit better. Wyoming is not as badly squashed as before, so more of its details are visible. All three of South Carolina's outliers on the original scale have been moved by the transformation and are no longer outliers on the square root scale. But more important, the variabilities are generally much closer to being equal from one state to another. However, some trend remains. The sparser states at the top definitely are less variable than the more populous states at the bottom.

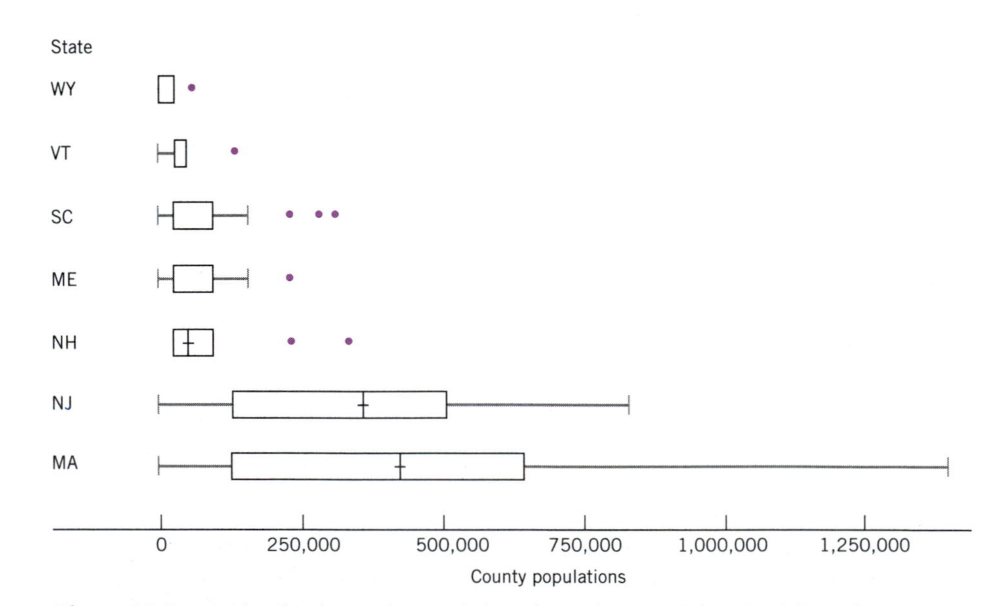

Figure 13.2 *Box plots for the populations of the counties of seven states, ordered by median county size for each state. Note that variability is systematically related to median: As medians increase (as you look from top to bottom) so does the interquartile range (the width of the box).*

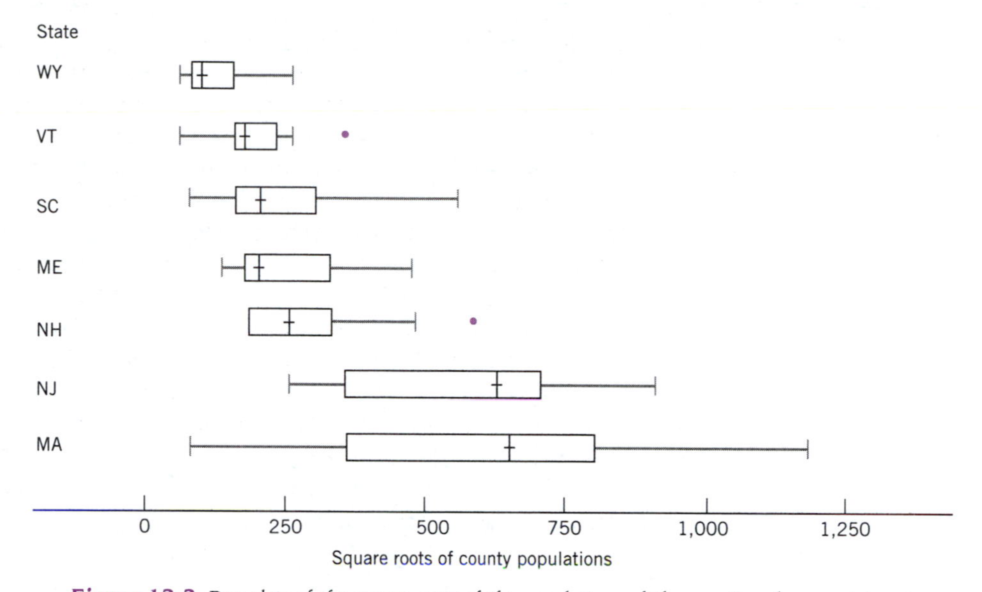

Figure 13.3 *Box plots of the square roots of the populations of the counties of seven states. A systematic relation still exists between level and variability, but less so.*

Let's try a stronger transformation. We will use logarithms to base 10, but logs to base e would provide equivalent results. Transforming the *original* data (*not* the square roots) and recomputing all of the box plots, we have Figure 13.4. The diagram is now very good. The variabilities have been stabilized by the log transformation. As the centers increase from top to bottom, the variabilities change a little, but not systematically. The variabilities of all of the states are approximately equal to each other with no trend relating to the centers.

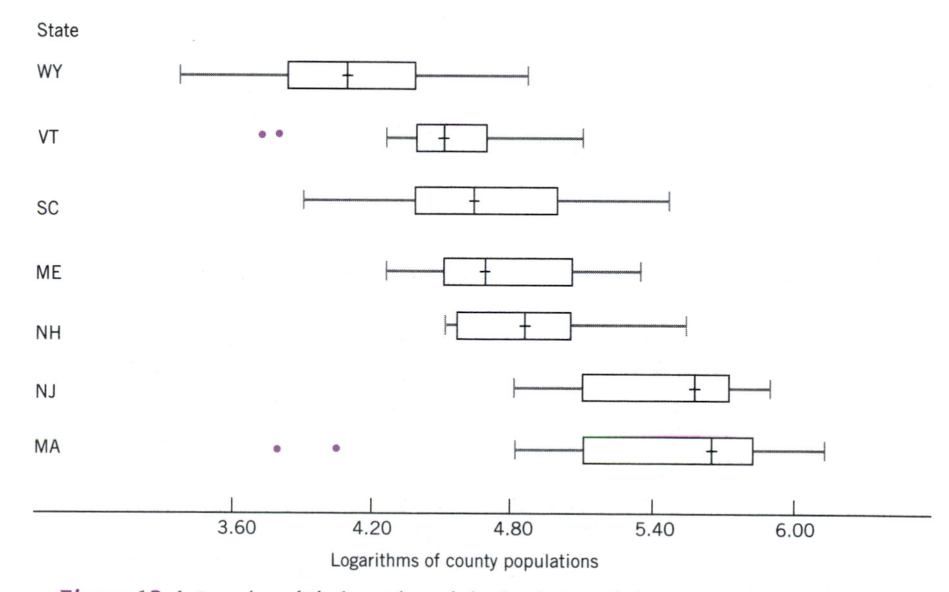

Figure 13.4 *Box plots of the logarithms of the populations of the counties of seven states. The variabilities have been stabilized: There is no longer any systematic increase or decrease in interquartile range as you look from top to bottom.*

Because the transformation has preserved the ordering, it is still clear that Massachusetts is the largest, Wyoming is the smallest, and so on. Certain details of the situation have become clearer, such as the existence of some extra low outliers on the log scale even though all of the high outliers on the original scale have now disappeared.

Transforming back from logarithms to numbers of people so we may report what we have learned, we see that the states are very similar in their county population distributions. Because an additive shift of the logarithms can change one distribution (approximately) into another here and logs change multiplication into addition, when we transform back we notice that one state's county sizes are approximately a fixed *multiple* of any other state's county sizes. For example, the county populations of New Jersey are like the county populations of Wyoming but with each one approximately 30 times larger. This factor was roughly estimated as the ratio of the original median population sizes for the two states: $395{,}066/12{,}373 = 31.9$, or approximately 30. This simple relationship—each state's county populations being like a constant multiple of another's—was not at all clear from the box plots (or histograms) on the original scale. But from simply looking at the distributions on the logarithmic scale, this relationship becomes immediately and plainly clear.

Exercise Set 13.1

1. Suppose you have two groups of perceived speed measurements, one taken from well-rested subjects 2 hours after they wake up, and the other taken from subjects who have had no sleep and nothing but coffee to eat or drink for 36 hours. The five-number summaries follow:

   Rested subjects: (1, 3, 4, 5, 7)

   Sleep-deprived subjects: (10, 30, 40, 50, 70)

   Find a transformation that makes the variability in the two sets of measures approximately equal.

2. Find a transformation that stabilizes the variability of measurements represented by these three five-number summaries:

   Control: (2, 4, 5, 6, 8)

   Treatment 1: (4, 16, 25, 36, 64)

   Treatment 2: (9, 25, 36, 64, 100)

13.2 Testing Whether Several Groups Are Different

When data have been collected for each of several groups or situations and the data values seem to be different from one group to another, we naturally wonder whether the differences are real or whether they could be merely the result of random sampling. After all, even if the groups had exactly the same basic characteristics, because of the presence of randomness (from measurement error or random sampling) we would not expect the measured data to look exactly alike from one group to another.

Our first inclination might be to look at groups two by two, using a series of t-tests to search for statistically significant differences. However, the significance level is based on the assumption that we are doing a single test. Even when there is nothing but random sampling variation, in 100 t-tests we would see about 5 statistically significant results. There are ways to correct for this, and we will introduce one way in Section 13.4. But first we want a way to test all the groups together to see whether there are any differences.

The ideas of populations and samples are important here. Each group is taken to represent a random sample from the population, and the result of the test is a decision about whether the population groups are different from each other. Because the sampled data will surely be at least a little different from one group to another, a formal statistical test of significance will be needed.

For example, suppose there are three experimental medical treatments for an illness that is not yet curable. A careful experiment would assign patients at random to one of the treatment groups, and their progress would be measured. One important goal of the experiment is to determine whether these treatments are any different from one another in their ability to help people in general who have the illness. This is a question about the population of potential people with the disease, which can be answered based on the sample of patients, together with statistical reasoning that will show whether the differences among treatments are detectable with respect to the usual patient-to-patient variability.

The Analysis of Variance

The analysis of variance, known as ANOVA (for ANalysis Of VAriance), is a very versatile and general collection of statistical techniques first suggested by Sir Ronald A. Fisher.[2] These techniques take the variability of a complex situation, as measured by the variance, and divide it into meaningful component parts.

For example, in an experiment measuring the behavior of people under various contexts and in different colored rooms, the overall variation in behavior would be divided into one component representing the variability from one context to another, another component representing the variability from one color to another, and a third component representing the remaining or "unexplained" variability. Statistical tests of whether the contexts made a difference or whether the colors made a difference would then be based directly on these variances; in other words, the variances are analyzed. Because variances are found from sums of squares computed from the data values, we will hear terms such as *sum of squares due to context* or *mean square* from time to time.

The Role and Limitations of the Analysis of Variance

The analysis of variance has a very specific role: to test definite hypotheses about whether differences exist under strict assumptions about the sampling situation. It can never substitute for a careful exploration of the basic properties of the data because its narrow role only provides yes-or-no answers to very specific questions. If not used carefully, it can provide exact answers to unimportant questions and leave us quite unaware of interesting features of the data. For example, it is possible to look at the results of an analysis of variance and never find out what the average values actually were in different situations.

But when used properly, in conjunction with other statistical methods, the analysis of variance fills the crucial role of setting the boundary between sampling variation and real differences in complicated situations.

13.3 One-Way Analysis of Variance and the F-Test

The one-way analysis of variance is the simplest form of the general ANOVA technique. We use it to test whether the population means of the groups are different. Thus, we use it to see whether the apparent differences in the averages computed for the groups are significantly different or whether those differences could be due to random sampling variability alone. This can be thought of as extending the two-sample *t*-test to more than two samples, but the procedures appear somewhat different.

Here is an outline of the procedures that will be involved in doing an *F*-test for differences among several groups. First, we check the necessary assumptions against the actual situation. Then we compute two kinds of averaged sums of deviation squares: the *between groups* and the *within groups* mean deviation squares. We need both of these because to see whether there is a difference among the groups, we must first have a baseline measure of underlying variability within the groups. The *F*-test is based on the ratio of these mean squares, which is compared to a critical value from a table to see whether the test is significant. Finally, if the test is significant, the detailed differences among groups may be studied further in a second stage of testing.

The Assumptions. To apply these procedures to our data, all three of the following assumptions should be satisfied, at least to a fair approximation.

1. Each group of data is normally distributed.
2. The variabilities are the same from one group to another.
3. The data were obtained by random sampling from a population. In particular, all data observations are independent of each other.

If the first two assumptions are not satisfied by the raw data, we would look for a transformation of the data to achieve normality and equal variability and then continue the analysis of variance using the transformed data. The third assumption is a fixed part of the experimental design, and usually nothing can be done once the data have been collected. Thus, a careful experimental design can make the difference between being able to reach valid conclusions from the data and not being able to validly apply statistical procedures. Of course, we need to carefully identify the appropriate population for a given data set so that this third assumption makes sense.

Note that there is no restriction on the number of data values in each group. Some of the more advanced ANOVA procedures do require that the experiment be balanced by having equal sample sizes. However, the simple one-way ANOVA does not have this restriction. All we need is at least two data values in each group to compute variances.

In the examples that follow, we will first develop ANOVA procedures for hand calculations in situations where the variabilities are roughly equal from group to group. (We would, as a rule of thumb, allow one SD to be up to twice as large as another.) We will show how to calculate a single *F*-statistic, how to evaluate it for statistical significance, and then how to isolate the groups that are different once we have established statistical significance. Finally, we will show computer calculations for the roughly equal variability situation. The unequal variability situation is beyond the scope of this course, unless a transformation can be found that produces roughly equal variability.

EXAMPLE 13.2 ANOVA Calculations (No Significant Difference): Seedlings

Suppose that because of your green thumb your friends have asked you for help in decorating their rooms with plants. You decide to perform an experiment while you are raising 24 seedlings, and you decide to test four fertilizers (A, B, C, and D), each on six seeds randomly selected from the package. During the experiment, only one mishap occurs (the cat knocks over one of the pots), but this actually presents no problem—we will analyze the remaining 23 data values.

The number of days between planting and sprouting is shown in Table 13.2. All three

Table 13.2 Sprouting time of seedlings (days)

	FERTILIZER			
	A	B	C	D
	3	4	2	5
	4	5	5	6
	4	5	3	3
	5	2	3	5
	2	3	4	3
	2	4		4
n	6	6	5	6
Average days to sprout	3.3333	3.8333	3.4000	4.3333
Standard deviation	1.2111	1.1690	1.1402	1.2111

assumptions are reasonably well satisfied here. Normality is difficult to verify with such small sample sizes, but at least there are no extreme outliers. The variabilities (as measured by the standard deviations) are fairly close to one another. Because seeds were randomly assigned to the fertilizers, the independence assumption is also satisfied.

We will use this example to illustrate the steps involved in a one-way ANOVA to answer the question "Can we detect a difference in the means from one fertilizer to another?"

The Between Groups Mean Square

When we only had two study groups, we could simply subtract one group mean from the other. But combining three or more group means requires more care to obtain a number that indicates generally how different they are from each other. The **between groups mean square** is that number—a measure of how different the groups are from one another. Essentially, this term will be a measure of the variability of the group averages, squaring and summing their deviations from an overall **grand mean,** almost like computing the variance of the group means.

PROCEDURE: FINDING THE BETWEEN GROUPS MEAN SQUARE

1. Find the mean, standard deviation, and sample size for each group.
2. Find the grand mean by multiplying each group mean by its sample size, summing these, and dividing by the sum of the sample sizes. Equivalently, find the grand mean by averaging all of the numbers in all of the groups.[3]
3. Subtract the grand mean from each group mean. Square each of these deviations and multiply each one by its respective sample size.
4. Add together the resulting terms to get the sum of squares.
5. Divide this result by its degrees of freedom, which is 1 less than the number of groups. You now have the between groups mean square.

Formulas

To represent this procedure by mathematical formulas, we will need the representation shown in Table 13.3 (page 436) for the summary values for each group mentioned in step 1. We then use these summary values to find the grand mean as defined in step 2.

Table 13.3 Summary values for between groups mean square

AVERAGE	STANDARD DEVIATION	SAMPLE SIZE	
$\bar{X}_1$	S_1	n_1	for the first group
$\bar{X}_2$	S_2	n_2	for the second group
⋮	⋮	⋮	
$\bar{X}_k$	S_k	n_k	for the last group (where k denotes the number of groups)

FORMULA

Formula for the grand mean

$$\textbf{grand mean} = \frac{n_1 \times \bar{X}_1 + n_2 \times \bar{X}_2 + \cdots + n_k \times \bar{X}_k}{n_1 + n_2 + \cdots + n_k}$$

or, in standard notation,

$$\bar{X} = \frac{\Sigma(n_i \times \bar{X}_i)}{\Sigma n_i}$$

We can now compute the between groups mean square and its degrees of freedom.

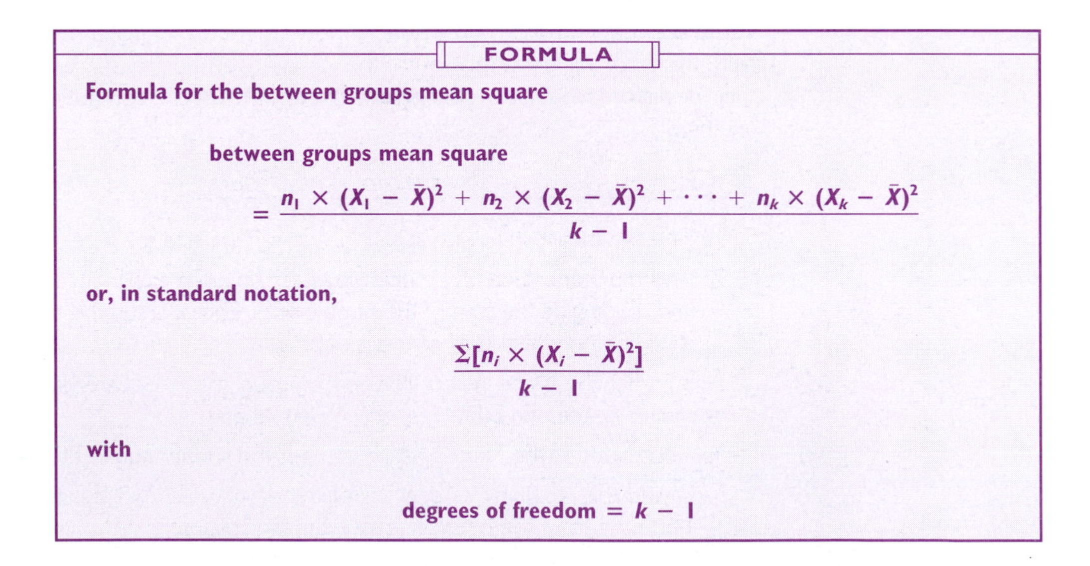

FORMULA

Formula for the between groups mean square

between groups mean square

$$= \frac{n_1 \times (X_1 - \bar{X})^2 + n_2 \times (X_2 - \bar{X})^2 + \cdots + n_k \times (X_k - \bar{X})^2}{k - 1}$$

or, in standard notation,

$$\frac{\Sigma[n_i \times (X_i - \bar{X})^2]}{k - 1}$$

with

$$\textbf{degrees of freedom} = k - 1$$

Note that, by convention, no square root will be taken. Instead, we work directly with the squares when doing ANOVA procedures. The purpose of multiplying by the sample sizes in step 3 is to give each group an importance, or weighting, in proportion to the information it brings to the situation. The degrees of freedom represent the number of items whose variability is being assessed (the number of group means) less 1 because the grand mean has been subtracted. For example, if there are 5 groups, each with 10 observations, then the degrees of freedom will be 5 − 1 = 4. Note that only the number of group means (5 because there are 5 groups) entered into this calculation.

The resulting between groups mean square is a measure of how different the group means are from one another. If they were all the same, the between groups mean square would be 0. The more different they are, the larger this term becomes.

For the seedling example, the calculations are as follows:

$$\text{grand mean} = \frac{6(3.3333) + 6(3.8333) + 5(3.4000) + 6(4.3333)}{6 + 6 + 5 + 6}$$
$$= \frac{85.9994}{23}$$
$$= 3.7391$$

This expresses the fact that when we place all 23 seedlings together, the average sprouting time is about $3\frac{3}{4}$ days.

between groups mean square

$$= \frac{6(3.3333 - 3.7391)^2 + 6(3.8333 - 3.7391)^2 + 5(3.4000 - 3.7391)^2 + 6(4.3333 - 3.7391)^2}{4 - 1}$$
$$= \frac{3.7347}{3}$$
$$= 1.2449$$

This value, 1.24, is a measure of how different the mean sprouting times of the various fertilizers are from the overall average. Note that this sum of squares has 3 degrees of freedom, corresponding to the 3 in the denominator.

Next, we will compute a reference value to see whether this term is larger than we should expect from sampling variation only.

The Within Groups Mean Square

The **within groups mean square** is also known as the *mean square due to error* or the *residual mean square.* Its purpose is to measure the variability of each individual group. However, because we have assumed that all groups are equally variable, we will combine the variation of all the groups to arrive at a good single estimate of this variability that combines the information from all of the groups. This term is also called the *pooled estimate of variance* and its square root the *pooled estimate of standard deviation,* because the information about variability has been placed in a common pool, so to speak.

PROCEDURE: FINDING THE WITHIN GROUPS MEAN SQUARE

1. Take the standard deviation for each group and square it.
2. Multiply each of these squares by 1 less than its respective sample size.
3. Add these to obtain a sum of squares.
4. Divide the resulting sum of squares by its number of degrees of freedom, which is the total number of data values in all groups (the sum of the sample sizes) minus the number of groups. The result is the within groups mean square.

Formula

The mathematical formulas for computing the within groups mean square and its degrees of freedom are as follows.

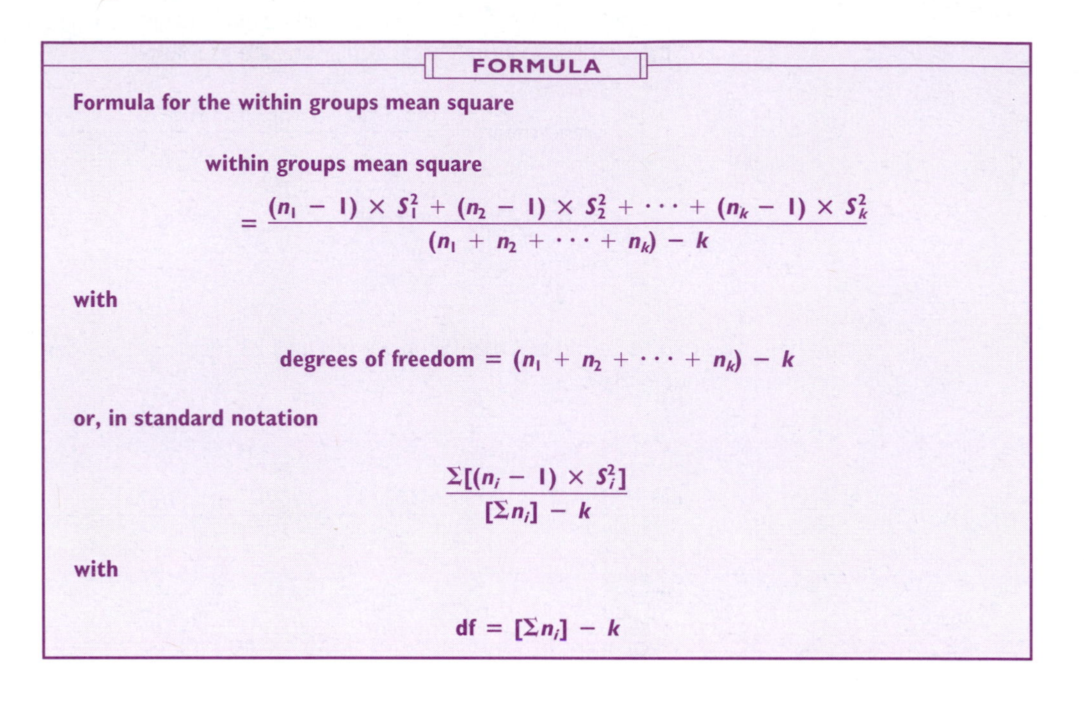

FORMULA

Formula for the within groups mean square

$$\text{within groups mean square} = \frac{(n_1 - 1) \times S_1^2 + (n_2 - 1) \times S_2^2 + \cdots + (n_k - 1) \times S_k^2}{(n_1 + n_2 + \cdots + n_k) - k}$$

with

$$\text{degrees of freedom} = (n_1 + n_2 + \cdots + n_k) - k$$

or, in standard notation

$$\frac{\Sigma[(n_i - 1) \times S_i^2]}{[\Sigma n_i] - k}$$

with

$$\text{df} = [\Sigma n_i] - k$$

Again, we see that some weighting has been done, in step 2, so that the greater information contributed by the larger groups is recognized. Note also that there is a second kind of degrees of freedom because each of the mean squares has its own number. The within groups number of degrees of freedom here is simply the sum of the usual separate degrees of freedom that we would have found for the variability of each separate group, and they represent the total number of data values less 1 for each group mean that is subtracted. For example, if there are 5 groups with 10 observations in each group, the number of degrees of freedom is $50 - 5 = 45$ because the total number of data values is $10 + 10 + 10 + 10 + 10 = 50$ for all 5 groups.

We now compute the within groups mean square for the seedling example:

$$\text{within groups mean square} = \frac{(6 - 1)(1.2111)^2 + (6 - 1)(1.1690)^2 + (5 - 1)(1.1402)^2 + (6 - 1)(1.2111)^2}{6 + 6 + 5 + 6 - 4}$$

$$= \frac{26.7007}{19}$$

$$= 1.4053$$

This value, 1.41, is a measure of how different the individual observations are from the average sprouting time for their fertilizer and provides an indication of the underlying randomness of the situation. This sum of squares has 19 degrees of freedom, corresponding to the denominator.

The F-Test

The F-test (named in honor of Sir Ronald A. Fisher[4]) is now very easily completed. We compute the F-value (or F-statistic) by dividing the between groups mean square by the within groups mean square:

$$F = \frac{\text{between groups mean square}}{\text{within groups mean square}}$$

This measures how different the group means are from one another (the numerator) with respect to the general overall amount of randomness in the situation (the denominator). Thus, it provides a measure of whether the group means are "more different from one another than is reasonable" for the situation if the groups were identically distributed.

We make the final decision as to significance by looking up a value in a table. As you might expect, the F-statistic behaves differently according to how many data values and groups there are. Fortunately, this only depends on the two degrees of freedom numbers (for the between groups and the within groups mean squares). If the exact numbers of degrees of freedom are not found along the side and top of the table, a safe conservative solution is to use the next smaller numbers. A more difficult, but more exact, procedure would interpolate between the numbers listed in the table.

After finding the critical value from the table, we will

decide that the differences are statistically significant if the F-value is larger than the critical value from the table.

Or, we will

decide that the differences are not statistically significant if the F-value is smaller than the critical value from the table.

If we reject the null hypothesis that "the means of the population groups are all identical," we conclude that there are indeed some differences among the groups. On the other hand, if the null hypothesis is accepted, we conclude that there is no strong evidence in favor of the groups being different.

To find the F-value for the seedling example, we find the ratio of the mean sums of squares computed earlier:

$$F = \frac{1.2449}{1.4053}$$

$$= .8859$$

The critical F-value with 3 and 19 degrees of freedom found from Table 13.4 (page 440) is between 3.10 (for 3 and 20 degrees of freedom) and 3.29 (for 3 and 15 degrees of freedom). Since our F-value (.8859) is smaller than this tabled critical value, we decide that the apparent differences among fertilizers are *not statistically significant.*

Thus, the apparent superiority of Fertilizer A, which resulted in the shortest observed mean sprouting time of 3.33 days, may not be real. In fact, because of the natural variation in sprouting time from one seed to another (even with the same fertilizer), we cannot detect any differences at all from one treatment to another.

Table 13.4 *F-table: 5% critical values*

BETWEEN GROUPS DEGREES OF FREEDOM (number of groups minus 1)	WITHIN GROUPS DEGREES OF FREEDOM (the total sample size for all groups, less the number of groups)							
	5	10	15	20	30	60	120	INFINITY
1	6.61	4.96	4.54	4.35	4.17	4.00	3.92	3.84
2	5.79	4.10	3.68	3.49	3.32	3.15	3.07	3.00
3	5.41	3.71	3.29	3.10	2.92	2.76	2.68	2.60
4	5.19	3.48	3.06	2.87	2.69	2.53	2.45	2.37
5	5.05	3.33	2.90	2.71	2.53	2.37	2.29	2.21
6	4.95	3.22	2.79	2.60	2.42	2.25	2.17	2.10
7	4.88	3.14	2.71	2.51	2.33	2.17	2.09	2.01
8	4.82	3.07	2.64	2.45	2.27	2.10	2.02	1.94
9	4.77	3.02	2.59	2.39	2.21	2.04	1.96	1.88
10	4.73	2.98	2.54	2.35	2.16	1.99	1.91	1.83
11	4.70	2.94	2.51	2.31	2.13	1.95	1.87	1.79
12	4.68	2.91	2.48	2.28	2.09	1.92	1.83	1.75
13	4.66	2.89	2.45	2.25	2.06	1.89	1.80	1.72
14	4.64	2.86	2.42	2.22	2.04	1.86	1.78	1.69
15	4.62	2.85	2.40	2.20	2.01	1.84	1.75	1.67
16	4.60	2.83	2.38	2.18	1.99	1.82	1.73	1.64
17	4.59	2.81	2.37	2.17	1.98	1.80	1.71	1.62
18	4.58	2.80	2.35	2.15	1.96	1.78	1.69	1.60
19	4.57	2.79	2.34	2.14	1.95	1.76	1.67	1.59
20	4.56	2.77	2.33	2.12	1.93	1.75	1.66	1.57
25	4.52	2.73	2.28	2.07	1.88	1.69	1.60	1.51
30	4.50	2.70	2.25	2.04	1.84	1.65	1.55	1.46
60	4.43	2.62	2.16	1.95	1.74	1.53	1.43	1.32
Infinity	4.36	2.54	2.07	1.84	1.62	1.39	1.25	1.00

Note that with a huge number of groups and a huge number of total observations, the critical value is approximately 1.00. In the rest of the table, the critical test values are larger due to the uncertainty of the estimates of the variances when the sample sizes or the group sizes are small.

EXAMPLE 13.3 ANOVA *Calculations (with Significant Difference): Chemical Stimulants and Reaction Times*

Medical researchers are interested in the effects of stimulants on reaction times. In one well-designed experiment[5] the investigators began by randomly dividing a group of 20-day-old rats into three groups. One group received ATRO (atropine) only, the second group received SPI (spiroperidol) only, and the third group received both in combination. One hour after the stimulant was administered, the time in seconds for each rat to move its raised front paws to the floor was measured. The smaller the number, the faster the reaction time. The results are shown in Table 13.5.

Before we test to see whether there are any differences among these three treatments, we should check the assumptions required by the one-way ANOVA. From the box plots for these three groups in Figure 13.5, we see that the groups appear quite different. We might well expect these apparent differences to be confirmed by statistical hypothesis testing. We also see from this illustration that assumption 2 for the one-way ANOVA is not satisfied by this data set: Note

Table 13.5 *Reaction time in seconds for three stimulants*

ATRO	SPI	COMBINATION
10.5	35.8	16.0
.8	10.5	5.9
.7	10.5	11.5
.7	5.2	4.4
.3	20.9	17.7
.7	44.2	13.5
.3	19.6	60.0
	20.7	2.3

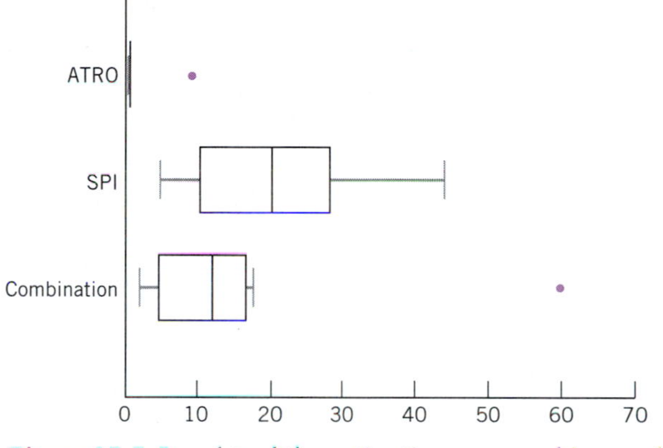

Figure 13.5 *Box plots of the reaction times, measured in seconds*

the extreme differences in variability from one group to another. Because it looks as though a fairly strong transformation will be needed to equalize these variabilities, we will try the log transformation. Table 13.6 shows the transformed data values (using natural logarithms). The box plots for this transformed data set (Figure 13.6, page 442) show much more reasonably equal variabilities. So we will proceed with the analysis, using logarithms.

Table 13.6 *Reaction time in log seconds for three stimulants*

ATRO	SPI	COMBINATION
2.35	3.58	2.77
−.22	2.35	1.77
−.36	2.35	2.44
−.36	1.65	1.48
−1.20	3.04	2.87
−.36	3.79	2.60
−1.20	2.98	4.09
	3.03	.83

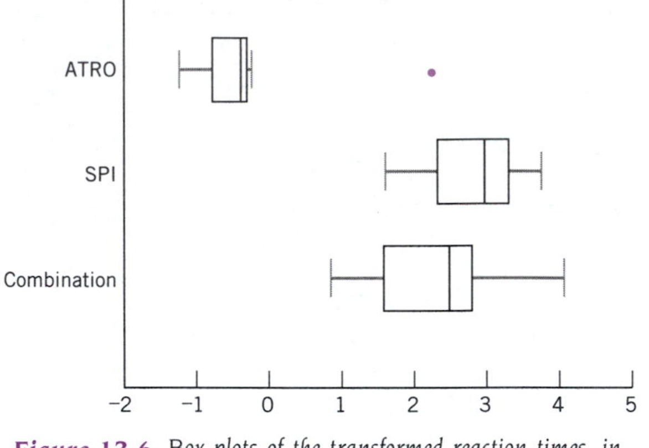

Figure 13.6 Box plots of the transformed reaction times, in log seconds

How will we interpret the results of an *F*-test on the log scale? Because the test looks for differences among the groups, and any differences (or similarities) on the original scale will still be there on the log scale, the basic conclusions will still apply. And because the three groups look similar to each other (except for center) on the log scale, it appears that perhaps the reaction time is affected by a constant multiple for each treatment. But are these differences from one treatment to another real at all?

The basic statistics needed for doing the *F*-test are shown in Table 13.7.

From these, we first find the grand mean:

$$\text{grand mean} = \frac{7(-.1928) + 8(2.8454) + 8(2.3594)}{7 + 8 + 8}$$

$$= \frac{40.2888}{23}$$

$$= 1.7517$$

This value represents the overall mean of reaction time, measured in log seconds. It may be transformed back to seconds by using the natural antilog function (taking e to the 1.7517 power), which would suggest a typical reaction time of 5.76 seconds, combining all three groups on the log scale.

The next step is to find the between and within groups mean squares.

between groups mean square

$$= \frac{7(-.1928 - 1.7517)^2 + 8(2.8454 - 1.7517)^2 + 8(2.3594 - 1.7517)^2}{3 - 1}$$

$$= \frac{26.4676 + 9.5694 + 2.9544}{2}$$

$$= \frac{38.9914}{2}$$

$$= 19.4957$$

Table 13.7 *Summary statistics for reaction times measured in log seconds*

ATRO	SPI	COMBINATION	
Mean	−.1928	2.8454	2.3594
Standard deviation	1.197	.7005	.9967
Sample size	7	8	8

This number, 19.4957, measures how different the group means are from the overall mean.

$$\begin{aligned}\text{within groups mean square} &= \frac{(7-1)1.197^2 + (8-1).7005^2 + (8-1).9967^2}{7+8+8-3} \\ &= \frac{8.5969 + 3.4349 + 6.9539}{20} \\ &= \frac{18.9857}{20} \\ &= .9493\end{aligned}$$

This number, .9493, measures the basic underlying variability. That it is so much smaller than the between groups mean square suggests that the means are more different than this variability would ordinarily allow and that maybe the means are, in fact, different. The *F*-test will decide this formally by finding the ratio of these two mean squares:

$$\begin{aligned}F &= \frac{19.4957}{.9493} \\ &= 20.5369\end{aligned}$$

Looking in Table 13.4 (page 440) for the *F*-critical value with 2 and 20 degrees of freedom (from the denominators of the mean square terms), we find that

$$\text{critical value} = 3.49$$

Because our computed *F*-value (20.5369) is larger than this tabled critical value, the differences among groups are indeed significant. Differences among treatments are highly unlikely to have occurred by chance, and therefore we reject the null hypothesis that the treatments all had equal effects.

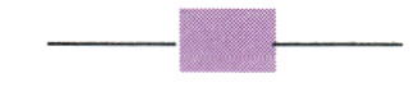

Exercise Set 13.2

3. Suppose you have measurements on three groups of trees: Four elms that average 20 feet tall with an SD of 5 feet, four vine maples that average 10 feet tall with an SD of 4 feet, and four firs that average 30 feet tall with an SD of 6 feet.

a. What is the between groups sum of squares?

b. What is the within groups sum of squares?

c. What is the value of F?

d. What is the tabled critical value of F for this number of groups and this number of observations?

e. Is there a statistically significant difference in the heights of the different types of trees?

4. Suppose you have three medical treatments: Standard drug, Experimental drug, and Combined standard and experimental. You give these to randomly assigned patients, and record the number of days of treatment required. (More days required means the treatment is less effective.) Your results follow:

   Standard drug: $N = 36$, average $= 7$, SD $= 2$

   Experimental drug: $N = 35$, average $= 9$, SD $= 2$

   Combined: $N = 30$, average $= 5$, SD $= 2$

   a. What is the between groups sum of squares?

   b. What is the within groups sum of squares?

   c. What is the value of F?

   d. What is the tabled critical value of F for this number of groups and this number of observations?

   e. Is there a statistically significant difference in the number of days it takes patients to recover?

5. Suppose you are trying three different fertilizers on a crop and measuring yield in units per acre. You get these results:

   Fertilizer One: $N = 4$, average yield $= 3.1$, SD $= .10$

   Fertilizer Two: $N = 4$, average yield $= 3.1$, SD $= .15$

   Fertilizer Three: $N = 4$, average yield $= 2.5$, SD $= .12$

   Is there a statistically significant difference for these three fertilizers?

Discussion: Interpreting the F-test

The results of an F-test are interpreted in much the same way as the other hypothesis tests we have seen. The null hypothesis is that the groups all have the same population mean. Because of the assumptions of normal distributions and equal variabilities, the null hypothesis actually assumes that the groups are identically distributed in all respects. The alternative hypothesis is that the groups are not all the same; that is, there are some differences in the groups. Note that the alternative hypothesis includes, for example, the possibility that two out of three groups are the same but the third one is different.

The Practical Interpretation. There are two possible interpretations in practice. First, if the null hypothesis is accepted because the computed F-value was smaller than the critical value from the table, we conclude that all of the groups are identical. This does not prove that they really are, only that no differences can be demonstrated based on the statistical evidence. The true situation might be either that there really are no differences (and so the decision was correct) or that there are differences but they are small and masked by variability perhaps due to small sample sizes (and so the decision was wrong). Regardless of the true situation, the investigation ends here for this analysis of the data set: There are no detectable differences, and therefore there is nothing to explain further based on the data set.

Second, if the null hypothesis is rejected because the computed F-value was larger than the tabled critical value, we conclude that there really are some differences from one group to another. The differences, apparent in the sample averages, are "certified" as statistically significant and are probably not due to sampling variability only. In this case, we have the right to examine the group differences in detail and a "license" to explain them. The second stage addresses some of these issues.

13.4 Identifying Which Groups Are Different

As we mentioned earlier, repeating *t*-tests on enough groups will eventually produce some "statistically significant" difference just by chance. We will show one way to correct for the repeated tests, but bear in mind that this procedure assumes that we first found a significant difference using ANOVA.

If the *F*-test is found to be significant, then we may pass along to the second stage of looking for which groups are different. (If the *F*-test was not significant, then we are not allowed to look further because, except for sampling variability, no real group differences in the population have been established.) The second stage pinpoints where the differences are by identifying specific pairs of groups, whereas the *F*-test was just a single overall procedure for all groups at once.

At the second stage, each group average is compared to every other group average using a modified *t*-test to see whether they are significantly different. These *t*-tests are similar to those that we used before in comparing just two groups, but this time we will use the pooled estimate of variability as estimated by all of the groups, even when we are just comparing two groups. The pooled variance estimate gives us a more sensitive test. We also modify the degrees of freedom calculation, which corrects for the repeated testing.

Thus, to test whether two particular group averages are significantly different, we will compute a pooled standard error and its appropriate number of degrees of freedom.

PROCEDURE: PERFORMING SECOND-STAGE ANALYSIS

1. Did the *F*-test find significant differences among the groups? If not, then do not proceed further because you already know that they are not significantly different. If yes, proceed to step 2.
2. Add the reciprocals of the sample sizes of the two groups in question.
3. Multiply by the pooled estimate of variance. (This is the same as the within groups mean square, which is the denominator of the *F*-statistic.)
4. Take the square root of the quantity from step 3. This is the standard error to use for the difference between the two group averages:

$$\text{standard error} = \sqrt{(\text{within groups mean square}) \times \left(\frac{1}{n_i} + \frac{1}{n_j}\right)}$$

   when comparing group i to group j.

5. Use the within groups degrees of freedom, which is the total sample size minus the number of groups:

$$\text{degrees of freedom} = (n_1 + n_2 + \cdots + n_k) - k$$
$$= [\Sigma n_i] - k$$

6. Compute the confidence interval as we did for the two-sample t-test on page 397, but be sure to use the standard errors and degrees of freedom that you have just computed. The confidence interval is then constructed to go from

$$\text{(difference in sample means)} - t \times \text{standard error}$$

to

$$\text{(difference in sample means)} + t \times \text{standard error}$$

where t is the value from the t-table with the appropriate number of degrees of freedom (from step 5) and the appropriate level for the test (95% confidence interval to test at level .05, for example).

7. If 0 is not within the confidence interval, then these two groups are declared to be significantly different. If 0 is within the interval, then there is no significant difference between the groups.

If we follow these steps, all comparisons can be done and a list of significantly different pairs of groups made. Occasionally, it might be that the F-test is significant, yet no individual pairs are significantly different. This would tell us that there are differences, but because of randomness it is difficult to pinpoint exactly where the differences are. On the other hand, it may happen that some individual pairs are significantly different, yet the overall F-test is not. This would tell us that there really are no significant differences anywhere at all because the individual t-tests are not valid unless the F-test is found to be significant.

EXAMPLE 13.4 *Second-Stage Analysis: Chemical Stimulants and Reaction Times*

Recall from Example 13.3 that we found a statistically significant difference between the different stimulants and their effects on reaction time. Because the differences are significant, we may start looking at the differences and explaining them. It certainly looks as though ATRO is very different from the other two. We might also wonder whether SPI is different from the combination. The second stage will help us answer these questions.

To test whether ATRO is different from SPI, the calculations are as follows. First we find the mean difference:

$$\begin{aligned}\text{mean difference} &= (-.1928) - 2.8454 \\ &= -3.0382\end{aligned}$$

We then compute the standard error for this difference using the within groups mean square and the sample sizes for these two groups:

$$\begin{aligned}\text{standard error} &= \sqrt{.9493\left(\frac{1}{7} + \frac{1}{8}\right)} \\ &= \sqrt{.2543} \\ &= .5043\end{aligned}$$

Next, we need the t-value for 20 degrees of freedom (the within groups degrees of freedom because we are using that mean square). From the t-table, we find a critical value of 2.086. Therefore, the confidence interval for the difference between ATRO and SPI goes from

$$-3.0382 - 2.086 \times .5043 = -4.09$$

to

$$-3.0382 + 2.086 \times .5043 = -1.99$$

This interval does not contain 0, so we decide that

ATRO and SPI are significantly different from each other.

Repeating this procedure for the comparison of ATRO with the combination, we find a confidence interval for this difference from

$$-2.5522 - 2.086 \times .5043 = -3.60$$

to

$$-2.5522 + 2.086 \times .5043 = -1.50$$

so we also decide that

ATRO and the combination are significantly different from each other.

For the last possible comparison—SPI against the combination—the standard error will be different because this time both groups have a sample size of 8, instead of one with 7 and one with 8. The standard error this time is slightly smaller, reflecting the greater precision in the group means that results from the additional observation. The confidence interval this time goes from

$$.4860 - 2.086 \times .4872 = -.53$$

to

$$.4860 + 2.086 \times .4872 = 1.50$$

This time, 0 is within the interval. We therefore decide that

SPI and the combination are not significantly different from each other.

The slight difference we see in the box plot comparison could easily be due to random sampling variation. In retrospect, we probably should have expected this.

The final conclusions are the following:

1. There are significant differences among these three treatments.
2. ATRO by itself results in a significantly faster reaction time than either of the other treatments.
3. SPI and the combination are not significantly different from each other.

We are now in an excellent position to explain various possible theories about how these chemicals actually achieve these statistically significant results.

Discussion

The particular procedure for the second stage of the analysis we use is just one of several choices of statistical procedures developed for the purpose of statistically comparing many groups simultaneously. This method is called the **least significant difference test,** attributed to Fisher and described in mathematical detail by R. Miller (1981).[6]

Just as the theoretical interpretation of hypothesis tests showed before, there are two types of errors possible in this analysis. A Type I error occurs when, in reality, there are no differences, but because of randomness we mistakenly decided that there were. This is the error controlled at 5%. Thus, when there are no true population differences, we will wrongly decide there are only 5% of the time.

A Type II error occurs when there are, in fact, true differences among the population groups and we mistakenly decide that there are none. This error is not easy to control because it depends on how different the population means actually are from one another, and this is information we usually do not have.

Because the second stage requires that the *F*-test find significance first, the combination of both procedures still controls the Type I error at the 5% level. That is, if there are in fact no differences among the population group means, then the conclusions for a given experiment will be wrong (and find significant differences) only 5% of the time. However, during that 5% of the time that the data mislead and suggest there are differences (by having a significant *F*-test), there may be many individual errors made when comparing all of the individual pairs of groups. Thus, the Type I error is controlled at 5% only to the extent that mistakes are or are not made for a given situation. The actual number of mistakes made when comparing many groups during the second stage (that is, when identifying which groups are different) is not controlled.

Exercise Set 13.3

6. Suppose you do an *F*-test on the results of a set of three drug treatments (call them A, B, and C) and, overall, there is no significant difference. The director of marketing at your drug company is upset by this result, and urges you to do just a *t*-test for a difference between drugs A and C. Is this the way *t*-tests were designed to be used? If not, where is the problem?

7. Recall the data on three (hypothetical) drug treatments from Exercise Set 13.2:

Standard drug: $N = 36$, average $= 7$, SD $= 2$

Experimental drug: $N = 35$, average $= 9$, SD $= 2$

Combined: $N = 30$, average $= 5$, SD $= 2$

If an *F*-test for a significant overall difference was statistically significant, go ahead with a second-stage analysis to identify which group-to-group differences are statistically significant.

8. Recall the data on three (hypothetical) fertilizers from Exercise Set 13.2:

Fertilizer One: $N = 4$, average yield $= 3.1$, SD $= .10$

Fertilizer Two: $N = 4$, average yield $= 3.1$, SD $= .15$

Fertilizer Three: $N = 4$, average yield $= 2.5$, SD $= .12$

If an *F*-test for a significant overall difference was statistically significant, go ahead with a second-stage analysis to identify which group-to-group differences are statistically significant.

EXAMPLE 13.5 Computer Analysis of Variance: Seedling Data

It is of course much simpler to do the arithmetic for ANOVA on a computer. We set up the seedling data from Example 13.2 on a computer file as shown in Table 13.8. The fertilizer type, which is given as A, B, C, or D in the original example, is now given by the numbers 1, 2, 3, and 4. These are in the first column, "Type." The number of days to sprouting are given in the second column, "Time." Now we use the ANOVA procedures in Minitab—other statistical packages have similar routines—to do the calculations. Bear in mind that although the computer does the arithmetic, it is still our responsibility to choose an appropriate analysis and to interpret the results.

Table 13.8

TYPE	TIME
1	3
1	4
1	4
1	5
1	2
1	2
2	4
2	5
2	5
2	2
2	3
2	4
3	2
3	5
3	3
3	3
3	4
4	5
4	6
4	3
4	5
4	3
4	4

Display 13.1 (page 450) is the ANOVA table produced by Minitab. There are a few differences between the terminology used in the table and what we used earlier, but all of the information is the same. Let's start with the two numbers at the right, which are the final results. We have an *F*-value of .89, which is essentially the same as the $F = .8859$ that we got by hand. When we did the analysis by hand we used a table to look up critical *F*-values for the test of statistical significance. The computer program instead gives us an observed probability, p, of .466. So we know that the *F*-value is not statistically significant (because $p > .05$), and we have a sense that *F*-values this large would be very common if there were no real differences between these fertilizers.

To the left of the *F*-value is a column labeled "MS," which stands for "Mean Square." The between groups mean square, which we calculated by hand at 1.2449, is shown on this output as a mean square of 1.24 whose source—the column on the left—is type of fertilizer. The

DISPLAY 13.1

```
MTB > Oneway Time Type.

ANALYSIS OF VARIANCE ON Time


SOURCE   DF   SS      MS     F      P
Type      3    3.73   1.24   0.89   0.466
ERROR    19   26.70   1.41
TOTAL    22   30.43
```

within groups mean square, which we calculated by hand as 1.4053, is shown on this output as a mean square of 1.41 whose source is ERROR.

The remaining two columns give the sums of squares (SS) and the degrees of freedom (DF) used to calculate the mean squares.

Exercise Set 13.4

9. Display 13.2 is an ANOVA table produced from an analysis of the grades of a few students being tutored by three different teachers.

DISPLAY 13.2

```
MTB > Oneway 'Grades' 'Teacher'.

ANALYSIS OF VARIANCE ON Grades

 SOURCE   DF SS       MS    F    P
 Teacher   2    51.5 25.7 0.28 0.763
 ERROR    13 1212.5 93.3
 TOTAL    15 1264.0
```

a. What is the value of *F*?

b. What is the observed probability of this *F*-value?

c. Are the differences in the students' grades statistically significant?

d. What is the mean square due to differences between teachers? (This is the "between groups mean square" in our terminology.)

e. What is the mean square due to ERROR? (This is the "within groups mean square" in our terminology.)

EXAMPLE 13.6 *Computer Analysis of Variance: Chemical Stimulants and Reaction Times*

Display 13.3 is an analysis of the logged reaction times. It includes a second-stage analysis to determine which groups are different. The ANOVA table gives an *F*-value of 20.53, with an observed probability of $p = 0.000$. There is no chance (to three decimal places, anyway) of seeing these results just on the basis of random sampling, if there is no real difference in the reaction times. But where are the differences?

DISPLAY 13.3

```
MTB > Oneway 'LnTime' 'Drug';
SUBC>   Fisher 5.

ANALYSIS OF VARIANCE ON LnTime

SOURCE     DF      SS          MS         F         P
Drug        2     38.992     19.496     20.53     0.000
ERROR      20     18.988      0.949
TOTAL      22     57.981

LEVEL       N      MEAN       STDEV       INDIVIDUAL 95 PCT CI'S FOR MEAN
   1        7    -0.1928      1.1972      BASED ON POOLED STDEV
   2        8     2.8454      0.7006      -------+---------+---------+---------
   3        8     2.3594      0.9967       (----*----)
                                                     (----*----)
                                                  (----*----)
POOLED STDEV = 0.9744                     -------+---------+---------+---------
                                                0.0       1.5       3.0
Fisher's pairwise comparisons

Family error rate = 0.118
Individual error rate = 0.0500

Critical value = 2.086

Intervals for (column level mean) - (row level mean)

            1            2
-------------------------------
2     -4.0902
      -1.9863

3     -3.6041      -0.5302
      -1.5002       1.5023
-------------------------------
```

The output includes a table of means and standard deviations for the three treatment groups, a calculated pooled standard deviation, and a graph of 95% confidence intervals for the three samples. Looking at the three confidence intervals, we see that there is no overlap between that of drug 1 (ATRO) and either of the other two. But the other two overlap quite a bit. This suggests that ATRO is significantly different from the other two, but they are not significantly different from each other.

The last part of the output includes a table of confidence intervals for the differences between average reaction times for the various combinations. We see that the confidence interval for the difference between drug 1 (ATRO) and drug 2 (SPI) runs from −4.0902 to −1.9863. This interval does not include 0, so we conclude that the observed difference is statistically significant. Similarly, the confidence interval for the difference between drug 1 (ATRO) and drug 3 (the combination) runs from −3.6041 to −1.5002. It does not include 0 either, so we conclude that the difference between ATRO and the combination is also statistically significant. Finally, we see that the interval for the difference between drugs 2 and 3 (SPI and the combination) runs from −.5302 to 1.5023. This interval does include 0, so we conclude that these are not significantly different.

Exercise Set 13.5

Display 13.4 shows the output for an analysis of student grades for three different tutors.

DISPLAY 13.4

```
MTB > Oneway 'Grades' 'Tutor';
SUBC>   Fisher 5.

ANALYSIS OF VARIANCE ON Grades

SOURCE      DF        SS          MS         F        P
Tutor        2       656.9       328.4      5.35     0.015
ERROR       18      1105.7        61.4
TOTAL       20      1762.6

                                          INDIVIDUAL 95 PCT CI'S FOR MEAN
                                          BASED ON POOLED STDEV
LEVEL      N      MEAN       STDEV     --+---------+---------+---------+----
   1       7     94.71       3.45                         (--------*--------)
   2       7     82.14      10.81      (--------*--------)
   3       7     83.71       7.45         (--------*--------)
                                       --+---------+---------+---------+----
POOLED STDEV =  7.84                   77.0      84.0      91.0      98.0

Fisher's pairwise comparisons

Family error rate = 0.118
Individual error rate = 0.0500

Critical value = 2.101

Intervals for (column level mean) − (row level mean)

          1          2
-------------------------
2      3.770
      21.373

3      2.198    −10.373
      19.802      7.230
-------------------------
```

10. a. What is the value of *F*?

b. What is the observed probability of this *F*-value?

c. Are there any statistically significant differences here?

11. **a.** Looking at the table of means, what are the mean grades earned by students of tutor 1?

b. What is the mean of the grades earned by students of tutor 2?

c. What is the mean of the grades earned by students of tutor 3?

12. Looking at the graph of confidence intervals, does one interval seem to be shifted away from the other two? If so, which one?

13. **a.** The last part of the output is labeled "Intervals for (column level mean) − (row level mean)." Tutor 1 is the first column. What is the confidence interval for the difference between the average grades of tutor 1's students and the average grades of tutor 2's students?

b. Are the average grades of the students of tutors 1 and 2 significantly different?

14. **a.** What is the confidence interval for the difference between the average grades of tutor 1's students and the average grades of tutor 3's students?

b. Are the average grades of the students of tutors 1 and 3 significantly different?

15. **a.** What is the confidence interval for the difference between the average grades of tutor 2's students and the average grades of tutor 3's students?

b. Are the average grades of the students of tutors 2 and 3 significantly different?

13.5 A Nonparametric Test of Several Medians

The Kruskal–Wallis test is the nonparametric equivalent of the one-way ANOVA. The procedure is similar to that used in the Mann–Whitney test, in the sense that all of the data are first ranked together (with tied scores sharing their average rank), and then calculations are done on the ranks for each group. In the Kruskal–Wallis test, the average rank for each group is calculated and then used as the basis for a test statistic H. The test statistic sums the deviations of each group's average rank from the overall average rank and corrects for group size and total sample size. This statistic has approximately a chi-square distribution (see Chapter 14) with degrees of freedom equal to 1 less than the number of groups, if each group has at least five members. A large value of H indicates that there are differences between at least some of the groups.

EXAMPLE 13.7 *Difference of Several Medians: Alfalfa Yields*

Recall the data on alfalfa yields, introduced in Chapter 2 (Table 2.18) and available as data set ALFALFA.MTW distributed by Minitab. These data give the yield for six different varieties of alfalfa, with each variety planted in each of four different fields. If we are reluctant to assume that these data meet the assumptions for an analysis of variance to see if the mean yields are different, we might use a Kruskal–Wallis test instead. We might, for instance, be in an argument with a colleague who denies that the yields are normally distributed, and we want to show this person that our results are not sensitive to this assumption. Or we might just be trying to be conservative.

We do the test first comparing the yields for the different fields, then comparing the yields for the different varieties. Display 13.5 (page 454) shows dot plots of yield for each field, followed by a Kruskal–Wallis test for a difference of medians. Looking first at the dot plots, we see that there is nothing particularly unusual about the data—no outliers, approximately equal variability, and an arguably normal distribution. (If it were not for the hypothetical argument with our colleague and our desire to be conservative, we could go ahead and use ANOVA with these data.) Further, by eye, the yields appear to be roughly equal from field to field.

DISPLAY 13.5

```
MTB > DotPlot 'YIELD';
SUBC>  Same;
SUBC>  By 'FIELD'.

FIELD
1                .         .                  ..         ..
      ---+---------+---------+---------+---------+---------+---YIELD

FIELD
2      .                        .               .           .      .   .
      ---+---------+---------+---------+---------+---------+---YIELD

FIELD
3       .           .                     .         .        ..
      ---+---------+---------+---------+---------+---------+---YIELD

FIELD
4      .                .                            .       .  ..
      ---+---------+---------+---------+---------+---------+---YIELD
       2.40      2.60      2.80      3.00      3.20      3.40

MTB > Kruskal-Wallis 'YIELD' 'FIELD'.

LEVEL       NOBS     MEDIAN     AVE. RANK     Z VALUE
    1          6      3.050          11.3       -0.47
    2          6      3.080          13.6        0.43
    3          6      3.020          12.3       -0.07
    4          6      3.145          12.8        0.10
OVERALL       24                     12.5

H = 0.31 d.f. = 3 p = 0.957
H = 0.32 d.f. = 3 p = 0.957 (adjusted for ties)
```

In the Kruskal–Wallis test we are testing whether the median yield is different from field to field. The four fields are numbered 1 to 4 in the LEVEL column. Each has 6 observations (NOBS), and the median yields are all a little over 3. Turning to the test of significance, we see that $p = .957$. There is better than a 95% probability that we will see differences this large or larger just on the basis of chance, even when there are no real differences between the fields. We cannot reject the hypothesis that the fields are really equal in their potential yields.

Now let us look at the yields of the different varieties. These are shown as dot plots, followed by a Kruskal–Wallis test, in Display 13.6. Looking at the dot plots, we now see that there appear to be substantial differences from one group to another. Varieties 4 and 6 seem to produce lower yields than the other varieties. Turning to the Kruskal–Wallis test, we see that the median yields for varieties 4 and 6 are lower than the medians for the other varieties. Looking at all of the groups together, we see that the probability of differences this large or larger, under the null hypothesis of no real differences, is only $p = .002$. We can reject the hypothesis that there are no real differences.

Note that we have a warning message on the output: "One or more small samples." All six of these groups have only four observations—one for each field. The rule of thumb is that the

DISPLAY 13.6

```
MTB > DotPlot 'YIELD';
SUBC>  Same;
SUBC>  By 'VARIETY'.

VARIETY
1                                                             . :  .
    ---+---------+---------+---------+---------+---------+---YIELD

VARIETY
2                                                    . .       .   .
    ---+---------+---------+---------+---------+---------+---YIELD

VARIETY
3                                                 .      ..    .
    ---+---------+---------+---------+---------+---------+---YIELD

VARIETY
4                             . ..      .
    ---+---------+---------+---------+---------+---------+---YIELD

VARIETY
5                                                        .    . .      .
    ---+---------+---------+---------+---------+---------+---YIELD

VARIETY
6       :.      .
    ---+---------+---------+---------+---------+---------+---YIELD
      2.40      2.60      2.80      3.00      3.20      3.40

MTB > Kruskal-Wallis 'YIELD' 'VARIETY'.

LEVEL       NOBS     MEDIAN    AVE. RANK    Z VALUE
   1          4      3.255       20.5         2.48
   2          4      3.120       15.0         0.77
   3          4      3.075       12.3        -0.08
   4          4      2.630        6.5        -1.86
   5          4      3.210       18.3         1.78
   6          4      2.375        2.5        -3.10
OVERALL      24                  12.5

H = 19.15 d.f. = 5 p = 0.002
H = 19.16 d.f. = 5 p = 0.002 (adjusted for ties)

* NOTE * One or more small samples
```

Kruskal–Wallis probability calculations are acceptably accurate if all the groups contain five or more observations. Since we only have four per group, the calculated probabilities are somewhat off. But the calculated probability is so small that we continue to feel secure that we can reject the null hypothesis.

Exercise Set 13.6

16. Suppose you have the prices (in thousands of dollars) for homes in three neighborhoods:

Capitol Hill: 124, 200, 222, 225, 250, 250, 279, 280, 300
North Shore: 85, 125, 150, 175, 180, 185, 195, 197, 198, 199
East Side: 270, 285, 350, 400, 450, 750, 850, 999, 1,895

Is there a statistically significant difference in median price?

Summary

A few new features arise when we go from two to several groups of numbers (in which all data are in the same measurement units so that they can be directly compared to each other). Exploring the data involves doing a box plot (or a histogram) for each group and aligning them all on the same scale so as to permit easy comparison of basic features such as center, variability, distribution shape, and outliers. If the variabilities are different and in particular related to the central values, a transformation may be used to stabilize these variabilities from one group to another. This can help with both the understanding of the data and the application of formal statistical testing procedures.

The *analysis of variance* (*ANOVA*) is a versatile, general set of statistical techniques that divide an overall variance into component parts in order to test for various effects in a complicated data situation. Its role is to test the significance of differences under very specific assumptions; it does not substitute for a careful exploration or a deeper understanding of the data.

In particular, the *one-way analysis of variance* is used to test for the statistical significance of differences among several groups of numbers. The *F-statistic* is the ratio of the *between groups* to the *within groups* mean square. Significance is declared if this computed *F*-value exceeds the value in the *F*-table, found by using two different numbers of degrees of freedom (one for each mean square). For the conclusions of this analysis of variance procedure to be valid, (1) each group of data must be normally distributed, (2) the variabilities must be equal from one group to another, and (3) the data must have an interpretation in which each group is a random sample from a population and all data values are independent of each other. For the conclusions to be only approximately valid, these assumptions need only be approximately satisfied.

If the *F*-test is not significant, then any apparent differences among the groups cannot be ascribed to differences in the population groups, but could be due to sampling variability alone. In this case, nothing further can be done. But if the *F*-test is significant, so that we decide some differences do indeed exist, then there is a second stage of analysis, the *least significant difference test*, during which *t-tests* (based on a pooled variability estimate with its degrees of freedom) are used to see which particular pairs of groups can be declared significantly different from each other.

The level of this *F*-test is controlled, for example, at 5%, which says that if there are in reality no differences among the groups, then we will make the wrong decision (of deciding that there are some differences) only 5% of the time in the long run for many repeated experiments. When this Type I error is made (wrongly rejecting the null hypothesis that there are no differences), however, the number of errors made at the second stage (identifying which groups are different) is not controlled to any further extent.

Often, an overall test like the *F*-test will be used to decide whether the experimental results should be explained or not, in the sense that if the *F*-test is not significant, then there are no established differences to explain. This mechanism helps prevent researchers from inventing elaborate theories to explain random, nonrepeatable results. This is why obtaining a statistically significant result is considered so important to an investigator.

If the populations do not have normal distributions, the Kruskal–Wallis test can be considered as an alternative to the one-way analysis of variance. This is a nonparametric procedure that tests whether there are significant differences among the medians of several independent samples.

Problems

1. For each of the following situations, find a transformation that stabilizes the variability from one group to another. Complete two displays of quick box plots: one for the original units of measurement and one for your chosen transformation (if it is different from the original units).

a. Five-number summaries[7] for the following four groups:

58	(1.20,	2.01,	2.72,	3.67,	6.69)
73	(2.23,	4.48,	5.75,	7.39,	13.46)
45	(4.95,	7.39,	9.97,	13.46,	27.11)
49	(16.44,	29.96,	40.45,	54.60,	99.48)

b. Five-number summaries for the following five groups:

35	(.35,	.56,	.74,	.92,	1.23)
14	(.03,	.12,	.21,	.30,	.48)
12	(.69,	.98,	1.18,	1.40,	1.60)
83	(.60,	.85,	1.07,	1.29,	1.64)
55	(.02,	.08,	.16,	.24,	.43)

c. Five-number summaries for the following four groups:

16	(3.1,	8.5,	12.1,	15.5,	22.3)
23	(7.2,	11.6,	17.2,	20.4,	28.1)
9	(2.8,	7.9,	12.4,	14.8,	22.1)
12	(18.3,	22.4,	28.6,	31.3,	39.2)

d. Five-number summaries for the following five groups:

40	(1.04,	1.13,	1.22,	1.32,	1.56)
40	(1.07,	1.17,	1.26,	1.38,	1.66)
40	(1.31,	1.53,	1.74,	2.10,	3.66)
40	(.90,	.96,	1.01,	1.06,	1.18)
40	(1.40,	1.66,	1.91,	2.45,	12.40)

2. Consider the groups of numbers shown in Table 13.9 (page 458), representing the average income per person in countries grouped by parts of the world, as measured by per capita

Table 13.9 Income data

NORTHERN AFRICA		SOUTHWEST ASIA		WESTERN EUROPE	
Algeria	650	Bahrain	2,250	Austria	4,050
Egypt	280	Cyprus	1,380	Belgium	5,210
Libya	3,360	Iraq	970	France	5,190
Morocco	430	Israel	3,380	W. Germany	5,890
Sudan	150	Jordan	400	Luxembourg	5,690
Tunisia	550	Kuwait	11,640	Netherlands	4,880
		Lebanon	1,080	Switzerland	6,650
		Oman	1,250		
		Qatar	5,830		
		Saudi Arabia	2,080		
		Syria	490		
		Turkey	690		
		UAR	13,500		
		Yemen	120		

gross national product expressed in U.S. dollars.[8] This data set is no longer current, but it has interesting features.

a. Do box plots for each group, placing them all on a common scale, and briefly describe what you see.

b. Which assumptions, if any, of the one-way analysis of variance are satisfied by the data in this situation?

c. Take the square roots of the data values and repeat parts (a) and (b).

d. Take the logarithms of the data values and repeat parts (a) and (b).

e. Which is your choice of the best transformation? Why?

3. Repeated measurements of radioactive counts (the number of particles emitted from radioactive material) in a 1-minute interval were done on several materials, as shown in Table 13.10. (These data were simulated using the Poisson distribution, a theoretical distribution that behaves very much like this kind of data. It represents the number of independent rare events that occur in a given time interval.)

a. Explore this data set using one box plot for each material and briefly describe what you see.

b. Repeat part (a) using the square root transformation of the data values.

c. Repeat part (a) using the logarithm transformation of the data values.

d. Which transformation is preferable? Why?

4. Suppose that a survey of four national forests resulted in the basic statistics shown in Table 13.11, based on measurements of the heights of randomly selected trees.

a. What additional information would you need to tell whether it would be valid to do a one-way analysis of variance for this data?

b. What would the one-way analysis of variance procedure be a test of in this situation?

c. Compute the grand mean.

d. Compute the between groups mean square and its degrees of freedom.

e. Compute the within groups mean square and its degrees of freedom.

Table 13.10 *Radioactive counts*

MATERIAL			
1	2	3	4
83	8	32	96
85	6	26	91
89	4	32	75
100	3	29	79
94	8	23	85
86	4	20	87
86	6	29	81
98	7	26	80
98	6	27	90
97	4	37	107
98	4	27	87
79	5	26	84
104	5	31	85
79	3	14	86

Table 13.11 *Tree heights*

	FOREST 1	FOREST 2	FOREST 3
Mean height in feet	60.2	82.5	73.4
Standard deviation	11.2	19.0	15.5
Sample size	493	810	335

f. Compute the *F*-value.

g. Find the appropriate critical value from the *F*-table.

h. Do the *F*-test and state its result.

i. Briefly interpret this result.

5. "Mary the Magnificent" has two magic wands that she claims can, among other things, extend the life of razor blades. To test this claim, 20 razor blades were randomly divided into four groups: the control group (nothing done), those that had the black magic wand (A) waved over them, those that had the blue magic wand (B) waved over them, and those that had both magic wands waved over them. Each blade was then used repeatedly, and the number of successful smooth shaves that each one produced was recorded. The summary statistics are shown in Table 13.12.

a. What would the one-way analysis of variance procedure be a test of in this situation?

b. Compute the between groups mean square and its degrees of freedom.

Table 13.12 *Useful life of razor blades*

	CONTROL GROUP	WAND A ONLY	WAND B ONLY	BOTH WANDS
Mean number of shaves	15.3	18.2	12.4	19.6
Standard deviation	2.9	3.5	1.8	2.3
Number of blades	5	5	5	5

c. Compute the within groups mean square and its degrees of freedom.

d. Compute the *F*-statistic and complete an *F*-test.

e. Briefly interpret the results.

f. In particular, have you proven that the magic wands are ineffective? (*Hint:* Think about this before answering; you might also want to refer to "the practical interpretation" material for interpreting the *F*-test on page 444.)

6. A study of treatments for edema involves patients under the care of three doctors. After 10 weeks of the study, the serum potassium in units of meq/l was found to be as shown in Table 13.13.

Table 13.13 *Treatments for edema*

DR. HYDE	DR. JEKYLL	DR. CALIGARI
3.3	3.7	3.9
3.4	3.1	3.3
3.9	4.1	3.0
3.5	3.5	2.9
3.5	3.3	3.6
3.6	4.5	3.3
3.7	3.6	3.5
	4.1	3.7
	5.0	3.9
	4.6	
	3.7	

a. Find the average potassium level for each doctor's patients.

b. Are these average values different from each other? (Please answer this question literally.)

c. Are these average values *significantly* different from each other?

d. Is there any statistical evidence here that being under the care of one doctor or another makes any difference?

7. Display 13.7 is a Minitab analysis of variance for the iris petal data presented earlier.

a. What is the value of *F*?

b. What is the probability of getting an *F*-value this large or larger, under the null hypothesis?

c. Write a sentence stating the null hypothesis being tested.

d. Write a sentence stating your conclusion regarding the null hypothesis.

8. Suppose you are interested in buying a used sailboat in the 30- to 35-foot range. Using the data in SAILSMLL.MTW, determine whether there is a statistically significant difference in price between manufacturers. Describe the difference(s), and support your description with appropriate statistics.

9. Using the data in IRIS.MTW giving the sepal and petal measurements of three different iris species, calculate the ratio of sepal length to petal length for each specimen. Then determine if there is a statistically significant difference in this ratio between species.

10. In a study of the effectiveness of cognitive therapy versus relaxation therapy versus a minimum-contact control in the treatment of panic attacks, Lapp et al. (1994) report the

DISPLAY 13.7

```
MTB > Retrieve 'IRIS.MTW'.
MTB > AOVOneway 'Setosa' 'Versiclr' 'Virginic'.

ANALYSIS OF VARIANCE
SOURCE      DF       SS          MS          F           p
FACTOR      2        437.103     218.551     1180.16     0.000
ERROR       147      27.223      0.185
TOTAL       149      464.325
                                        INDIVIDUAL 95 PCT CI'S FOR MEAN
                                        BASED ON POOLED STDEV
LEVEL        N       MEAN        STDEV  ---------+---------+---------+-------
Setosa       50      1.4620      0.1737 (*)
Versiclr     50      4.2600      0.4699                    (*-)
Virginic     50      5.5520      0.5519                                  (*)
                                        ---------+---------+---------+-------
POOLED STDEV = 0.4303                          2.4       3.6       4.8
```

results of a questionnaire designed to measure fear of being in open places, administered after treatment. Data modeled after their results are presented in Table 13.14.[9] Analyze these data and report the results. (Higher numbers indicate greater fear.)

Table 13.14 *Results of agoraphobia fear scale after cognitive therapy* (CT), *relaxation therapy* (RT), *or minimum contact control* (MCC)

CT	RT	MCC
5	0	8
7	0	9
4	32	7
8	14	5
0	4	5
7	7	15
6	0	22
1	6	14
4	4	19
7	23	18
0	15	0
0	12	17
5	0	6
9	0	21
7	12	14
2	15	15
3	18	3
	13	1
	0	7
		0
		2
		5

11. Recall the series of three (hypothetical) studies presented in Problems 12–14 of Chapter 5, in which rats were exposed to electric shock paired with three different "unconditioned stimuli." The length of time over 5 seconds that it took the rats to complete 5 seconds of licking at a water bottle, in the presence of each stimuli, was then measured. These data were presented in Table 5.9 and data file RATLICKS.MTW. Is there a statistically significant difference between the conditioning effects achieved with the three different conditioned stimuli?

12. Beecher (1966)[10] compared morphine to a placebo as a way to control pain. He used a "double blind" procedure in which neither the subject not the researcher interacting with the subject knew whether the "treatment" was morphine or the placebo. His results showed that for low levels of pain there was effectively no difference between morphine and the placebo. For stronger pain, morphine was more effective than the placebo.

Suppose you decide to replicate this study. You induce various levels of pain in your volunteer (and reasonably well-paid) subjects, and then have a nurse (who does not know

Table 13.15 *Pseudodata for pain relief for various levels of pain after receiving morphine or a placebo*

SUBJECT	TREATMENT	PAIN LEVEL	RELIEF	NUMBER OF DOSES
1	1	1	1	1
2	1	1	1	2
3	1	1	1	2
4	1	1	1	1
5	1	1	1	2
6	1	2	0	2
7	1	2	1	2
8	1	2	1	2
9	1	2	1	3
10	1	2	1	4
11	1	3	0	6
12	1	3	0	6
13	1	3	0	6
14	1	3	0	6
15	1	3	0	6
16	2	1	1	1
17	2	1	1	2
18	2	1	1	2
19	2	1	1	1
20	2	1	1	2
21	2	2	1	2
22	2	2	1	2
23	2	2	1	2
24	2	2	1	3
25	2	2	1	3
26	2	3	1	4
27	2	3	1	3
28	2	3	1	4
29	2	3	1	6
30	2	3	0	6

Treatment: 1 = placebo, 2 = morphine
Pain level: 1 = mild, 2 = moderate, 3 = severe
Relief: 0 = no, 1 = yes

which preparation is being used) administer either morphine or the placebo. The nurse waits 90 seconds, then asks whether the pain has been relieved. If not, another small dose is administered. This procedure is repeated until either pain relief is achieved or the dose level reaches a medically and ethically established limit at, say, six very small doses.

The nurse records two pieces of data—whether pain relief was achieved, and the number of doses administered—as shown in Table 13.15 and data file PLACEBO.MTW. What are the results?

13. *Computer problem:* Using the data in data file SAILSMLL.MTW, determine whether there are statistically significant differences in average length by different motor type. Describe the differences, if any. Comment on how these differences (if any) may have come about.

14. Display 13.8 gives the results of an analysis of variance of the alfalfa yields in different *fields* from the data file ALFALFA.MTW. Write a brief paragraph giving the highlights of the results.

DISPLAY 13.8

```
MTB > Retrieve 'ALFALFA.MTW'.

MTB > Oneway 'YIELD' 'FIELD'.

ANALYSIS OF VARIANCE ON YIELD
SOURCE      DF        SS          MS          F         p
FIELD        3     0.019       0.006       0.05     0.985
ERROR       20     2.572       0.129
TOTAL       23     2.591
                                              INDIVIDUAL 95 PCT CI'S FOR MEAN
                                              BASED ON POOLED STDEV
                                              ---------+---------+---------+-------
LEVEL      N     MEAN        STDEV             (--------------*--------------)
1          6     2.9400      0.3026                 (--------------*--------------)
2          6     2.9983      0.3859           (--------------*--------------)
3          6     2.9233      0.3674              (--------------*--------------)
4          6     2.9600      0.3729           ---------+---------+---------+-------
                                                     2.80      3.00      3.20
POOLED STDEV = 0.3586
```

15. Display 13.9 (page 464) gives the results of an analysis of variance of the alfalfa yields of different *varieties* of alfalfa from the data file ALFALFA.MTW. Write a brief paragraph giving the highlights of the results.

16. Recall that data set CARSPEED.MTW (first discussed in Example 12.6) includes recorded car speeds under four different intervention methods designed to cause drivers to slow down. Analyze these data and report your results.

17. For the Whistler Mountain ski area data (Table 2.8, data file WHISTLER.MTW), is there a statistically significant relationship between type of ski lift and the vertical rise of the lift? Report your results in a brief paragraph, with supporting computations and/or graphics.

18. Recall the data on the effects of two sleeping medications presented in Table 12.11. Make up a third column of data that you think might represent the effects of two cups of coffee taken at bedtime. Conduct an analysis of the resulting data set and report your results.

DISPLAY 13.9

```
MTB > Oneway 'YIELD' 'VARIETY'.

ANALYSIS OF VARIANCE ON YIELD
SOURCE       DF       SS           MS           F         p
VARIETY       5     2.43507      0.48701      56.22     0.000
ERROR        18     0.15593      0.00866
TOTAL        23     2.59100

                                       INDIVIDUAL 95 PCT CI'S FOR MEAN
                                       BASED ON POOLED STDEV
LEVEL    N     MEAN      STDEV     ----+---------+---------+---------+--
1        4    3.2600    0.0374                               (---*--)
2        4    3.1250    0.1318                          (--*--)
3        4    3.0625    0.0998                        (--*--)
4        4    2.6500    0.0698            (--*---)
5        4    3.2325    0.1223                              (---*--)
6        4    2.4025    0.0585     (--*--)
                                   ----+---------+---------+---------+--
POOLED STDEV = 0.0931                 2.40      2.70      3.00      3.30
```

19. Suppose you are trying to develop a theory of the evolutionary significance of different taste sensitivities in animals. As part of the development of this theory, you need to test the effects of flavoring an already attractive food with a second flavor that you think the animal will find unpleasant.[11] You want to use an animal other than the white rat, so you decide to use Mongolian gerbils as your study animal and peanuts as the already-attractive food. For flavoring agents, you choose sucrose octa-acetate (SOA) and denotomium benzoate (Bitrex), both of which are extremely bitter to humans.

Ten gerbils are randomly assigned to each of three experimental conditions: natural (unflavored), SOA-flavored, and Bitrex-flavored. You measure the length of time it takes for the animals to eat their peanut, either natural or flavored. (Somewhat to your surprise, you find that the animals do eat all the peanuts.) You repeat the experiment for 7 days, after which your lab assistant refuses to handle Bitrex ever again. Table 13.16 shows your data for the first and last days of the study. Describe the results, including tests for statistical significance.

20. The WISC-R is an individually administered IQ test that generates three summary scores: a "verbal" IQ, a "performance" IQ, and a "full scale" IQ. The first two scales are made up of different items; the third scale combines items from the first two. In an effort to explore the effects of cultural background and motivation on test scores, Devers et al. (1994)[12] administered the WISC-R to 52 Native American students. A randomly assigned half of the students were given a poker chip (which could be exchanged for various prizes, including cash, at 5 cents a chip) each time they answered a question correctly. The control group was not told when they answered questions correctly. The basic results are in Table 13.17.

a. Describe any patterns you see in the data.

b. Speculate on why these patterns might exist.

c. Using the techniques we have studied so far, how would you go about analyzing the data to test for statistical significance?

Table 13.16 *Preferences for three tastes of peanuts at two time intervals (Data file GERBILS.MTW)*

TASTE	LATENCY1	LATENCY7
1	7.1	4.7
1	6.2	6.4
1	3.7	8.7
1	7.7	4.1
1	6.4	5.6
1	3.7	3.7
1	5.1	3.8
1	4.9	9.3
1	6.3	6.4
1	5.3	7.0
2	21.4	5.0
2	20.4	8.3
2	22.3	1.5
2	22.9	4.2
2	25.0	4.6
2	27.4	7.1
2	18.1	5.0
2	23.5	1.8
2	23.1	9.5
2	24.7	4.2
3	55.6	62.7
3	59.2	69.0
3	57.3	64.1
3	61.3	65.7
3	58.1	61.8
3	61.5	62.4
3	60.1	62.5
3	61.4	64.5
3	60.1	62.2
3	57.9	68.1

Taste condition: 1 = plain peanut, 2 = peanut + SOA, 3 = peanut + Bitrex. Latencies are at day 1 and day 7.

Table 13.17 *Means and standard deviations from a study comparing feedback versus no feedback on the three WISC-R scores*

GROUP	VERBAL IQ MEAN (SD)	PERFORMANCE IQ MEAN (SD)	FULL SCALE IQ MEAN (SD)
Control	88.8 (6.3)	101.2 (10.0)	94.4 (6.9)
Experimental	102.3 (12.8)	109.9 (13.2)	106.3 (13.1)

21. Some things are easier to learn than others. Humans, for instance, seem to have more trouble recognizing, naming, and remembering odors than they do colors. Bowers et al. (1994)[13] studied the ease with which students could learn to remember pairs of colors, pairs of odors, and pairs with one color and one odor. The colors were yellow, blue, and red; the odors were strawberry, banana, and orange. In the experiment, students were

Table 13.18 Percent of correct matches when students are learning color–color, odor–odor, color–odor, and odor–color pairs (Data modeled after Bowers et al., 1994)

ORDER	RIGHT %	ORDER	RIGHT %
1	78.5	3	76.2
1	69.0	3	69.5
1	67.2	3	77.3
1	72.3	3	72.5
1	70.6	3	61.2
1	72.1	3	66.2
1	70.1	3	77.1
1	74.6	3	74.1
1	67.7	3	81.9
1	65.6	3	69.2
2	58.0	4	61.6
2	60.7	4	66.3
2	58.1	4	62.1
2	61.9	4	71.2
2	48.9	4	49.4
2	60.3	4	54.6
2	68.2	4	56.1
2	51.0	4	65.4
2	62.3	4	60.9
2	60.0	4	61.5

Order code: 1 = color–color, 2 = odor–odor, 3 = color–odor, 4 = odor–color.

presented with the stimulus—either an odor or a color—and then had to select the correct match. This was repeated 45 times for each student. The outcome measure was the percent of times the student picked the correct match. Notice that if we treat all colors as equal and all odors as equal, we have four different experimental conditions: color–color, odor–odor, color–odor, and odor–color. The basic finding was that it was easier to learn pairs when the color was presented first.

You decide to replicate the experiment, and you get the data in Table 13.18 (data file ODORCOLR.MTW). Describe these results, including tests for statistical significance.

22. Suppose you review the research literature on learning odor–color pairs, including your own replication of the Bowers et al. experiment in Problem 21. You speculate that perhaps not all odors are equal: Perhaps the three odors used in the earlier experiment (strawberry, banana, and orange) were all perceived and remembered simply as "food" odors. You decide to do a systematic replication of the odor–color part of the experiment, but you choose the three odors "strawberry," "banana," and "bull elephant." Your results are in Table 13.19 as well as data file ODORMORE.MTW. Describe and analyze these results.

23. Recalling the name of a person to whom we have just been introduced can be a problem at any age, and it has been suggested that it gets harder with age.[14] Suppose that you want to examine this phenomenon. You recruit subjects in four different age groups: 5–10, 20–25, 25–35, and 60–79. You then use a standardized group of photographs and name labels to introduce the subjects to a series of people. Later, you show the subjects the photographs and ask them to recall the names. Your data consist of the number of names correctly recalled. The data are in Table 13.20 (data file NameTest.mtw). Describe and analyze your results.

***Table* 13.19** *Percent of correct matches when students are learning odor–color pairs, using different odors*

ORDER	RIGHT %	ORDER	RIGHT %
1	69.7	2	61.7
1	52.7	2	45.8
1	60.6	2	62.1
1	60.9	2	58.1
1	62.4	2	62.3
1	53.1	3	75.6
1	59.3	3	73.9
1	58.1	3	83.7
1	63.1	3	71.4
1	62.3	3	71.2
2	62.6	3	76.1
2	50.3	3	74.8
2	62.2	3	75.4
2	55.6	3	70.7
2	59.0	3	71.8

Odor code: 1 = strawberry, 2 = banana, 3 = bull elephant.

***Table* 13.20** *Pseudodata on recall of faces by age group* (*Data file NameTest.mtw*)

AGE GROUP	NUMBER FACES RECALLED	AGE GROUP	NUMBER FACES RECALLED
1	11	3	8
1	3	3	17
1	7	3	11
1	8	3	13
1	4	3	9
1	8	3	11
1	10	3	11
1	5	3	11
1	10	3	13
1	4	3	15
2	16	4	13
2	9	4	9
2	11	4	3
2	16	4	8
2	12	4	7
2	15	4	9
2	14	4	6
2	14	4	3
2	16	4	7
2	10	4	2

Age groups: 1 = 2–10, 2 = 20–25, 3 = 25–35, 4 = 60–79.

24. Research on the effect of birth order on personality and achievement dates at least to Galton's *English Men of Science,* published in 1874. One current speculation (Claxton, 1994) is that first-born children may receive more feedback from their parents about their task performance.[15] As a master's thesis project, you decide to replicate part of Claxton's study. You interview 152 people about the parental feedback they received on their task performance while growing up, assign scores using an established protocol, and classify your respondents as firstborn, middle born, or lastborn. Your data are in Table 13.21 and data file BrthOrdr.mtw. Describe and analyze your results.

Table 13.21 *Levels of parental feedback, by birth order* (*Pseudodata modeled after Claxton,* 1994)

FIRSTBORN		MIDDLE	LASTBORN	
10.6	15.3	17.4	11.3	18.7
19.5	9.8	15.3	14.8	13.5
17.4	17.4	13.0	15.7	15.2
17.6	16.6	10.8	12.5	14.6
13.7	13.2	15.7	11.8	10.2
14.5	14.1	14.6	11.5	9.3
14.2	13.9	14.8	9.9	13.0
15.2	19.0	13.5	6.4	12.2
14.9	13.0	14.6	12.0	18.4
14.3	11.8	13.2	12.0	17.9
15.4	11.4	13.6	10.7	16.2
15.2	14.0	12.4	17.6	14.2
14.0	17.8	12.3	16.1	8.8
11.2	11.2	16.0	13.9	15.5
13.7	14.3	19.0	13.2	13.5
13.5	15.4	11.8	13.3	6.3
15.7	11.6	18.5	17.5	13.1
16.3	17.0	15.1	11.8	16.2
11.8	17.2	14.5	11.9	11.4
11.2	20.0	13.3	12.7	13.1
13.2	8.2		13.0	11.6
14.0	17.4		14.5	11.1
16.1	16.8		13.4	17.4
12.3	18.8		11.4	13.5
14.2	15.1		14.7	16.4
11.3	14.3		16.7	16.9
18.5	11.0		15.7	12.1
16.7	12.2		12.4	14.4
16.2	18.7		16.6	17.4
12.9	16.0		10.8	11.7
10.0	18.0		13.2	15.5
17.4	12.4		17.5	16.7
9.2	12.5		14.1	
12.6				

chapter 14

Categorical Data and Chi-Square Analysis

So far, our statistical inference has primarily been based on numerical measurements. We have also learned how to work with percentage data based on binary (yes/no) situations where there are only two categories. In this chapter we show how to do statistical inference with categorical variables that have multiple categories.

Chi-square tests are based on the differences between observed counts and those we would have expected if the null hypothesis were true. One chi-square test tells us whether data from a categorical variable could reasonably have come from a population with given hypothesized percentages. Another chi-square test tells us whether two categorical variables are independent of one another; if they are not independent, then we conclude that the two variables are related or associated.

14.1 Frequencies for a Single Categorical Variable

We will start with tests for a single categorical variable. The general procedure is this: We count the number of cases in each category. We then ask how many cases we would "expect" in each category—where "expect" is used in the probabilistic sense—if the cases were randomly selected from a given population. We then compare the number of cases we actually got with the number we expect. If the numbers are very different, we decide that we can reject the null hypothesis that the actual numbers could be the result of a random sample. To decide whether they are "very different," we calculate a statistic called **chi-square.**

EXAMPLE 14.1 *Describing a Single Categorical Variable: Dice*

Suppose we want to decide whether a die is "fair," or whether it produces some results more frequently than others. We toss a six-sided die over and over again. The resulting data values represent the sample. The population may be thought of as a very large number of dice with each one already tossed, and with $\frac{1}{6}$ of them showing 1 dot, $\frac{1}{6}$ showing 2 dots, and so on. The sample results of 60 tosses are summarized in Table 14.1.

Table 14.1 *Data for die tossing*

NUMBER OF DOTS VISIBLE	NUMBER OF TOSSES	PERCENT
One	5	8.3
Two	12	20.0
Three	9	15.0
Four	9	15.0
Five	14	23.3
Six	11	18.3

For a fair die (one that is not "loaded"), we would expect about $\frac{1}{6}$ of the tosses to be in each category. Due to the random uncertainty involved, we do not obtain exactly equal representation of 10 tosses or 16.7% in each category. Although we already know how to test any *particular* percentage to see whether it is significantly different from the hypothesized value of 16.7%, such a test will not answer the question correctly for all six percentages simultaneously. We will soon see how to test the null hypothesis that "this is a fair die" against the alternative hypothesis "this die is unfairly loaded" using the *chi-square test,* which does take account of all six categories.

EXAMPLE 14.2 *Describing a Single Categorical Variable: Legislative Representation*

Consider the ethnic backgrounds of a hypothetical elected legislature of 123 people shown in Table 14.2. In this case, it is probably not reasonable to expect the percentages to be equal

Table 14.2 *Hypothetical legislature*

ETHNIC BACKGROUND	NUMBER OF LEGISLATORS	PERCENT IN LEGISLATURE
White	93	75.6
Black	15	12.2
Hispanic	9	7.3
Asian	6	4.9

Table 14.3 *Hypothetical population*

ETHNIC BACKGROUND	PERCENT IN POPULATION
White	62.1
Black	18.3
Hispanic	12.9
Asian	6.7

from one category to another. However, if the legislature is supposed to be representative of the population at large, then these percentages should be approximately close to the census population values (shown in Table 14.3).

A quick comparison of the percentages shows first that the legislative percentages are not equal to the population percentages, and second, that whites are overrepresented in the legislature and all other groups are underrepresented. Although this might look suspicious, we must keep in mind that some variation about the population percentages is perfectly normal and to be expected whenever any group is chosen. We might ask the question "Is there statistical evidence that the legislative percentages are more than just randomly different from the population percentages?" and find the answer using the chi-square test.

14.2 Testing Against Hypothesized Percentages: The Chi-Square Test

To test whether the observed data could have come from a hypothesized set of population percentages, we compare the observed number in each category to the number we would expect if we matched the population percentages exactly. To specify precisely what we are doing, it is conventional to define the hypotheses to be tested as follows. The **null hypothesis, H_0,** is that the actual population percentages are exactly equal to the hypothesized values. The **alternative hypothesis, H_1,** is that the actual population percentages are different from the hypothesized values. And when we say "expected," we mean expected under the null hypothesis.

For the die-tossing example, we would expect on average (for the null hypothesis, which says that the die is fair) to find that $\frac{1}{6}$ of the 60 tosses (that is, 10 tosses) will occur for each side of the die. We show this in Table 14.4 (page 472). There are certainly some discrepancies between the observed numbers of tosses and those expected. Should we conclude that the

Table 14.4 *Expected data for die tossing*

NUMBER OF DOTS VISIBLE	OBSERVED NUMBER OF TOSSES	EXPECTED NUMBER OF TOSSES
One	5	10
Two	12	10
Three	9	10
Four	9	10
Five	14	10
Six	11	10

die is "loaded" so that 1 is unlikely and 5 is favored? Or could this much discrepancy be due to ordinary, usual randomness? The **chi-square test** is used to measure the combined discrepancies between the observed and expected numbers of tosses.

PROCEDURE: CALCULATING CHI-SQUARE

1. Compute the *expected* number for each category by multiplying the hypothesized population proportion by the total sample size:

$$\text{expected number} = \text{population proportion} \times \text{sample size}$$

2. For each category, subtract the expected number from the observed number, then square the result. This is a measure of the discrepancy between the data and the hypothesized population percentages.
3. For each category, divide the result of step 2 by the expected number. This has the effect of adjusting for the fact that when larger numbers are expected, larger deviations also generally occur.
4. Sum the resulting values from step 3. This is the **chi-square statistic.** The larger this number is, the more different the observed numbers are from the hypothesized population proportions.
5. Find the degrees of freedom, defined as the number of categories minus 1.
6. Compare the computed chi-square statistic (from step 4) to the value in the chi-square table (Table 14.5) with the appropriate degrees of freedom (from step 5) and decide the results of the test as follows.
   - If the computed chi-square statistic is *larger* than the value in the table, then we *reject* the null hypothesis that the population percentages could have been the hypothesized values and *accept* the alternative hypothesis that the actual population percentages are different from the hypothesized values. We would also say that the observed percentages are significantly different from the hypothesized values.
   - If the computed chi-square statistic is *smaller*, then we *accept* the null hypothesis that the population percentages could have been the hypothesized values and say that the observed discrepancy is within the level of ordinary and usual randomness associated with random sampling.

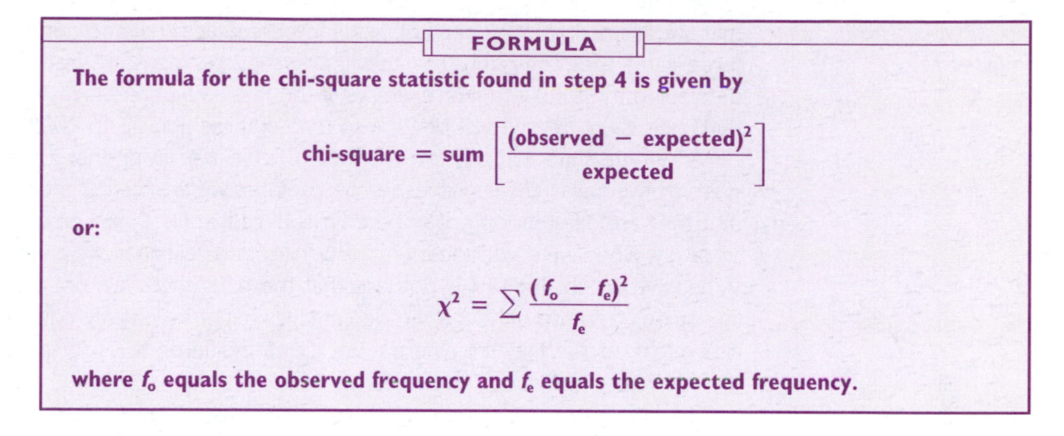

FORMULA

The formula for the chi-square statistic found in step 4 is given by

$$\text{chi-square} = \text{sum}\left[\frac{(\text{observed} - \text{expected})^2}{\text{expected}}\right]$$

or:

$$\chi^2 = \sum \frac{(f_o - f_e)^2}{f_e}$$

where f_o equals the observed frequency and f_e equals the expected frequency.

Note that it is the observed and expected numbers of *counts* that are used, *not* the percentages. That is, we would use 9 to represent the observed number of threes, rather than the 15.0% that it represents as a percentage.

The Chi-Square Distribution and Table

If the (unknown) population percentages are indeed equal to the hypothesized values, then the chi-square statistic will have a sampling distribution that is approximated by a distribution called the **chi-square distribution**. There is a different chi-square distribution for each number of degrees of freedom. Since the chi-square statistic tends to be large when the observed percentages are different from the hypothesized values, we will reject the null hypothesis (that they are equal) when the chi-square statistic is "larger than reasonable." The chi-square table (Table 14.5) provides a critical value that controls the Type I error (wrongly deciding significance) at level .05, as we know is traditional in general for hypothesis tests.

Technical Assumptions. For the chi-square test to be valid, the following two conditions must hold.

1. The data must be a random sample from a large population.
2. The expected number in each category should not be too small. A rule of thumb is to demand that at least 5 counts be expected in each category.

Discussion

We use the counts, rather than the percentages, in calculating chi-square because the percentage values do not "know" how many cases they are based on and so do not carry along the necessary information about how much random variation they contain. The numbers of counts do allow this important information about the extent of random variation to enter into the chi-square formula.

We have defined the number of degrees of freedom as the number of categories minus 1 for the following reasons. The randomness in each category may generally be thought of as contributing one "piece of information" or 1 degree of freedom to the chi-square statistic, with the exception of the last category. The last one is not free because the total sample size is fixed. Thus, all categories except one are free to vary. Another way to think about this is to realize that we are really testing percentages, and percentages have to add up to 100%. So if we have three categories, then only, say, the first two percentages are free to vary because they then determine the third one. For example, if the first two percentages are 30% and 50%, then the third one is *not free*; it *must* be 20% so that all three add up to 100%.

When we state that this test has level .05, we are saying that if the population percentages were truly equal to the hypothesized percentages we are testing, then we would wrongly decide that they are different only 5% of the time. Because this Type I error is controlled and happens so rarely, when we do decide there are significant differences, we are making a strong statement. When we accept the null hypothesis that the differences are not significant, the conclusion is less strong because, although the percentages might be exactly equal to those hypothesized, it also might be that they are different but that the difference is too small to measure reliably with the (perhaps small) amount of data we have.

The chi-square test is not completely exact: It is an approximation. It is a very good approx-

Table 14.5 *Chi-square table: 5% critical values*

DEGREES OF FREEDOM	CRITICAL VALUE	DEGREES OF FREEDOM	CRITICAL VALUE	DEGREES OF FREEDOM	CRITICAL VALUE
1	3.841	35	49.80	68	88.25
2	5.991	36	51.00	69	89.39
3	7.815	37	52.19	70	90.53
4	9.487	38	53.38	71	91.67
5	11.07	39	54.57	72	92.81
6	12.59	40	55.76	73	93.94
7	14.07	41	56.94	74	95.08
8	15.51	42	58.12	75	96.22
9	16.92	43	59.30	76	97.35
10	18.31	44	60.48	77	98.48
11	19.67	45	61.66	78	99.62
12	21.03	46	62.83	79	100.7
13	22.36	47	64.00	80	101.9
14	23.68	48	65.17	81	103.0
15	25.00	49	66.34	82	104.1
16	26.30	50	67.50	83	105.3
17	27.59	51	68.67	84	106.4
18	28.87	52	69.83	85	107.5
19	30.14	53	70.99	86	108.6
20	31.41	54	72.15	87	109.8
21	32.67	55	73.31	88	110.9
22	33.92	56	74.47	89	112.0
23	35.17	57	75.62	90	113.1
24	36.41	58	76.78	91	114.3
25	37.65	59	77.93	92	115.4
26	38.88	60	79.08	93	116.5
27	40.11	61	80.23	94	117.6
28	41.34	62	81.38	95	118.8
29	42.56	63	82.53	96	119.9
30	43.77	64	83.67	97	121.0
31	44.98	65	84.82	98	122.1
32	46.19	66	85.96	99	123.2
33	47.40	67	87.11	100	124.3
34	48.60				

imation when the sample size is large. The theory the test is based on says that as the sample size becomes very large (that is, tends to infinity), the computed chi-square statistic follows more and more closely the theoretically ideal chi-square distribution when the null hypothesis is indeed true. However, this approximation is imperfect in the real world of finite and small amounts of data. When the sample size is very small, the chi-square test should not be used because it is not valid in the sense that its Type I error is not guaranteed to be .05. The rule of thumb is to try to avoid using the chi-square test when fewer than 5 cases are expected in any category. This problem of small sample size can sometimes be avoided by lumping similar categories together, adding up the counts in each one.

Table 14.5 is the table of critical values, calculated so that the probability of a Type I error is approximately .05.

EXAMPLE 14.3 Calculating Chi-Square: Fair and Unfair Dice

Table 14.6 *Chi-square calculation for die tossing*

NUMBER OF DOTS VISIBLE	f_o	f_e	$(f_o - f_e)$	$(f_o - f_e)^2$	$\frac{(f_o - f_e)^2}{f_e}$
One	5	10	−5	25	2.5
Two	12	10	2	4	.4
Three	9	10	−1	1	.1
Four	9	10	−1	1	.1
Five	14	10	4	16	1.6
Six	11	10	1	1	.1
				Chi-square statistic:	4.8
				Degrees of freedom:	6 − 1 = 5
				Critical value from chi-square table:	11.07

The calculations for the die-tossing example we considered earlier are shown in Table 14.6. Because the computed chi-square statistic (4.8) is less than the value in the chi-square table (11.07) for 5 degrees of freedom, we conclude that

The die's differences from complete fairness are not statistically significant.

That is, a fair die with equal probabilities of $\frac{1}{6}$ for all six faces could very reasonably have produced the kind of results we see here. There is no reason to suspect that 1 is especially unlikely to come up in future tosses or that 5 will continue to occur most often (although it might if the die is, in fact, loaded[1]). (Of course, there is also no reason to suppose that 1 will have to "make up" for its earlier absence by coming up more often later on, because the die has no memory of previous rolls.) Although we have not proven that the die is fair, we accept this possibility because we lack convincing evidence to the contrary.

However, should this pattern of apparently favored and unfavored outcomes continue with many more tosses, then the additional evidence might be enough to allow us to conclude that the die is indeed loaded. For example, suppose that the next 600 tosses of the die produced results that continued the same pattern seen before. Then, this unfair (loaded) die might produce the data in Table 14.7 (page 476).

Table 14.7 *Loaded die, 600 tosses*

NUMBER OF DOTS VISIBLE	OBSERVED NUMBER OF TOSSES	EXPECTED NUMBER OF TOSSES IF FAIR
One	45	100
Two	116	100
Three	96	100
Four	94	100
Five	137	100
Six	112	100

With this additional information (600 as compared to only 60 tosses before), we would expect that the chi-square test would be able to detect that this is, in fact, not really a fair die. The chi-square test does indeed do this, as the calculations in Table 14.8 show. Because the computed chi-square statistic (48.46) is larger than the table value (11.07) with 5 degrees of freedom, we conclude that based on this data set

the die's differences from fairness are statistically significant.

With the additional information, it is now quite clear that the percentages are not all equal.

Table 14.8 *Chi-square calculation for loaded die, 600 tosses*

NUMBER OF DOTS VISIBLE	f_o	f_e	$(f_o - f_e)$	$(f_o - f_e)^2$	$\frac{(f_o - f_e)^2}{f_e}$
One	45	100	−55	3,025	30.25
Two	116	100	16	256	2.56
Three	96	100	−4	16	.16
Four	94	100	−6	36	.36
Five	137	100	37	1,369	13.69
Six	112	100	12	144	1.44
				Chi-square statistic:	48.46
				Degrees of freedom:	6 − 1 = 5
				Critical value from chi-square table:	11.07

Let's try one more experiment. We will toss a *fair* die 600 times to see what happens to the chi-square statistic when the die is indeed fair. The results for this fair die are shown in Table 14.9.

What will happen to the chi-square statistic? Its job is to assess the differences between the observed and expected numbers and come out with a standard measure of discrepancy that takes advantage of the additional information contained in the larger sample size. The results are shown in Table 14.10.

These are very satisfying results. The computed chi-square statistic (6.22) is less than the tabled value (11.07) for 5 degrees of freedom, telling us to accept the null hypothesis that

this is a fair die.

Table 14.9 *Fair die, 600 tosses*

NUMBER OF DOTS VISIBLE	OBSERVED NUMBER OF TOSSES	EXPECTED NUMBER OF TOSSES IF FAIR
One	97	100
Two	105	100
Three	113	100
Four	81	100
Five	107	100
Six	97	100

Table 14.10 *Chi-square calculation for fair die, 600 tosses*

NUMBER OF DOTS VISIBLE	f_o	f_e	$(f_o - f_e)$	$(f_o - f_e)^2$	$\frac{(f_o - f_e)^2}{f_e}$
One	97	100	−3	9	.09
Two	105	100	5	25	.25
Three	113	100	13	169	1.69
Four	81	100	−19	361	3.61
Five	107	100	7	49	.49
Six	97	100	−3	9	.09
				Chi-square statistic:	6.22
				Degrees of freedom:	6 − 1 = 5
				Critical value from chi-square table:	11.07

More precisely, there is not enough evidence to rule out the possibility that all six faces have equal chances of coming up, based on these 600 tosses.[2]

EXAMPLE 14.4 Calculating Chi-Square: Legislative Representation

The calculations for the legislative representation example are shown in Table 14.11. The expected numbers are found by multiplying 123 (the sample size) by the population proportion in each case. For White, this is 123 × .621 = 76.38. Other expected frequencies are computed similarly.

Table 14.11 *Chi-square calculation for hypothetical legislature*

ETHNIC BACKGROUND	f_o	f_e	$(f_o - f_e)$	$(f_o - f_e)^2$	$\frac{(f_o - f_e)^2}{f_e}$
White	93	76.38	16.62	276.22	3.62
Black	15	22.51	−7.51	56.40	2.51
Hispanic	9	15.87	−6.87	47.20	2.97
Asian	6	8.24	−2.24	5.02	.61
				Chi-square statistic:	9.71
				Degrees of freedom:	4 − 1 = 3
				Critical value from chi-square table:	7.815

Because the computed chi-square statistic (9.71) is larger than the tabled chi-square critical value (7.815) with 3 degrees of freedom, we reject the null hypothesis that the legislature could be a random sample from the population at large with the same ethnic representation. In particular,

There is a statistically significant misrepresentation in the legislature as compared to the population.

Discussion

What have we really done here? It is important to be careful with the interpretation. The chi-square test requires the assumption that the data being tested are a random sample from the target population. In government legislatures, representatives are not chosen by a deliberately random mechanism. However, the usefulness of the chi-square test in such a situation is that it provides an objective benchmark from which we can measure the significance of the deviation from perfect ethnic representation. That is, we ask the *hypothetical* question "Could the representation be this unbalanced had we chosen our legislators at random from the population?" The answer is not based on the chi-square test here. Since the ethnic backgrounds do not look like a random sample from the population, it follows that there is something systematic (non-random) about the selection process that causes certain ethnic backgrounds to be favored and others to be underrepresented. Note further that statistics makes no value judgments as to whether this is due to unfair discrimination or something else (perhaps minority voters actually favored the elected candidates, for example). Statistics merely provides an answer to the question "Could this have happened at random?"

Exercise Set 14.1

1. Suppose you roll a die 60 times, with the results shown in Table 14.12.

***Table* 14.12**

FACE	FREQUENCY
One	10
Two	8
Three	12
Four	11
Five	9
Six	10

You want to see whether these results are consistent with the hypothesis that this is a fair die.

a. What are the expected frequencies, assuming a fair die?

b. Calculate chi-square for these data. (*Hint:* It is easiest to set up a work table like Table 14.6.)

c. How many degrees of freedom are there for this chi-square?

d. What is the critical value for this chi-square, at a significance level of .05?

e. What do you conclude with respect to the fairness of the die?

2. The U.S. Bureau of the Census has estimated the number of children and young adults there will be in the United States in 2020. They projected 18 million children under 5 (19%), 31.9 million 5- to-13-year-olds (34%), 15 million 14- to-17-year-olds (16%), and 27.2 million 18- to-24-year-olds (30%).[3]

Suppose you survey the 100 children and young adults in your neighborhood and get the results shown in Table 14.13. You want to see whether this age distribution is significantly different from the distribution projected by the Bureau of the Census for the country as a whole.

Table 14.13

AGE	FREQUENCY
Under 5	36
5 to 13	34
14 to 17	20
18 to 24	10

a. What are the expected frequencies, assuming this neighborhood can be viewed as a random sample from the projected population?

b. Calculate chi-square for these data.

c. How many degrees of freedom are there for this chi-square?

d. What is the critical value for this chi-square test, at a significance level of .05?

e. What do you conclude?

14.3 Two-Way Classification: Bivariate Categorical Data

When each object has been classified into two different sets of categories, the result is **bivariate categorical data,** often referred to as a **two-way classification** because each object is classified in two ways. Although we may, of course, apply the techniques we just learned for analyzing each of the two categorical variables separately, we can gain much more by also analyzing the two variables together to measure their *association* with each other. Before we learn the chi-square test for association between two categorical variables, we will first look at some examples of two-way classifications.

First, consider the classification of projectiles commonly used in competitive sports, as shown in Table 14.14. Analysis of bivariate categorical data usually begins by expressing the data set in the form of a two-way table of counts. To do this, we count the number of objects with each combination of attributes (in this case, color and shape) and arrange these counts in appropriate places in the table. For example, there are no objects that are conical and brown, but there are two objects that are round and white.

Table 14.14 *Classification of sports projectiles*

SPORTS PROJECTILE (OBJECT)	COLOR (FIRST CATEGORICAL VARIABLE)	SHAPE (SECOND CATEGORICAL VARIABLE)
Baseball	white	round
Football	brown	ellipsoidal
Hockey puck	black	cylindrical
Basketball	brown	round
Badminton birdie	white	conical
Golf ball	white	round

Table 14.15 *Two-way table for sports projectiles*

SPORTS PROJECTILES	BLACK	BROWN	WHITE
Conical	0	0	1
Cylindrical	1	0	0
Ellipsoidal	0	1	0
Round	0	1	2

The completed table is shown in Table 14.15. Note that each object is represented just once in the table and that the sum of all the numbers in the table (six in this example) corresponds to the total number of objects in the original listing.

For another example of data from a two-way classification, consider the data on the number of applicants to an advanced degree program at a major university shown in Table 14.16. This is the simplest possible table, with just two categories in each of the two classifications—a two-by-two table. Such a table will result whenever two yes/no variables are observed for each subject; thus, questionnaires and surveys typically result in many possible two-by-two tables of counts.

The two-by-two table is also common in medical studies. Typically, one of the variables is something observable early on (such as a symptom or the result of a test) and the other category is the end result, possibly observable only later on in the future (for example, whether the patient was cured or not).

Consider the results of a study of the effectiveness of surgery in controlling cancer of the colon.[4] When cancerous tumors are removed, it is not always possible to remove all cancerous cells without removing too much of the patient's vital organs. This is measured by the row variable in the two-by-two table of Table 14.17.

Table 14.16 *Two-way table for degree applicants*

NUMBER OF APPLICANTS	MALE	FEMALE
1994 year	64	25
1995 year	49	16

Table 14.17 *Two-by-two table for cancer treatment*

NUMBER OF CASES		WAS CANCER CONTROLLED? YES	NO
Was cancer present at the edge of the surgery?	Yes	28	11
	No	182	6

■ Exercise Set 14.2

3. Construct a two-way table for the data shown in Table 14.18. We suggest setting it up like Table 14.19.

Table 14.18

VOTER	REGION OF BIRTH	PARTY AFFILIATION
Sam	West	Owl
Mary	Northeast	Elk
Frank	South	Elk
Fran	North Central	Owl
Jack	West	Owl
Jill	Northeast	Elk
Art	South	Owl
Alice	North Central	Elk
Bary	West	Elk
Betty	Northeast	Owl

Table 14.19

	PARTY AFFILIATION	
REGION	OWL	ELK
North Central		
Northeast		
South		
West		

4. Suppose you have these random sample data on the genders and party affiliations of registered voters:

CASE	GENDER	PARTY
1	M	E
2	F	O
3	M	E
4	F	O
5	M	E
6	F	O
7	M	E
8	F	O
9	M	E
10	F	O
11	M	O
12	F	E

Codes: M = male, F = female; O = Owl, E = Elk.

Use these data to construct a two-by-two table.

Exploring Bivariate Categorical Data

There are three different ways to express percentages for a two-way table of counts, each method highlighting a different aspect of the data.

1. Joint percentages
2. Row percentages
3. Column percentages

The method (or methods) we use will depend on whether we want to compare the table entries directly to one another or to consider one of the variables as a "baseline" and compute the percentages of the other variable separately within each level of the baseline variable. For example, we might want to compute the percentage of males separately for the 1994 and 1995 college applications to see how it has changed from one year to the other, using the year as a baseline variable.

EXAMPLE 14.5 Exploring Bivariate Categorical Data: Artists' Selection and Rejection

Artists often submit slides of their work to be reviewed by judges who then decide which artists' work will be selected for an exhibition. In the 1980 Marietta College Crafts National Exhibition, a total of 1,099 artists applied to be included in a national exhibit of modern crafts. If we classify each artist according to the two categorical variables (1) was the artist selected or not, and (2) the artist's region of residence, then we arrive at the two-way classification shown in Table 14.20.[5]

The first step is to adjoin totals to the table—row totals, column totals, and an overall total—as in Table 14.21. Here, the row totals (which form a column at the far right) are simply the sum of everything in their row; for example, 63 + 299 = 362 artists applied from the North Central region. Similarly, the column totals (which form a row at the bottom) represent the total in the column above them; for example, 63 + 55 + 44 + 54 = 216 artists were selected to appear in the exhibition.

The overall total in the lower right-hand corner represents the total number of artists who applied to the exhibition, from all regions and whether selected or rejected. The overall total may be found in three ways: either directly (by adding up all of the entries to the original table), by adding up the row totals, or by adding up the column totals. This gives us a way to check

Table 14.20 *Two-way table for artist selection*

	NUMBER OF ARTISTS	
	SELECTED	REJECTED
North Central	63	299
Northeast	55	207
South	44	208
West	54	169

Table 14.21 *Two-way table for artist selection, with totals*

	NUMBER OF ARTISTS		
	SELECTED	REJECTED	TOTAL
North Central	63	299	362
Northeast	55	207	262
South	44	208	252
West	54	169	223
Total	216	883	1,099

our arithmetic: Compare the sum of the row totals to the sum of the column totals to be sure that they are the same. In this case, we find

$$\text{sum of row totals} = 362 + 262 + 252 + 223$$
$$= 1{,}099$$
$$\text{sum of column totals} = 216 + 883$$
$$= 1{,}099$$

These are equal, as they must be.

Exercise Set 14.3

5. For the data used in Exercise Set 14.2, what are the totals for:
   a. voters born in the West?
   b. voters born in the Northeast?
   c. voters born in the South?
   d. voters born in the North Central region?
   e. members of the Owl party?
   f. members of the Elk party?
6. Use the totals calculated in Exercise 5 to make a table like Table 14.22, but with numbers where the question marks are.

Table 14.22

	PARTY AFFILIATION		
REGION	OWL	ELK	TOTAL
North Central	?	?	?
Northeast	?	?	?
South	?	?	?
West	?	?	?
Total	?	?	?

Joint Percentages

To express a table of counts in terms of **joint percentages,** we divide each entry in the table of counts by the overall total. The top left entry will then be

$$\frac{63}{1{,}099} = .057, \quad \text{or } 5.7\%$$

This tells us that 5.7% of all applicants were from the North Central region and were selected to appear in the exhibition. The percentages for all other categories (including the row and column totals) are shown in Table 14.23 (page 484).

Table 14.23 *Joint percentages for artist selection*

	SELECTED	REJECTED	OVERALL
North Central	5.7%	27.2%	32.9%
Northeast	5.0%	18.8%	23.8%
South	4.0%	18.9%	22.9%
West	4.9%	15.4%	20.3%
Overall	19.7%	80.3%	100.0%

Discussion

There is a lot of information in Table 14.23. Besides giving us the percentage of artists in any given category of both variables (for example, 18.9% of applicants were from the South and were rejected), it also gives us the simple percentages of each classification separately in the Overall row and column. If we concentrated on the region of origin and forgot momentarily about selection or rejection, then the right-hand column would give us the percentages obtained by classifying artists by region only; for example, 32.9% of all applicants came from the North Central region. Similarly, the bottom Overall row gives us the percentage of artists selected without regard to which region they were from; for example, 80.3% of all applicants were rejected.

Note, however, that some comparisons are not easily apparent from this display. If we wanted to compare the overall selection rate of 19.7% to the selection rate in the South, for example, we would have to do further computation. Or, if we wanted to compare the home regions of the selected artists to those of the rejected artists (for example, 32.9% overall came from the North Central region; what percentage of the *selected* artists came from this region?), this would also require further effort. These questions and others like them are answered by the other two kinds of percentage displays—row percentages and column percentages—which are discussed in the next sections.

Exercise Set 14.4

7. Calculate the joint percentages for the first table worked out in Exercise Set 14.3. Your table should look like Table 14.24.

Table 14.24

	PARTY AFFILIATION		
REGION	OWL	ELK	TOTAL
Northeast	?	?	?
South	?	?	?
North Central	?	?	?
West	?	?	?
Total	?	?	100%

8. Calculate joint percentages for the table of gender and party affiliation that you worked out in Exercise Set 14.2.

Row Percentages

To express a table of counts as **row percentages,** we divide each entry in the table of counts by the total for its row. The top left entry will then be

$$\frac{63}{362} = .174, \quad \text{or } 17.4\%$$

This tells us that 17.4% of the applicants from the North Central region were selected to appear in the exhibition. Because the rows are regions in this table, the row percentages will give us percentages of selection and rejection for each region separately so that the regions can be compared to one another. Table 14.25 is the complete display of row percentages.

Table 14.25 *Row percentages for artist selection*

	SELECTED	REJECTED	OVERALL
North Central	17.4%	82.6%	100.0%
Northeast	21.0%	79.0%	100.0%
South	17.5%	82.5%	100.0%
West	24.2%	75.8%	100.0%
Overall	19.7%	80.3%	100.0%

This is basically a table of selection rates (and of rejection rates) arranged by region together with the overall selection and rejection rates. This allows an easy comparison of regions. The North Central region has the lowest selection rate: Only 17.4% of its applicants were selected to appear in the exhibition. The West has the highest rate, with 24.2% being selected. All of these selection rates may be compared to the overall selection rate of 19.7%. Some are higher and some are lower. This is expected, because the overall selection rate is a weighted average of the selection rates for the regions. (It is not a simple average because some regions have different numbers of applicants than others.)

The 100% values listed in the Overall column are there to remind us that each percentage is on the basis of its row only; that is, each row contains 100% of the data for the calculation.

For this data set, row percentage display is probably the most useful if we think of the regions as "given" and the judges' selection decisions as perhaps depending in part on regional characteristics. It is easier to think about regions possibly determining artistic characteristics (and therefore selection or rejection) than it is to think of rejection determining the region of origin of an artist. This choice of a stimulus/response pair of variables (region is the stimulus and selection is the response here) is exactly analogous to the assignment of numerical bivariate data to X and Y, where Y is thought of as being predicted from X. The difference is that now we are predicting (percentages of) categories of selection from categories of region rather than predicting numbers from numbers.

Now that we have examined selection rates and compared them by region, we might ask whether the apparent differences from one region to another are the kind of differences we would expect if decisions were made randomly (or, equivalently, if artwork was similar from one region to another) or whether there is any evidence here that the western United States might be significantly favored in the selection process. This question will soon be answered by a second kind of chi-square test. But first let's complete the summary picture with the third and last kind of percentage display.

Exercise Set 14.5

9. Calculate the row percentages for the table worked out in Exercise Set 14.3. Your table should look like Table 14.26.

Table 14.26

	PARTY AFFILIATION		
REGION	OWL	ELK	TOTAL
North Central	?	?	100%
Northeast	?	?	100%
South	?	?	100%
West	?	?	100%
Total	?	?	100%

10. Which of the following questions can we answer just by reading this table?
    a. What percent of Owl party members were born in the West?
    b. What percent of those born in the West join the Owl party (assuming that Owl and Elk are the only choices)?
11. Calculate row percentages for the gender and party affiliation data from Exercise Set 14.2.
12. In one or two short sentences, describe your results above.

Column Percentages

To express a table of counts as **column percentages,** we divide each entry in the table of counts by the total for its column. The top left entry will then be

$$\frac{63}{216} = .292, \quad \text{or } 29.2\%$$

This tells us that of the applicants selected to appear in the exhibition, 29.2% came from the North Central region. Because the columns represent the selection/rejection decision, the column percentages will give us the geographical distribution (in terms of percent from each region) separately for those selected and those rejected. Thus, we may compare the geographical distribution of those selected to that of those rejected, as shown in Table 14.27.

This is basically a table of geographical distributions. The first column is the geographical distribution of those selected (29.2% from North Central, 25.5% from Northeast, and so on).

Table 14.27 *Column percentages for artist selection*

	SELECTED	REJECTED	OVERALL
North Central	29.2%	33.9%	32.9%
Northeast	25.5%	23.4%	23.8%
South	20.4%	23.6%	22.9%
West	25.0%	19.1%	20.3%
Overall	100.0%	100.0%	100.0%

The second column is the geographical distribution for those rejected, and the last column is the geographical distribution for everybody. The largest percentage from the first column, 29.2%, indicates that the North Central region had the largest representation in the show (compared to the other regions, in terms of the actual number of artists with no adjustment for population differences among regions).

The overall percentage for the West, indicating that 20.3% of all applicants were from the West, is a weighted average of the 25.0% of those selected from the West together with the 19.1% of those rejected from the West. Note that 20.3% is much closer to 19.1% than 25.0%, reflecting the fact that many more artists were rejected than selected, and so the overall geographical distribution should follow that of the larger group of rejected artists more closely.

The 100% values listed in the Overall row are there to remind us that each percentage is on the basis of its column; that is, these are column percentages.

We might wonder whether there is any significant difference between those selected and those rejected in terms of their geographical distribution. That is, are these three columns basically the same except for randomness? The chi-square test of the next section will show us how to answer this. Curiously, the very same test works for equality of row percentages as well as for equality of column percentages.

Exercise Set 14.6

13. Calculate the column percentages for the table worked out in Exercise Set 14.3. Your table should look like Table 14.28.

***Table* 14.28**

	PARTY AFFILIATION		
REGION	OWL	ELK	TOTAL
North Central	?%	?%	?%
Northeast	?%	?%	?%
South	?%	?%	?%
West	?%	?%	?%
Total	100%	100%	100%

14. Which of these questions can we answer just by reading this table?

a. What percent of Owl party members were born in the West?

b. What percent of those born in the West join the Owl party (assuming that Owl and Elk are the only choices)?

15. Calculate column percentages for the gender and party affiliation data from Exercise Set 14.2.

16. In one or two short sentences, describe your results above.

14.4 Testing Independence in a Two-Way Table: The Chi-Square Test

What Is Independence?

The **chi-square test for a two-way table** is a test for *independence* or *lack of association* between the row and column factors in a two-way table of counts arising from bivariate categorical data. It is, at the same time, (1) a test for equality of row percentages from one row to another and (2) a test for equality of column percentages from one column to another. This allows us to use our choice of row or column percentages, whichever makes sense to us for the data, and then test for differences using the same chi-square test regardless of the choice.

For example, if a table were based on gender (male or female) and salary level (low, medium, or high), then a chi-square test would tell whether there is any statistically significant association between gender and salary level—that is, whether there is any evidence of potential salary discrimination based on gender. The concept of independence is that knowledge about one variable does not help us predict the outcome for the other variable. In this case, independence would say the knowledge that a person is male (or female) does not affect expectations regarding his (or her) salary. Likewise, the knowledge that a person receives a high (or low) salary does not tell us that the person is more likely to be either male or female. Independence is a hypothesis, which can be either true or false in the real world. The chi-square test helps us decide when a set of data is not compatible with the hypothesis of independence.

It is useful to understand what independence is *not*. Because it is a property of the relationship between the two factors (row and column), it has nothing to do with each variable separately. It is *not* a test of whether more people are male than female. It is *not* a test of whether more people make lower salaries than higher salaries. It *is* a test of the interaction, or dependence, between these factors; it is a test of whether gender and salaries are linked.

The Expected Table

The chi-square test begins with the observed table of counts and then constructs from it a hypothetical table of counts called the **expected table.** The expected table shows how the numbers would have been if the row and column factors were independent of each other. It is constructed by multiplying the row sum by the column sum and dividing by the overall sum.

$$\text{expected} = \frac{(\text{row sum}) \times (\text{column sum})}{\text{overall sum}}$$

This formula is applied to each and every entry in the observed table. The resulting expected table is the table that we would have expected (on average) for a perfectly independent table with no randomness. It may be interpreted as saying that of all the cases in this row (that is, of the row sum) take a portion according to the proportion in this column (that is, multiply by the column sum divided by the overall sum). In other words, if 35 objects are classified as being in a certain row and one-fifth of all objects are in a certain column, then we would expect (under independence that objects within this row are assigned to columns in the same way as objects are overall) that $35 \times (\frac{1}{5}) = 7$ objects would fall into this row and this column. This formula may equivalently be interpreted as saying that of all the cases in this column, take a portion according to the proportion in this row.

The observed table of counts for the earlier example of the selection of artists' works to appear in an exhibition is as shown in Table 14.29. The expected number of artists selected from the North Central region is

$$\begin{aligned} 362 \times \frac{216}{1{,}099} &= 362 \times .196542 \\ &= 71.148 \end{aligned}$$

We included the intermediate calculation to show that, in effect, we are answering the question "Of the 362 artists from the North Central region, based on the overall selection rate of .197 or 19.7% for everyone, how many of these North Central artists would have been selected if there were no regional differences?"

Table 14.29 *Observed table for artist selection*

	NUMBER OF ARTISTS		
	SELECTED	REJECTED	TOTAL
North Central	63	299	362
Northeast	55	207	262
South	44	208	252
West	54	169	223
Total	216	883	1,099

Continuing our calculations, we find that the expected number of artists rejected from the South is

$$252 \times \frac{883}{1{,}099} = 252 \times .803458$$
$$= 202.471$$

Gathering together all of the calculated expected values, we find the expected table shown in Table 14.30. By comparing this expected table to the observed table, we can tell how different the observed data are from perfect independence. In this example, independence of region and selection means that the percent selected in the different regions is the same except for sampling randomness; that is, there are no systematic differences among regions. Independence also means that the regional distribution of those selected is the same as that for those rejected.

Table 14.30 *Expected table for artist selection*

	EXPECTED NUMBER OF ARTISTS		
	SELECTED	REJECTED	TOTAL
North Central	71.148	290.852	362
Northeast	51.494	210.506	262
South	49.529	202.471	252
West	43.829	179.171	223
Total	216	883	1,099

Exercise Set 14.7

17. Given the data in Table 14.31, what are the expected cell frequencies?

Table 14.31

	PARTY AFFILIATION		
REGION	OWL	ELK	TOTAL
North Central	20	10	30
Northeast	10	20	30
South	10	10	20
West	10	10	20
Total	50	50	100

18. Work out the expected cell frequencies for the gender and party affiliation data originally presented in Exercise Set 14.2.

Discussion

To see more clearly just what the expected table represents, first notice that the row and column totals (and the overall total, also) are the same as for the original observed table. Thus, we have preserved the basic properties of the row variable (the geographical distribution) and the column variable (the selection rate). However, the row percentages computed from the expected table have a special pattern due to independence (see Table 14.32).

Table 14.32 *Row percentages from expected table for artist selection*

	NUMBER OF ARTISTS (*row percentages*)		
	SELECTED	REJECTED	OVERALL
North Central	19.7%	80.3%	100.0%
Northeast	19.7%	80.3%	100.0%
South	19.7%	80.3%	100.0%
West	19.7%	80.3%	100.0%
Overall	19.7%	80.3%	100.0%

It is quite clear from Table 14.32 that the selection rates are equal among all regions: Each has a rate of 19.7%, the overall selection rate. Similarly, the geographical distributions are seen to be equal if we compute column percentages, as shown in Table 14.33. Again, it is clear that the regional distribution is the same regardless of selection or rejection. For example, in the expected table, 32.9% of those selected as well as 32.9% of those rejected come from the North Central region, and both of these correspond to the fact that in the original observed data table, 32.9% of all artists came from this region.

Table 14.33 *Column percentages from expected table for artist selection*

	NUMBER OF ARTISTS (*column percentages*)		
	SELECTED	REJECTED	OVERALL
North Central	32.9%	32.9%	32.9%
Northeast	23.8%	23.8%	23.8%
South	22.9%	22.9%	22.9%
West	20.3%	20.3%	20.3%
Overall	100.0%	100.0%	100.0%

The Chi-Square Test for a Two-Way Table

The chi-square statistic is a single number that summarizes how different the observed table is from the expected table (which expresses independence). It is computed and tested in almost exactly the same way as the chi-square for a single categorical variable.

PROCEDURE: COMPUTING AND TESTING CHI-SQUARE FOR A TWO-WAY TABLE

1. Construct the expected table whose entry, for a given row and column, is

$$\text{expected} = \frac{(\text{row sum}) \times (\text{column sum})}{\text{overall sum}}$$

2. For each entry in the table (not including the Total row and column at the right and bottom), subtract the expected number from the observed value, then square the result. This is a measure of the discrepancy between the data and what we would have seen under the hypothesis of independence.
3. For each entry in the table, divide the result of step 2 by the expected number. This has the effect of adjusting for the fact that when larger numbers are expected, larger deviations also generally occur.
4. Sum the values resulting from step 3 to get the chi-square statistic. The larger this number is, the more different the observed table is from what we would have seen under the hypothesis of independence.
5. Find the degrees of freedom, defined as the product of (1 less than the number of rows) times (1 less than the number of columns).
6. Compare the computed chi-square statistic (from step 4) to the critical value in the chi-square table with the appropriate degrees of freedom (from step 5). Then, decide the results of the test as follows:
   - If the computed chi-square statistic is *larger* than the value in the table, we *reject* the null hypothesis that the population percentages could have satisfied the hypothesis of independence and *accept* the alternative hypothesis that there is a link between the row and the column factors. We would say that the table exhibits statistically significant dependence (or association) between row and column factors.
   - If the computed chi-square statistic is *smaller*, we *accept* the null hypothesis that the population percentages could have satisfied the independence assumption. Any apparent link between the row and column factors is within the level of ordinary and usual sampling randomness. There is no significant association between the row and column factors.

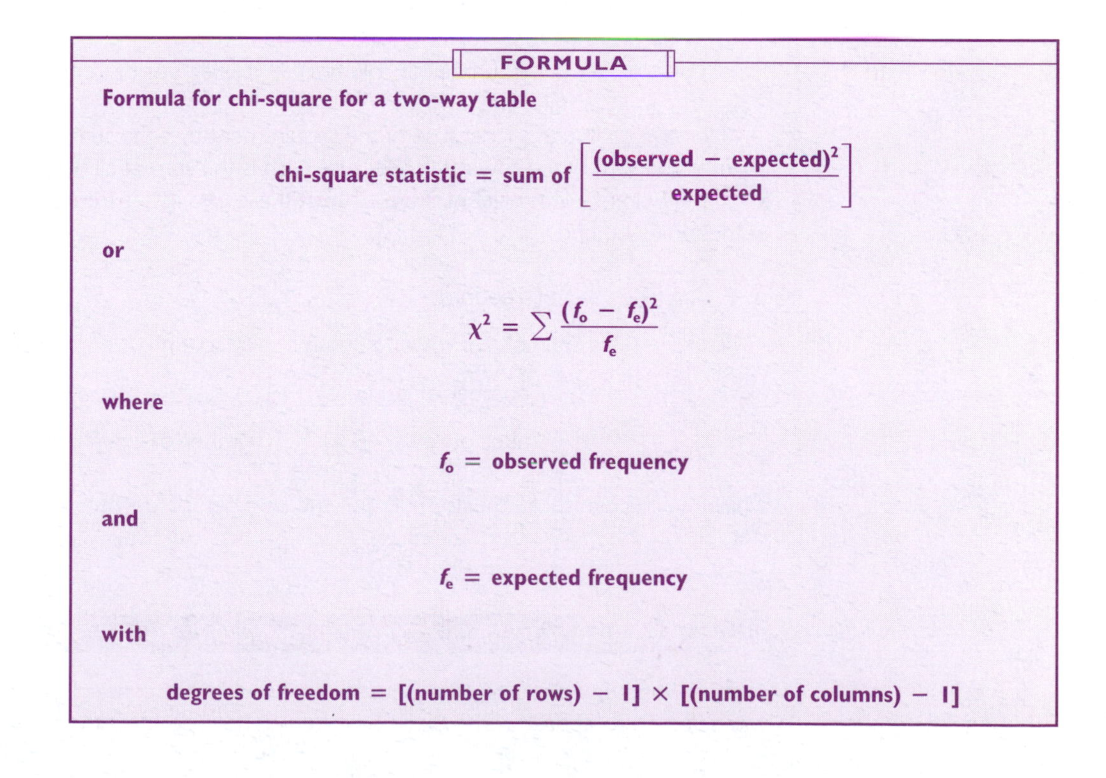

FORMULA

Formula for chi-square for a two-way table

$$\text{chi-square statistic} = \text{sum of} \left[\frac{(\text{observed} - \text{expected})^2}{\text{expected}} \right]$$

or

$$\chi^2 = \sum \frac{(f_o - f_e)^2}{f_e}$$

where

$$f_o = \text{observed frequency}$$

and

$$f_e = \text{expected frequency}$$

with

$$\text{degrees of freedom} = [(\text{number of rows}) - 1] \times [(\text{number of columns}) - 1]$$

Discussion

It is important again to note that it is the observed and expected *numbers of counts* that are used, not the percentages, because only the numbers of counts allow important information about the extent of random variation to enter into the chi-square formula. It is not possible to examine a table of percentages and test whether there is significant dependence unless we know how many cases those percentages were based on.

It is also important that the numbers not be too small: The chi-square test is only approximately valid according to theory, becoming more and more correct as the numbers increase. One rule of thumb is to require that all the expected numbers of counts be at least 5. If some numbers are small, but we still would like to perform a test, we could merge some categories of one or the other variable (or both). For example, suppose the rows in the artists' selection and rejection example had originally been defined by state rather than region. With many more states than regions, there would have been too few counts in some table entries to test the table. Grouping nearby states into regions is one way to avoid this problem while still permitting a reasonable interpretation of the results.

Where do the degrees of freedom come from? Each number being added in step 4 contributes a potential degree of freedom to the chi-square statistic, and there are

$$(\text{number of rows}) \times (\text{number of columns})$$

such numbers because this is the number of entries in the entire table. However, by computing the expected table under the assumption of independence, we lose some degrees of freedom by having estimated the row and column totals. Because the row and column totals must be equal (for both the observed and the expected tables), first we lose (for the rows):

$$(\text{number of rows}) \text{ degrees of freedom}$$

and then we lose (for the columns):

$$[(\text{number of columns}) - 1] \text{ degrees of freedom}$$

We have subtracted 1 because of the requirement that the sum of the column totals be the same as the sum of the row totals (because each is the sum of all numbers in the original table). With the help of a little algebra, if we subtract these two losses from the total number of entries in the table, we find

$$\begin{aligned}
&\text{degrees of freedom}\\
&\quad = (\text{number of rows}) \times (\text{number of columns}) - (\text{number of rows})\\
&\qquad - [(\text{number of columns}) - 1]\\
&\quad = [(\text{number of rows}) - 1] \times [(\text{number of columns}) - 1]
\end{aligned}$$

which we recognize as the formula for the number of degrees of freedom we defined in step 5.

EXAMPLE 14.6 *Chi-Square Test of Independence: Artists' Selection and Rejection*

To test whether there is any link between where an artist comes from and whether he or she was selected to appear in the exhibition, we will now perform the chi-square test. For step 1, the observed table together with the expected table we constructed earlier is shown in Table 14.34.

Table 14.34 *Observed and expected table for artist selection*

	OBSERVED TABLE: NUMBER OF ARTISTS		EXPECTED TABLE: NUMBER OF ARTISTS	
	SELECTED	REJECTED	SELECTED	REJECTED
North Central	63	299	71.148	290.852
Northeast	55	207	51.494	210.506
South	44	208	49.529	202.471
West	54	169	43.829	179.171

For step 2, we square the difference between observed and expected for each entry in the table. For example, for the top-left entry (Selected, North Central region), we find $(63 - 71.148)^2 = 66.390$. (This differs slightly from Table 14.34 because of rounding.)[6] Proceeding similarly for the rest of the table, we find what is shown in Table 14.35.

Table 14.35 *(Observed − Expected)^2 for artist selection*

	NUMBER OF ARTISTS	
	SELECTED	REJECTED
North Central	66.395	66.395
Northeast	12.291	12.291
South	30.566	30.566
West	103.451	103.451

Next, for step 3, we adjust for the fact that some of these entries may be large simply because more counts were expected. We divide each of these by the expected number; for example, the top-left entry will be $66.395/71.148 = .933$, and the bottom-right entry will be $103.451/179.171 = .577$. The complete table is shown in Table 14.36.[7]

Table 14.36 *(Observed − Expected)^2/Expected for artist selection*

	NUMBER OF ARTISTS	
	SELECTED	REJECTED
North Central	.933	.228
Northeast	.239	.058
South	.617	.151
West	2.360	.577

There is a useful interpretation for this table, whose entries will be summed to find the chi-square statistic. Each entry is an indication of "how far out of line" from independence the corresponding counted value is in the following statistical sense: The square roots of these values indicate approximately how many standard deviations away from the expected number the observed value has fallen. Thus, values that are very much larger than about 4 might be indicating something of interest because the square root of 4 is 2, and we recall that two standard deviations from the mean is beginning to be excessive.

In Table 14.36, the largest value is 2.360 for the selection from the West. Remember that when we first looked at the table, we noted that the selection rate seemed high for the West. Is it statistically significant? The chi-square test will now give us the formal answer.

The chi-square statistic is, from step 4, the sum of the entries in the table of (Observed − Expected)^2/Expected values.

$$\text{chi-square statistic} = .933 + .228 + .239 + .058 + .617 + .151 + 2.360 + .577$$
$$= 5.163$$

We must now find the degrees of freedom, step 5, so we may find the appropriate value in the chi-square table:

$$\begin{aligned}\text{degrees of freedom} &= [(\text{number of rows} - 1)] \times [(\text{number of columns}) - 1] \\ &= (4 - 1) \times (2 - 1) \\ &= 3 \times 1 \\ &= 3\end{aligned}$$

The critical value from the chi-square table with 3 degrees of freedom is 7.815.

The result of the chi-square test, step 6, is as follows. Because the computed chi-square statistic, 5.163, is *smaller* than the value in the chi-square table (7.815), we *accept* the null hypothesis that the population percentages could have satisfied the independence assumption (that is, that region of origin is unrelated to selection). In particular,

There is no evidence that judges are discriminating by region.

and

The apparent variations in selection rate by region are not statistically significant.

In particular, our observation that selection rates are highest in the West could easily be due to sampling randomness. We do not need to look further to explain the observed differences. In fact, we have no license to explain the differences and should accept the "innocent because not proven guilty" hypothesis that the selection rates are equal across regions. Remember that we have not *proven* this hypothesis (because the selection rates might indeed be very slightly different from one region to another); however, we have no statistical evidence against independence.

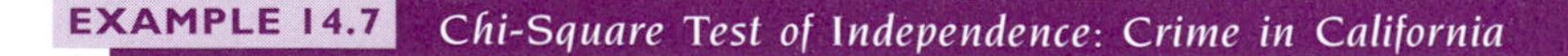

EXAMPLE 14.7 Chi-Square Test of Independence: Crime in California

Not only do different cities have different crime rates, they also show differences in the *types* of crimes committed. The techniques we have learned can help us analyze these differences. Table 14.37 represents the incidence of selected crimes in some California cities during the first 6 months of 1982.[8] Dividing each of the numbers in Table 14.37 by the grand total of 1,822,

Table 14.37 *Crime in California*

	CRIME INCIDENCE, FIRST HALF OF 1982			
	HOMICIDE	RAPE	ARSON	TOTAL
Berkeley	7	36	28	71
Oakland	44	192	208	444
San Francisco	63	294	209	566
San Jose	13	214	514	741
Total	127	736	959	1,822

Table 14.38 *Joint percentages for crime in California*

	HOMICIDE	RAPE	ARSON	OVERALL
Berkeley	.4%	2.0%	1.5%	3.9%
Oakland	2.4%	10.5%	11.4%	24.4%
San Francisco	3.5%	16.1%	11.5%	31.1%
San Jose	.7%	11.7%	28.2%	40.7%
Overall	7.0%	40.4%	52.5%	100.0%

we find the joint percentages that show the proportion of all crimes in all four cities represented by each category (Table 14.38).

From the Overall column in this display of joint percentages, we see that the cities are not all the same, with San Jose accounting for 40.7% compared to Berkeley with only 3.9% of these crimes. Some of this variation is likely to be explained by the fact that Oakland has a much larger population than Berkeley. Similarly, from the Overall row, we see that homicide is a relatively rare crime (only 7.0% of all these crimes), whereas arson is the most common, at 52.6%. However, our focus here will be more on the *independence* (or nonindependence) of the type of crime and city. That is, we will ask whether the types of crimes committed in these four cities differ from one city to the other after adjusting for the facts that some cities have more crime than others and that some crimes are more common than others.

Let's look at row percentages to see how the crimes committed in each city divide up among the three categories (Table 14.39). From this table, it is easy to see in what ways these cities are unlike each other. For example, San Jose has a strikingly low homicide rate (1.8%, compared to about 10% for the other cities). San Jose also stands out as having a much higher rate of arson (69.4%, compared to the 52.6% overall).

Table 14.39 *Crime incidence in California*

	ROW PERCENTAGES			
	HOMICIDE	RAPE	ARSON	OVERALL
Berkeley	9.9%	50.7%	39.4%	100.0%
Oakland	9.9%	43.2%	46.8%	100.0%
San Francisco	11.1%	51.9%	36.9%	100.0%
San Jose	1.8%	28.9%	69.4%	100.0%
Overall	7.0%	40.4%	52.6%	100.0%

Are these apparent city-to-city differences statistically significant? First, to see how much of each crime in each city we would expect if crimes were truly independent of cities, we construct the expected table shown in Table 14.40.

Table 14.40 *Expected table for crime in California*

	HOMICIDE	RAPE	ARSON	TOTAL
Berkeley	4.949	28.681	37.370	71
Oakland	30.948	179.355	233.697	444
San Francisco	39.452	228.637	297.911	566
San Jose	51.650	299.328	390.021	741
Total	127	736	959	1,822

We then calculate the (Observed - Expected)^2/Expected values (which will be added up to find the chi-square statistic), as shown in Table 14.41. Summing all these values, we find the chi-square statistic and degrees of freedom.

$$\text{computed chi-square statistic} = 166.222$$
$$\text{degrees of freedom} = (\text{number of rows} - 1) \times (\text{number of columns} - 1)$$
$$= 6$$

Looking in the chi-square table under 6 degrees of freedom, we find that the critical value from the chi-square table is 12.59.

Table 14.41 *(Observed − Expected)^2/Expected for crime in California*

	HOMICIDE	RAPE	ARSON
Berkeley	.850	1.868	2.350
Oakland	5.504	.892	2.826
San Francisco	14.055	18.686	26.535
San Jose	28.922	24.324	39.410

Because the computed chi-square statistic (166.222) is larger (in fact, much larger) than the tabled critical value (12.59) for 6 degrees of freedom, we conclude that we have a significant result, that is, the observed table shows statistically significant deviations from independence. We may now say that

Incidence of different types of crimes differs significantly from one city to another.

Because the result is significant, we now have a license to explain the apparent differences in the table (which we would not have had if the computed chi-square statistic had been smaller than 12.59). Looking back at the row percentages, we now have a right to say that some of these numbers are different within columns. For example, we might begin searching for reasons to explain why San Jose has lower homicide and higher arson rates than the other cities.

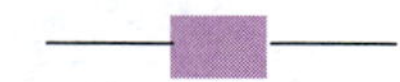

It is common to use the chi-square test like this to first reject, if possible, the null hypothesis of independence so that some more interesting theories may be applied that will explain the link between the row and the column factors. By using a chi-square test first, we are (at least partially) protected from the human tendency to see interesting structures (such as the associations between factors) where in fact only randomness may exist.

Exercise Set 14.8

19. To calculate a chi-square test of independence for the hypothetical data on region of birth and party affiliation (Table 14.18 on page 481), answer the questions below.

a. How many degrees of freedom are there for a table with four rows and two columns?

b. What is the critical value of chi-square at a .05 significance level and 3 degrees of freedom?

c. Complete the calculation of chi-square for the hypothetical political parties, started in Table 14.42.

Table 14.42 *Work table*

	OBSERVED	EXPECTED	DIFFERENCE	SQUARED DIFFERENCE	SQUARE/ EXPECTED
W, Owl	20	15	5	25	1.67
NE, Owl	10	15	−5	25	1.67
S, Owl	10	10	0	0	?
NC, Owl	10	10	0	?	?
W, Elk	10	15	?	?	?
NE, Elk	20	?	?	?	?
S, Elk	?	?	?	?	?
NC, Elk	?	?	?	?	?
Chi-square = SUM =					?

d. On the basis of the chi-square value you calculated in part (c), can you reject the hypothesis that party affiliation and region of birth are independent?

20. Is there a statistically significant relationship between gender and party affiliation for the data originally presented in Exercise Set 14.2?

21. In a brief paragraph, describe the relationship between gender and party affiliation in the data originally presented in Exercise Set 14.2. Include a statement about the statistical significance (if any) of the relationship. (Assume that your audience has not seen the data or worked the problem themselves.)

14.5 Other Significance Levels

In the crime in California example, the chi-square statistic (166.222) was *much* larger than the tabled value (12.59) for 6 degrees of freedom. In a case like this, we are entitled to feel very strong and confident about our conclusion that crime rates have different patterns in different cities.

The Level .01 Chi-Square Table: Is It Highly Significant?

The formal procedure for deciding whether we have a *highly significant result* rather than just a significant one is to use a more demanding chi-square table. The table we have been using has

Table 14.43 *Chi-square table: 1% critical values*

DEGREES OF FREEDOM	CRITICAL VALUE	DEGREES OF FREEDOM	CRITICAL VALUE	DEGREES OF FREEDOM	CRITICAL VALUE
1	6.635	35	57.34	68	98.03
2	9.209	36	58.62	69	99.23
3	11.34	37	59.89	70	100.4
4	13.28	38	61.16	71	101.6
5	15.08	39	62.43	72	102.8
6	16.81	40	63.69	73	104.0
7	18.47	41	64.95	74	105.2
8	20.09	42	66.20	75	106.4
9	21.66	43	67.46	76	107.6
10	23.21	44	68.71	77	108.8
11	24.72	45	69.96	78	110.0
12	26.22	46	71.20	79	111.1
13	27.69	47	72.44	80	112.3
14	29.14	48	73.68	81	113.5
15	30.58	49	74.92	82	114.7
16	32.00	50	76.15	83	115.9
17	33.41	51	77.38	84	117.1
18	34.80	52	78.61	85	118.2
19	36.19	53	79.84	86	119.4
20	37.56	54	81.07	87	120.6
21	38.93	55	82.29	88	121.8
22	40.29	56	83.51	89	122.9
23	41.64	57	84.73	90	124.1
24	42.98	58	85.95	91	125.3
25	44.31	59	87.16	92	126.5
26	45.64	60	88.38	93	127.6
27	46.96	61	89.59	94	128.8
28	48.28	62	90.80	95	130.0
29	49.59	63	92.01	96	131.1
30	50.89	64	93.22	97	132.3
31	52.19	65	94.42	98	133.5
32	53.48	66	95.62	99	134.6
33	54.77	67	96.83	100	135.8
34	56.06				

been computed so that the probability of Type I error (wrongly deciding significance) is 5%. If we reduce this probability to 1%, then the table values are somewhat larger, reflecting the fact that it should be less likely that we would declare significance purely by random chance. To do this, we use the level .01 chi-square table. Table 14.43 lists the critical values that are calculated so that the probability of a Type I error is approximately[9] .01.

Returning to the crime in California example, we see that only the chi-square tabled value has changed.

computed chi-square statistic = 166.222 (same as before)
degrees of freedom
= (number of rows − 1) × (number of columns − 1)
= 6 (same as before)

Looking in this new level .01 chi square table under 6 degrees of freedom, we find that the

critical value from chi-square table = 16.81

which is larger than the 12.59 we used before for the level .05 test.

Because the computed chi-square statistic (166.222) is still larger than the tabled value (16.81) for 6 degrees of freedom, we conclude that we have a highly significant result; that is, the observed table shows statistically significant deviations from independence even at the more demanding 1% level. We may now say that

Incidence of different types of crimes differs highly significantly from one city to another.

***p*-Value Notation.** Sometimes, we use *p-value notation* to distinguish between these levels of significance. We would write

significant ($p < .05$)

to express the fact that the chi-square test found significance at level .05. If the result was also significant using the level .01 chi-square table, then we would write

significant ($p < .01$)

We may also use p-value notation with other statistical tests.

Computer Applications

As usual, computers can be used to perform the routine work of counting and tabulating and doing arithmetic, leaving us with the more interesting work of collecting or locating the data, deciding what tables or calculations the computer should perform, and interpreting the tables and calculations.

EXAMPLE 14.8 *Computer Tables and the Chi-Square Test: Smoking and Activity*

We develop this example using the data file PULSE.MTW, distributed by Minitab. The two variables included in the data set concern smoking and routine physical activity. The variables are named "SMOKES" and "ACTIVITY," and the possible answers are coded as numbers. The interpretations of the codes are:

SMOKES: 1 = smokes regularly, 2 = does not smoke regularly

ACTIVITY: Usual level of physical activity: 1 = slight, 2 = moderate, 3 = a lot

There are 92 cases in the data set. We print dot plots of the smoking and activity variables in Display 14.1 (page 500).

Looking at the dot plots, we notice that one case has been reported as 0. But the possible codes—the ones that have a defined meaning—are 1, 2, and 3. Oops. If we leave this coding mistake in the data set, it can seriously affect the analysis. If we had collected the data ourselves, we could go back to the original data forms and see if we could find the mistake. Since we cannot do that, we have to make a reasonable decision and go on with the analysis.

We have two reasonable options: We can discard the case that was miscoded, or we can

DISPLAY 14.1

DOT PLOTS OF SMOKES AND ACTIVITY

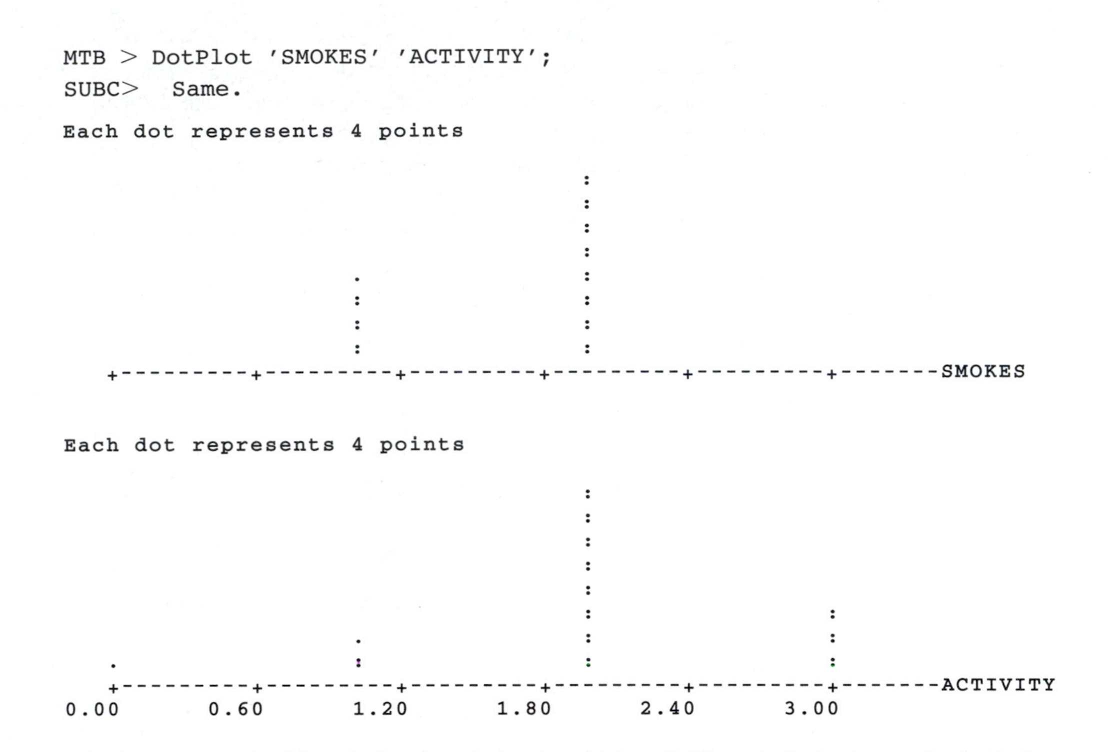

substitute a meaningful code for the obviously mistaken 0. The whole business of substituting meaningful codes is obviously open to abuse, and we have to be careful that we do not invent data that support our beliefs. But in this case, it seems reasonable that the interviewer may have deliberately recorded a 0 because the study participant answered "none" when asked about usual level of physical activity. (Well-trained interviewers use only the codes on the form, but occasionally even they will try to improve the instrument by inventing new codes.) Given that we have a reasonable argument about how the error occurred, we will go ahead and change this code to "1," for "slight" physical activity. The analysis follows.

DISPLAY 14.2

TABLE OF SMOKES AND ACTIVITY

```
MTB > Table 'SMOKES' 'ACTIVITY';
SUBC>  Counts.

ROWS: SMOKES     COLUMNS: ACTIVITY

             1        2        3      ALL

   1         4       19        5       28
   2         6       42       16       64
ALL         10       61       21       92

CELL CONTENTS --
                    COUNT
```

All of the information of a standard table is contained in Display 14.2, but the labeling of the table is perhaps not as informative as we would like. Notice that the heading of the table identifies the row variable (SMOKES) and the column variable (ACTIVITY). There are two rows, labeled "1" and "2." These are the codes for the SMOKES variable. We recall that "1" means "smokes regularly" and "2" means "does not smoke regularly." Unless we have been handling this data set for so long that this interpretation is automatic, it is worth writing these labels on the output by hand. The columns are headed "1," "2," and "3," which we recall mean "slight," "moderate," and "a lot" of activity, respectively. Each entry in the body of the table indicates the number of individuals with that level of smoking and that level of physical activity. There are 4 individuals who smoke regularly and have slight physical activity, 19 who smoke regularly and have moderate physical activity, and so on. The numbers at the ends of the rows give the row totals (there are 28 regular smokers, versus 64 who do not smoke regularly). The numbers at the bottoms of the columns give the column totals.

The program indicates that the cells—the intersections of rows and columns—contain only the counts. It is hard to directly interpret cell counts—what does it mean that there are 6 nonsmokers who have slight activity, compared to 4 smokers? The problem is that these numbers are not standardized so that they can be directly compared. Therefore, we will convert the counts to percentages, so the numbers can be directly compared.

In Display 14.3 we have asked the program to calculate row percentages and to show these along with the cell counts. Similar subcommands can be used to calculate column percentages or joint percentages. You can even ask the program to calculate and report all three percentages simultaneously.

Now we can say that 14% of the smokers report slight physical activity, compared with 9% of the nonsmokers. Looking at the high end of the activity variable, we see that 18% of the smokers report a lot of physical activity, compared to 25% of the nonsmokers. It appears that smoking may be associated with activity level, with smokers less likely to be active. We will test for the significance of this possible association next, using the chi-square test.

DISPLAY 14.3

TABLE PRODUCED BY THE MINITAB TABLES COMMAND, WITH COUNTS AND ROW PERCENTAGES

```
MTB > Table 'SMOKES' 'ACTIVITY';
SUBC>  Counts;
SUBC>  RowPercents.

ROWS: SMOKES      COLUMNS: ACTIVITY

              1         2         3       ALL

  1           4        19         5        28
          14.29     67.86     17.86    100.00

  2           6        42        16        64
           9.37     65.63     25.00    100.00

ALL          10        61        21        92
          10.87     66.30     22.83    100.00

CELL CONTENTS —
                COUNT
                % OF ROW
```

DISPLAY 14.4

CHI-SQUARE TEST OF ASSOCIATION BETWEEN SMOKING AND ACTIVITY

```
MTB > Table 'SMOKES' 'ACTIVITY';
SUBC>  ChiSquare 2.

ROWS: SMOKES      COLUMNS: ACTIVITY

                1         2         3       ALL
  1             4        19         5        28
             3.04     18.57      6.39     28.00

  2             6        42        16        64
             6.96     42.43     14.61     64.00

ALL            10        61        21        92
            10.00     61.00     21.00     92.00

CHI-SQUARE =   0.882   WITH D.F. =   2

CELL CONTENTS --
                COUNT
                EXP FREQ
```

The basic layout of Display 14.4 is the same as before. Each cell now contains the observed count and the expected count, under the null hypothesis of no association (independence). The program reports that chi-square equals .882, with 2 degrees of freedom. This particular program does not report the probability of a chi-square this large or larger, nor does it provide an indication of statistical significance. Using the tables, we see that with 2 degrees of freedom, we need a chi-square of 5.991 or larger to reject the null hypothesis at the 5% significance level. So, in spite of the reasonableness of supposing that there is an association between smoking and level of physical activity, and in spite of the glimmer of support we saw looking at the row percentages, on the basis of these data we cannot reject the hypothesis that the apparent association is simply the result of chance.

Minitab does include a way to find the probability of a particular chi-square value, using variations on the PDF or PROBABILITY DENSITY FUNCTION command. In Display 14.5, we use CDF to find the cumulative probability that, under the null hypothesis, we would see a chi-square this large or smaller. Notice that this is not the probability we look at for a test of significance. We want the probability that the value is this large or larger, so we have to subtract this value from 1. Rounding, we subtract .36 from 1 and get .74, which is clearly not significant.

DISPLAY 14.5

```
MTB > CDF .882;
SUBC>  Chisquare 2.

    0.8820   0.3566
```

Summary

Categorical data or *qualitative data* arise when measurement assigns a category, rather than a number, to each object. To *summarize* a categorical variable, we report the percentage of the total number of objects belonging to each category. It is also important to report the total number of objects so we know the basis for these percentages. Because each of the categories generates a yes/no variable, we may use our earlier techniques for assigning a standard error to each percentage individually.

A *chi-square test* can be used to test whether an observed set of percentages computed from a sample could have come from a population with a set of fixed, given, hypothesized population percentages. The general procedure for performing a chi-square test involves computing the number of counts that would have been expected in each case if the hypothesized percentages had held exactly. Adding up the terms "observed minus expected, squared, divided by expected," we compute the chi-square statistic. The formula is

$$\text{chi-square statistic} = \text{sum}\left[\frac{(\text{observed} - \text{expected})^2}{\text{expected}}\right]$$

The degrees of freedom, with one variable, is 1 less than the number of categories. The computed chi-square statistic is then compared to the value in the chi-square table with these degrees of freedom. If the computed value is larger than the value in the table, the sample values show significant differences from the hypothesized population values. If the computed value is smaller, we accept the null hypothesis that the actual population percentages could reasonably have been the hypothesized percentages.

A *two-way classification* arises when we have bivariate categorical data—that is, when we measure two different categorical variables on each object under study. We explore the data using the basic table of counts (with totals adjoined as the last row and column) and also with *joint percentages, row percentages,* and *column percentages.* The *chi-square test* for a two-way table is a test of *independence* versus *dependence*; that is, it is a test of whether there is any association between the row factor and the column factor. Thus, independence involves the interaction of the two variables under study, rather than their individual, separate properties. The computations involve first computing the *expected table,* which represents what we would expect had the two variables been perfectly independent, with entries given by

$$\text{expected} = \frac{(\text{row total}) \times (\text{column total})}{\text{overall total}}$$

We then compute the chi-square statistic as the sum of $(\text{observed} - \text{expected})^2/\text{expected}$ values, with one term in the sum for each entry in the original data table, excluding the totals. The degrees of freedom for a table of counted data is

$$\text{degrees of freedom} = [(\text{number of rows}) - 1] \times [(\text{number of columns}) - 1]$$

We compare the computed chi-square statistic to the tabled value with these degrees of freedom. If the computed chi-square statistic is larger than the value in the table, we conclude that the observed counts show significant association (that is, we reject the null hypothesis of the independence of row and column factors). If the computed value is smaller, we accept the null hypothesis that the factors could be truly independent in the population and that any deviations observed in the data could reasonably be due to randomness of sampling rather than anything else.

Both of the chi-square tests (for one or for two variables) follow the usual conventions of statistical hypothesis testing. In particular, the Type I error is controlled so that there is only a 5% probability of wrongly declaring significance (rejecting the null hypothesis when it is in fact true). Certain *technical assumptions* must be satisfied for this to hold: (1) the data must arise as a random sample from a population, and (2) we should require at least 5 counts in each category because the chi-square test is based on an approximation that breaks down when there are very small numbers.

Note also that it is important that the chi-square tests be done on *counts,* not percentages. Although percentages are very useful to us for purposes of interpretation, we absolutely must use counts for our observed and expected values. The nature of randomness depends critically on how many objects fall into each category, not just on the percentage.

The chi-square tests may be performed at levels other than .05 by using an appropriate chi-square table. All other computations remain the same. Using *p-value notation,* we would write ($p < .05$) to express the fact that the result was significant at level .05, and we would write ($p < .01$) to express the fact that the result was significant at level .01.

Problems

Basic Reviews

Problems 1–4 apply to Table 14.44, a hypothetical collection of fish.

Table 14.44 Numbers of fish

	MALE	FEMALE	TOTAL
Salmon	300	300	600
Flounder	100	200	300
Tuna	20	80	100
Totals	420	580	1,000

1. **a.** What percent of all the fish are flounder males?
   b. What percent of the males are flounders?
   c. What percent of the flounders are males?
2. **a.** Calculate a table of row percentages from these data.
   b. Looking at the table of row percentages from part (a), does the percent of males seem to be constant from one species to another, or does there seem to be an association between species and percent male?
3. **a.** If there were no association between species and percent male, how many male flounders would you expect to see in this sample?
   b. If there were no association between species and percent male, how many male salmon would you expect to see in this sample?

c. If there were no association between species and percent male, how many male tuna would you expect to see in this sample? (*Hint/pedagogical clue:* Notice that the total number of males is fixed at 420, so you can get the number of male tuna by subtracting the number of male salmon and male flounders from 420.)

d. How many female salmon would you expect to see in this sample, if species and gender were independent? (*Hint/pedagogical clue:* Notice that the total number of salmon is fixed at 600, so you can get the number of female salmon by subtracting the number of male salmon from 600.)

4. a. Calculate chi-square for Table 14.44.

b. How many degrees of freedom are there for chi-square for Table 14.44?

c. What is the critical value for chi-square at the 5% significance level with 2 degrees of freedom?

d. Based on the results of your chi-square analysis, can you reject the null hypothesis that there is no association between species and percent male?

Analysis

5. In an introductory statistics class, there are 13 freshmen, 52 sophomores, 22 juniors, and 6 seniors.

a. Identify the categorical variable here.

b. Identify the objects being classified.

c. Summarize this data set using percentages.

d. Find the standard error of the percentage of sophomores. Interpret this standard error, viewing the class as a random sample from the population of students who might have taken statistics during the time this data set was collected.

e. It certainly looks as though there are more sophomores than anyone else in the course. Before long-range plans are put into effect placing some additional emphasis on material most appropriate to sophomores, a committee would like to be sure that this is not just a "fluke" of this particular year's class. Perform a chi-square test to see whether data like these could have been generated as a random sample from a population of students in which all four classes have equal representation.

f. Interpret the results of your test in part (e) in a brief paragraph.

6. An anthropologist is using the coding scheme shown in Table 14.45 for ancient artifacts found during a parking lot excavation project in South America. As objects are unearthed, they are coded, resulting in the following data set for the first 20 objects.

5 5 2 2 1 3 2 4 1 4 3 2 5 4 2 1 4 3 3 5

Table 14.45

OBJECT	CODE NUMBER
Bangle	1
Ring	2
Spindle	3
Coin	4
Dish	5

a. The anthropologist's field computer has just calculated that the average code is 2.90 for this data set. Should this number have been calculated? Does this number have any meaning?

b. Summarize this data set using percentages.

c. Based on large-scale field studies, the usual tribe in the area has been characterized by the percentages shown in Table 14.46.

Table 14.46

OBJECT	POPULATION PERCENTAGE
Bangle	11%
Ring	23%
Spindle	14%
Coin	38%
Dish	14%

d. Would it be correct to perform a chi-square test on this data set to see whether the observed 20 objects could have come from a population with these percentages? Why or why not? (*Hint:* Think about the technical assumptions.)

e. Later, when 200 objects were gathered, the sample included the numbers of objects shown in Table 14.47. Test whether this sample could have been obtained randomly from a population with the percentages of the usual tribe from part (c).

Table 14.47

OBJECT	NUMBER
Bangle	27
Ring	42
Spindle	32
Coin	81
Dish	18

f. How strongly does your result from part (e) help support the claim that this excavation is from the usual tribe in the area and not from a different, visiting group? In particular, is your result absolute proof?

7. Consider terrorist activity against businesses during the first three quarters of 1985.[10] In Latin America there were 443 bombings, 56 attacks on installations, and 7 assassinations. In Europe during this same time period, there were 101 bombings, 6 attacks on installations, and 10 assassinations.

a. This is a bivariate categorical data set. Identify the two variables and the objects being classified.

b. Construct a two-way table of counts using geographical region to define the columns.

c. Explain what is meant by the hypothesis of independence for this particular data set.

d. Compute the column percentages and write a paragraph interpreting and comparing them.

e. Compute the expected table, under the assumption that each kind of terrorist activity occurs the same percentage of the time in both regions.

f. Perform a chi-square test of independence.

g. Write a paragraph interpreting the results of part (f). In particular, are patterns of terrorism the same or different on these two continents?

8. Compute the column percentages for the crime in California example. What do they tell you? If you were in charge of law enforcement for all four cities, which percentages (joint, row, or column) would be most useful to you in deciding:

a. How many homicide detectives to station in each city.

b. How many of your San Jose people to assign to arson investigation.

c. How much of your total force to assign to rape prevention in San Francisco.

9. Consider the two-by-two table of number of applicants for an advanced degree program at a major university, shown in Table 14.48.

***Table* 14.48**

	NUMBER OF APPLICANTS	
	MALE	FEMALE
1994 year	64	25
1995 year	49	16

a. Which would be most appropriate to the understanding of female representation: joint percentages, row percentages, or column percentages? Compute for your choice and interpret the results.

b. If there had been, in fact, no fundamental change in female representation from one year to the next, how many of each gender would we have expected to apply in each year?

c. Perform a chi-square test for independence at level .05.

d. Perform a chi-square test for independence at level .01.

e. Interpret the results.

10. Consider the two-by-two table representing the effectiveness of surgery in controlling cancer of the colon shown in Table 14.49.[11]

***Table* 14.49**

NUMBER OF CASES		WAS CANCER CONTROLLED? YES	NO
Was cancer present at the edge of the surgery?	Yes	11	28
	No	6	182

a. In what percentage of cases was cancer controlled overall?

b. Of those cases with cancer present at the edge of surgery, in what percentage of these cases was cancer controlled?

c. Of those cases with cancer not present at the edge of surgery, in what percentage of these cases was cancer controlled?

d. Perform a chi-square test of independence to decide whether the presence or absence of cancer at the edge of surgery gives any indication of whether the cancer is likely to be controlled by the operation.

e. Interpret the results of the test in part (d).

11. Large systems often have some problems, and the United States space program has had a few. The heat-protective tiles of the space shuttle *Columbia* had a tendency to fall off the vehicle during the stress of reentry into the earth's atmosphere. Consider the data set in Table 14.50, which shows what happened to each tile during two different flights.[12]

Table 14.50

	NUMBER OF TILES	
	FELL OFF	STAYED ON
April	164	30,836
November	12	30,988

a. Compute the percentage of tiles that fell off during each of the two flights. During which flight (April or November) did a larger percentage of tiles fall off the shuttle?

b. Suppose tiles tended to fall off randomly and that there was, in fact, no real difference between these two flights. How many tiles would we have expected to see in each place of the table with no differences and no randomness?

c. Perform the chi-square test of independence at level .05.

d. Perform the chi-square test of independence at level .01.

e. Interpret the results of parts (c) and (d). In particular, is there strong evidence that the situation had improved from April to November?

12. Gregor Mendel, an Austrian monk, proposed the genetic theory of inheritance in 1866. He supported his theory with the results of 8 years of experiments breeding peas and examining the inheritance of seven different characteristics.

One characteristic studied was plant height. Mendel crossed purebred tall plants (6 to 7 feet tall) with purebred short plants (.75 to 1.5 feet tall) to get hybrid plants, all of which appeared to be tall but all of which (according to his theory) carried the short form of the gene as well as the tall form. He proposed that when he bred these hybrids, one-quarter of their progeny would be pure tall, one-half would be mixed and appear tall, and one-quarter would be pure short. Since there is no way to distinguish the hybrids that appear tall from the pure tall, in total three-quarters of the plants would be tall and one-quarter short. Mendel reports doing this experiment, and that of 1,064 plants, 787 were tall and 277 were short.[13]

How well do these data fit Mendel's theory?

13. Bronowski (1973)[14] includes a quote from Mendel giving the exact results for the breeding experiments with height and summarizing the results for all seven characteristics. Mendel reports that the ratio of the dominant forms to recessive forms is 2.98 to 1, very close to the 3 to 1 that he predicted. Assume that there were approximately 1,000 plants (as in the height experiment) and that the seven characteristics were observed on each of these plants. How well do the results fit Mendel's model? What is the probability of seeing a *larger* difference between model and results, just on the basis of chance, when the model is true?

14. The data file PULSE.MTW, distributed by Minitab and used here with permission, contains data from a simple demonstration experiment in which students were asked to report certain background information and then record their pulse rates two times. Half of the students, randomly picked, ran in place between the two recordings. The background information includes the gender of the participants (1 = male, 2 = female) and whether they smoke regularly (1 = smokes regularly, 2 = does not smoke regularly). Are gender and smoking independent in these data?

15. There is a traditional belief that creative individuals use psychoactive substances (for example, alcohol or marijuana) to enhance their creativity. Does this really work? Lapp, Collins, and Izzo (1994)[15] report an experiment where subjects were given alcohol or a placebo, and either were told they were drinking alcohol or not drinking alcohol. They measured creativity with a picture-arranging problem. Subjects who thought they had drunk alcohol were found to be more creative (whether or not they really drank alcohol), but alcohol itself did not make them more creative.

Suppose you try to replicate this experiment. You give everyone a creativity test. Half the participants are then given a drink with alcohol, and the other half are given a placebo. (Both drinks are flavored so that it is not clear whether they contain alcohol or not.) Half of the subjects are told that they drank alcohol, and half are told that they did not. But half of the subjects are being misled and are told either that they did drink alcohol when they did not or that they did not when they did.

After a suitable interval, everyone is given another creativity test. Half the subjects show enhanced creativity. The data are in Table 14.51. Is enhanced creativity associated with actually drinking alcohol, with the belief that one has drunk alcohol, or with neither?

Table 14.51 *Effects of alcohol versus nonalcohol consumption and "belief" on a measure of creativity*

ALCOHOL DATA				BELIEF DATA			
		Creativity improved?				*Creativity improved?*	
		yes	no			yes	no
Alcohol	yes	18	22	Told	yes	26	14
	no	22	18		no	14	26

16. Competition for National Science Foundation fellowships for graduate study is strong, and the pool of qualified applicants is large. To achieve equitable geographic representation, some individuals who would otherwise get honorable mentions are given awards. This introduces an element of randomness, albeit incomplete, between awardees and honorable mentions. Chapman and McCauley (1993)[16] make this assumption and go on to argue that any differences in the subsequent careers of these awardees compared to honorable mentions might be explained in terms of the effect receiving the award has on people's expectations. They find no evidence of a difference between awardees and honorable mentions following graduate school, but they do find that there is a statistically significant effect on the likelihood that an awardee will finally receive a Ph.D. Of 2,423 awardees, 1,689 received a Ph.D.; of 1,885 honorable mentions, 1,190 received a Ph.D.

a. Show that this difference is statistically significant.

b. Is the difference necessarily due to the prestige and enhanced expectation associated with receiving the award? Formulate an alternative explanation.

17. To study what types of psychotherapy work best for different mental or behavioral problems, Beck et al. (1994)[17] compared the effectiveness of three methods in the treatment of panic disorder. (Panic disorder is well named: It consists of a panic reaction—intense fear, pounding heart, and so forth—to ordinary experiences, such as leaving the house.) The three "treatments" tested were cognitive therapy (CT), where the patient was taught about the relationship of maladaptive thoughts and anxiety and helped to recognize panic-generating thoughts and replace them with alternative thoughts; relaxation therapy (RT), where patients were trained in relaxation techniques; and a minimum contact control (MCC), who received no formal treatment but were interviewed once a week about their current situations and provided with a sympathetic listener.

a. At the end of treatment, the numbers of patients who were free of panic attacks were as shown in Table 14.52. Is there a statistically significant difference?

Table 14.52 *Number of patients free of panic attacks*

	PANIC FREE?	
	YES	NO
Cognitive therapy	11	6
Relaxation therapy	9	10
Minimum contact control	8	14

b. The authors also classified the patients as to whether they had improved, even if not panic free. These results were as shown in Table 14.53. Is there a statistically significant difference?

Table 14.53 *Number of patients showing decreased panic attacks*

	PANIC FREE?	
	YES	NO
Cognitive therapy	14	3
Relaxation therapy	13	6
Minimum contact control	8	14

18. Angina pectoris is chest pain associated with heart conditions. In the 1950s a new operation led to many apparent successes in treatment, but when a careful double-blind study was done it turned out that the really important component was the belief of the surgeons and their patients (Beecher, 1961).[18] It turned out that simply making skin incisions and sewing them back up was just as effective as the real operation, providing that the attending physician believed in the efficacy of the operation and believed that it had been performed.

Sketch out, in a paragraph, how you would replicate this study. Construct a data set that you might get from such a study and analyze it. Report your results.

19. Suppose you have detailed information on 360 school children. They are equally divided among three "tracks"—fast, regular, and slow. Seventeen percent of the pupils are of Mexican descent, the remainder are non-Mexican. Of the pupils in the fast track, 6% are of Mexican descent, while 29% of the slow-track students are of Mexican descent. What is the probability of this happening by chance? [Data and ethnic terms are modeled after material in Rosenthal and Jacobsen (1968, p. 63).][19]

20. Suppose you have detailed information on 300 students in a school that divides pupils into three equal size tracks—fast, regular, and slow. Fifty-three percent of the students are boys. Only 38% of the fast-track pupils are boys, while 69% of the slow-track pupils are boys. What is the probability of this happening by chance? [Data modeled after material in Rosenthal and Jacobsen (1968, p. 63).][20]

21. Drawing and interpreting graphs is an integral part of math and science, and there has been a lot of interest in using computers as an aid in teaching these skills. One problem is to evaluate how well students have mastered graphing. The cheapest way is to use multiple-choice tests. How well do they work? One problem used in this research is called the "walk to the wall"—students are asked to graph an individual's distance from the starting point, over time, walking to the wall and back.[21] The correct graph is shown in Figure 14.1.

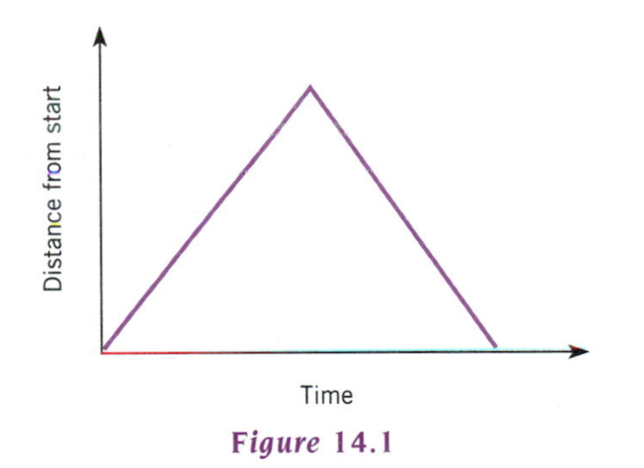

***Figure* 14.1**

Suppose you give this problem to 1,000 students in grades 8 through 12. Half of them are allowed to draw their own graph (the free-response, or F-R, condition), and half are given a multiple-choice test. Fifty-six percent of the kids in the free-response condition draw a correct graph, compared to 37% in the multiple-choice condition. Is this statistically significant?

22. Continuing Problem 21 above, you also ask your 1,000 students to draw or identify the graph of the speed of a ball rolling down a hill, up a small incline, then onto a flat surface. The results are in Table 14.54 (page 512).[22] Describe and report these results.

23. Continuing your research from Problem 22, you compare the correct responses of younger students to those of older students. The data are in Table 14.55 (page 512).[23]

Table 14.54 Numbers of students correctly graphing speed of ball rolling down hill, up incline, and onto plane. [Modeled after Berg and Smith (1994)]

	CORRECT	WRONG
Free-response	275	225
Multiple-choice	250	250

Table 14.55 Percent correct responses on ball/hill graph, lower grades versus upper grades [Modeled after Berg and Smith (1994)]

	GRADES	
	8 AND 9	10, 11, AND 12
Free-response	50% of 200	58% of 300
Multiple-choice	38% of 200	59% of 300

a. Is there a statistically significant difference by grade level for the free-response graphs?

b. Is there a statistically significant difference by grade level for the multiple-choice graphs?

24. Data file GENETICS.MTW includes the phenotypes of 200 randomly selected pea plants that were raised in an altered atmosphere. If the atmospheric alteration had no effect, each phenotype should appear with about the probability given in Table 14.56. Is this a good model, or does the observed distribution of phenotypes differ significantly?

Table 14.56 Expected distribution of phenotypes for Minitab data file GENETICS.MTW

TYPE	PROBABILITY
1 = smooth-yellow	.5625
2 = smooth-green	.1875
3 = wrinkled-yellow	.1875
4 = wrinkled-green	.0625

chapter 15

Bivariate Data and Regression

In this chapter we consider how two measurements might be related to one another. We will explore the different possible relationships and how to identify them. Linear, or straight-line, relationships play a special role in statistics. We will learn how to measure the strength of a linear relationship and how to find the formula for the line that does the best job of predicting one variable from the other.

15.1 What Are Bivariate Data?

In our next big step to a new kind of data structure, we consider two related or paired groups of numbers, for which we will no longer require that the two groups have the same units of measurement. The big payoff will be our ability to exploit the relationship between the two variables to predict the value of one variable given a value for the other. We have in mind a situation for which each person, object, or experimental unit contributes two numbers to the data set. There are several somewhat different situations that all fall under this category of data structure: data from a stimulus/response experiment, time-series data, and (more generally) data on any two variables that can each be measured and whose relationship is of interest.

The Stimulus/Response Experiment

We use the **stimulus/response experiment** to generate bivariate data in an effort to understand the effect, if any, of one thing on another. For this kind of experiment, the stimulus is under the control of the experimenter and can be set at various levels. An observation (the response) would be made for each of several levels of the stimulus. The experiment would be repeated, and each time that it is performed, a new pair of numbers (amount of stimulus, size of response) joins the data set. Some examples include the following.

1. To limit the amount of resources expended by a health maintenance organization on visits that do not really require medical treatment, some managements impose a small "nuisance payment" that people must pay each time they use the facility. To study the effect of this method, we might perform an experiment where people would be assigned (randomly, of course, so that we could exclude the influence of other factors) a certain amount of nuisance payment, which would be the *stimulus factor.* Over the next year, the extent to which these same individuals used the facility for nonemergency care would be monitored, representing the *response factor.* Each person in the study would therefore contribute a pair of numbers—nuisance payment size (stimulus) and utilization (response)—and we would then have a bivariate data set. We would have more than just two separate groups of numbers because nuisance payments and utilization measures are linked in natural pairs, with each person contributing one of each to the data set. The objectives of this study would include answering questions like:
   - Are nuisance payments effective in reducing the demand for unnecessary services?
   - What is the nature of the relationship between payment and demand?
   - What would be a good choice of amount for the nuisance payment?
2. To produce sheet metal in the most economical way, the conditions under which the process operates may be controlled. In particular, the temperature can be adjusted (this is the stimulus because it is under our control) and has an effect on production. Some particular measure of the quality of the process, such as the amount of acceptable material produced in 1 hour, can be used as a measure of the response factor. We would then collect data at various settings of temperature, with each trial contributing one pair of data values: the temperature (stimulus) and the amount of acceptable material (response). Questions that could be studied include:
   - Does temperature have an effect on production?
   - How large is this effect and what is its nature?
   - What is the best setting for the temperature?

3. An experimental medical treatment for acne is being tested. Dose levels (amount of the drug, the stimulus) are assigned randomly to teenagers who are participating in the experiment. The extent of the improvement in their situation is measured as the response. Each person contributes this pair of numbers (stimulus level and measured improvement) to the data set. Questions to be examined include:
   - Does the drug work?
   - How well does it work?
   - What is the effect of dose level?

Time-Series Data

Many things change as time progresses. A collection of data values measured at successive intervals of time is referred to as a **time series.** Such a data structure consists of two groups of numbers: the time at which each measurement was taken and the measurement itself. There are many examples of this kind of data set.

1. The fields of economics and business furnish many examples. For instance, interest rates change over time. A bivariate data set could be collected that would list the prime rate (the interest rate that commercial banks charge on loans to their most financially healthy customers) each week for the past year. Although this may first look like only one group of numbers (the interest rates themselves), each number actually represents a pair of data values in which the first number is the time (week of the year, say) and the second is the interest rate. Many time-series data sets represent pairs of numbers in this implicit sense. Questions that might be answered by analyzing this data set include:
   - Are interest rates changing?
   - What are the trends, if any, in interest rates?
   - What would we forecast for future interest rates?

   This last question, prediction, is very difficult to answer with certainty. A good prediction can be worth a lot to an individual, corporation, or government agency, but it is unfortunately not easy to arrive at. Even the experts are usually wrong, to some extent, with their economic predictions, but this should not surprise those of us who recognize that the world in general (and economic activity in particular) includes a strong component of randomness.
2. In the fields of geology and paleontology, time is a very important parameter as the history of the earth can be traced over presumably hundreds of millions of years. For example, the number of species that were living at any given time can be estimated based on the classification of fossils found all over the earth. The resulting time series (with pairs of numbers representing time period and number of species) could be analyzed in order to provide quantitative insight into the development of life on earth. We might ask:
   - Is the number of species nearly constant?
   - Are there sudden drops in the number of species, showing mass extinctions?
   - If there are mass extinctions, how do the numbers of species change as conditions stabilize?
3. A census provides demographic information about populations of people and indicates how they change over time. An excellent collection of statistical time-series information is published by the U.S. Bureau of the Census.[1] Included is information on population size, migration, size of the labor force, natural resources, energy, communications, agriculture,

and many other changing aspects of the United States. Many questions can be addressed based on these kinds of information, including:

- How fast is the population growing?
- What has been happening to natural resources?
- How many sheep and goats have traditionally been permitted to graze on U.S. National Forest System lands and what trends are there (if any) over time?

Observing the Relationship Between Two Variables

We can analyze the nature of a relationship, if any, between two variables of interest in any situation for which data are available. This is more general than the previous two kinds of examples (stimulus/response and time-series) because nothing special needs to be assumed about the two variables. A procedure may result in two measurements with neither being controlled (so that neither may be considered the stimulus) and neither representing time (so that it is not a time series). Yet such situations can still be viewed and analyzed as bivariate data. Here are some examples.

1. In a survey using questionnaires, each completed form provides not just one but many numbers for each person. There is one number provided by each question on the questionnaire. In this case, we could go beyond studying the questions separately, one at a time, by looking for relationships between pairs of questions. For example, rather than just looking at the distribution of income separately from the distribution of age, we could search for a connection between the two. This connection can be studied only if we have pairs of data values (age and income or education and income) measured for each person in the sample. For instance:
   - Does income increase with age?
   - Does income increase with education?
   - Does income increase with education among people with bachelor's and higher degrees?
2. Pollution is a complex issue. There are many different kinds of chemical contaminants that find their way into the air and can be measured in different cities. Thus, a data set could be constructed that would consist of a pair of numbers for each of a group of cities, such as measurements of the yearly average amount for both carbon monoxide and sulfur oxides. Analyzing such a data set of pairs of numbers would provide insight into questions such as:
   - Are these two kinds of pollution related?
   - What is the nature of the relationship, if any?
   - Are there typical trends and variations from one city to another?

Recognizing a Bivariate Situation in a Table of Data

We must sometimes use imagination to recognize a set of bivariate data hidden within a larger, more complicated data structure. But often a table of data organized as rows and columns can be analyzed by looking at a pair of columns (or alternatively, a pair of rows if this makes more sense), using the methods for analyzing paired data that we will cover in this chapter.

Exercise Set 15.1

1. Suppose a law school wants to use an aptitude test score to predict their applicants' first-year grade point averages. What are the two variables here?
2. Suppose you have the data shown in Table 15.1. What are the two variables here?

Table 15.1

	CASE				
	1	2	3	4	5
Father's height	60	63	66	69	72
Son's height	62	65	68	71	61

3. Classify each of the following studies as a stimulus/response experiment, a time series, or an observational study.
   - **a.** A study of temperature changes in the Antarctic over the course of the year
   - **b.** A study of how randomly assigned patients with a certain condition respond to two different therapies
   - **c.** A study analyzing telephone survey data that include questions on education level and political party affiliation
4. In one or two sentences, describe the critical difference between a stimulus/response experiment and an observational study that is neither a stimulus/response experiment nor a time series.

15.2 Exploring Bivariate Data

Once we have data to look at, the first step should always be to explore the data, using graphical pictures, so that we can examine the basic structures and properties. Each of the two variables that make up a bivariate data set can, of course, be studied on its own using the methods that we have learned (such as histograms and box plots) for exploring a single group of numbers. The important new aspect is that bivariate data transcends these separate considerations, so we can study how the two factors interact with or relate to each other. To effectively study and explore these interrelations, we will use a bivariate plot of the data.

The Bivariate Plot

The **bivariate plot** has several other names: It is known as a *scatterplot,* an X-Y plot, and a *two-dimensional plot,* among other names. Whatever we call it, it is the single most important tool we have when trying to understand the structure in a bivariate data set, and it should always be examined whenever we work with this kind of data.

Each pair of numbers in the data set will contribute one point to this bivariate plot. By convention, the first number of each pair is called X and will correspond to horizontal distance. The second number in each pair is called Y and corresponds to vertical distance. Each pair of data values therefore specifies a horizontal and a vertical distance. At the place where they meet, we will put a symbol such as a circle or a cross (sometimes these are connected by a line instead). The resulting bivariate plot is often referred to as a plot of Y against X.

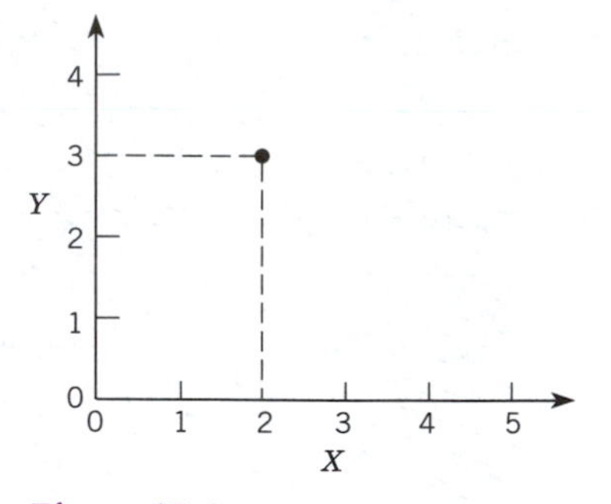

Figure 15.1 One data pair becomes one point in a bivariate plot, as with X = 2 and Y = 3 here.

For example, the pair of numbers (2, 3) would be represented as a point 2 units to the right of 0 and 3 units high, as shown in Figure 15.1.

Discussion: Which One Is X? Which One Is Y?

In real data situations, there are traditional conventions to help us decide which variable (or factor) of the pair to call X and which to call Y. If one factor is under our control (as, for example, in a stimulus/response experiment), then the controllable factor (the stimulus) would be called X and the other factor (the response) Y. Sometimes, we refer to X as the **independent variable** and Y as the **dependent variable** (because it depends on X). Thus, when examining a growing plant's response to a controlled amount of fertilizer, the amount of fertilizer would be X and the amount of growth Y.

If neither factor is under our control, there are still some guidelines to help us decide which one should be X and which should be Y. If one factor is an indicator or predictor of the other, then we assign the factors so that Y is predicted from X. For example, although we cannot (currently) control the weather, we do think of barometric pressure as a predictor of precipitation. Thus, given paired measurements of the morning's barometric pressure and the afternoon's rainfall, we would designate the pressure as X and the rainfall as Y.

In a time-series situation, X will usually denote the time at which the observation was made, and Y will denote the observation itself.

Exercise Set 15.2

5. Suppose you have the data shown in Table 15.2.

a. Plot these data on an X-Y coordinate system. (If you do not have graph paper handy, you can use a ruler to lay out a grid on ordinary paper. That will be accurate enough for our purposes.)

b. Does it seem more natural to predict son's height from father's height, or the other way around?

c. If you use father's height as the predictor variable, do you call it X or Y?

Table 15.2

	CASE				
	1	2	3	4	5
Father's height	60	63	66	69	72
Son's height	62	65	68	71	61

6. Refer to Exercise 1 in Exercise Set 15.1.

a. Which is the predictor, or X, variable?

b. Which is the response, or Y, variable?

EXAMPLE 15.1 *Exploring Bivariate Data: Ages and Heights of Children*

Consider the ages and heights of children shown in Table 15.3. From this table, we see only some lists of numbers. To obtain a better idea of how heights and ages might be related to each other, we will plot height against age. We do it in this order because it is easier to think of height depending on age (that is, height is predicted by age) rather than vice versa. Age is therefore X, and height is Y. The bivariate plot is shown in Figure 15.2.

Table 15.3 *Ages and heights of children*

CHILD NUMBER	AGE (YEARS)	HEIGHT (CENTIMETERS)
1	1.7	72
2	3.2	88
3	3.1	85
4	.6	63
5	1.2	77
6	1.8	84
7	3.8	100

Except for some scattering due to random variation, the points generally become higher as they move to the right. This kind of behavior in a bivariate plot indicates that Y increases as X increases. In the context of the example, this says that older children are generally taller than younger ones because a larger height corresponds to a larger age, which is what we know from our personal experience of watching children grow. The relationship is not perfect, particularly because of the natural variation from one child to another that is expressed by the presence of some random scattering in the distribution of the points.

Aside from this important indication of the relationship between the two factors, we can

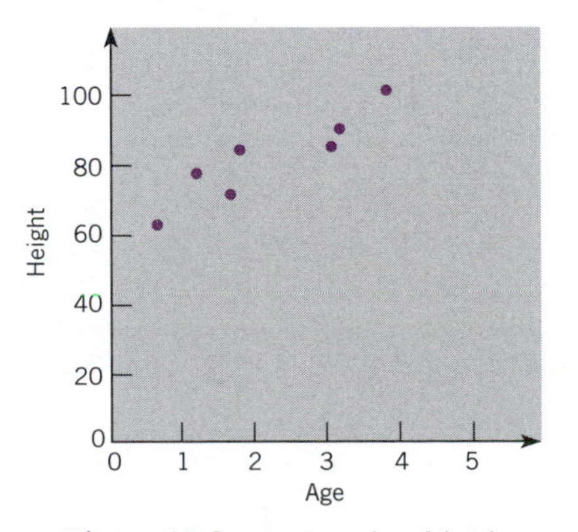

Figure 15.2 *Bivariate plot of height against age for seven children*

also see something of the distributions of height and age separately. The age distribution appears fairly evenly distributed from about $\frac{1}{2}$ to 4 years, as we see by looking at the horizontal locations of the data points and referring to the horizontal scale. The height distribution seems also fairly evenly distributed, extending from about 60 to 100 cm, as we see by looking at the vertical locations of the data points and referring to the vertical scale.

Although details of the individual distributions of two factors would be better studied by using separate histograms, the bivariate plot nonetheless provides some useful indication of these separate distributions. The main strong point of the bivariate plot, however, is its ability to show us the relationship between the two variables.

No Relationship Between *X* and *Y*

We will now look at how to identify the various kinds of structures that often arise in bivariate data. It may be that the two factors X and Y are not at all related to each other (even though it is usually more interesting when they are). When they are unrelated, the distribution of Y (by itself) is the same regardless of what the value of X happens to be. Figures 15.3, 15.4, and 15.5 represent some examples in which the average and variability of Y are the same regardless of the value of X, in the sense that the heights of the points on the left (those with small values of X) have approximately the same distribution as the heights of the points on the right (those with large values of X). Because the distribution of Y does not generally depend on the value of X, we say that these two variables are not related.

These illustrations show some of the ways a lack of relationship between the variables X and Y can be expressed. The situation depicted in Figure 15.3 might best be described as a shapeless cloud of points, except perhaps for some randomness. (We should probably resist the temptation to read too much into this random variation.) Figure 15.4 seems to show an elongated structure but because it is basically flat and untilted (except for random variation), the distribution

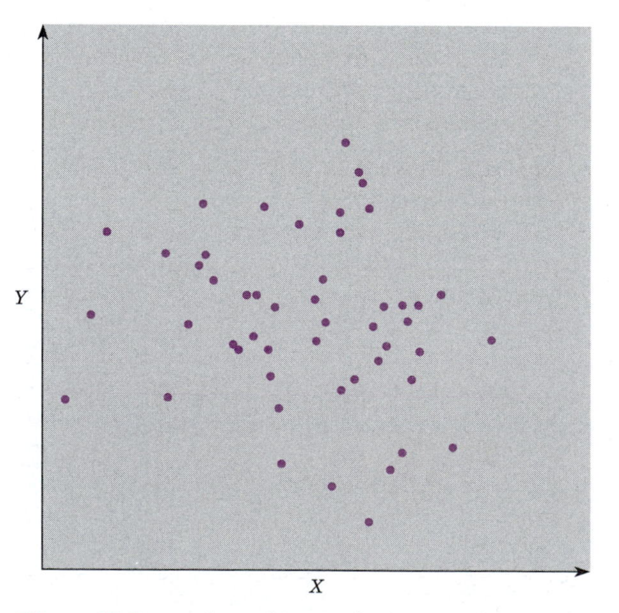

Figure 15.3 *No relationship exists between X and Y. It is basically a circular cloud of points.*

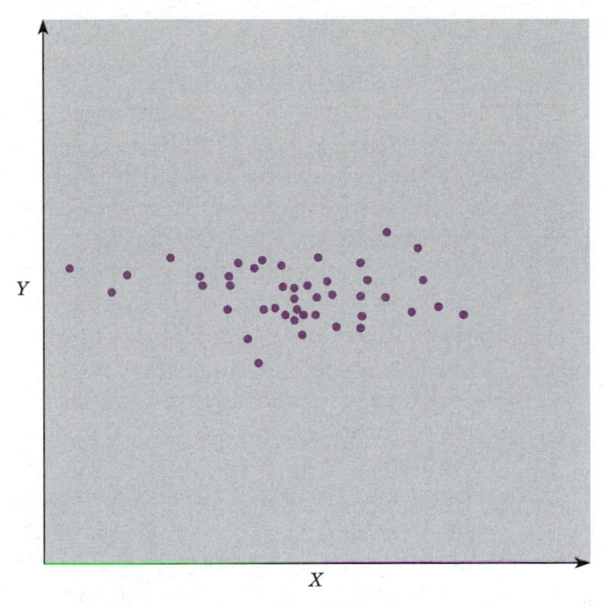

Figure 15.4 *No relationship exists between X and Y. Do not be deceived by a stretching of the X scale; there is no "tilt" to the data.*

of Y stays about the same regardless of which horizontal region of the X values we consider. This flat elongation is due to the fact that Y, as displayed, has less variability than X. If Y had more variability than X, we would obtain a result like that in Figure 15.5. The vertically oriented elongated structure there still indicates no relationship between the two variables because smaller values of X correspond to the same distribution of Y values as larger values of X do.

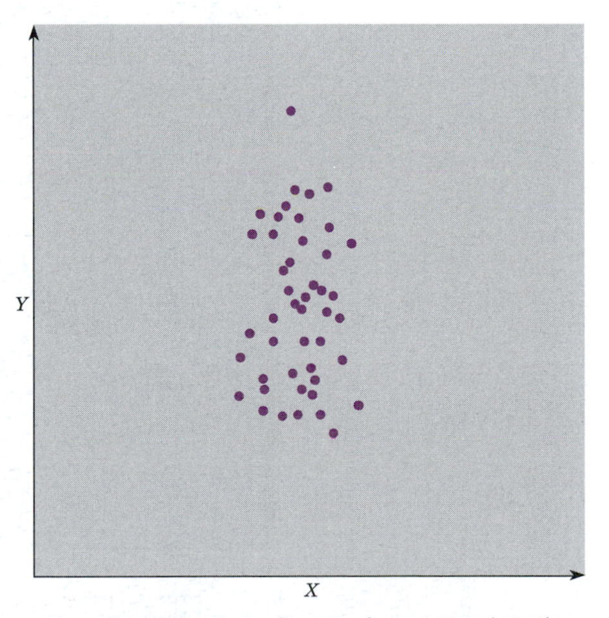

Figure 15.5 *No relationship exists between X and Y. This time, the Y scale has been stretched, but there is still no tilt.*

A Linear Trend

One very satisfying kind of structure to find in a bivariate data set is one in which the Y values trace out a straight line as X changes, except for some random variation about that line. Because it is the simplest kind of structure to quantify and interpret, we will give this kind of situation special attention later on. This sort of straight-line behavior is called **linear** structure (you can see the root word *line* within this technical term). Figures 15.6 and 15.7 represent some examples of bivariate data sets with linear structure.

A linear trend can be either increasing (as in Figure 15.6, where increasingly larger values of X correspond to increasingly larger values of Y) or decreasing (as in Figure 15.7, where increasing values of X correspond to decreasingly smaller values of Y) and can have a small or large amount of variability around the imaginary straight line that could pass through the data. Some of the cases of "no relationship" also indicate a linear trend, with the special case of a 0 trend (indicated earlier in Figure 15.4).

An example with real data, rather than computer-simulated data, is shown in Figure 15.8, indicating the relationship between the age and height of a sample of young children.[2] The relationship depicted is fairly linear over most of the range of age and height in the data set. Perhaps the overly rapid descent in the lower-left part of the data is not linear, where the very youngest children appear to have heights so small that these points seem to be below the line suggested by the rest of the data. However, throughout most of the data, the relationship is linear.

Of course, we know that outside the range of the data, the relationship cannot continue always to be linear because people eventually stop growing. Fortunately, a linear relationship need hold only in some particular region of interest (such as from 1 to 6 years in this example) for it to be worthy of study. The linear relationship in this example may be interpreted as saying that (to a good approximation) children of this age group grow at a constant rate.

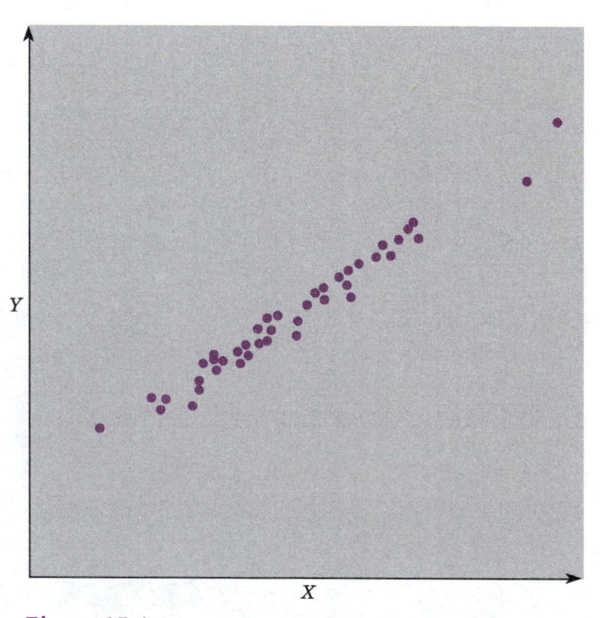

Figure 15.6 *An increasing linear trend: Larger (higher) Y values are associated with larger (farther right) X values, and the increase is like a straight line but with some randomness.*

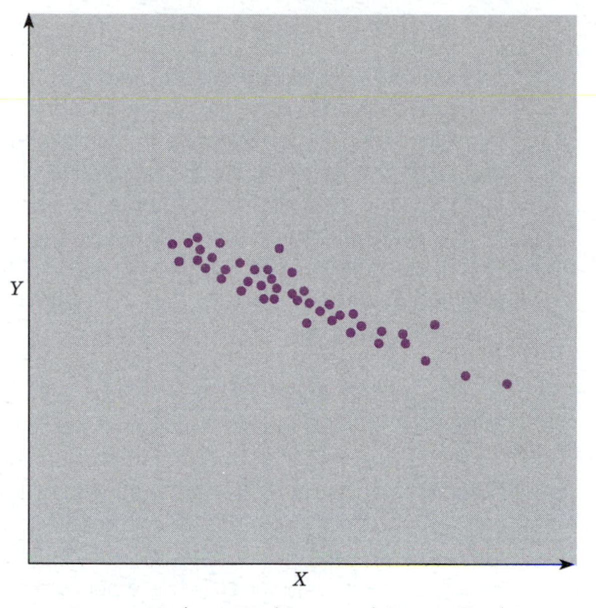

Figure 15.7 *A decreasing linear trend: Larger Y values are associated with smaller X values.*

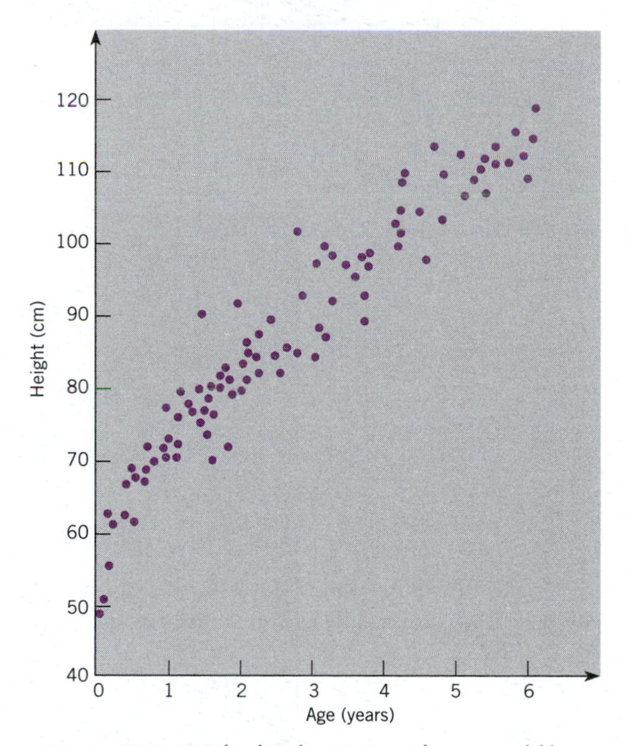

Figure 15.8 *Height plotted against age for young children. There is a fairly linear relationship here.*

A Nonlinear Trend

As we have seen throughout this text, the real world does not always conform to our hopes and expectations. In place of pleasing linear structure, we often see nonlinear structure in which the values of Y trace a curve as X changes. **Nonlinear** literally means anything other than a straight line. Perhaps the easiest way to deal with a nonlinear trend is to transform the data so that the trend becomes linear, if this is possible. We will see how to do this later on.

It is important to be aware when a nonlinear trend exists; this is one of the reasons for exploring the data as the first step of analysis. If methods of linear analysis are used on data with nonlinear structure, the results will not be altogether valid. Of course, a slight amount of curvature would not cause any serious problems.

Some examples of nonlinear trends are shown in Figures 15.9, 15.10, and 15.11. A generally increasing trend is depicted in Figure 15.9. Because it increases more and more slowly and seems to level off as X increases, the slope (or tilt) does not stay the same and it is therefore nonlinear. Figure 15.10 shows a nonlinear trend that first decreases and then increases. This same data set is displayed in Figure 15.11, using a connected line instead of individual dots for each data point. Notice how much difference the method of display makes in how the trends look, even though the same data are used in both cases. The connected line has the effect of making it look as though there is less scatter, while emphasizing the individual changes in Y as X changes. The connected line is a good choice for displaying a time series.

EXAMPLE 15.2 *Nonlinear Structure: The Stock Market*

Figures 15.12 and 15.13 (page 526) show a time series with nonlinear structure: the behavior of the stock market in the 20th century, as measured by Standard and Poor's index of industrial

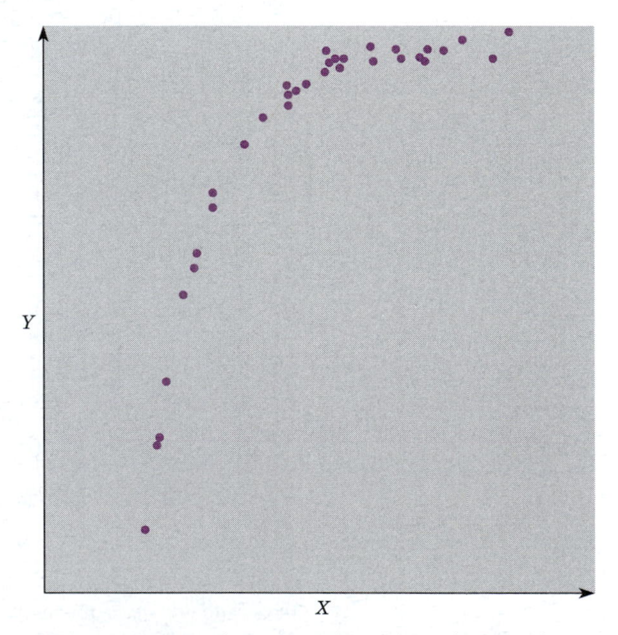

Figure 15.9 *A nonlinear (curved) trend. In this example, larger Y values are associated with larger X values but the relationship is not linear (not a straight line).*

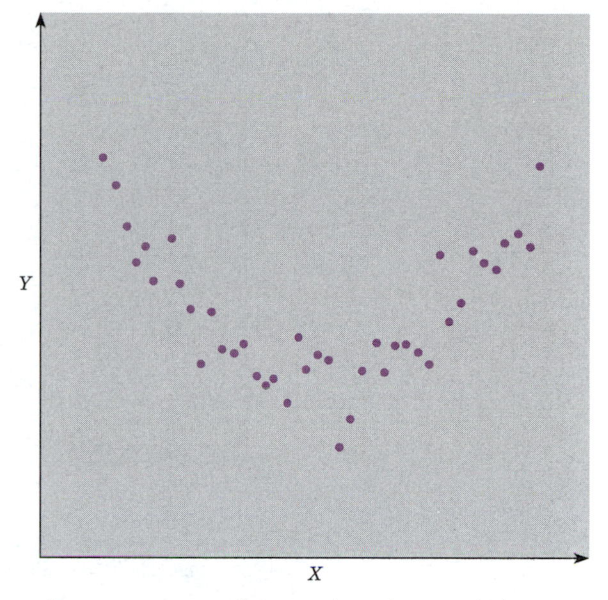

Figure 15.10 *A nonlinear trend. In this example, larger Y values are associated with either small or large X values.*

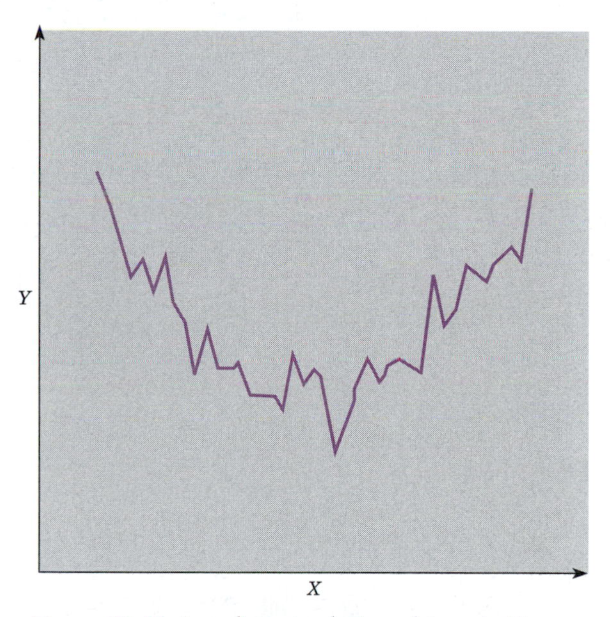

Figure 15.11 *A nonlinear trend: Same data as in Figure 15.10, but displayed using connected lines instead of separate symbols.*

stocks.[3] Again, the scatterplot and the connected line plot are provided for comparison. In practice, of course, we would only use one of them, probably Figure 15.13 because this connected line display conveys the time trend more directly.

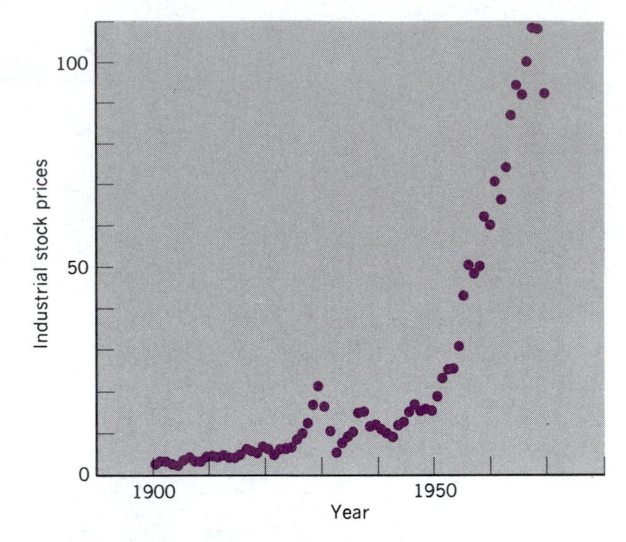

Figure 15.12 *A time series with nonlinear structure: Standard and Poor's index of industrial stock prices, 1900–1970.*

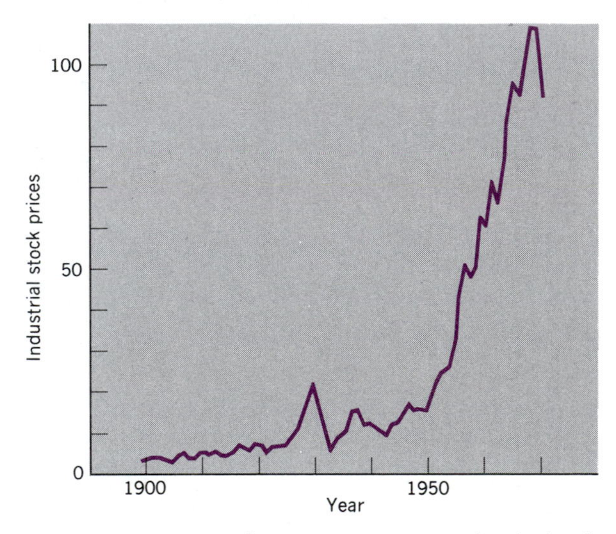

Figure 15.13 *Same data as in Figure 15.12, but displayed using connected lines. This is often preferable in displaying a time series.*

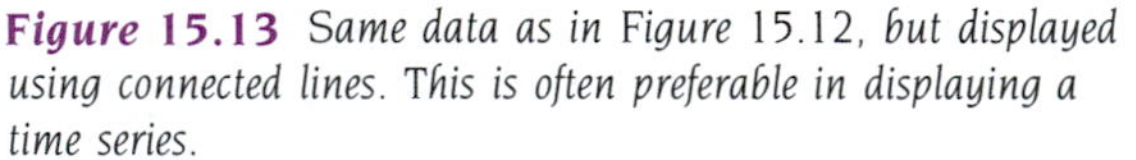

EXAMPLE 15.3 Nonlinear Structure: Paying Questionnaires

Figure 15.14 shows the nonlinear result of an experiment to measure the effect of money on the probability of a questionnaire being answered and returned.[4] Questionnaires were sent to randomly selected people, and varying amounts of money were also included, so that 50 people received each amount of money. This is a stimulus/response experiment. The stimulus X is the amount of money (which was controlled by the researchers), and the response Y is the percentage of questionnaires that were returned. The data are reported in Table 15.4.

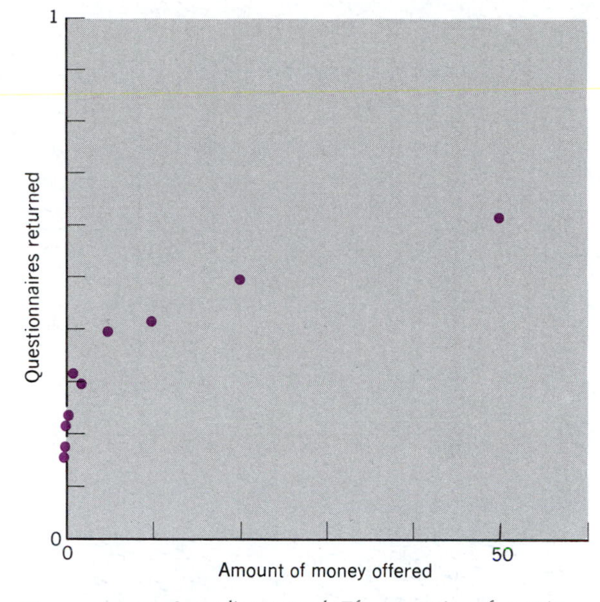

Figure 15.14 A nonlinear trend. The proportion of questionnaires returned is plotted against the amount of money (in French francs) included in the envelope.

Table 15.4 Results of a study of questionnaire response when payments are enclosed; each percentage is based on a different random sample of 50 people.

PAYMENT ENCLOSED (FRENCH FRANCS)	PERCENT OF FORMS RETURNED
.00	16%
.10	18%
.20	22%
.50	24%
1.00	32%
2.00	30%
5.00	40%
10.00	42%
20.00	50%
50.00	62%

The nonlinear structure of this plot shows us that small amounts of money have a large effect on the percentage of forms returned, but that as more and more money is involved, a point of diminishing returns is reached. The effect of an additional amount of money on the number of forms returned is less and less as the amount of money grows larger and larger, resulting in the curved leveling off of the relationship.

Exercise Set 15.3

7. Seven bivariate plots are shown in Figure 15.15. Which panel shows:
   a. a perfect increasing linear relationship?
   b. a perfect decreasing linear relationship?
   c. an increasing linear relationship with random error?

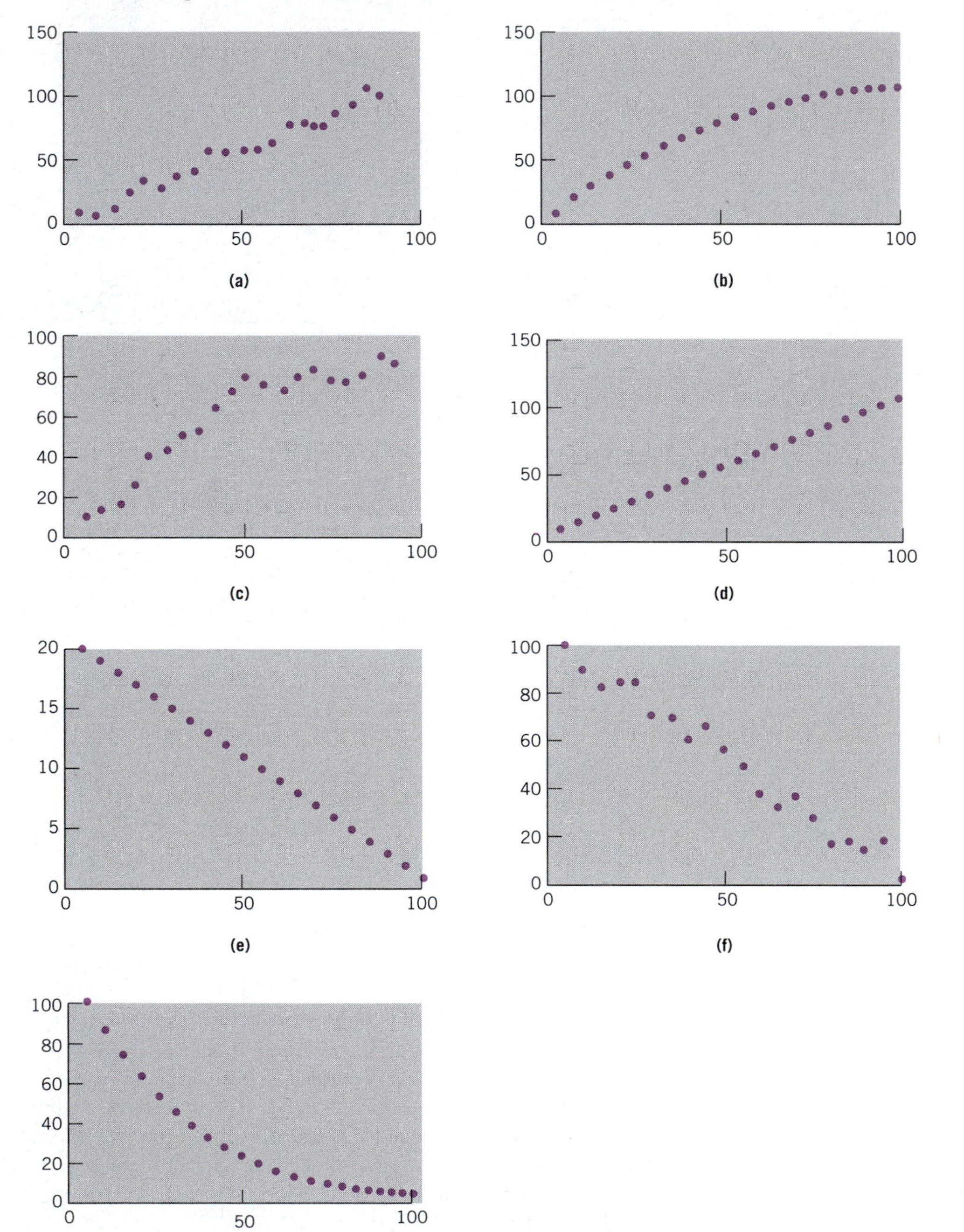

Figure 15.15

d. a decreasing linear relationship with random error?

e. a smoothly increasing curvilinear relationship?

f. a smoothly decreasing curvilinear relationship?

g. an increasing curvilinear relationship with random error?

8. Suppose you have these data on education (X) and income (Y):

CASE	ONE	TWO	THREE	FOUR
Years of school	8	10	12	16
Annual Income (thousands of $)	6	9	12	18

a. Plot the data.

b. In a sentence, describe the apparent relationship.

Clustering

Sometimes a data set should not all be analyzed at once. If there are several distinct clusters visible in the bivariate plot, we should consider either analyzing each cluster separately or completing an analysis that takes account of the different groups. At the very least, we should make note of the clustering so that we can consider possible reasons for it.

EXAMPLE 15.4 *Clustering: Life Expectancy and Infant Mortality*

Consider the bivariate plot of life expectancy against the infant mortality rate, with each data point representing one country, shown in Figure 15.16.[5] This does not look like just one group of numbers. There seem to be two separate clusters: one with high infant mortality and a shorter life expectancy, and the other with less infant mortality and a longer life expectancy. In

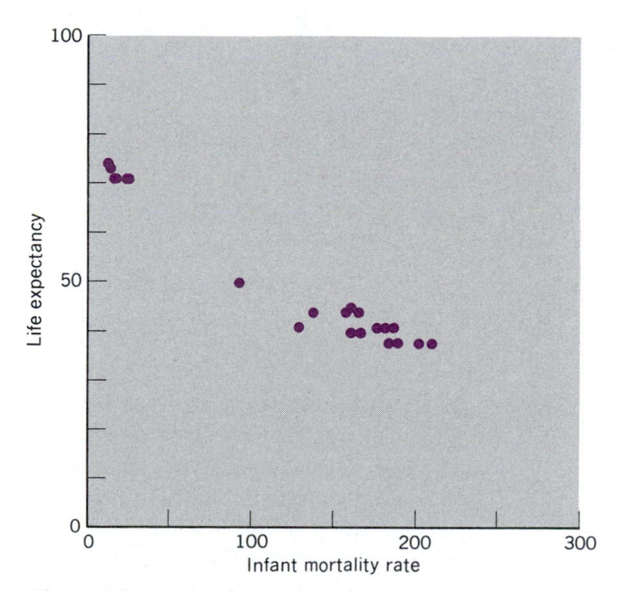

Figure 15.16 *An example of clustering: Note that there are basically two different groups of data here. Life expectancy is plotted against infant mortality for each of a group of countries.*

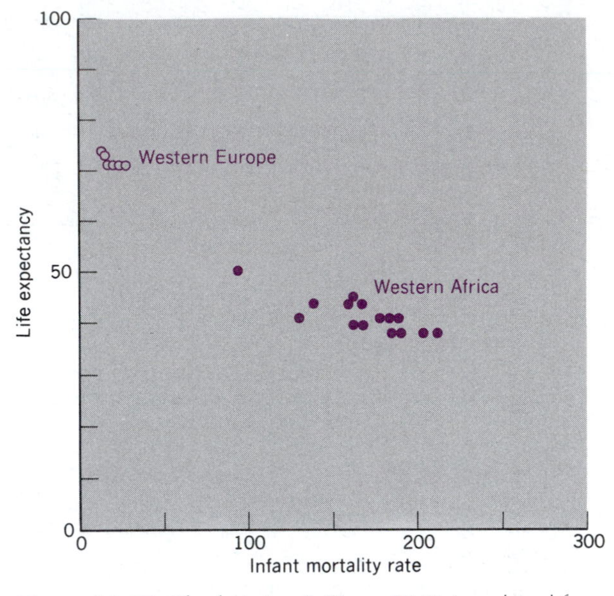

Figure 15.17 *The clustering in Figure 15.16 is explained by indicating Western European countries separately from West African countries.*

fact, the countries represented within each cluster are also related by geographical area. Figure 15.17 reveals that the countries involved are from Western Europe and Western Africa and that by using a different plotting symbol for the two areas, it is possible to identify the two clusters in a natural way.

In the preceding example, although there was clustering, the whole data set seemed to show the same relationship between the two factors as did the individual groups. Western Europe seemed to be almost an extension of the trend among the Western African nations, rather than completely separate.

This simple behavior, in which the parts behave as the whole, will not always hold. In Figure 15.18, three distinct clusters are shown. Although each of the three clusters individually exhibits an increasing linear trend, this trend does not apply to all three together. When we see something like this in our data, we really should find out about it; otherwise, we might reach the wrong conclusion and leave unnoticed some important structure.

Outliers

As in our previous study, one or more data points may not seem to belong with the rest of the data. One warning: An outlier in a bivariate data set may or may not be an outlier in one or both of the coordinates separately. Thus, an examination of separate histograms or box plots will not be enough to locate outliers. Figure 15.19 illustrates a bivariate outlier that would not be an outlier in either of the coordinates (X or Y) separately. However, because it is quite different from the trend in the bivariate data, it is nonetheless an outlier.

Outliers cause varying amounts of trouble, depending on how extreme they are and whether they are outliers in X alone. Certainly, an outlier should always be questioned to see whether it was plotted correctly, typed in correctly, recorded correctly, and so on. Then robust methods

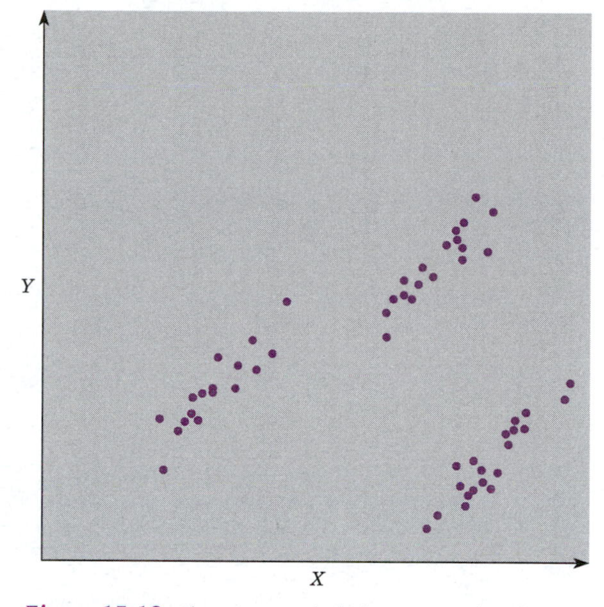

Figure 15.18 *Clustering in which the parts (each with a strong increasing linear trend) behave differently from the whole (which has no overall tilt).*

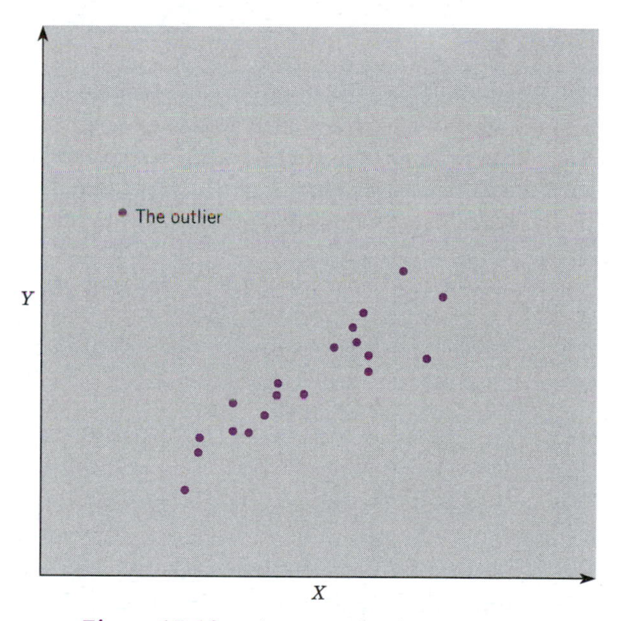

Figure 15.19 *A data set with a bivariate outlier*

can be used to limit the effects of outliers on the analysis, so that the conclusions would not be overly dependent on just one or a few data values.

Unequal Variability

The pattern in the way that Y changes as X does is most easily analyzed when the variability of Y stays fairly constant regardless of the value of X. Unfortunately, this does not always happen.

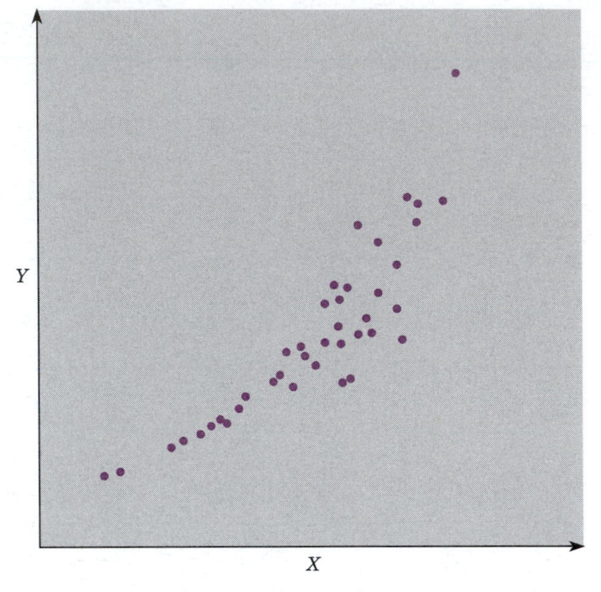

Figure 15.20 *An example of unequal variability. Note that the randomness of "scatter" in the data is stronger when X is large (to the right) than when X is small.*

Figure 15.20 shows one way in which the variability in *Y* can change as *X* does. Note that *Y* is more variable (and less predictable) when *X* is large (to the right), but that *Y* shows very little variability (as is indicated by the thin curved line that the points trace) when *X* is small (to the left). This kind of *inhomogeneity of variance* or *heterogeneity of variance* (two equivalent ways of saying the same thing) often occurs with real data sets. Two examples are illustrated in Figures 15.21 and 15.22.[6]

Unequal variability can be a problem with certain kinds of analysis. Sometimes, it can be controlled or corrected by using a transformation of *Y* that compresses the scale where the variability is large, helping to equalize it across the range of *X* values. Sometimes, it is also necessary or useful to transform *X*.

15.3 Analyzing a Linear Relationship: Correlation

Now let's consider bivariate data sets that have linear structure; that is, the plot of the data shows basically a straight-line trend except for some randomness. The key words in the analysis of linear structure are *correlation* and *regression*. **Correlation analysis** provides a measure of how strong the (linear) relationship is between the two factors. **Regression analysis** provides a method for drawing a good straight line through the data points in order to summarize this linear structure. In this section we concentrate on correlation analysis; we will cover regression analysis in Section 15.4.

Correlation: A Measure of Association

Correlation is a measure of the strength of linear association between two variables, *X* and *Y*. If *X* and *Y* tend either to increase or to decrease together, then they are said to be correlated. If

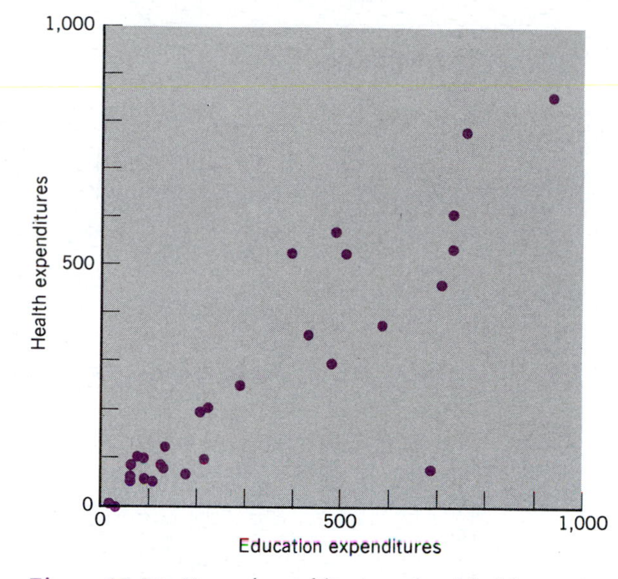

Figure 15.21 *Unequal variability in a plot of health expenditures against education expenditures, on a per capita basis, for European countries. The general trend is increasing (i.e., countries that spend more on one also usually spend more on the other), but the variation in dollar amounts is greater among countries that spend relatively larger amounts.*

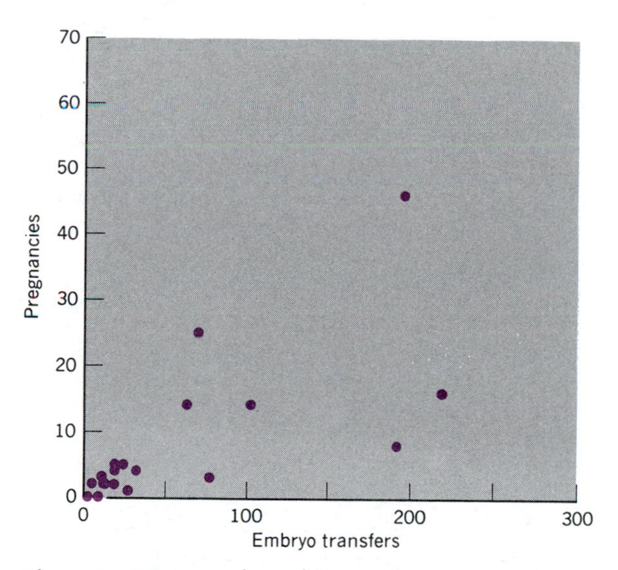

Figure 15.22 *Unequal variability in data on external fertilization in medical clinics (test-tube babies). We plot, for each clinic, the number of successes (pregnancies) against the number of attempts (embryo transfers).*

the values of Y are about the same regardless of the value of X, then they are not correlated. For example, we know that taller children tend to weigh more than shorter children, so we can say that *height* and *weight* are *correlated.*

The most standard measure of the strength of this relationship is the **correlation coefficient,** sometimes called the *product-moment correlation.* It can be computed in several equivalent ways, but we chose the method to be outlined next because it relies directly on averages and standard deviations.

PROCEDURE: COMPUTING THE CORRELATION COEFFICIENT

1. Compute the average and standard deviation in the usual way for the X values only, then compute these summaries for the Y values only.
2. Multiply each X value by the corresponding Y value and find the average of the resulting products.
3. From the average found in step 2, subtract the product of the two averages found in step 1. (This is a delicate step in which round-off error can cause problems if you did not use enough digits of precision in the previous steps.)
4. Divide the value found in step 3 by the standard deviation of X, then divide the result by the standard deviation of Y.
5. Multiply the value found in step 4 by the sample size, then divide the result by 1 less than the sample size. This is the correlation coefficient.

   Check: Look at the result. If it is larger than 1 or smaller than −1, then it is definitely wrong. Start over and check each step.

FORMULA

Formula for the correlation coefficient

A formula that expresses the above steps in standard notation is as follows.

$$r = \frac{(\overline{XY}) - \overline{X} \times \overline{Y}}{S_x \times S_y} \times \frac{n}{n-1}$$

where $\overline{X}$ and $\overline{Y}$ are the averages, S_x and S_y are the standard deviations, and $(\overline{XY})$ is the average of the products $X_1 \times Y_1, X_2 \times Y_2, \ldots, X_n \times Y_n$.

As usual, we represent the data as shown in Table 15.5, where n is the sample size.

Table 15.5

X	Y
X_1	Y_1
X_2	Y_2
⋮	⋮
X_n	Y_n

EXAMPLE 15.5 *Computing a Correlation: Neanderthal Bones*

Table 15.6 shows the measurements of fossil bones of five Neanderthal humans.[7] We would expect there to be an association between the arm length and the leg length because of our personal observation that bigger people are bigger all over, whereas smaller people are generally smaller all over. Looking at Figure 15.23, we see that there does seem to be a linear relationship. The correlation coefficient will measure the extent to which arm and leg lengths are associated with each other.

Table 15.6 *Measurements of fossil Neanderthal arm and leg bones, in mm*

SPECIMEN	ARM (HUMERUS)	LEG (FEMUR)
A	312	430
B	335	458
C	286	407
D	312	440
E	305	422
Average	310.00	431.40
Standard deviation	17.56	19.15

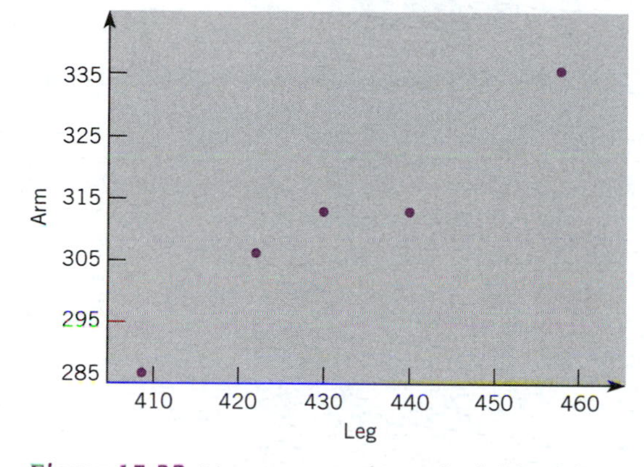

Figure 15.23 *Measurements of Neanderthal fossil bones*

To compute the correlation, we will make a new column whose entries are the product of arm length times leg length for each specimen. For example, the first entry will be $312 \times 430 = 134{,}160$ (see Table 15.7).

Table 15.7

SPECIMEN	ARM × LEG
A	134,160
B	153,430
C	116,402
D	137,280
E	128,710
Average	133,996.40

Combining these numbers, we compute the correlation:

$$\text{correlation} = \frac{133{,}996.40 - 310.00 \times 431.40}{17.56 \times 19.15} \times \frac{5}{5-1}$$

$$= \frac{133{,}996.40 - 133{,}734.00}{336.274} \times \frac{5}{4}$$

$$= \frac{262.4}{336.274} \times 1.25$$

$$= .78032 \times 1.25$$

$$= .975$$

The correlation of .975 provides a strong answer to our question. Because it is very close to the highest possible correlation (1.00), we conclude that arm and leg lengths are highly associated with each other. Because it is a positive number, it reflects the fact that those individuals with longer arms also tended to have longer legs.

Exercise Set 15.4

9. Find the correlation of X and Y for the data in Table 15.8.

Table 15.8

X	Y
1	3
2	4
3	5
4	6
5	7

10. Find the correlation of X and Y for the data in Table 15.9.

Table 15.9

X	Y
1	4
2	3
3	5
4	7
5	6

11. Find the correlation of X and Y for the data in Table 15.10.

Table 15.10

X	Y
1	4
2	7
3	5
4	3
5	6

12. Find the correlation of X and Y for the data in Table 15.11.

Table 15.11

X	Y
1	6
2	7
3	5
4	3
5	4

Discussion: Interpreting the Correlation Coefficient

The correlation coefficient is always a number between -1 and 1. A correlation of exactly 1 can happen only when the data points fall exactly along an upward-tilting straight line for which Y increases as X increases. A value of exactly -1 can result only when the data points fall exactly on a downward-tilting straight line for which Y decreases as X increases.

A value near 0 indicates that there is little or no linear association between X and Y and suggests that X and Y are independent of each other. However, a look at the bivariate plot is advised to confirm this, because nonlinear association could be present.

A value between 0 and 1 indicates a generally increasing trend. The closer the correlation is to 1, the stronger the linear trend. A value between -1 and 0 indicates a generally decreasing trend. The closer the correlation is to -1, the stronger the linear trend. Note that we are not necessarily measuring how steep the line is, but are measuring instead how tightly the points cluster around the line.[8]

All bets are off if the data set is not linear. For example, an outlier can distort the correlation, and the correlation can be misleading when the data set has nonlinear structure. But when the data set is well behaved and linear, the correlation coefficient does an excellent job of summarizing the strength of the relationship between X and Y.

In Figure 15.24 (page 538), we see a series of examples showing bivariate plots and their corresponding correlation values, in the case of linear data. This should help you develop an understanding of what the correlation is measuring.

Figure 15.25 (page 539) shows a series of examples to indicate what can go wrong with the usual interpretation of the correlation coefficient when outliers or curvature are present. The first row of Figure 15.25 illustrates that a single outlier can cause the correlation to look very impressive ($r = .84$, which is close to 1.00, suggesting strong association) even though the rest of the data exhibit very little correlation ($r = .05$). The fact that the correlation coefficient responds so strongly to even a single outlier is an indication that it is not robust and does not necessarily summarize the structure contained in most of the data. The second row shows how an outlier can change an underlying situation with strong negative correlation into either a situation with positive correlation or a situation with almost no correlation.

The last two rows in Figure 15.25 show the effects of curvature. The next-to-last row illustrates two situations with no correlation: $r = 0$. Ordinarily, this would be interpreted as saying that there is no association between X and Y. However, in these cases, there is a strong association (pattern) but it is not a straight line. A single-number summary, like the correlation coefficient, cannot be expected to adequately describe every complex situation. That is why there is no substitute for looking at the data. Finally, the last row shows how an increasing trend exhibits less correlation as it becomes more curved, even though the relationship is still very strong.

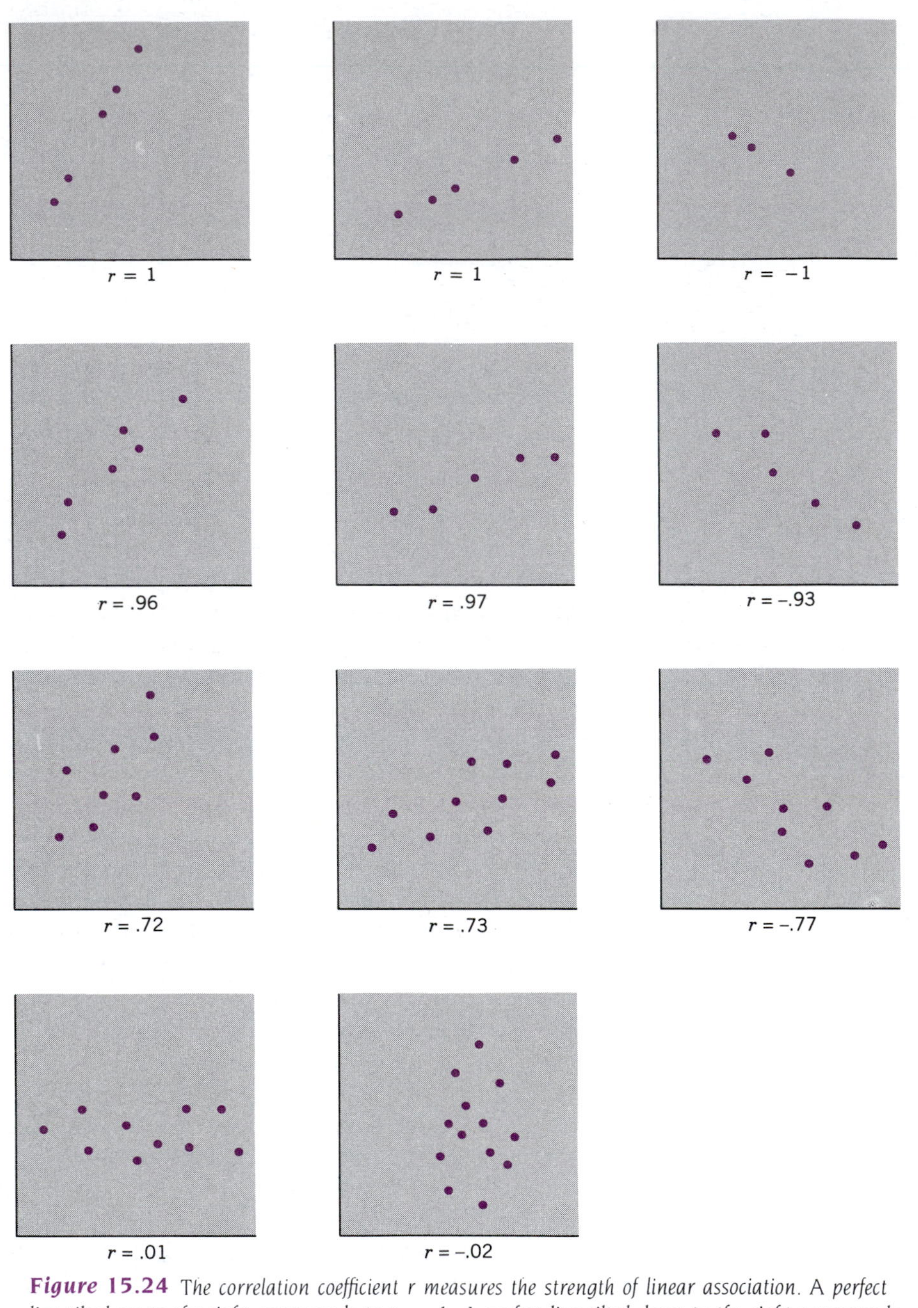

Figure 15.24 *The correlation coefficient r measures the strength of linear association. A perfect line tilted up to the right corresponds to $r = 1$. A perfect line tilted down to the right corresponds to $r = -1$. Here we see some intermediate cases as well.*

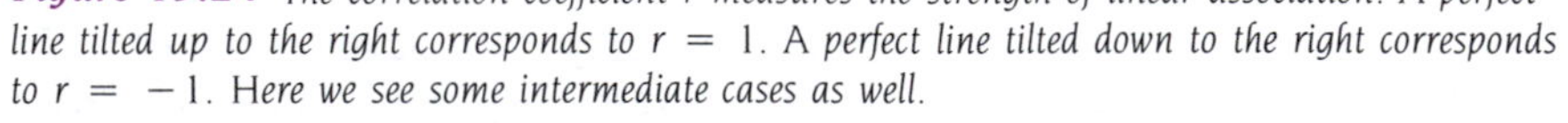

EXAMPLE 15.6 Interpreting Correlation: Pollution Sources

Consider the amount of pollution produced by utilities and the amount produced by industries. Would you expect states with a large amount of pollution by utilities also to have a high level of industrial pollution? Or would you think that these two sources were not related? The correlation coefficient can help answer this.

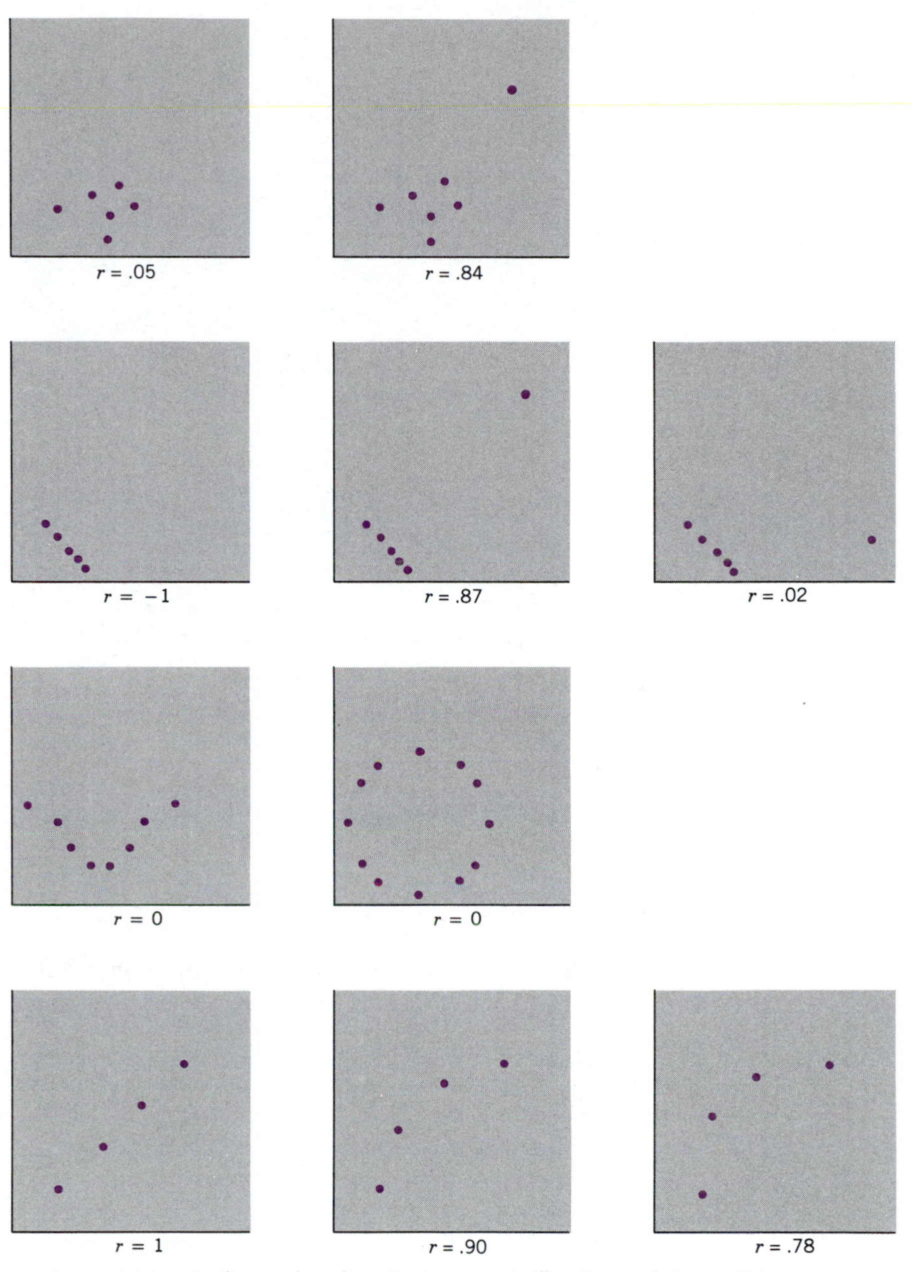

Figure 15.25 *Outliers and nonlinearity (curvature) affect the correlation coefficient in unexpected ways.*

As we have noted, it is always a good idea to look at the data first to make sure that it looks reasonably linear and that there are no extreme outliers. This seems to be the case with the data displayed in Figure 15.26 (page 540). The data values represent the average sulfur dioxide emission rates for each state of the United States from utility and industrial boilers, measured in units of lb/mm Btu.[9]

The correlation coefficient for this data set is $r = .42$, indicating some relationship (but far from a perfect association) between industrial and utility pollution. This confirms our visual

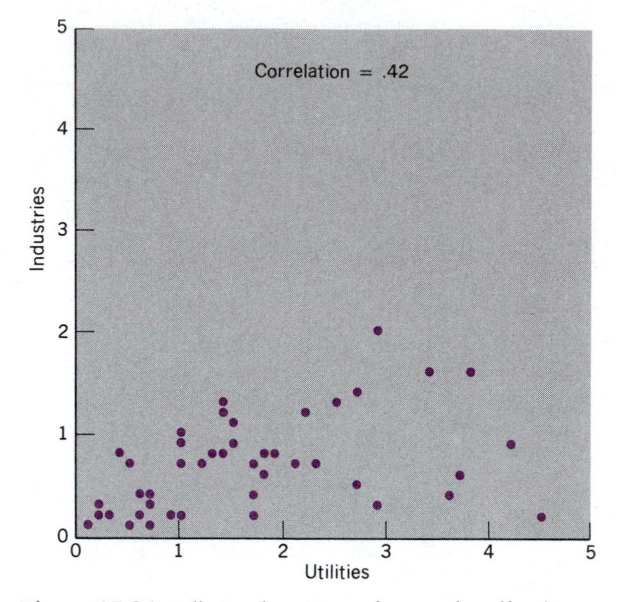

Figure 15.26 *Pollution data. For each state, the sulfur dioxide emissions of its industries are plotted against the emissions of its utilities. The correlation of $r = .42$ indicates some relationship (but far from a perfect association) bertween pollution by industries and pollution by utilities.*

assessment of the data plot, which seems quite random but has a hint of a steady trend of larger industrial pollution levels for states with more pollution by utilities.

Discussion (A Warning): Correlation Is Not Causation

Statistics can only do so much. Because the correlation coefficient is based on two columns of numbers, it is a useful measure of the linear association between those columns of numbers. But be careful not to go beyond this interpretation! Any hint of one factor "causing" another goes well beyond statistics and requires an argument based in the field that generated the data (and perhaps an argument based on the philosophy of just what "causation" is).

For example, consider cigarettes and cancer. Statistical reasoning has played a part in research that has demonstrated convincingly that smoking causes cancer, but it does not do so alone. A high statistical correlation between smoking and cancer, based on a random sample of people in the population, might be caused by several factors. It might be that smoking causes cancer and that this is why an association was seen. However, it might also be that having cancer causes people to smoke. The statistics would not be able to differentiate between these two possibilities. Or it could be that some third factor underlies both of these: Perhaps urban dwellers tend to smoke more and also tend to develop more cancer. It takes more than just statistical reasoning to sort out these possibilities.

As another example, consider a teaching situation. There is a high degree of association between your being in class and the teacher's being there. However, this correlation does not in any sense prove that your act of walking to class causes the teacher to go there also. Although these events are highly related, they probably do not cause each other.

Exercise Set 15.5

13. Suppose you are reading some articles in an area of interest, and you notice that a series of correlation coefficients is reported. The various values are 1.0, .95, .75, .33, .05, −.3, −.8 and −1.33. Pick the value that is best described by each of the following statements.

a. A perfect positive correlation

b. A very strong positive correlation

c. A strong positive correlation

d. A moderate positive correlation

e. Almost no correlation

f. A moderate negative correlation

g. A strong negative correlation

h. An obvious mistake, since the claimed value is impossible

14. Some friends of yours are taking an introductory statistics class. As a class assignment, they are to see whether there is a relationship between two variables. They plug the data into the formulas, get a correlation coefficient of 0, and decide that there is no relationship. They ask you whether they did the problem correctly. You look at the data and decide to plot it out. The plot is shown in Figure 15.27. Which of the following do you tell your friends?

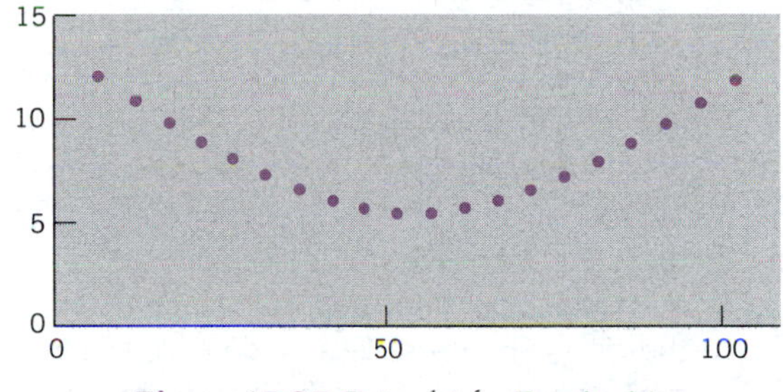

Figure 15.27 *Data plot for Exercise 14*

a. You are right; there is no relationship.

b. The correlation coefficient only measures linear relationships, so it is 0 here, but there appears to be a clear curvilinear relationship.

15. Suppose that over a 3-month period in spring and early summer, two things happen: Corn grows taller and taller, and people wear lighter and lighter clothing. That is to say, there is a correlation between the height of the corn and the weight of the clothing. Does this correlation help prove that changes in the height of corn cause changes in the clothing people wear? Why or why not?

Testing the Significance of a Correlation

If we compute a correlation of, say, $r = .63$ based on a random sample of size 20 from a population, what can we conclude about the situation? In particular, can we conclude that X and Y are associated in the population? Where do we draw the line between deciding that there is an association or not?

There is a simple test of hypothesis that will decide whether the correlation is significantly different from 0. If the correlation is significant, we conclude that the two factors are indeed associated. But if the correlation is found to be nonsignificant, we conclude that the two factors might not be related at all and that the correlation that we think we see could reasonably be due to sampling randomness.

This hypothesis test compares the observed sample correlation to the hypothesized value of 0, which expresses a lack of association, to see whether the difference is larger than what chance alone would have produced. The null hypothesis is that there is no association (the population correlation is 0), whereas the alternative hypothesis is that there is some association (the population correlation is not 0).

The test is performed by seeing how the absolute value of the correlation (this means ignoring any minus sign) compares to a value in a critical value table such as Table 15.12. If the absolute value of the correlation is larger, then the correlation is statistically significant; we reject the null hypothesis and accept the alternative hypothesis to conclude that there is indeed correlation in the population. If the tabled value is larger, then the computed correlation is not statistically significant and we do not reject the null hypothesis that there is no correlation in the population.

For example, the correlation coefficient of .975 that we computed for the fossil Neanderthal arm and leg bones was computed from a sample of size 5. The value in the table for a sample of this size, at the conventional 5% testing level, is .8783. Because the observed correlation is larger than the tabled value, we conclude that it is statistically significant. This is statistical inference, and our conclusion is that in the population of Neanderthals, there was indeed association between arm and leg length. A correlation this high is probably not just an instance of sampling randomness.

In the example of industrial pollution, we found a correlation of $r = .42$ for a sample of size 44 (data were not available for all of the states). Comparing this to the value in the table (.3120 for 40, or .2940 for 45), we see that the connection between industrial and utility pollution is indeed statistically significant, even though the situation includes a strong component of randomness. We conclude that we can detect some association, even though the association may not be very strong. The assumptions need to be interpreted properly here: The test is whether these pollution levels by state could have reasonably come from an independent random sample from an uncorrelated population (see the more detailed assumptions presented next). Although the states are not really a random sample from any population, we are free to think of them as such provided we make this assumption explicitly.

Assumptions Needed for the Test of Significance. The following conditions need to be true for this test of significance to be valid. The first condition guarantees that the sample represents the population, whereas the second one is technical and is needed for the particular table of critical values (Table 15.12) to apply.

1. The data must be obtained from a random sample selected from the population of interest.
2. The distribution of data points must be *bivariate normal.* Such a distribution will have a bivariate plot that looks linear except for randomness. This rules out outliers and nonlinear data sets. Furthermore, the individual distributions of each of the two factors should look normal.

For sample sizes not listed in the table, a conservative test may be used by selecting the critical value for the next smaller sample size listed. Such a test is conservative in the sense that the chance of a Type I error is even less than the specified level. For larger sample sizes, an approximate critical value may be found by taking the square root of the sample size, finding its reciprocal, and multiplying by the appropriate multiplier at the bottom of the column.

Table 15.12 *Testing the correlation coefficient*

SAMPLE SIZE	CRITICAL VALUES			
	10% LEVEL	5% LEVEL	1% LEVEL	.1% LEVEL
3	.9877	.9969	.9999	1.0000
4	.9000	.9500	.9900	.9990
5	.8054	.8783	.9587	.9911
6	.7293	.8114	.9172	.9740
7	.6694	.7545	.8745	.9509
8	.6215	.7067	.8343	.9247
9	.5822	.6664	.7977	.8982
10	.5494	.6319	.7646	.8721
11	.5214	.6020	.7348	.8470
12	.4973	.5760	.7079	.8233
13	.4762	.5529	.6835	.8009
14	.4575	.5324	.6614	.7796
15	.4409	.5140	.6411	.7600
16	.4259	.4973	.6226	.7416
17	.4124	.4821	.6055	.7244
18	.4000	.4683	.5897	.7082
19	.3887	.4555	.5750	.6929
20	.3783	.4438	.5614	.6786
21	.3687	.4329	.5487	.6650
22	.3598	.4227	.5368	.6522
23	.3515	.4132	.5256	.6401
24	.3438	.4044	.5151	.6286
25	.3365	.3961	.5051	.6176
26	.3297	.3882	.4958	.6072
27	.3233	.3809	.4869	.5973
28	.3172	.3739	.4785	.5879
29	.3115	.3673	.4705	.5788
30	.3061	.3610	.4629	.5702
35	.2826	.3338	.4296	.5321
40	.2638	.3120	.4026	.5006
45	.2483	.2940	.3801	.4741
50	.2353	.2787	.3610	.4514
55	.2241	.2656	.3445	.4316
60	.2144	.2542	.3301	.4142
65	.2058	.2441	.3173	.3988
70	.1982	.2352	.3059	.3849
75	.1914	.2272	.2957	.3724
80	.1852	.2199	.2864	.3610
85	.1796	.2133	.2780	.3507
90	.1745	.2072	.2702	.3411
95	.1698	.2017	.2631	.3323
100	.1654	.1966	.2565	.3242
multiplier	1.645	1.960	2.576	3.291

Exercise Set 15.6

16. a. Suppose we have four cases (Jack, Jim, Sally, and Sue) selected randomly from a larger group. We get data on two variables for each of them, and each of the variables appears to be normally distributed. We calculate a correlation coefficient and get a value of .97. Is this statistically significant at the 5% level?

    b. Suppose we have the same situation as in part (a), but this time we get a correlation coefficient of −.97. Is this statistically significant at the 5% level?

17. a. Suppose we have measurements on two normally distributed variables for 30 randomly selected cases and the correlation coefficient works out to .25. Is this statistically significant at the 5% level?

    b. Suppose instead that the correlation coefficient works out to −.25. Is this statistically significant at the 5% level?

15.4 Analyzing a Linear Relationship: Regression

If a bivariate plot of data seems to have linear (straight-line) structure, we can summarize the basic relationship between the two variables X and Y by choosing a particular straight line that passes through the data This line can then be used to estimate (or predict) a value of Y for a given value of X. This process is called **linear regression analysis.** The idea is illustrated in Figure 15.28.

Many bivariate situations have linear structure, at least approximately. Recall that this means that for two different values of X, we expect the difference between the Y values (on the average) to be proportional to the difference in the X values. Here are some examples of linear structure.

- Suppose X represents the number of cans of tuna fish purchased at the supermarket, and Y represents the total price paid at the cash register. Then for two shopping trips, the difference in Y will be the price of one can times the difference in X. That is, for each unit increase in X, we expect Y to increase a fixed amount.

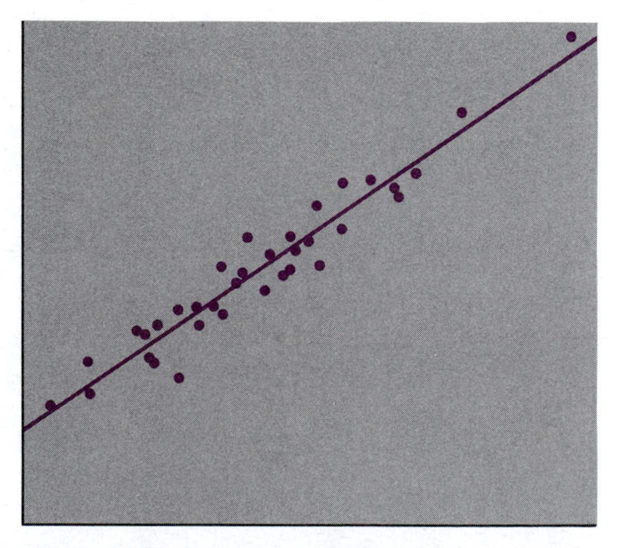

Figure 15.28 *A straight line may be used to summarize linear structure in bivariate data.*

- Suppose X represents the number of purple martins with nests near a field, and Y represents a measure of mosquito activity. Because purple martins eat mosquitoes, we would expect to see increases in X (birds) associated with decreases in mosquito activity. In some situations, we might even expect to see a linear relationship, expressing the fact that each additional bird consumes a certain number of mosquitoes, causing a proportionate decrease in activity.

The notion of linear structure is captured mathematically by a straight line drawn on a bivariate plot of X and Y. The straight line represents the case in which Y changes proportionately to changes in X.

Straight Lines

The equation of a straight line is

$$Y = a + b \times X$$

The coefficient b is the slope, measuring how steep the line is. The coefficient a is the *Y-intercept,* measuring how high the line is when X is 0. These concepts are illustrated in Figure 15.29. Note that b is the change in height (or Y value) that accompanies a horizontal move of 1 unit. If the slope b is positive, then Y increases as X does. A flat horizontal line has a slope of 0. A negatively sloped line tilts downward to the right, indicating a decreasing relationship between X and Y. All of this is illustrated in Figure 15.30 (page 546).

Suppose we had the line

$$Y = 5 + 3 \times X$$

To interpret this line, we notice that (1) when X is equal to 0, Y is equal to 5, and (2) each time X changes by 1 unit, Y changes by 3 units. If Y represented length in centimeters and X represented time in months, then this equation could describe an animal that was 5 cm long at birth and grew 3 cm each month. Or if Y represented thousands of dollars and X represented the number of robots manufactured in a day, then this equation could describe the daily cost of running a robot manufacturing plant where the fixed costs were \$5,000 per day (even if nothing were manufactured) and the incremental unit costs for materials were \$3,000 per robot manufactured.

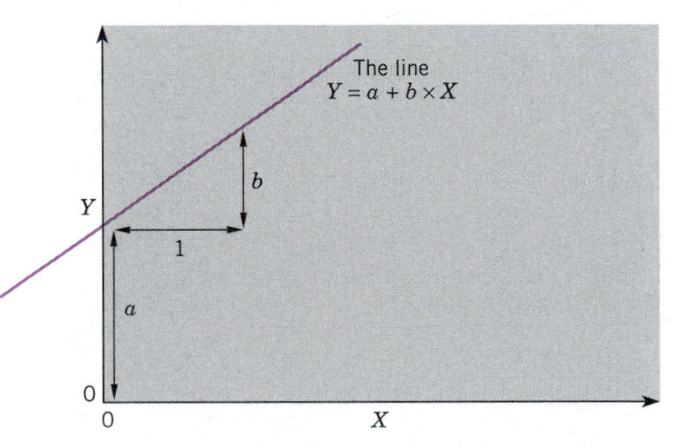

Figure 15.29 *The line $Y = a + b \times X$. The Y-intercept is a, where the line hits the vertical Y-axis. The slope is b, indicating the rise (or fall if negative) of the line as it moves 1 unit to the right.*

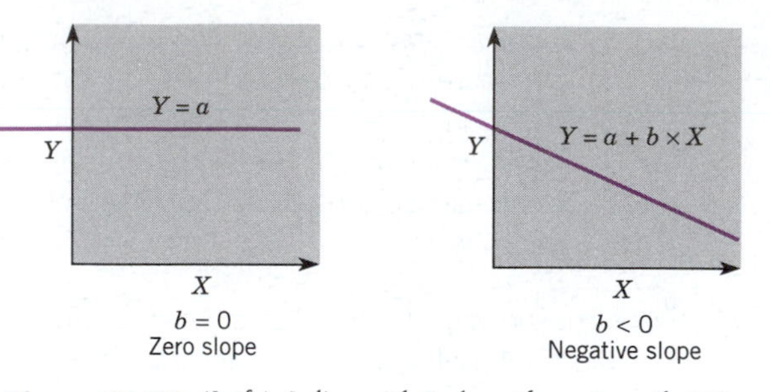

Figure 15.30 (Left) *A line with 0 slope ($b = 0$ so that $Y = a$ is the equation for the line).* (Right) *A line with negative slope ($b < 0$) so that it tilts down to the right.*

To draw a line for which we know the equation, we plot at least two points (preferably three or more to guard against errors) and then connect them with a ruler. We pick a value (any value) of X, then compute $a + b \times X$ to find the Y value to plot with this X. For example, using the line $Y = 5 + 3 \times X$, with the choice $X = 8$, we find $Y = 5 + 3 \times 8 = 5 + 24 = 29$. Thus, we know that the point (8, 29) falls on the line. Similarly, the points (4, 17) and (0, 5) are also on the line.

Using Two Typical Points

Now we want to start with the data points and find a way to write an equation for a line that predicts Y from X. The simplest and fastest way to calculate the equation of a line through a data set is to look at the bivariate plot, choose two points (one toward the left and the other toward the right), and use the straight line that connects them (see Figure 15.31). From the two typical points, denoted (X', Y') and (X'', Y''), the values of the slope b and the intercept a are easily computed:

$$b = \frac{Y'' - Y'}{X'' - X'}$$

$$a = \frac{X'' \times Y' - X' \times Y''}{X'' - X'}$$

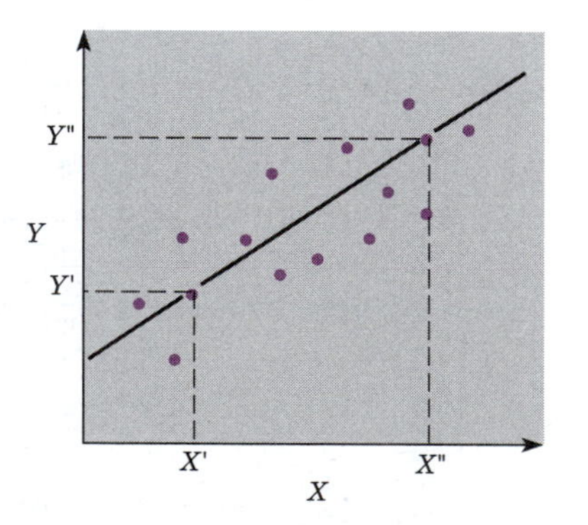

Figure 15.31 *Using the two typical points (X', Y') and (X'', Y'') to draw a straight line through a data set*

For example, if the two typical points are $(X', Y') = (1, 7)$ and $(X'', Y'') = (3, 4)$, then we would find that

$$b = \frac{4 - 7}{3 - 1} = -1.5$$

is the slope of the line and

$$a = \frac{3 \times 7 - 1 \times 4}{3 - 1} = 8.5$$

is the Y-intercept. These points and their line are illustrated in Figure 15.32.

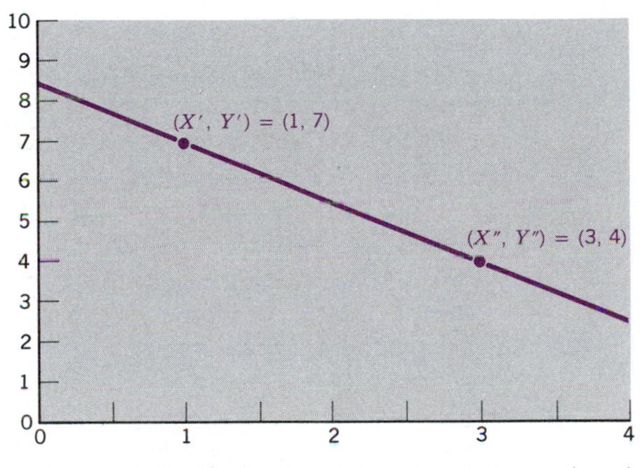

Figure 15.32 *The line $Y = 8.5 - 1.5 \times X$ goes through the two points (1, 7) and (3, 4).*

Discussion

The "two typical points" line is a very subjective method because it depends on our particular choice of points. Thus, we should try to choose the points so that the line is an effective summary of the structure in the entire data set. To do this, we may sometimes need to choose a typical point in the middle of a "cloud" of data where there is no actual data point.

Another way to choose a line by eye is to use a rubber band or a string, which is moved about until it seems to fit the data closely. A black straight line drawn on a transparent plastic sheet also works well as a line that can be adjusted until it looks appropriate with respect to the data. But the subjective nature of these methods makes them useful primarily for informal calculation. If someone were to dispute our particular choice of line, there would be no solid basis on which to stand. The least-squares line, discussed next, gives us a more solid foundation.

The Least-Squares Regression Line

The most standard method of estimating a straight line is by the method of **least squares.** The formula for this method is easily done by computer. When the data are well behaved (linear with no outliers), the method of least squares chooses the line that comes closest to the data in a well-defined sense.

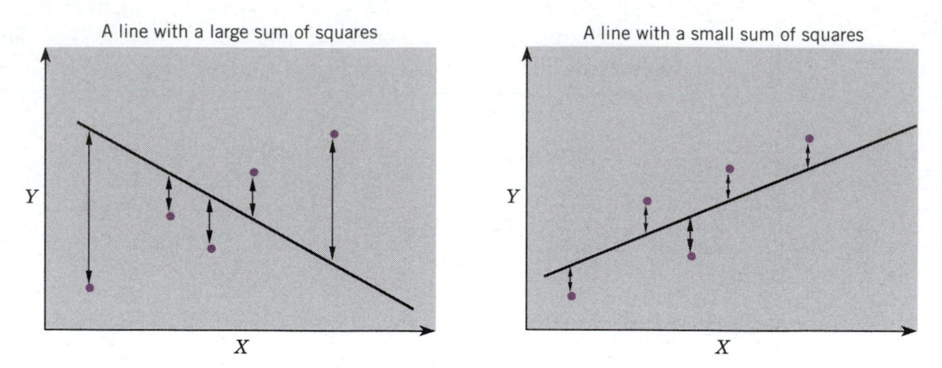

Figure 15.33 (Left) *The line here has a large sum of squares, is not very close to the data, and is not a good choice of summary. The term* sum of squares *refers to the sum of the squares of the vertical distances indicated here.* (Right) *The line here has a small sum of squares and is a better choice of summary.*

To understand what the least-squares line represents, we need to understand some of the theory behind it. The least-squares line actually minimizes the discrepancy between the points and the line in the following way. If we measure the vertical distance from each point to the line and add the squares of these distances, then we have a measure of how far the line is from the points. This *sum of squares* criterion is illustrated in Figure 15.33. The line that comes closest to all of the points, in the sense of making the sum of these squared vertical distances as small as possible, is defined to be the least-squares line. The least-squares line is the best possible line with respect to this sum of squared differences.

By taking the vertical distances instead of the horizontal distances or the direct perpendicular distances from each point to the line, we obtain a line that is best for the purpose of *predicting Y from X*. We have minimized the prediction error of the line in the Y direction. This means that the least-squares line for predicting Y is different from the least-squares line for predicting X. This is good if our purpose is really to predict Y (or understand Y) based on X.

Computing the Least-Squares Line

Fortunately, we do not have to consider all possible lines and then select the one that makes the sum of squared vertical distances as small as possible each time we analyze a data set. There is a formula that gives us the answer directly, where the slope is expressed in terms of the correlation between X and Y and the standard deviations of X and Y.

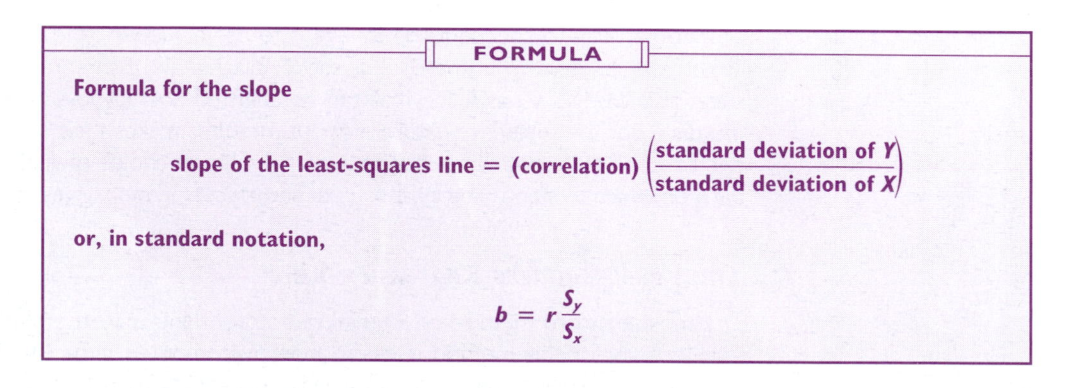

FORMULA

Formula for the slope

$$\text{slope of the least-squares line} = (\text{correlation})\left(\frac{\text{standard deviation of } Y}{\text{standard deviation of } X}\right)$$

or, in standard notation,

$$b = r\frac{S_Y}{S_X}$$

It happens that the least-squares line *always* goes through the point defined by the average value of X and the average value of Y. So we can easily find the intercept once the slope has been computed, by using the averages of X and Y to find a point we know is on the line. The intercept for the least-squares line is computed as follows.

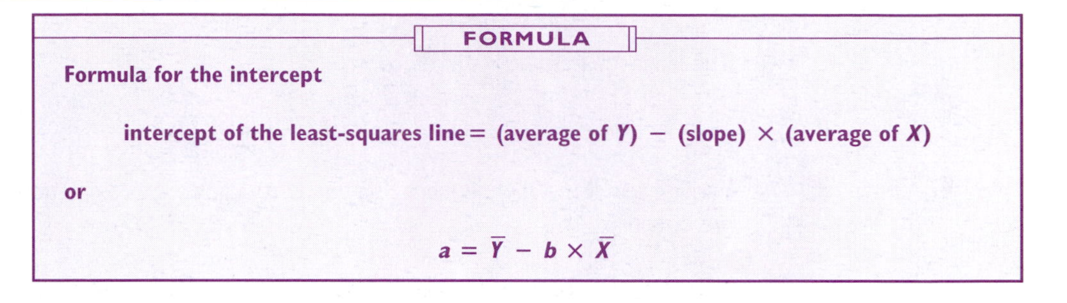

FORMULA

Formula for the intercept

intercept of the least-squares line = (average of *Y*) − (slope) × (average of *X*)

or

$$a = \overline{Y} - b \times \overline{X}$$

Once the slope and the intercept are found, we have found the line itself and can draw it in the bivariate plot of the data.

EXAMPLE 15.7 *Computing the Least-Squares Line: Professional Degrees Awarded to Women*

Women made significant gains in the 1970s in terms of their acceptance into professions that had been traditionally populated by men. To measure just how big these gains were, we will compare the percentage of professional degrees awarded to women in 1973–1974 to the percentage awarded in 1978–1979 for selected fields of study.[10] (See Table 15.13.) We have chosen X and Y in the way shown because it is easiest to think of the past (the earlier years) as predicting the future (the later years). Because the least-squares technique tries to predict Y from X, this choice is only natural.

Table 15.13 *Percentage of professional degrees awarded to women in each of eight fields*

FIELD	*X*: DEGREES IN 1973–1974	*Y*: DEGREES IN 1978–1979
Dentistry	2.0%	11.9%
Law	11.5	28.5
Medicine	11.2	23.1
Optometry	4.2	13.0
Osteopathic medicine	2.8	15.7
Podiatry	1.1	7.2
Theology	5.5	13.1
Veterinary medicine	11.2	28.9
Average	6.188	17.675
Standard deviation	4.437	8.127
Correlation	.9412	

Note the high correlation, .9412, indicating a strong association between the percentages of degrees to women by field in 1973–1974 and the same kind of percentages in 1978–1979. Based on these summary statistics, we can easily compute the coefficients of the least-squares line:

$$\text{slope} = .9412 \times \frac{8.127}{4.437} = 1.7239, \quad \text{about } 1.72$$

$$\text{intercept} = 17.675 - 1.7239 \times 6.188 = 7.008, \quad \text{about } 7.01$$

The least-squares line is therefore

$$Y = 7.01 + 1.72 \times X$$

Writing this in more understandable terms, to a good approximation as a summary of the data, we obtain

$$(\text{percentage in } 1978\text{–}1979) = 7.0 + 1.7 \times (\text{percentage in } 1973\text{–}1974)$$

We can now interpret the coefficients of this least-squares line. The intercept term, 7.0, says that even those professions with very few women increased their representation by about 7 percentage points. The slope term, 1.7, says that those fields that already had some representation by women tended to increase in representation even faster, nearly doubling (because 1.7 is nearly 2) what they had before and then adding the intercept term of 7% to that. This is a useful summary of the behavior of these data points and describes the trends in the acceptance of women in these professions.

The least-squares line is shown together with a bivariate plot of this data set in Figure 15.34. The data points have been labeled for ease of recognition of these fields, a luxury that we have when the sample size is not too large.

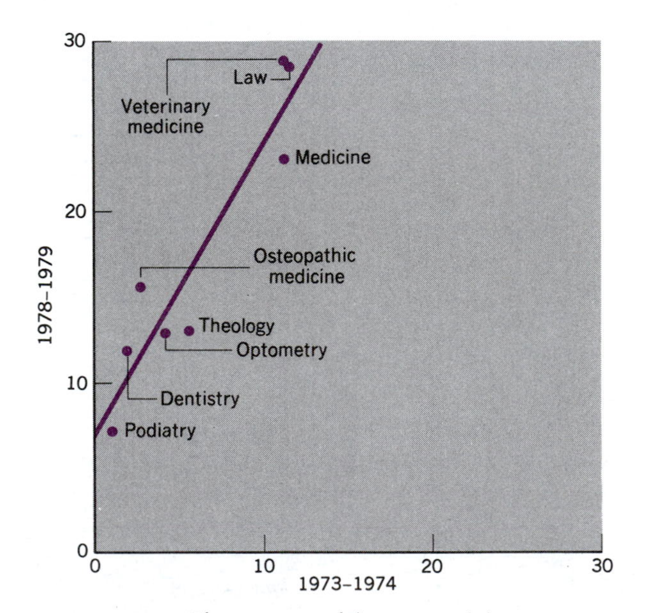

Figure 15.34 *The percentage of degrees awarded to women for several fields of study. More recent data (1978–1979) are plotted against earlier data (1973–1974) so that we can see the changes over this 5-year period. The least-squares line is also indicated.*

Predicted Values

Once the least-squares line has been computed based on the data, it is straightforward to find the **predicted value** of Y corresponding to any value of X. This is given by the point on the line.

$$\text{predicted value} = (\text{intercept}) + (\text{slope})X$$

or

$$\hat{Y} = a + b \times X$$

The standard notation for predicted Y value is $\hat{Y}$, read "Y-hat." The formula above expresses the value of Y we would *expect* at that value of X, based on the data but without the randomness of the actual situation. Note that we can find a predicted value for any value of X either corresponding to an actual data value or not. Naturally, if we try an X value that is very far from most of the data, the predicted value may not be reasonable because we are predicting beyond the information in the data.

For our example, here are some predicted values for the data in Table 15.13:

$$\begin{aligned} &\text{predicted value for dentistry in 1978–1979} \\ &\quad = 7.00 + 1.72 \times 2.0 \\ &\quad = 10.4 \end{aligned}$$

$$\begin{aligned} &\text{predicted value for medicine in 1978–1979} \\ &\quad = 7.00 + 1.72 \times 11.2 \\ &\quad = 26.3 \end{aligned}$$

The actual value for dentistry (11.9%) is slightly above but fairly close to the predicted value of 10.4%. This relationship may be seen visually in Figure 15.34, where the data point for dentistry is just *above* but close to the line.

The actual value for medicine (23.1%) is below the predicted value of 26.3%. This corresponds to the data value in Figure 15.34 being below the line. We might interpret this relationship as indicating that the actual percentage for women in medicine was slightly below what we might have expected, given the gains that women have achieved in their representation in other fields. However, it is not clear whether this apparent lag in the field of medicine is significant or just part of the natural randomness of the situation.

Exercise Set 15.7

18. Table 15.14 (page 552) shows the four bivariate data sets first seen in Exercise Set 15.4.

Note 1: You already calculated the correlation coefficients for each set, and on the way to the correlation coefficient you calculated the mean and the SD for each variable. (We hope you can still find your work from Exercises 9–12.)

Note 2: The best-fitting prediction line, or regression line, or least-squares line is found by using the formulas on pages 548–549.

a. What is the least-squares prediction line for data set A?

b. What is the least-squares prediction line for data set B?

Table 15.14

A		B		C		D	
X	Y	X	Y	X	Y	X	Y
1	3	1	4	1	4	1	6
2	4	2	3	2	7	2	7
3	5	3	5	3	5	3	5
4	6	4	7	4	3	4	3
5	7	5	6	5	6	5	4

c. What is the least-squares prediction line for data set C?

d. What is the least-squares prediction line for data set D?

19. Given the regression equation

$$\hat{Y} = 5 + 2 \times X$$

what value of y is predicted for $x = 10$?

20. Plot the line defined by $\hat{Y} = 10 + 3 \times X$.

Residuals

After a line is estimated to represent the primary trend in the data, any remaining structure is captured by the **residuals.** Examination of residuals is an important step in hierarchical modeling, because they will help decide whether the current model (such as a straight line) is adequate or if more attention to detail is needed. In general, residuals are defined by

$$\text{residual} = \text{data} - \text{summary}$$

In the case of linear regression, we think of Y as being predicted (fit, summarized, or explained) by X. Therefore, in this case we define the residuals from the regression line to be

$$\begin{aligned}\text{residual} &= (\text{actual value of } Y) - (\text{predicted value of } Y)\\ &= \text{vertical distance to the line}\end{aligned}$$

Some residuals, indicated as vertical distances from points to the line, for the data set on women in the listed professions are illustrated in Figure 15.35. The residuals are computed by the formula

$$\text{residual} = Y - (a + b \times X)$$

Or, if you have already calculated a predicted value $\hat{Y}$,

$$\text{residual} = Y - \hat{Y}$$

There is one residual for each of the original data points. To examine the residuals for any additional remaining structure, we plot them against the original X values, forming a **residual plot.**

Here are the computations needed for the degree data to find the residuals of the data from the least-squares line, which we recall was $Y = 7.01 + 1.72 \times X$. The first residual is $11.9 - (7.01 + 1.72 \times 2.0) = 11.9 - 10.45 = 1.45$. The other residuals are computed similarly, with the results shown in Table 15.15.

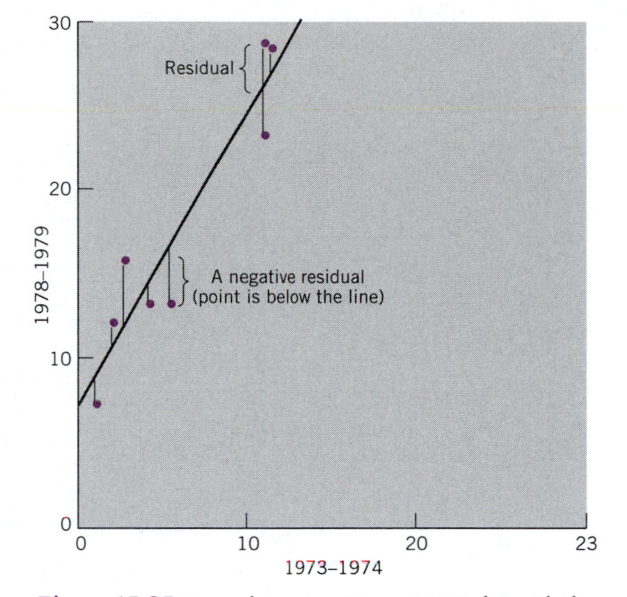

Figure 15.35 Same data set as Figure 15.34, but with the residuals indicated. For each data point, the size of its residual is the vertical distance to the line (as indicated). The residual is positive if the point is above the line and negative if the point is below.

Table 15.15 Percentage of professional degrees awarded to women and residuals of 1978–1979 regressed on 1973–1974

X (1973–1974)	Y (1978–1979)	$\hat{Y}$	RESIDUALS $Y - (7.01 + 1.72 \times X)$
2.0	11.9	10.45	1.45
11.5	28.5	26.79	1.71
11.2	23.1	26.27	−3.17
4.2	13.0	14.23	−1.23
2.8	15.7	11.83	3.87
1.1	7.2	8.4	−1.70
5.5	13.1	16.47	−3.37
11.2	28.9	26.27	2.63

Plotting the residual column against the X column, we get the residual plot, in which we see the remaining variation in the percentage of women in 1978–1979 that could not be explained by the linear trend based on the percentage in 1973–1974 (see Figure 15.36, page 554). This is a good residual plot because it looks random, indicating that there is no further structure we could extract from the situation. The straight line has already summarized it all.

Discussion: Interpreting a Residual Plot

The height of the residuals should be centered around 0 and the plot should not show any extreme tilting to one side or the other. Such patterns should have already been accounted for by the regression line and subtracted when the residuals were calculated. This provides a check to see whether the regression line is the best fit for the data.

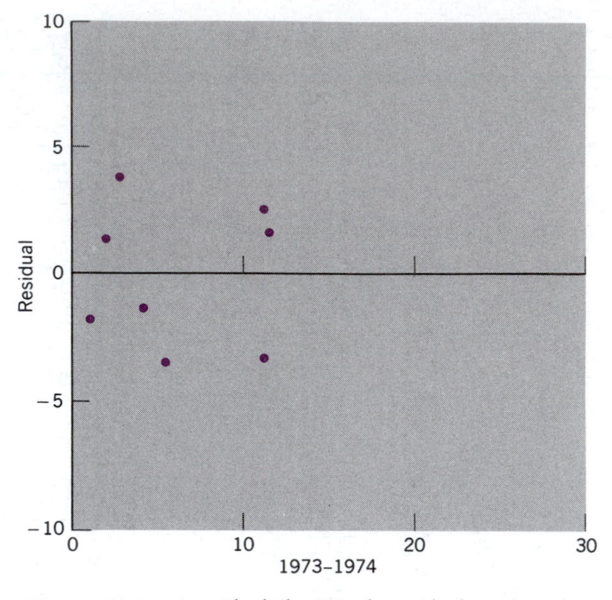

Figure 15.36 *A residual plot. We plot residuals against the original X values. Ordinarily, we expect the plot to look random and flat, as it does here.*

The best structure to see in a residual plot is no structure at all, just a flat scattering of points, as we saw in Figure 15.36. This tells us that the straight line has captured all of the important structure in the data, leaving only randomness, and that the regression line is a good description of the trends in the data.

A curved pattern in the residual plot indicates that there was some curvature in the original data. Because we can expand the vertical scale in the residual plot, we can often see curved patterns in residuals that come from a data set that looked straight in its original form. Any slight curvature can be magnified in the residual plot and subtle remaining structures may be found.

EXAMPLE 15.8 Residuals of a Curvilinear Relation: Average Heights and Ages of Girls

Consider the average heights of girls at ages 2 through 11.[11] This plot (Figure 15.37) looks like a fairly straight line with very little scatter or randomness. The least-squares line is

$$\text{height} = 76.641 + 6.366 \times \text{age}$$

which we can easily interpret. About 6.366 cm of growth are added each year, with an initial height at birth (age 0) extrapolated to be 76.641 cm (which does not necessarily imply that this is a good estimate of height at birth because we have no data at age 0). The computed residuals are shown in Table 15.16.

Already, we can see structure in the residuals column: The pattern of plus and minus signs is not random, but progresses from minus to plus to minus. The residual plot clearly shows that a lot of structure still remains in these residuals (see Figure 15.38, page 556). The slight curvature in the original data has been magnified many times in this plot because the strong straight component of the original data has been subtracted. Note that the vertical scale has been

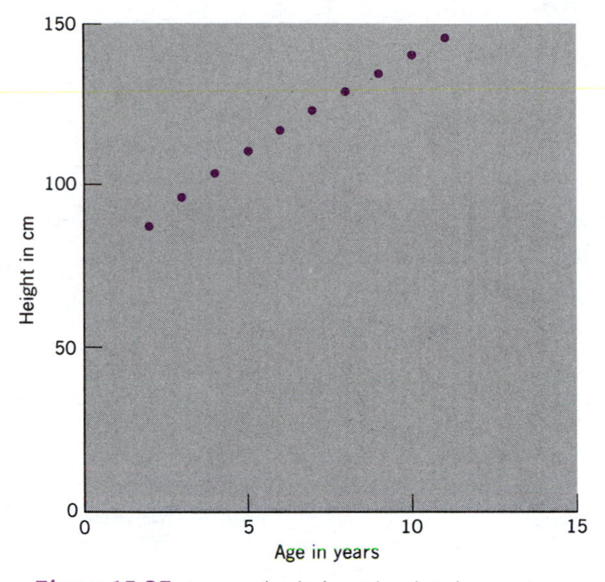

Figure 15.37 *Average height for girls, plotted against age. This looks linear with very little randomness.*

Table 15.16 *Age and height of girls*

AGE	HEIGHT (cm)	RESIDUALS: HEIGHT − (76.641 + 6.366 × AGE)
2	86.5	−2.87
3	95.5	−.24
4	103.0	.89
5	109.8	1.33
6	116.4	1.56
7	122.4	1.20
8	128.2	.63
9	133.8	−.14
10	139.6	−.70
11	145.0	−1.67

expanded many times in the residual plot as compared to the original data plot. The curvature is due to the fact that the growth rate is not constant (as estimated, 6.366 centimeters per year). In fact, younger children grow slightly faster than older children over this range from 2 to 11 years.

Generally, when the residual plot is only slightly curved, we might simply make a note of it but still use the straight-line summary of the data. One way to handle more extensive curvature is with a transformation. We would transform the original data, then estimate a straight line for the transformed data. The other way to handle extensive curvature is to pass a curve rather than a straight line through the data. This is called *nonlinear regression* and is an advanced topic that is beyond the scope of this text. (More details may be found in a book on regression, such as the one by N. Draper and H. Smith.[12])

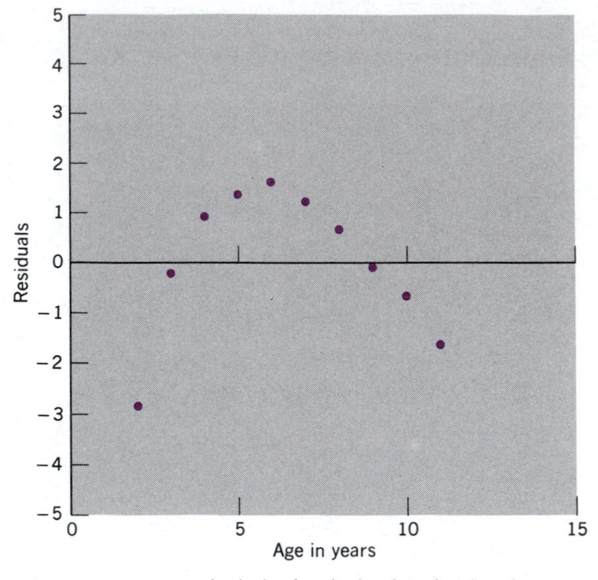

Figure 15.38 Residual plot for the heights of girls. This is highly curved, indicating nonlinear structure in the original data. Note the power of the residual plot in exposing this curvature, which is almost imperceptible in the original data (Figure 15.37).

Exercise Set 15.8

21. Find the residuals when

a. $\hat{Y} = 5 + 2 \times X$, and you have an observed value of $Y = 20$ at $X = 10$.

b. $\hat{Y} = 15 - 2 \times X$, and you have an observed value of $Y = 5$ at $X = 5$.

22. *True or false:* If the linear model does a complete job of summarizing the relationship between X and Y, the residual plot will look random.

15.5 How Much Is Explained, and Is It Significant?

We can think of X as explaining some of the variability of Y in the following sense. There is variability in the heights (Y values) of children. We can account for (or explain) some of this variability by noticing that older children are taller because *if we know the age* (X value) of a child, we can predict their height more accurately than if we ignore the age and simply use the average height $\bar{Y}$ as a predictor.

Usually X explains some, but not all, of the variability in Y. To see this, note that two children with the same age (same X value) do not always have the same height (Y value). However, the variation in heights for children of the same age will be less than for children in general.

To measure how much of the variability of Y is explained by X, we use the **coefficient of determination,** denoted ***r*-squared** or r^2, which is simply the square of the correlation coefficient r, interpreted as a percentage. The closer the correlation coefficient is to 1 or -1, the stronger the association is between X and Y because the r-squared is closer to 100%. If the correlation actually is $r = 1$ or $r = -1$, then the r-squared is 100%. For these perfect relationships (with all points on a straight line) Y can be predicted perfectly from X (using the least-squares line) and 100% of the variability in Y is explained by X. On the other hand, if the correlation is

$r = 0$ then the r-squared is 0%, suggesting that there is no relationship between X and Y and that none of the variability of Y is explained by X. In general, the r-squared is actually the percentage of the variance of Y explained by X (it is not a ratio of standard deviations).

One way to measure the reduced variability of Y is to use S, the **estimated standard deviation about the regression line,** which is reported as part of computer regression results and may be interpreted as the standard deviation of the residuals (prediction errors). To actually see that the variability of Y is reduced when information about X is used, we may compare S_y (the ordinary sample standard deviation of Y, which does not use X) to S. We may interpret S_y as telling us the typical error made using $\bar{Y}$ to predict the height of a new child. Likewise, we may interpret S as telling us the typical error made using the predicted value (from the least-squares line) to predict height (Y) from the age (X). When S is less than S_y, it looks like regression is achieving its goal of reducing error. But is it significantly less?

When we ask whether this explanation (of Y by X) is statistically significant, we are asking whether the apparent relationship between X and Y (in the sample data) might just be a random coincidence or an artifact of sampling. Just as there is a test for the significance of a relationship in bivariate data using the correlation, there is also a test that uses regression. The answer is always the same (even the p-value is identical) whether we test the correlation coefficient or the regression coefficient because both approaches test the same question: "Is there a relationship?"

Computer Analysis: Correlation and Regression

Computers simplify both the plotting of bivariate data and the computation of correlation and regression equations. In this section, we will use Minitab first to conduct an analysis of the age and height data that parallels the analysis we did by hand. Then we will analyze data on chocolate milk shakes from fast-food restaurants.

There are three programs involved: "Plot" or "Gplot" for the graphics, "Correlation" for the correlation coefficient, and "Regress" for the regression equation and tests of significance.

EXAMPLE 15.9 A Significant Correlation: Age and Height

The first thing we do is plot the data, as in Figure 15.39. Clearly, there is a strong relationship between the variables, and at first glance it appears to be almost perfectly linear. Using the Correlation program, we find that the correlation coefficient has a value of .997—which is very

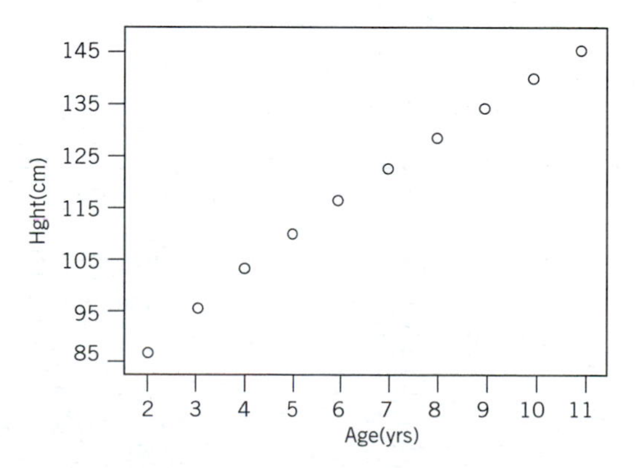

Figure 15.39 *Age and height data for Example 15.9*

close to perfect and (by the table—the program does not print this) statistically significant for a random sample of size 3 or more. Sample output is shown in Display 15.1.

DISPLAY 15.1

```
MTB > Correlation 'Age(yrs)' 'Hght(cm)'.
Correlation of Age(yrs) and Hght(cm) = 0.997
```

A separate program calculates the regression equation. It also allows us to save the residuals for later analysis and calculates some tests of significance. In the output shown in Display 15.2, the first command names columns 3 and 4 as the places where the residuals and the regression equation coefficients should be stored. The commands for the Regress program itself, which start on the next line, ask for a regression of height on age (for example, they ask the computer to find an equation that predicts height from age) and for the residuals and coefficients to be stored in the columns named RESI1 and COEF1.

DISPLAY 15.2

```
MTB > Name c3 = 'RESI1' c4 = 'COEF1'

MTB > Regress 'Hght(cm)' 1 'Age(yrs)';

SUBC>  Residuals 'RESI1';

SUBC>  Coefficients 'COEF1'.

The regression equation is
Hght(cm) = 76.6 + 6.37 Age(yrs)

PREDICTOR     COEF      STDEV     T-RATIO    P
Constant      76.641    1.188     64.52      0.000
Age(yrs)      6.3661    0.1672    38.08      0.000

s = 1.518    R-sq = 99.5%    R-sq(adj) = 99.4%

Analysis of Variance

SOURCE        DF    SS        MS        F         P
Regression    1     3343.5    3343.5    1450.45   0.000
Error         8     18.4      2.3
Total         9     3361.9

Unusual Observations

Obs.Age(yrs)   Hght(cm)   Fit Stdev.   Fit     Residual   St.Resid
1 2.0          86.500     89.373       0.892   -2.873     -2.34R

R denotes an obs. with a large st. resid.
```

Let's look at each section of the output in Display 15.2. It starts with a nice declarative statement of the main result.

```
The regression equation is

Hght(cm) = 76.6 + 6.37 Age(yrs)
```

This is what we got when we did the problem by hand, and it has the same interpretation: To predict the average heights of girls age 2 to 11 (which is the range of the original data), we start with 76.6 cm and add 6.37 cm for each year of age.

Next, the output shows a table of coefficients.

PREDICTOR	COEF	STDEV	T-RATIO	P
Constant	76.641	1.188	64.52	0.000
Age(yrs)	6.3661	0.1672	38.08	0.000

These columns repeat some of the information already given in the regression equation, with a little more detail: The constant (intercept) in the equation is now 76.641 cm and the slope is 6.3661. They also give standard deviations for both coefficients. The standard deviation for the constant is 1.188 cm, and for the slope it is .1672 cm/yr. These have the same sort of interpretation we have been giving other standard deviations, and they can be used to calculate confidence intervals for these coefficients. The next to last column gives the *t*-ratio, which is the number of standard deviations separating the observed coefficient value from a hypothesized value of 0.

In this example the *t*-ratios are quite large. The probability of observing coefficients at least as extreme as those observed if the null hypothesis is true works out to 0.000 (that is, it is less than .0005, which would round to .001). Thus we would reject the null hypothesis and accept the alternative hypotheses, namely, that on average girls are estimated to have some height even at age 0 and that (at least within the limits of these data) the older they are the taller they are. The conclusion is that there is a significant relationship between age and height.

On the next line of the output, we see three values.

```
s = 1.518   R-sq = 99.5%   R-sq(adj) = 99.4%
```

The "s" value is the estimated standard deviation of the observed values around the regression line. It is, in other words, the sample standard deviation of the residuals (dividing by $n - 2$ instead of $n - 1$). As such, it has the usual interpretation (about two thirds of the observations should be within 1 standard deviation of the predicted value, and so on) and can be used to calculate the probability that a randomly selected observation will be a given distance from the prediction line. This *S* value of 1.518 indicates the estimated variability about the regression line. This is considerably less than the standard deviation of height (about 20) for all of these data values (representing different ages), indicating the substantial improvement in prediction we get by using age to explain or predict height.

The "R-sq" value is the square of the correlation coefficient. (If we squared the *r* value of .997 we got with the Correlation program, we would get an *r*-squared of .994009, rather than .995. The difference, of course, is due to rounding error.) This tells us that 99.5% of the variability in average height, from one data value to another, can be explained by age (using the least-squares prediction equation). The remaining .5% of variability reflects the fact that these predictions, while very good, are not quite perfect. Why does the program report the square of the correlation coefficient, rather than the coefficient itself? It happens that *r*-squared is the proportion of the total variance in *Y*—the average squared deviations around the mean of *Y*—that is removed when we use the regression line to predict *Y*. It is a direct measure of the improvement in our prediction, provided that we square our errors.

The correlation coefficient itself does not have an interpretation that directly reflects the amount of improvement in our predictions. Instead, it has an interpretation in terms of the amount of change we would expect to see in the response variable, *Y*, with a given change in the predictor variable, *X*, provided that both are measured in standard deviations. If there is a perfect correlation ($r = 1$), then the predicted value of *Y* goes up 1 standard deviation each time the value of X goes up 1 standard deviation. If $r = .5$, Y goes up (on average) half a standard deviation every time X goes up a standard deviation. If $r = .0$, Y does not change (on average) when X goes up a standard deviation. Likewise, if $r = -.5$, Y goes down (on average) half a standard deviation every time X goes up a standard deviation, and if $r = -1.0$, Y goes down a full standard deviation every time X goes up a standard deviation.

The important point here is that both r and r-squared each have precise, though different, interpretations. You may see either value reported, and you can move freely from one to the other by squaring or taking square roots (except that the negative square root is used for a decreasing relationship).

The "R-sq(adj)" (adjusted R-squared) value on the output uses a degrees-of-freedom computation to correct for some possible distortions. In particular, the program used here has the capability of using more than one predictor variable. (That general technique is called *multiple regression.*) The adjustment is important as predictor variables are added. We are using only one predictor variable, and the difference shows only in the third decimal place, which may just be rounding error.

The next section on the output is labeled "Analysis of Variance" and is another test of significance. (See Chapter 13.) Notice that it produces a single observed probability for the whole equation, which is identical to that of the regression coefficient. The analysis of variance is particularly useful when more than one predictor variable is being used.

The regression line does indeed fit the observed data nicely, as shown in Figure 15.40. Using the Plot program on the residuals stored in column 3, we get Figure 15.41. We see that there

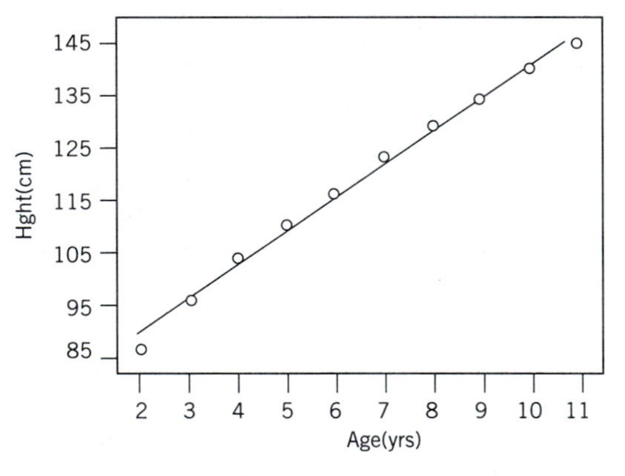

Figure 15.40

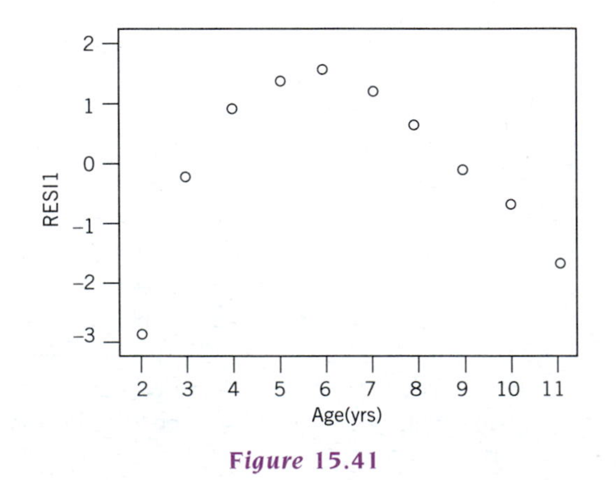

Figure 15.41

is indeed still some structure in the data. It is more conspicuous than in the last figure because we were able to use a larger scale.

EXAMPLE 15.10 A Correlation That Is Not Significant: Energy in Milk Shakes

Here is an example of a correlation that is not statistically significant. While there does appear to be a relationship in the data, the number of data points is small enough that the apparent relationship could be explained by chance.

Our data, shown in Table 15.17, come from a review of chocolate milk shakes sold in fast-food restaurants published in *Consumer Reports.*[13] They include the percent of total calories that come from fat, the total calories, the total grams of fat, price, and number of ounces in a serving.

Table 15.17 *Chocolate shake data*

	FAT%CAL	CAL	GRAMSFAT	PRICE	OUNCES
McD Lowfat	5	320	2	1.11	10
Burger King	20	400	9	1.10	11
Arby's	23	451	12	1.07	12
Hardee's	23	390	10	1.16	12
Wendy's	25	460	13	1.01	11

Let's look at the relationship between serving size and total number of calories. We plot the data in Figure 15.42 and we see that while it is not perfect, there does appear to be a relationship between the size of the shake and the number of calories it contains. (If we were concerned with different-sized servings of the same brand of shake, we would expect a perfect relationship, but we have five different brands in this data set.)

DISPLAY 15.3

```
MTB > Correlation 'Calories' 'Ounces'.

Correlation of Calories and Ounces = 0.622
```

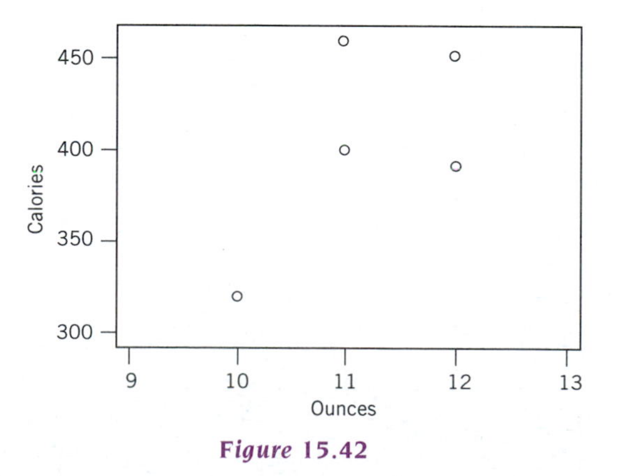

Figure 15.42

Running the Minitab Correlation program, we get what is shown in Display 15.3. To find out whether the correlation of .622, which appears to be substantial, is statistically significant we run the Regression program and get the results shown in Display 15.4. We see that the *t*-ratio for "ounces" is 1.37, with a *p*-value of .263. This is not statistically significant. Thus, on the basis of this data set we cannot reject the null hypothesis that there is no relationship between the size of a fast-food chain's standard-sized chocolate shake and the number of calories it contains.

DISPLAY 15.4

```
MTB > Regress 'Calories' 1 'Ounces'.

The regression equation is
Calories = - 63 + 41.7 Ounces

PREDICTOR        COEF          STDEV          T-RATIO          P
Constant         -63.0         340.7          -0.18            0.865
Ounces           41.71         30.36          1.37             0.263

s = 50.80   R-sq = 38.6%   R-sq(adj) = 18.2%

Analysis of Variance

SOURCE         DF     SS        MS       F        P
Regression     1      4872      4872     1.89     0.263
Error          3      7741      2580
Total          4      12613
```

Exercise Set 15.9

These exercises are based on the data on percent of degrees awarded to women in various professional fields, comparing academic year 1972–1973 to 1978–1979. In all cases, the percent of degrees awarded to women in 1972–1973 is used to predict the percent in 1978–1979. We start with a simple plot of the data, given in Figure 15.43.

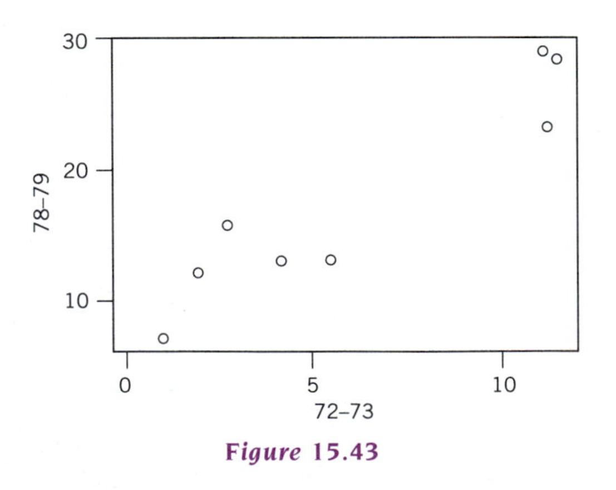

Figure 15.43

23. Overall, does there appear to be a relationship between the percent of degrees awarded to females in 1972–1973 (the horizontal axis) and the percent awarded in 1978–1979?

24. Which of the following statements is true?

a. In general, the greater the percent of degrees awarded to females in 1972–1973, the greater the percent awarded to females in 1978–1979.

b. In general, the percent of degrees awarded to females in 1972–1973 tells nothing about the percent awarded to females in 1978–1979.

c. In general, the greater the percent of degrees awarded to females in 1972–1973, the smaller the percent awarded to females in 1978–1979.

Now we calculate the correlation coefficient, as shown in Display 15.5.

DISPLAY 15.5

```
MTB > Correlation '72-73' '78-79'.

Correlation of 72-73 and 78-79 = 0.941
```

25. What is the value of the correlation coefficient?

The results of the Regress program are shown in Display 15.6.

DISPLAY 15.6

```
MTB > Regress 'F%78-79' 1 'F%72-73' 'SRES2' 'FITS2';

SUBC>  Residuals 'RESI2';

SUBC>  Coefficients 'COEF2'.

The regression equation is
F%78-79 = 7.01 + 1.72 F%72-73

PREDICTOR    COEF      STDEV     T-RATIO    P
Constant     7.007     1.882     3.72       0.010
F%72-73      1.7241    0.2527    6.82       0.000

s = 2.966  R-sq = 88.6%  R-sq(adj) = 86.7%

Analysis of Variance

SOURCE       DF    SS        MS        F        P
Regression   1     409.60    409.60    46.57    0.000
Error        6     52.77     8.80
Total        7     462.38
```

26. What is the formula for predicting the percent of female graduates expected in 1978–1979 from a program that had 8.0% female graduates in 1972–1973?

Figure 15.44 (page 564) shows the plot of the data points with the regression line drawn through them.

27. Put tracing paper (or something similar) over the page and draw a vertical line connecting each observation with the prediction line.

a. About how far off is the first observation? What is the square of this value?

b. About how far off is the second observation? What is the square of this value?

c. About how far off is the third observation? What is the square of this value?

d. Can you find a line where the sum of the squared deviations is smaller than it is with this line?

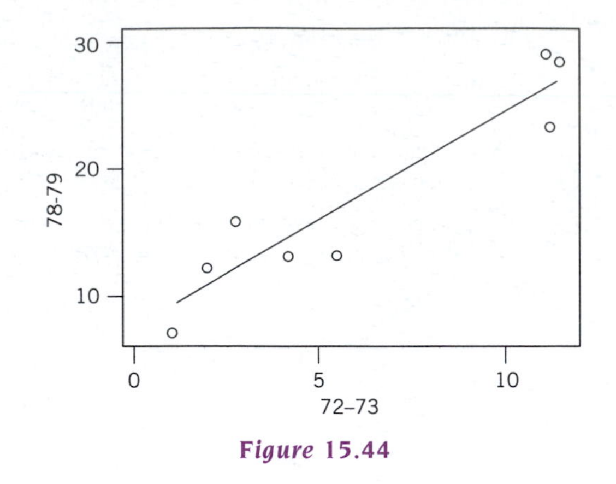

Figure 15.44

Figure 15.45 is a scatterplot of the residuals.

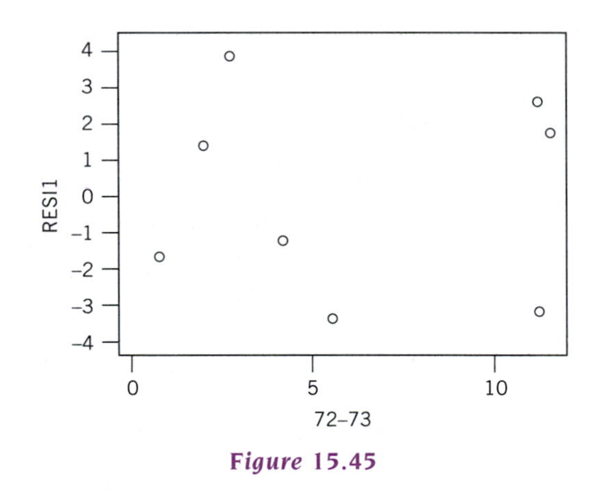

Figure 15.45

28. a. What is the approximate value of the first residual?

b. What is the approximate value of the second residual?

c. Do the residuals appear to have a clear shape and direction (for example, "up and to the right," or "down and to the right"), or do they look more like a formless cloud?

15.6 A Nonparametric Correlation: Spearman's Rho

Spearman's rho is a nonparametric alternative to the standard correlation coefficient. Like many of the other nonparametric techniques, it is robust in the sense that it is not particularly sensitive to the shape of the distribution or the presence of outliers. Further, it only requires that the data be measured at the level of ordered categories, whereas the regular correlation coefficient requires measured numerical data.

PROCEDURE: CALCULATING SPEARMAN'S RHO

1. Convert each X value to a rank value among the X's and convert the corresponding Y values to rank values among the Y's.
2. Calculate the usual correlation coefficient using these ranks rather than the original values.

Since the correlation is based on ranks, it is often called a **rank order correlation.**

EXAMPLE 15.11 *Rank Order Correlation: Exam Grades*

We developed this example using data file EXAM.MTW. It includes first and second exam grades for the eight students in a graduate statistics class. The raw data are retrieved and displayed first as numbers, then as dot plots, in Display 15.7.

DISPLAY 15.7

```
MTB > Retrieve 'C:\MTBWIN\STUDENT1\EXAM.MTW'.
MTB > Print 'EXAM 1' 'EXAM 2'.

ROW    EXAM 1    EXAM 2
 1       89        83
 2       56        77
 3       78        91
 4       88        87
 5       94        99
 6       87        80
 7       96        85
 8       72        75

MTB > DotPlot 'EXAM 1' 'EXAM 2';
SUBC>  Same.

  .                    .       .        ...   . .
 -+---------+---------+---------+---------+---------+-----EXAM 1

                       . .   .   . . .   .        .
 -+---------+---------+---------+---------+---------+-----EXAM 2
56.0      64.0      72.0      80.0      88.0      96.0
```

Looking at the dot plots, we notice that there is one score that is substantially lower than the others. While it is probably not an outlier in a technical sense, and we would be willing to argue that the data meet the assumptions for a regular correlation analysis, we decide that we are really most interested in the extent to which we can predict a student's position in the class (rather than the student's test grade). This is a natural problem for a rank order correlation.

We use the Minitab Rank command to find each student's rank in the class (1 is lowest) for each exam, then we do dot plots and a scatterplot of the results to examine our data (Display 15.8, page 566). Looking at the ranks and the dot plots, there are no surprises. The scatterplot of the ranked data seems to show a clear positive correlation.

We now calculate the correlation coefficient and the regression statistics using the ranked

DISPLAY 15.8

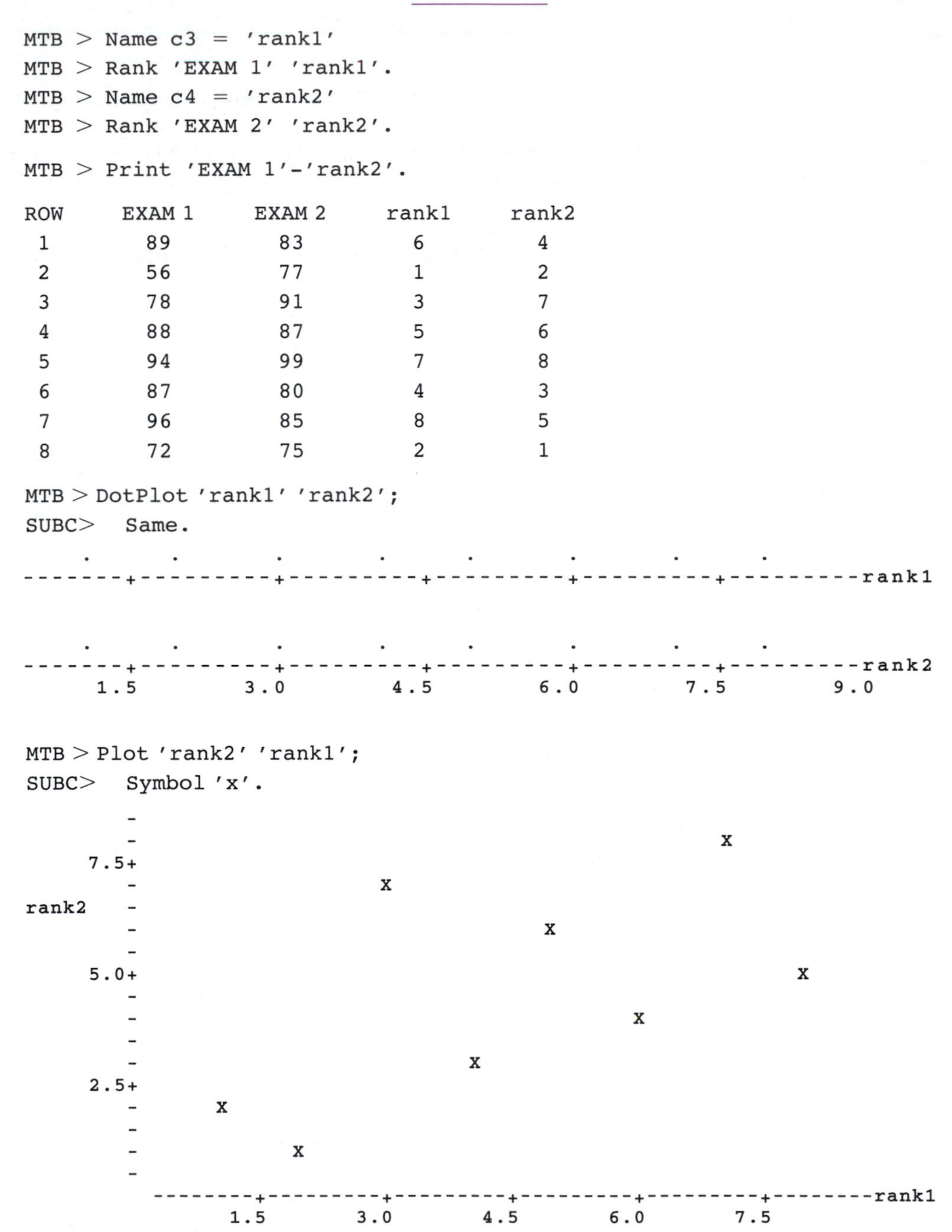

```
MTB > Name c3 = 'rank1'
MTB > Rank 'EXAM 1' 'rank1'.
MTB > Name c4 = 'rank2'
MTB > Rank 'EXAM 2' 'rank2'.

MTB > Print 'EXAM 1'-'rank2'.

ROW    EXAM 1    EXAM 2    rank1    rank2
 1       89        83        6        4
 2       56        77        1        2
 3       78        91        3        7
 4       88        87        5        6
 5       94        99        7        8
 6       87        80        4        3
 7       96        85        8        5
 8       72        75        2        1

MTB > DotPlot 'rank1' 'rank2';
SUBC>   Same.

       .      .       .       .      .       .      .      .
-------+---------+---------+---------+---------+---------rank1

       .      .       .       .      .       .      .      .
-------+---------+---------+---------+---------+---------rank2
      1.5       3.0       4.5       6.0       7.5       9.0

MTB > Plot 'rank2' 'rank1';
SUBC>   Symbol 'x'.

           -
           -                                         x
      7.5+
           -                  x
rank2      -
           -                            x
           -
      5.0+                                               x
           -
           -                                  x
           -
           -                       x
      2.5+
           -     x
           -
           -          x
           -
             -------+---------+---------+---------+---------+--------rank1
                   1.5       3.0       4.5       6.0       7.5
```

data (Display 15.9). We find that the rank order correlation coefficient is .595, and, looking at the *P*-value for the analysis of variance, we see that on the basis of this sample alone, we cannot reject the hypothesis that there is no association between class rank on the first exam and class rank on the second exam.

This is, of course, not what we expected. We have a substantial amount of personal experience that would lead us to think that class rank *is* generally associated from one test to another.

DISPLAY 15.9

```
MTB > Correlation 'rank1' 'rank2'.

Correlation of rank1 and rank2 = 0.595

MTB > Regress 'rank2' 1 'rank1'.

The regression equation is
rank2 = 1.82 + 0.595 rank1

Predictor       Coef        Stdev       t-ratio      p
Constant        1.821       1.657       1.10         0.314
rank1           0.5952      0.3280      1.81         0.120

s = 2.126   R-sq = 35.4%   R-sq(adj) = 24.7%

Analysis of Variance

SOURCE        DF    SS        MS        F       p
Regression    1     14.881    14.881    3.29    0.120
Error         6     27.119    4.520
Total         7     42.000
```

However, on the basis of this set of data alone we cannot claim to have evidence supporting that association. The apparent association we see in these particular data could be explained by chance.

Exercise Set 15.10

29. Suppose you think there is a relationship between typical house prices in neighborhoods, and the general desirability of the neighborhood. For one city, you have a panel of homeowners rank the desirability of different neighborhoods, and then you calculate the median price of the homes for sale in each neighborhood. You get these results:

	MOST DESIRABLE				LEAST DESIRABLE
Neighborhoods	East Side	North End	Capitol Hill	Central	South End
Median price	\$350,000	\$250,000	\$150,000	\$125,000	\$85,000

Without doing any arithmetic, what is the value of the rank order correlation coefficient for desirability and median price?

30. Suppose someone proposes that, among humans, unusual abilities in one area will be "balanced" by weaknesses in other areas. In an effort to test this model, and after extensive negotiations, you manage to get aptitude test scores for a few randomly selected grade-school students. You then have a panel of classmates rank these students in terms of "social skills and ability to get along with others." Higher aptitude scores are better. Social skills are ranked with 1 being best. Here are your results:

Rank on Social Skills	1	2	3	4	5	6
Aptitude Score	16	14	10	12	8	9

a. Calculate the rank order correlation coefficient.

b. In a brief paragraph, describe the results and state whether they are consistent with the balancing model.

Summary

Bivariate data sets occur whenever we have two groups of numbers arranged as pairs—for example, when each case (or experimental unit) contributes two numbers to the data set. Typical examples include the stimulus/response experiment, time-series data, and almost any study that looks for a relationship between two things. One of the two will be called X and the other Y; usually, we think of the variable Y as being predicted from the variable X.

The first step in analyzing bivariate data is to display the data in a bivariate plot. If there is no relationship between the two variables, the Y values of the points in the plot will have the same distribution regardless of the X values. The easiest structure to quantify is the *linear* trend, in which the points fall along a straight line except for randomness. In a *nonlinear* trend, the points fall along a curved line. *Clustering* is when the data points seem to fall into more than one distinct group, and an effort should be made to find out why it happens. Sometimes, it is appropriate to isolate the clusters and analyze each one separately. Outliers, points that are far away from the rest of the data, should be examined and possibly set aside (if a solid reason can be found for doing so). If there is *unequal variability* in Y near different values of X, analysis is more difficult but a transformation might help equalize the variability.

There are two main approaches to the analysis of linear (straight-line) structure in a bivariate data set of (X, Y) pairs: correlation and regression. *Correlation* is a measure of the strength of linear association between X and Y and is a number from -1 to 1 that is computed from the data. A value near 1 indicates a strong increasing linear trend, whereas a value near -1 indicates a strong decreasing linear trend. Values of the correlation coefficient nearer to 0 indicate somewhat weaker trends. A correlation near 0 is usually interpreted as indicating that there is no association between X and Y, but in fact there still might be some interesting nonlinear structure visible in the bivariate plot of the data.

Correlation is a measure of association between the data values of X and Y. Any conclusions about one causing the other may use statistics as supporting evidence, but require additional logic from outside the realm of statistics. The correlation quantifies the extent of the association, but usually cannot indicate why things are related.

A correlation coefficient may be tested to see whether it is significantly different from 0 (that is, to see whether X and Y are significantly correlated, as opposed to being independent of each other) by comparing the computed correlation coefficient to the proper number in a table of critical values. For the test to be valid, the data must be a random sample from a population and must be normally distributed in a bivariate sense.

Linear regression is the process of estimating a good straight line that describes and summarizes the trends in the data. This is done by estimating the slope b and the intercept a in the equation of a straight line, $Y = a + b \times X$. One informal, subjective method for estimating a line is to pick two typical points from the bivariate plot. Another is to move a string or another indicator of a line until it looks like a good summary of the data. The most standard method of estimating a straight line is the method of *least squares,* which chooses the line that minimizes the sum of the squared deviations in the Y direction and therefore produces a line that predicts Y from X. A *predicted value* for Y may be computed for any given value of X, simply by evaluating the equation of the line. This predicted value indicates the value of Y expected for that value of X. Predicted values may be computed for any X, data value or not. Care must be used when predicting beyond the range of the data.

Residuals represent any structure remaining in the data after a line has been subtracted. They are found by subtracting the height of the line at X, namely, $a + b \times X$, from Y for each data point. These residuals may be plotted against the original X values to form a residual plot that can be examined for further structure. A flat, random pattern of residuals centered around 0 indicates that only randomness remains in the data. Any slight curvature in the data will be magnified by the expanded scale of the residual plot and will be more easily detected so that it can be dealt with either by transformation of the original data or by methods of nonlinear regression.

Computerized regression analysis typically provides standard errors and hypothesis tests for the regression coefficients. The hypothesis test of the slope coefficient b indicates whether X and Y are significantly related. The result of this test is always the same as the result of the test for significance of the correlation coefficient, because they are both tests of relationship. The squared correlation, *r-squared,* indicates the proportion of the variance of Y that is removed by, or explained by, X. The estimated standard deviation in regression, S, indicates the typical size of the residuals.

When the data do not meet the assumptions required for a standard regression analysis (for example, one or both variables are not normally distributed, there are outliers, or the data have been measured as ordered categories), we can use a nonparametric alternative. The rank order correlation coefficient, Spearman's rho, is calculated by first ranking the data, and then calculating the correlation coefficient using the ranks, rather than the original data values.

Problems

Basic Reviews

1. For each of the following situations, state whether the correlation is significant.
   - **a.** Correlation = .63, sample size = 20
   - **b.** Correlation = .63, sample size = 10
   - **c.** Correlation = −.27, sample size = 25
   - **d.** Correlation = .02, sample size = 100
   - **e.** Correlation = −.86, sample size = 15
2. Plot each of the following straight lines and in each case say whether the relationship between X and Y is increasing, decreasing, or flat.
   - **a.** $Y = 3 + 5 \times X$
   - **b.** $Y = 4 - 2 \times X$
   - **c.** $Y = -10 + 3 \times X$
   - **d.** $Y = -5 - X$
   - **e.** $Y = 2 \times X$
   - **f.** $Y = 8$
3. For each of the following pairs of typical points, find the formula for the line that connects them.
   - **a.** (0, 0) and (1, 1)
   - **b.** (1, 2) and (5, 7)
   - **c.** (0, −1) and (4, 8)
   - **d.** (1.4, 3.6) and (10.2, 1.5)
   - **e.** (−1, 8) and (17, 8)
   - **f.** (−2, −3) and (−14, 15)

4. Find the least-squares regression line under each of the following circumstances.
   a. Average of X is 56.3, average of Y is 41.2, standard deviation of X is 5.7, standard deviation of Y is 3.9, and correlation is .62.
   b. Average of X is -14, average of Y is 104, standard deviation of X is 1.1, standard deviation of Y is 23.2, and correlation is $-.84$.
5. Find the predicted value of Y in each of the following cases.
   a. Prediction equation is $Y = 17.3 - 5.1 \times X$, at $X = 6$
   b. Prediction equation is $Y = -36 + 18 \times X$, at $X = -5$
6. Find the residual values from the line $Y = 173 + 54 \times X$ for each of the following data points.
   a. $X = 10$, $Y = 693$
   b. $X = 10$, $Y = 751$
   c. $X = 15$, $Y = 872$

Analysis

7. Consider the data in Table 15.18 on monthly normal temperatures.[14]

Table 15.18

	JANUARY	MARCH
Albany, NY	22	33
Albuquerque, NM	35	46
Anchorage, AK	12	24
Asheville, NC	38	46
Atlanta, GA	42	51
Baltimore, MD	42	53
Birmingham, AL	44	53
Bismarck, ND	8	25
Boise, ID	29	41
Boston, MA	29	38

   a. Do a bivariate plot of this data set.
   b. Describe any structure that you see.
   c. Compute the correlation coefficient.
   d. Find the regression equation.
   e. Plot the regression line.
   f. Describe the relationship between the two variables, in words.
8. Consider the data set in Tables 15.19a and 15.19b, consisting of the speed (in mi/hr) of the winner of the Indianapolis 500 race, for two time periods.[15]
   a. Plot this data set.
   b. Describe any structure that you see.

Table 15.19a

YEAR	SPEED
1967	151
1968	152
1969	157
1970	156
1971	158
1972	163
1973	159
1974	159

Table 15.19b

YEAR	SPEED
1986	171
1987	162
1988	145
1989	168
1990	186
1991	176
1992	134
1993	157

c. Compute the correlation coefficients for the two time periods, and for the data as a whole.

d. Find the regression equation for each time period, and for the data as a whole.

e. Plot the regression lines.

f. Describe the relationship between the two variables, in words, for the two time periods and for the data as a whole.

g. Speculate on what is going on here.

9. Table 15.20 shows some yearly revenues (money received in billions of dollars) and the number of telephones (in millions) for 10 of the 21 companies that were divested by AT&T as part of a court settlement around early 1982.[16]

Table 15.20

	TELEPHONES	REVENUES
Bell Telephone of PA	8.0	1.97
Chesapeake and Potomac Telephone (MD)	3.6	.96
Diamond State Telephone	.5	.14
Indiana Bell Telephone	2.7	.75
Mountain States Telephone	7.9	2.69
New Jersey Bell Telephone	6.6	1.91
Northwestern Bell Telephone	5.7	1.82
Pacific Northwest Bell Telephone	3.9	1.39
South Central Bell Telephone	10.6	3.58
Southwestern Bell Telephone	16.7	5.84

a. Would it be more natural to think of the number of telephones as predicting revenues or vice versa? Why?

b. Plot this data set.

c. What kind of structure do you see?

10. Consider the hypothetical data in Table 15.21 (page 572) from an experiment conducted at a brewery to determine the best aging time for a particular kind of beer.

Table 15.21

AGING TIME	QUALITY OF BREW
1	10.2
2	12.6
3	14.3
4	15.1
5	15.0
6	15.3
7	14.5
8	13.6
9	12.8

a. What general kind of experiment is this?

b. Which variable should be X and which should be Y? Why?

c. Plot the data.

d. What kind of structure do you see?

11. Consider the data set shown in Table 15.22, which reports the verbal and quantitative scores of a group of students.

Table 15.22

VERBAL ABILITY	QUANTITATIVE ABILITY
536	541
625	613
473	495
721	726
753	732
597	625
462	473
530	799
675	693

a. Plot the data and describe the structure that you see.

b. Complete a box plot of the verbal abilities, treating this column as a single group of numbers.

c. Do a box plot of the quantitative abilities.

d. Given that your answers to parts (b) and (c) showed no outliers when just one kind of score is considered, can you conclude that there are no outliers in the bivariate data set?

12. Tape recordings of bird songs have been analyzed as part of a zoology project, and some of the (hypothetical) data appears in Table 15.23.[17]

a. Plot the data.

b. Does it look like one cohesive group? If not, how would you describe its structure?

Table 15.23

LENGTH OF BIRD SONG	AVERAGE PITCH OR FREQUENCY
8.3	173
4.6	493
4.5	503
8.1	133
2.3	255
2.5	201
9.1	152
3.9	472
1.6	162
1.8	193
4.7	418
4.9	403
7.9	151

c. Given that birds use different kinds of songs for different purposes, can you explain why a data set like this one would be reasonable?

13. Consider the data shown in Table 15.24 on motor vehicle traffic deaths in 1979 and 1980 for northwestern states.[18]

Table 15.24

STATE	1979	1980
Washington	1,032	985
Oregon	675	646
Idaho	333	329
Montana	332	325
Wyoming	244	244
Nevada	365	346

a. Do a bivariate plot of this data set and describe its structure.

b. Compute the correlation coefficient.

c. Interpret this correlation coefficient. In particular, does it correspond to the plot of the data?

d. If we think of this data set as a random sample from an abstract population, is the correlation significant? Interpret your answer.

e. Find the equation of the least-squares line for this data set.

f. Interpret this line as a way of predicting fatalities in 1980 based on their levels in 1979.

g. Plot the data and draw the least-squares line. Describe what you see.

h. Find the predicted values for all states.

i. Compute the residuals from the least-squares line.

j. Do a residual plot and comment on what you see.

14. A hypothetical manufacturing plant has gathered data on weekly production, as shown in Table 15.25.

Table 15.25

NUMBER OF STEREOS PRODUCED	COST OF OPERATIONS
125	18,900
193	28,500
223	30,000
103	16,200
150	22,300

a. Find the least-squares line that predicts the cost of producing a given number of stereos.

b. Estimate the fixed costs of this plant (this is the *Y*-intercept).

c. Estimate the marginal unit costs of producing an extra stereo.

d. What is the predicted cost of producing 135 stereos?

15. Reconsider the data from Problem 9 on the number of telephones and the revenues of some telephone companies.

a. Find the correlation coefficient.

b. Find a prediction equation that estimates revenues based on the number of telephones.

c. What revenues would you expect from a telephone company with 5 million telephones?

16. Reconsider the data from Problem 10, on the quality of brew as a function of aging time.

a. What is the correlation coefficient?

b. Is it significant?

c. Is there an association between aging time and quality of brew? Why or why not? (Be careful! You may want to look at a plot of the data to be sure.)

17. Reconsider the data on chocolate milk shakes presented in Table 15.17. Do a correlation and regression analysis of the relationship between grams of fat (GRAMSFAT) and total calories (CAL). By "do a correlation and regression analysis," we mean:

a. Plot the data.

b. Calculate the correlation coefficient.

c. Test the correlation coefficient for statistical significance.

d. Find the formula for the best-fitting prediction line.

e. Construct a plot of the residuals.

You may want to use a computer.

18. For the data in Table 15.17:

a. Calculate a new variable (perhaps called UnitPric) giving the cost per ounce of the various shakes.

b. Do a correlation and regression analysis of the relationship between unit price and percent of calories as fat. (See Problem 17 for an outline of the analysis.)

19. For the data in Table 15.17, do a correlation and regression analysis of the relationship between total calories and percent of calories as fat (Fat%Cal). (See Problem 17 for an outline of the analysis.)

20. Tournament chess players are assigned ratings based on the difficulty of the tournaments they have won. Under the United States Chess Federation rating system, an estimated "average" player would be assigned a rating of 800. A rating of 2,200 makes one a U.S. Master: Bobby Fisher rated 2,785, and in 1990 Gary Kasparov, the world champion, rated about 2,900.

Computer chess programs have steadily been getting better. The top rankings earned by computers in various years are given in Table 15.26.[19] A plot of these results is given in Figure 15.46 and the results of a regression run are given in Display 15.10 (page 576).

a. What is the correlation between year and rating?

b. Is the correlation statistically significant?

c. What is the regression equation predicting rating from year?

d. Using the regression equation, what rating is predicted for a computer program in 1993?

e. Using the regression equation, what rating is predicted for a computer program in 1997?

Table 15.26 *Top computer chess rankings, by year*

YEAR	RATING
67	1,640
74	1,730
77	2,136
77	2,271
80	2,168
81	2,258
83	2,363
85	2,530
88	2,745

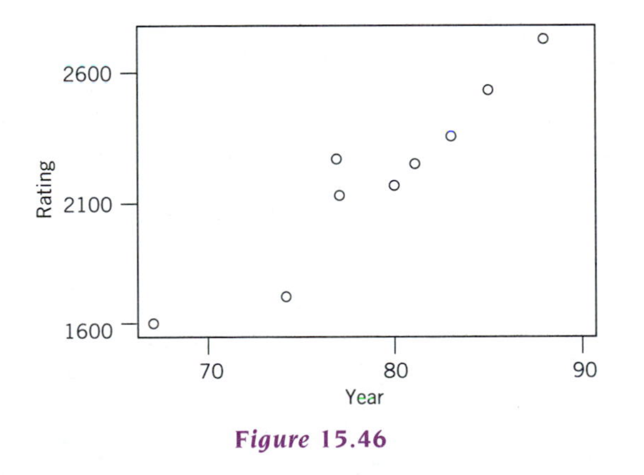

Figure 15.46

21. Molly Martin[20] assembled a group of ten friends and asked them to sample and rate 25 "energy bars." The ratings were summarized as thumbs up, so-so, and thumbs down. She then compiled the ratings with nutritional information, as shown in Table 15.27 (page 576).[21] Explore these data, formulate three (or more) hypotheses about variables that influence rating, and formally test those hypotheses.

DISPLAY 15.10

```
MTB > Retrieve 'CHESS.MTW';

MTB > Correlation 'Year' 'Rating'.

Correlation of Year and Rating = 0.949

MTB > Regress 'Rating' 1 'Year'.

The regression equation is
Rating = -1984 + 52.9 Year

PREDICTOR     COEF       STDEV     T-RATIO    P
Constant      -1984.1    527.8     -3.76      0.007
Year          52.947     6.653     7.96       0.000

s = 118.1   R-sq = 90.0%   R-sq(adj) = 88.6%

Analysis of Variance

SOURCE        DF    SS        MS        F        P
Regression    1     882757    882757    63.33    0.000
Error         7     97575     13939
Total         8     980332
```

Table 15.27 *Ratings and nutritional information for energy bars*

RATING	BRAND	COST	OUNCES	CALO-RIES	PRO-TEIN	CARBOS	FAT	CHOCO-LATE?
SS	Annie's Anytime Bar	.75	1.75	245	9	27	12	N
TD	Balance	2.75	1.65	180	14	18	7	N
TU	Bear Valley Pemmican	.99	3.75	420	17	59	13	N
TD	BTU Stoker	1.69	2.6	252	10	46	3	Y
TU	Clif Bar	1.50	2.4	250	6	50	2	N
SS	Edge Bar	1.50	2.5	234	10	44	<2	Y
TD	Fast Burners	1.69	2.0	155	5	27	2.5	N
SS	Gatorbar	1.69	1.17	110	1	25	1	N
TD	Glenny's Brown Rice	.95	1.14	120	1	28	1	N
SS	Hard Body Energy Bar	1.69	2.5	300	9	48	9	N
TU	Hoffman's Energy Bar	2.75	2	296	7	34	16	N
TD	Hot Stuff Fitness Bar	1.98	2.5	270	15	46	2	N
TD	Meal Pack	.99	3.75	410	17	57	13	N
TD	Nookies	.75	1.15	138	2	18	3	N
TD	PowerBar	1.49	2.25	225	10	42	2	N
TD	Pro-Amino	1.52	2.75	296	18	40	7	N
TD	ProSports Performance	1.69	2.25	240	15	40	<2	Y
SS	Pure Power Energy Bar	1.69	2.36	240	12	42	3	Y
TD	Spirulina Bee Bars	1.14	2.0	200	6	37	3	Y
SS	Steel Bar	1.75	3.0	368	16	68	4	N
SS	Super Cuts	1.69	2.0	155	5	27	2.5	N
TD	Tiger Sport	1.15	2.3	130	11	40	2	N
TD	TwinLab Amino Fuel	1.49	2.75	285	15	50	3	Y
TD	XTRNR	1.69	2.25	245	11	46	2	N
TU	Yohimbe Bar	1.69	2.0	225	6	37	4	Y

Note: Protein, carbohydrates, and fat in grams. "Chocolate?" is yes or no to "contains chocolate," even if not exclusively chocolate. (Spirulina Bee Bars contain carob and are classified as chocolate flavor here.)

22. Table 15.28 gives the 1991 population, in thousands, and the number of state employees, also in thousands, for the 50 states.[22] Explore these data and describe the relationship between the two variables.

Table 15.28 *Number of state employees and population for the 50 states, in 1991*

STATE	POPULATION (in thousands)	EMPLOYEES (in thousands)	STATE	POPULATION (in thousands)	EMPLOYEES (in thousands)
Alabama	4,091	82	Montana	809	17
Alaska	570	22	Nebraska	1,593	29
Arizona	3,748	51	Nevada	1,283	19
Arkansas	2,373	43	New Hampshire	1,104	16
California	30,380	325	New Jersey	7,753	113
Colorado	3,378	53	New Mexico	1,549	41
Connecticut	3,289	58	New York	18,055	269
Delaware	680	20	North Carolina	6,736	108
Florida	13,266	163	North Dakota	635	15
Georgia	6,623	112	Ohio	10,941	141
Hawaii	1,137	51	Oklahoma	3,175	68
Idaho	1,040	18	Oregon	2,922	55
Illinois	11,541	141	Pennsylvania	11,958	124
Indiana	5,610	90	Rhode Island	1,005	20
Iowa	2,795	56	South Carolina	3,560	81
Kansas	2,495	49	South Dakota	704	13
Kentucky	3,713	76	Tennessee	4,953	76
Louisiana	4,254	88	Texas	17,348	228
Maine	1,234	21	Utah	1,770	39
Maryland	4,859	87	Vermont	567	13
Massachusetts	5,996	88	Virginia	6,280	114
Michigan	9,380	139	Washington	5,012	96
Minnesota	4,432	66	West Virginia	1,803	34
Mississippi	2,593	47	Wisconsin	4,956	69
Missouri	5,157	74	Wyoming	460	11

23. Using Table 2.20, data file CAL_VT.MTW, find the correlation between calories per week used in exercise and change in ventilatory threshold. Does it appear that exercise increases ventilatory threshold? Based on these data, is the change statistically significant?

24. Using the data in Table 2.22, data file HEART.MTW, describe the relationship between heart fill rate and oxygen utilization capacity in younger men, in older men, and both groups combined. Are there apparently different relationships? Try to make sense of the patterns you see. What do you think is real? Why?

25. As you pondered whale ecology at the beach one day, you speculated that younger whales may have different feeding requirements than older whales, hence their mouths may make up a different relative proportion of their total length. Using the data from WHALES.MTW, check whether there is a correlation between total length and relative mouth size.

26. Suppose a random sample of girls produced the data shown in Table 15.29 (page 578). Calculate the correlation coefficient and the regression line predicting height from age. Do these appear to be substantially different from what we obtained in Example 15.8? If yes, explain the differences.

Table 15.29 *Age and height of a random sample of girls*

AGE (years)	HEIGHT (inches)
2	36.6
3	39.6
4	43.0
5	45.1
6	49.0
7	48.8
8	53.0
9	60.1
10	56.7
11	58.9

27. Table 6.11 and data file STATES.MTW contain data on the density of lawyers (nonlawyers per lawyer) and on the highest point in feet of each of the 50 states and the District of Columbia. Analyze and discuss the relationship between lawyer density and the highest point in the state or district.

28. It may be that individuals who actively seek information about new situations adjust to them more quickly than those who wait for things to be explained to them. For instance, Morrison (1993)[23] studied 135 new accountants and found that those who most frequently sought specific information were quicker to learn their job and to become socially integrated.

Table 15.30 *Frequency with which new employees ask questions, their job ratings, and number of people in work group who "like" them* (Pseudodata modeled after Morrison, 1993)

ASKS	RATING	LIKES
0	4	5
0	3	6
0	2	2
1	1	5
2	2	8
3	3	3
4	2	5
3	3	8
10	3	7
15	2	10
20	4	9
13	4	8
8	3	5
7	2	7
6	1	4

Rating: 4 = outstanding, 3 = excellent, 2 = good, 1 = needs improvement.

Suppose you wish to replicate this finding. You follow a group of 15 new research interviewers through their training and first 3 months on the job, noting (on a sample basis) how often each asks a job-skill-related question. At the end of 3 months you review their supervisor's evaluations to get a measure of job competence. You also have each person in the group rate all the others as someone they like, do not like, or do not know. You use each individual's total number of "likes" as an indication of that individual's integration into the group. Your results are in Table 15.30 (data file INFOSEEK.MTW). Do these data support the hypothesis?

29. Table 5.10, data file CELLS.MTW, contains information on the numbers of cells in growth solutions over 13 days. Treat the number of days as the predictor variable and find formulas predicting the number of cells. Comment on the adequacy of these predictor equations, using appropriate graphics.

Appendices

A Probabilities for a Standard Normal Random Variable

Probability that a standard normal random variable will be less than a given value. Special probabilities are indicated by double asterisks

VALUE	PROBABILITY	VALUE	PROBABILITY
.00	.500**	−.00	.500**
.01	.504	−.01	.496
.02	.508	−.02	.492
.03	.512	−.03	.488
.04	.516	−.04	.484
.05	.520	−.05	.480
.06	.524	−.06	.476
.07	.528	−.07	.472
.08	.532	−.08	.468
.09	.536	−.09	.464
.10	.540	−.10	.460
.11	.544	−.11	.456
.12	.548	−.12	.452
.13	.552	−.13	.448
.14	.556	−.14	.444
.15	.560	−.15	.440
.16	.564	−.16	.436
.17	.567	−.17	.433
.18	.571	−.18	.429
.19	.575	−.19	.425
.20	.579	−.20	.421
.21	.583	−.21	.417
.22	.587	−.22	.413
.23	.591	−.23	.409
.24	.595	−.24	.405
.25	.599	−.25	.401
.26	.603	−.26	.397
.27	.606	−.27	.394
.28	.610	−.28	.390
.29	.614	−.29	.386
.30	.618	−.30	.382
.31	.622	−.31	.378
.32	.626	−.32	.374
.33	.629	−.33	.371
.34	.633	−.34	.367
.35	.637	−.35	.363
.36	.641	−.36	.359
.37	.644	−.37	.356
.38	.648	−.38	.352
.39	.652	−.39	.348
.40	.655	−.40	.345
.41	.659	−.41	.341
.42	.663	−.42	.337
.43	.666	−.43	.334
.44	.670	−.44	.330
.45	.674	−.45	.326

(continued)

Continued

VALUE	PROBABILITY	VALUE	PROBABILITY
.46	.677	−.46	.323
.47	.681	−.47	.319
.48	.684	−.48	.316
.49	.688	−.49	.312
.50	.691	−.50	.309
.51	.695	−.51	.305
.52	.698	−.52	.302
.53	.702	−.53	.298
.54	.705	−.54	.295
.55	.709	−.55	.291
.56	.712	−.56	.288
.57	.716	−.57	.284
.58	.719	−.58	.281
.59	.722	−.59	.278
.60	.726	−.60	.274
.61	.729	−.61	.271
.62	.732	−.62	.268
.63	.736	−.63	.264
.64	.739	−.64	.261
.65	.742	−.65	.258
.66	.745	−.66	.255
.67	.749	−.67	.251
.674	.750**	−.674	.250**
.68	.752	−.68	.248
.69	.755	−.69	.245
.70	.758	−.70	.242
.71	.761	−.71	.239
.72	.764	−.72	.236
.73	.767	−.73	.233
.74	.770	−.74	.230
.75	.773	−.75	.227
.76	.776	−.76	.224
.77	.779	−.77	.221
.78	.782	−.78	.218
.79	.785	−.79	.215
.80	.788	−.80	.212
.81	.791	−.81	.209
.82	.794	−.82	.206
.83	.797	−.83	.203
.84	.800	−.84	.200
.85	.802	−.85	.198
.86	.805	−.86	.195
.87	.808	−.87	.192
.88	.811	−.88	.189
.89	.813	−.89	.187
.90	.816	−.90	.184
.91	.819	−.91	.181
.92	.821	−.92	.179
.93	.824	−.93	.176
.94	.826	−.94	.174
.95	.829	−.95	.171

Continued

VALUE	PROBABILITY	VALUE	PROBABILITY
.96	.831	−.96	.169
.97	.834	−.97	.166
.98	.836	−.98	.164
.99	.839	−.99	.161
1.00	.841	−1.00	.159
1.01	.844	−1.01	.156
1.02	.846	−1.02	.154
1.03	.848	−1.03	.152
1.04	.851	−1.04	.149
1.05	.853	−1.05	.147
1.06	.855	−1.06	.145
1.07	.858	−1.07	.142
1.08	.860	−1.08	.140
1.09	.862	−1.09	.138
1.10	.864	−1.10	.136
1.11	.867	−1.11	.133
1.12	.869	−1.12	.131
1.13	.871	−1.13	.129
1.14	.873	−1.14	.127
1.15	.875	−1.15	.125
1.16	.877	−1.16	.123
1.17	.879	−1.17	.121
1.18	.881	−1.18	.119
1.19	.883	−1.19	.117
1.20	.885	−1.20	.115
1.21	.887	−1.21	.113
1.22	.889	−1.22	.111
1.23	.891	−1.23	.109
1.24	.893	−1.24	.107
1.25	.894	−1.25	.106
1.26	.896	−1.26	.104
1.27	.898	−1.27	.102
1.28	.900	−1.28	.100
1.282	.900**	−1.282	.100**
1.29	.901	−1.29	.099
1.30	.903	−1.30	.097
1.31	.905	−1.31	.095
1.32	.907	−1.32	.093
1.33	.908	−1.33	.092
1.34	.910	−1.34	.090
1.35	.911	−1.35	.089
1.36	.913	−1.36	.087
1.37	.915	−1.37	.085
1.38	.916	−1.38	.084
1.39	.918	−1.39	.082
1.40	.919	−1.40	.081
1.41	.921	−1.41	.079
1.42	.922	−1.42	.078
1.43	.924	−1.43	.076
1.44	.925	−1.44	.075
1.45	.926	−1.45	.074

(continued)

Continued

VALUE	PROBABILITY	VALUE	PROBABILITY
1.46	.928	− 1.46	.072
1.47	.929	− 1.47	.071
1.48	.931	− 1.48	.069
1.49	.932	− 1.49	.068
1.50	.933	− 1.50	.067
1.51	.934	− 1.51	.066
1.52	.936	− 1.52	.064
1.53	.937	− 1.53	.063
1.54	.938	− 1.54	.062
1.55	.939	− 1.55	.061
1.56	.941	− 1.56	.059
1.57	.942	− 1.57	.058
1.58	.943	− 1.58	.057
1.59	.944	− 1.59	.056
1.60	.945	− 1.60	.055
1.61	.946	− 1.61	.054
1.62	.947	− 1.62	.053
1.63	.948	− 1.63	.052
1.64	.949	− 1.64	.051
1.645	.950**	− 1.645	.050**
1.65	.951	− 1.65	.049
1.66	.952	− 1.66	.048
1.67	.953	− 1.67	.047
1.68	.954	− 1.68	.046
1.69	.954	− 1.69	.046
1.70	.955	− 1.70	.045
1.71	.956	− 1.71	.044
1.72	.957	− 1.72	.043
1.73	.958	− 1.73	.042
1.74	.959	− 1.74	.041
1.75	.960	− 1.75	.040
1.76	.961	− 1.76	.039
1.77	.962	− 1.77	.038
1.78	.962	− 1.78	.038
1.79	.963	− 1.79	.037
1.80	.964	− 1.80	.036
1.81	.965	− 1.81	.035
1.82	.966	− 1.82	.034
1.83	.966	− 1.83	.034
1.84	.967	− 1.84	.033
1.85	.968	− 1.85	.032
1.86	.969	− 1.86	.031
1.87	.969	− 1.87	.031
1.88	.970	− 1.88	.030
1.89	.971	− 1.89	.029
1.90	.971	− 1.90	.029
1.91	.972	− 1.91	.028
1.92	.973	− 1.92	.027
1.93	.973	− 1.93	.027
1.94	.974	− 1.94	.026
1.95	.974	− 1.95	.026

Continued

VALUE	PROBABILITY	VALUE	PROBABILITY
1.960	.975**	−1.960	.025**
1.97	.976	−1.97	.024
1.98	.976	−1.98	.024
1.99	.977	−1.99	.023
2.00	.977	−2.00	.023
2.326	.990**	−2.326	.010**
2.500	.994	−2.500	.006
2.576	.995**	−2.576	.005**
3.000	.9987	−3.000	.0013
3.090	.9990**	−3.090	.0010**
3.291	.9995**	−3.291	.0005**
3.500	.9998	−3.500	.0002
4.000	.99997	−4.000	.00003

B Random Uniform Numbers Between 0 and 1

Table B.1 *Random uniform numbers between 0 and 1*

.4515	.8189	.7099	.7394	.3260	.6457	.0190	.7942	.9487	.4110
.9048	.6239	.5973	.3555	.4236	.8021	.0872	.1237	.4427	.9700
.9431	.4670	.0624	.2899	.0712	.4469	.5378	.3983	.9310	.1380
.7005	.8589	.2070	.8313	.8808	.1156	.9805	.7372	.3293	.9461
.4020	.1886	.1391	.4144	.8624	.3443	.6671	.3245	.6654	.3916
.4566	.1733	.1991	.2501	.6586	.6034	.6079	.7740	.9538	.3567
.9646	.4560	.9749	.7274	.9419	.6301	.2279	.5151	.9845	.1747
.5140	.4777	.9811	.7156	.8522	.2848	.4360	.3295	.9003	.9709
.1410	.7468	.8771	.8881	.3476	.0549	.9846	.1598	.4385	.9012
.3293	.6333	.0997	.1692	.3445	.3331	.9922	.3170	.7120	.4144
.9240	.8098	.7841	.6155	.5221	.5935	.5052	.2117	.4040	.6627
.8340	.4676	.8499	.2908	.6652	.5911	.9745	.7332	.2515	.1858
.3988	.4301	.1251	.2046	.4221	.7103	.5202	.4026	.6937	.9935
.2943	.2901	.3849	.4940	.3505	.5863	.8608	.4238	.5590	.1716
.8524	.6954	.1793	.5413	.1280	.1252	.7819	.7587	.6658	.1364
.6701	.9079	.3249	.9623	.8021	.0472	.0179	.3496	.2128	.6622
.3824	.7611	.9355	.6188	.9531	.0790	.4241	.6165	.8332	.8083
.4745	.1411	.8671	.7793	.6470	.9186	.3114	.3515	.1879	.8676
.0074	.1160	.4514	.5527	.0527	.0994	.8213	.1378	.9730	.4659
.1334	.8388	.3943	.7215	.5967	.8771	.2145	.4163	.2159	.9326
.2753	.3483	.7127	.0968	.5340	.2658	.9482	.3260	.7358	.0713
.8444	.3943	.5852	.5238	.8064	.3468	.4180	.4428	.6437	.4528
.3733	.4106	.9928	.6587	.0653	.9761	.2387	.5414	.7552	.3575
.9367	.7613	.0229	.6455	.7345	.0153	.9392	.5066	.7937	.4901
.7373	.3861	.7123	.8154	.9852	.2112	.9459	.5168	.2569	.5941
.7301	.9688	.3048	.5353	.5021	.8486	.7313	.6266	.8226	.0884
.1610	.7540	.7977	.3291	.9120	.3014	.0008	.8730	.1684	.8532
.8945	.1375	.6518	.3976	.7523	.6075	.7933	.0298	.0809	.2891
.4053	.1550	.5415	.6625	.8533	.1917	.3721	.1350	.7283	.7745
.3092	.9946	.0179	.4575	.0096	.8625	.7786	.9177	.9725	.6467

C *t*-Tables

Table C.1 *t-table for 80% confidence intervals: Two-sided 20% critical values (also one-sided 10% critical values for 90% one-sided confidence intervals)*

DEGREES OF FREEDOM	CRITICAL VALUE	DEGREES OF FREEDOM	CRITICAL VALUE	DEGREES OF FREEDOM	CRITICAL VALUE
1	3.078	11	1.363	21	1.323
2	1.886	12	1.356	22	1.321
3	1.638	13	1.350	23	1.319
4	1.533	14	1.345	24	1.318
5	1.476	15	1.341	25	1.316
6	1.440	16	1.337	26	1.315
7	1.415	17	1.333	27	1.314
8	1.397	18	1.330	28	1.313
9	1.383	19	1.328	29	1.311
10	1.372	20	1.325	30	1.310
				40	1.303
				50	1.299
				60	1.296
				120	1.289
				Infinity	1.282

Table C.2 *t-table for 90% confidence intervals: Two-sided 10% critical values (also one-sided 5% critical values for 95% one-sided confidence intervals)*

DEGREES OF FREEDOM	CRITICAL VALUE	DEGREES OF FREEDOM	CRITICAL VALUE	DEGREES OF FREEDOM	CRITICAL VALUE
1	6.314	11	1.796	21	1.721
2	2.920	12	1.782	22	1.717
3	2.353	13	1.771	23	1.714
4	2.132	14	1.761	24	1.711
5	2.015	15	1.753	25	1.708
6	1.943	16	1.746	26	1.706
7	1.895	17	1.740	27	1.703
8	1.860	18	1.734	28	1.701
9	1.833	19	1.729	29	1.699
10	1.812	20	1.725	30	1.697
				40	1.684
				50	1.676
				60	1.671
				120	1.658
				Infinity	1.645

Table C.3 *t-table for 95% confidence intervals: Two-sided 5% critical values (also one-sided 2.5% critical values for 97.5% one-sided confidence intervals)*

DEGREES OF FREEDOM	CRITICAL VALUE	DEGREES OF FREEDOM	CRITICAL VALUE	DEGREES OF FREEDOM	CRITICAL VALUE
1	12.706	11	2.201	21	2.080
2	4.302	12	2.179	22	2.074
3	3.182	13	2.160	23	2.069
4	2.776	14	2.145	24	2.064
5	2.571	15	2.131	25	2.060
6	2.447	16	2.120	26	2.056
7	2.365	17	2.110	27	2.052
8	2.306	18	2.101	28	2.048
9	2.262	19	2.093	29	2.045
10	2.228	20	2.086	30	2.042
				40	2.021
				50	2.009
				60	2.000
				120	1.980
				Infinity	1.960

Table C.4 *t-table for 98% confidence intervals: Two-sided 2% critical values (also one-sided 1% critical values for 99% one-sided confidence intervals)*

DEGREES OF FREEDOM	CRITICAL VALUE	DEGREES OF FREEDOM	CRITICAL VALUE	DEGREES OF FREEDOM	CRITICAL VALUE
1	31.820	11	2.718	21	2.517
2	6.964	12	2.681	22	2.508
3	4.540	13	2.650	23	2.500
4	3.747	14	2.624	24	2.492
5	3.365	15	2.602	25	2.485
6	3.143	16	2.583	26	2.478
7	2.998	17	2.567	27	2.473
8	2.896	18	2.552	28	2.467
9	2.821	19	2.539	29	2.462
10	2.764	20	2.528	30	2.457
				40	2.423
				50	2.403
				60	2.390
				120	2.358
				Infinity	2.326

Table C.5 *t-table for 99% confidence intervals: Two-sided 1% critical values (also one-sided .5% critical values for 99.5% one-sided confidence intervals)*

DEGREES OF FREEDOM	CRITICAL VALUE	DEGREES OF FREEDOM	CRITICAL VALUE	DEGREES OF FREEDOM	CRITICAL VALUE
1	63.605	11	3.106	21	2.831
2	9.920	12	3.055	22	2.818
3	5.840	13	3.012	23	2.807
4	4.604	14	2.977	24	2.797
5	4.031	15	2.947	25	2.787
6	3.707	16	2.920	26	2.779
7	3.499	17	2.898	27	2.770
8	3.355	18	2.878	28	2.763
9	3.250	19	2.861	29	2.756
10	3.169	20	2.845	30	2.750
				40	2.704
				50	2.678
				60	2.660
				120	2.617
				Infinity	2.576

Table C.6 *t-table for 99.8% confidence intervals: Two-sided .2% critical values (also one-sided .1% critical values for 99.9% one-sided confidence intervals)*

DEGREES OF FREEDOM	CRITICAL VALUE	DEGREES OF FREEDOM	CRITICAL VALUE	DEGREES OF FREEDOM	CRITICAL VALUE
1	317.781	11	4.024	21	3.526
2	22.296	12	3.929	22	3.504
3	10.212	13	3.852	23	3.484
4	7.170	14	3.787	24	3.466
5	5.893	15	3.733	25	3.449
6	5.203	16	3.684	26	3.434
7	4.783	17	3.644	27	3.420
8	4.500	18	3.609	28	3.407
9	4.296	19	3.578	29	3.395
10	4.143	20	3.550	30	3.384
				40	3.306
				50	3.261
				60	3.231
				120	3.159
				Infinity	3.090

Table C.7 *t-table for 99.9% confidence intervals: Two-sided .1% critical values (also one-sided .05% critical values for 99.95% one-sided confidence intervals)*

DEGREES OF FREEDOM	CRITICAL VALUE	DEGREES OF FREEDOM	CRITICAL VALUE	DEGREES OF FREEDOM	CRITICAL VALUE
1	636.012	11	4.437	21	3.818
2	31.463	12	4.312	22	3.791
3	12.915	13	4.216	23	3.766
4	8.602	14	4.137	24	3.744
5	6.868	15	4.070	25	3.724
6	5.952	16	4.012	26	3.706
7	5.405	17	3.963	27	3.689
8	5.040	18	3.920	28	3.673
9	4.780	19	3.882	29	3.658
10	4.586	20	3.848	30	3.645
				40	3.550
				50	3.496
				60	3.460
				120	3.373
				Infinity	3.291

D F-Tables

Table D.1 F-table: 10% critical values

"BETWEEN GROUPS" DEGREES OF FREEDOM (number of groups minus 1)	"WITHIN GROUPS" DEGREES OF FREEDOM (the total sample size for all groups, less the number of groups)							
	5	10	15	20	30	60	120	INFINITY
1	4.06	3.28	3.07	2.97	2.88	2.79	2.75	2.71
2	3.78	2.92	2.70	2.59	2.49	2.39	2.35	2.30
3	3.62	2.73	2.49	2.38	2.28	2.18	2.13	2.08
4	3.52	2.61	2.36	2.25	2.14	2.04	1.99	1.94
5	3.45	2.52	2.27	2.16	2.05	1.95	1.90	1.85
6	3.40	2.46	2.21	2.09	1.98	1.87	1.82	1.77
7	3.37	2.41	2.16	2.04	1.93	1.82	1.77	1.72
8	3.34	2.38	2.12	2.00	1.88	1.77	1.72	1.67
9	3.32	2.35	2.09	1.96	1.85	1.74	1.68	1.63
10	3.30	2.32	2.06	1.94	1.82	1.71	1.65	1.60
11	3.28	2.30	2.04	1.91	1.79	1.68	1.63	1.57
12	3.27	2.28	2.02	1.89	1.77	1.66	1.60	1.55
13	3.26	2.27	2.00	1.87	1.75	1.64	1.58	1.52
14	3.25	2.26	1.99	1.86	1.74	1.62	1.56	1.50
15	3.24	2.24	1.97	1.84	1.72	1.60	1.55	1.49
16	3.23	2.23	1.96	1.83	1.71	1.59	1.53	1.47
17	3.22	2.22	1.95	1.82	1.70	1.58	1.52	1.46
18	3.22	2.22	1.94	1.81	1.69	1.56	1.50	1.44
19	3.21	2.21	1.93	1.80	1.68	1.55	1.49	1.43
20	3.21	2.20	1.92	1.79	1.67	1.54	1.48	1.42
25	3.19	2.17	1.89	1.76	1.63	1.50	1.44	1.38
30	3.17	2.16	1.87	1.74	1.61	1.48	1.41	1.34
60	3.14	2.11	1.82	1.68	1.54	1.40	1.32	1.24
Infinity	3.10	2.06	1.76	1.61	1.46	1.29	1.19	1.00

Table D.2 *F-table: 5% critical values*

"BETWEEN GROUPS" DEGREES OF FREEDOM (number of groups minus 1)	"WITHIN GROUPS" DEGREES OF FREEDOM (the total sample size for all groups, less the number of groups)							
	5	10	15	20	30	60	120	INFINITY
1	6.61	4.96	4.54	4.35	4.17	4.00	3.92	3.84
2	5.79	4.10	3.68	3.49	3.32	3.15	3.07	3.00
3	5.41	3.71	3.29	3.10	2.92	2.76	2.68	2.60
4	5.19	3.48	3.06	2.87	2.69	2.53	2.45	2.37
5	5.05	3.33	2.90	2.71	2.53	2.37	2.29	2.21
6	4.95	3.22	2.79	2.60	2.42	2.25	2.17	2.10
7	4.88	3.14	2.71	2.51	2.33	2.17	2.09	2.01
8	4.82	3.07	2.64	2.45	2.27	2.10	2.02	1.94
9	4.77	3.02	2.59	2.39	2.21	2.04	1.96	1.88
10	4.73	2.98	2.54	2.35	2.16	1.99	1.91	1.83
11	4.70	2.94	2.51	2.31	2.13	1.95	1.87	1.79
12	4.68	2.91	2.48	2.28	2.09	1.92	1.83	1.75
13	4.66	2.89	2.45	2.25	2.06	1.89	1.80	1.72
14	4.64	2.86	2.42	2.22	2.04	1.86	1.78	1.69
15	4.62	2.85	2.40	2.20	2.01	1.84	1.75	1.67
16	4.60	2.83	2.38	2.18	1.99	1.82	1.73	1.64
17	4.59	2.81	2.37	2.17	1.98	1.80	1.71	1.62
18	4.58	2.80	2.35	2.15	1.96	1.78	1.69	1.60
19	4.57	2.79	2.34	2.14	1.95	1.76	1.67	1.59
20	4.56	2.77	2.33	2.12	1.93	1.75	1.66	1.57
25	4.52	2.73	2.28	2.07	1.88	1.69	1.60	1.51
30	4.50	2.70	2.25	2.04	1.84	1.65	1.55	1.46
60	4.43	2.62	2.16	1.95	1.74	1.53	1.43	1.32
Infinity	4.36	2.54	2.07	1.84	1.62	1.39	1.25	1.00

Table D.3 *F-table: 1% critical values*

"BETWEEN GROUPS" DEGREES OF FREEDOM (number of groups minus 1)	"WITHIN GROUPS" DEGREES OF FREEDOM (the total sample size for all groups, less the number of groups)							
	5	10	15	20	30	60	120	INFINITY
1	16.25	10.04	8.68	8.09	7.56	7.08	6.85	6.63
2	13.27	7.56	6.36	5.85	5.39	4.98	4.79	4.60
3	12.06	6.55	5.42	4.94	4.51	4.13	3.95	3.78
4	11.39	5.99	4.89	4.43	4.02	3.65	3.48	3.32
5	10.97	5.64	4.56	4.10	3.70	3.34	3.17	3.02
6	10.67	5.38	4.32	3.87	3.47	3.12	2.96	2.80
7	10.46	5.20	4.14	3.70	3.30	2.95	2.79	2.64
8	10.29	5.06	4.00	3.56	3.17	2.82	2.66	2.51
9	10.16	4.94	3.89	3.46	3.07	2.72	2.56	2.41
10	10.05	4.85	3.80	3.37	2.98	2.63	2.47	2.32
11	9.96	4.77	3.73	3.29	2.91	2.56	2.40	2.25
12	9.89	4.71	3.67	3.23	2.84	2.50	2.34	2.18
13	9.82	4.65	3.61	3.18	2.79	2.44	2.28	2.13
14	9.77	4.60	3.56	3.13	2.74	2.39	2.23	2.08
15	9.72	4.56	3.52	3.09	2.70	2.35	2.19	2.04
16	9.68	4.52	3.49	3.05	2.66	2.31	2.15	2.00
17	9.64	4.49	3.45	3.02	2.63	2.28	2.12	1.97
18	9.61	4.46	3.42	2.99	2.60	2.25	2.09	1.93
19	9.58	4.43	3.40	2.96	2.57	2.22	2.06	1.90
20	9.55	4.40	3.37	2.94	2.55	2.20	2.03	1.88
25	9.45	4.31	3.28	2.84	2.45	2.10	1.93	1.77
30	9.38	4.25	3.21	2.78	2.39	2.03	1.86	1.70
60	9.20	4.08	3.05	2.61	2.21	1.84	1.66	1.47
Infinity	9.02	3.91	2.87	2.42	2.01	1.60	1.38	1.00

Table D.4 *F-table: 0.1% critical values*

"BETWEEN GROUPS" DEGREES OF FREEDOM (number of groups minus 1)	"WITHIN GROUPS" DEGREES OF FREEDOM (the total sample size for all groups, less the number of groups)							
	5	10	15	20	30	60	120	INFINITY
1	47.16	21.03	16.56	14.81	13.29	11.97	11.38	10.83
2	37.10	14.90	11.33	9.95	8.77	7.77	7.31	6.91
3	33.19	12.55	9.33	8.10	7.05	6.17	5.78	5.42
4	31.07	11.26	8.25	7.09	6.12	5.30	4.95	4.62
5	29.74	10.46	7.56	6.46	5.53	4.76	4.42	4.10
6	28.82	9.91	7.09	6.02	5.12	4.37	4.04	3.74
7	28.15	9.50	6.74	5.68	4.82	4.09	3.76	3.47
8	27.64	9.19	6.47	5.43	4.58	3.86	3.55	3.26
9	27.23	8.94	6.25	5.23	4.39	3.68	3.38	3.10
10	26.90	8.74	6.08	5.07	4.24	3.54	3.24	2.96
11	26.63	8.58	5.93	4.94	4.11	3.42	3.12	2.84
12	26.40	8.43	5.81	4.82	4.00	3.31	3.01	2.74
13	26.21	8.31	5.71	4.72	3.91	3.23	2.93	2.66
14	26.04	8.21	5.61	4.64	3.82	3.15	2.85	2.58
15	25.90	8.12	5.53	4.56	3.75	3.08	2.78	2.51
16	25.77	8.04	5.46	4.49	3.69	3.01	2.72	2.45
17	25.66	7.97	5.39	4.43	3.63	2.96	2.67	2.40
18	25.55	7.91	5.34	4.38	3.58	2.91	2.62	2.35
19	25.46	7.85	5.29	4.33	3.53	2.87	2.57	2.31
20	25.38	7.80	5.24	4.29	3.49	2.83	2.53	2.27
25	25.07	7.60	5.07	4.12	3.33	2.67	2.37	2.10
30	24.86	7.46	4.95	4.00	3.22	2.55	2.26	1.99
60	24.32	7.12	4.64	3.70	2.92	2.25	1.95	1.66
Infinity	23.79	6.75	4.31	3.37	2.59	1.89	1.54	1.00

E The Correlation Coefficient

Testing the correlation coefficient

SAMPLE SIZE	CRITICAL VALUES			
	10% LEVEL	5% LEVEL	1% LEVEL	0.1% LEVEL
3	0.9877	0.9969	0.9999	1.0000
4	0.9000	0.9500	0.9900	0.9990
5	0.8054	0.8783	0.9587	0.9911
6	0.7293	0.8114	0.9172	0.9740
7	0.6694	0.7545	0.8745	0.9509
8	0.6215	0.7067	0.8343	0.9247
9	0.5822	0.6664	0.7977	0.8982
10	0.5494	0.6319	0.7646	0.8721
11	0.5214	0.6020	0.7348	0.8470
12	0.4973	0.5760	0.7079	0.8233
13	0.4762	0.5529	0.6835	0.8009
14	0.4575	0.5324	0.6614	0.7796
15	0.4409	0.5140	0.6411	0.7600
16	0.4259	0.4973	0.6226	0.7416
17	0.4124	0.4821	0.6055	0.7244
18	0.4000	0.4683	0.5897	0.7082
19	0.3887	0.4555	0.5750	0.6929
20	0.3783	0.4438	0.5614	0.6786
21	0.3687	0.4329	0.5487	0.6650
22	0.3598	0.4227	0.5368	0.6522
23	0.3515	0.4132	0.5256	0.6401
24	0.3438	0.4044	0.5151	0.6286
25	0.3365	0.3961	0.5051	0.6176
26	0.3297	0.3882	0.4958	0.6072
27	0.3233	0.3809	0.4869	0.5973
28	0.3172	0.3739	0.4785	0.5879
29	0.3115	0.3673	0.4705	0.5788
30	0.3061	0.3610	0.4629	0.5702
35	0.2826	0.3338	0.4296	0.5321
40	0.2638	0.3120	0.4026	0.5006
45	0.2483	0.2940	0.3801	0.4741
50	0.2353	0.2787	0.3610	0.4514
55	0.2241	0.2656	0.3445	0.4316
60	0.2144	0.2542	0.3301	0.4142
65	0.2058	0.2441	0.3173	0.3988
70	0.1982	0.2352	0.3059	0.3849
75	0.1914	0.2272	0.2957	0.3724
80	0.1852	0.2199	0.2864	0.3610
85	0.1796	0.2133	0.2780	0.3507
90	0.1745	0.2072	0.2702	0.3411
95	0.1698	0.2017	0.2631	0.3323
100	0.1654	0.1966	0.2565	0.3242
multiplier	1.645	1.960	2.576	3.291

For sample sizes not listed within this table, a conservative test may be used by selecting the critical value for the next smaller size listed. This is conservative in the sense that the chance of a Type I error is even less than the specified level (e.g. 5%).

For larger sample sizes, an approximate critical value may be found by taking the square root of the sample size, finding its reciprocal, and multiplying by the appropriate multipler at the bottom of the column.

F Chi-square Tables

Chi-square table: 10% critical values

DEGREES OF FREEDOM	CRITICAL VALUE	DEGREES OF FREEDOM	CRITICAL VALUE
1	2.706	51	64.30
2	4.605	52	65.42
3	6.251	53	66.55
4	7.779	54	67.67
5	9.236	55	68.80
6	10.64	56	69.92
7	12.02	57	71.04
8	13.36	58	72.16
9	14.68	59	73.28
10	15.99	60	74.40
11	17.27	61	75.51
12	18.55	62	76.63
13	19.81	63	77.75
14	21.06	64	78.86
15	22.31	65	79.97
16	23.54	66	81.09
17	24.77	67	82.20
18	25.99	68	83.31
19	27.20	69	84.42
20	28.41	70	85.53
21	29.61	71	86.64
22	30.81	72	87.74
23	32.01	73	88.85
24	33.20	74	89.96
25	34.38	75	91.06
26	35.56	76	92.17
27	36.74	77	93.27
28	37.92	78	94.37
29	39.09	79	95.48
30	40.26	80	96.58
31	41.42	81	97.68
32	42.58	82	98.78
33	43.74	83	99.88
34	44.90	84	101.0
35	46.06	85	102.1
36	47.21	86	103.2
37	48.36	87	104.3
38	49.51	88	105.4
39	50.66	89	106.5
40	51.80	90	107.6
41	52.95	91	108.7
42	54.09	92	109.8
43	55.23	93	110.8
44	56.37	94	111.9
45	57.50	95	113.0
46	58.64	96	114.1
47	59.77	97	115.2
48	60.91	98	116.3
49	62.04	99	117.4
50	63.17	100	118.5

Chi-square table: 5% critical values

DEGREES OF FREEDOM	CRITICAL VALUE	DEGREES OF FREEDOM	CRITICAL VALUE
1	3.841	51	68.67
2	5.991	52	69.83
3	7.815	53	70.99
4	9.487	54	72.15
5	11.07	55	73.31
6	12.59	56	74.47
7	14.07	57	75.62
8	15.51	58	76.78
9	16.92	59	77.93
10	18.31	60	79.08
11	19.67	61	80.23
12	21.03	62	81.38
13	22.36	63	82.53
14	23.68	64	83.67
15	25.00	65	84.82
16	26.30	66	85.96
17	27.59	67	87.11
18	28.87	68	88.25
19	30.14	69	89.39
20	31.41	70	90.53
21	32.67	71	91.67
22	33.92	72	92.81
23	35.17	73	93.94
24	36.41	74	95.08
25	37.65	75	96.22
26	38.88	76	97.35
27	40.11	77	98.48
28	41.34	78	99.62
29	42.56	79	100.7
30	43.77	80	101.9
31	44.98	81	103.0
32	46.19	82	104.1
33	47.40	83	105.3
34	48.60	84	106.4
35	49.80	85	107.5
36	51.00	86	108.6
37	52.19	87	109.8
38	53.38	88	110.9
39	54.57	89	112.0
40	55.76	90	113.1
41	56.94	91	114.3
42	58.12	92	115.4
43	59.30	93	116.5
44	60.48	94	117.6
45	61.66	95	118.8
46	62.83	96	119.9
47	64.00	97	121.0
48	65.17	98	122.1
49	66.34	99	123.2
50	67.50	100	124.3

Chi-square table: 1% critical values

DEGREES OF FREEDOM	CRITICAL VALUE	DEGREES OF FREEDOM	CRITICAL VALUE
1	6.635	51	77.38
2	9.209	52	78.61
3	11.34	53	79.84
4	13.28	54	81.07
5	15.08	55	82.29
6	16.81	56	83.51
7	18.47	57	84.73
8	20.09	58	85.95
9	21.66	59	87.16
10	23.21	60	88.38
11	24.72	61	89.59
12	26.22	62	90.80
13	27.69	63	92.01
14	29.14	64	93.22
15	30.58	65	94.42
16	32.00	66	95.62
17	33.41	67	96.83
18	34.80	68	98.03
19	36.19	69	99.23
20	37.56	70	100.4
21	38.93	71	101.6
22	40.29	72	102.8
23	41.64	73	104.0
24	42.98	74	105.2
25	44.31	75	106.4
26	45.64	76	107.6
27	46.96	77	108.8
28	48.28	78	110.0
29	49.59	79	111.1
30	50.89	80	112.3
31	52.19	81	113.5
32	53.48	82	114.7
33	54.77	83	115.9
34	56.06	84	117.1
35	57.34	85	118.2
36	58.62	86	119.4
37	59.89	87	120.6
38	61.16	88	121.8
39	62.43	89	122.9
40	63.69	90	124.1
41	64.95	91	125.3
42	66.20	92	126.5
43	67.46	93	127.6
44	68.71	94	128.8
45	69.96	95	130.0
46	71.20	96	131.1
47	72.44	97	132.3
48	73.68	98	133.5
49	74.92	99	134.6
50	76.15	100	135.8

Chi-square table: 0.1% critical values

DEGREES OF FREEDOM	CRITICAL VALUE	DEGREES OF FREEDOM	CRITICAL VALUE
1	10.83	51	87.96
2	13.81	52	89.27
3	16.26	53	90.57
4	18.46	54	91.87
5	20.51	55	93.16
6	22.45	56	94.46
7	24.32	57	95.75
8	26.12	58	97.03
9	27.87	59	98.32
10	29.58	60	99.60
11	31.26	61	100.9
12	32.90	62	102.2
13	34.52	63	103.4
14	36.12	64	104.7
15	37.69	65	106.0
16	39.25	66	107.3
17	40.79	67	108.5
18	42.31	68	109.8
19	43.82	69	111.1
20	45.31	70	112.3
21	46.79	71	113.6
22	48.26	72	114.8
23	49.72	73	116.1
24	51.17	74	117.3
25	52.62	75	118.6
26	54.05	76	119.8
27	55.47	77	121.1
28	56.89	78	122.3
29	58.30	79	123.6
30	59.70	80	124.8
31	61.09	81	126.1
32	62.48	82	127.3
33	63.87	83	128.6
34	65.24	84	129.8
35	66.61	85	131.0
36	67.98	86	132.3
37	69.34	87	133.5
38	70.70	88	134.7
39	72.05	89	136.0
40	73.40	90	137.2
41	74.74	91	138.4
42	76.08	92	139.7
43	77.41	93	140.9
44	78.75	94	142.1
45	80.07	95	143.3
46	81.40	96	144.6
47	82.72	97	145.8
48	84.03	98	147.0
49	85.35	99	148.2
50	86.66	100	149.4

G Nonparametric Confidence Interval and Sign Test

Table of ranks for nonparametric confidence interval and the sign test for the median

2-sided confidence: 1-sided confidence: 2-sided test level: 1-sided test level:	90% 95% 10% 5%	95% 97.5% 5% 2.5%	98% 99% 2% 1%	99% 99.5% 1% 0.5%	99.8% 99.9% 0.2% 0.1%	99.9% 99.95% 0.1% 0.05%
Sample size	rank	rank	rank	rank	rank	rank
7	1	1	1	1	1	1
8	2	1	1	1	1	1
9	2	2	1	1	1	1
10	2	2	1	1	1	1
11	3	2	2	1	1	1
12	3	3	2	2	1	1
13	4	3	2	2	1	1
14	4	3	3	2	2	1
15	4	4	3	3	2	2
16	5	4	3	3	2	2
17	5	5	4	3	2	2
18	6	5	4	4	3	2
19	6	5	5	4	3	3
20	6	6	5	4	3	3
21	7	6	5	5	4	3
22	7	6	6	5	4	4
23	8	7	6	5	4	4
24	8	7	6	6	5	4
25	8	8	7	6	5	5
26	9	8	7	7	5	5
27	9	8	8	7	6	5
28	10	9	8	7	6	6
29	10	9	8	8	6	6
30	11	10	9	8	7	6
31	11	10	9	8	7	7
32	11	10	9	9	7	7
33	12	11	10	9	8	7
34	12	11	10	10	8	8
35	13	12	11	10	9	8
36	13	12	11	10	9	8
37	14	13	11	11	9	9
38	14	13	12	11	10	9
39	14	13	12	12	10	9
40	15	14	13	12	10	10
41	15	14	13	12	11	10
42	16	15	14	13	11	11
43	16	15	14	13	12	11
44	17	16	14	14	12	11
45	17	16	15	14	12	12
46	17	16	15	14	13	12
47	18	17	16	15	13	12
48	18	17	16	15	13	13
49	19	18	16	16	14	13
50	19	18	17	16	14	14
Multiplier[1]	1.645	1.960	2.326	2.576	3.090	3.291

[1]For sample sizes larger than 50, a good approximation to this rank may be found by taking the square root of the sample size, multiplying by the appropriate multiplier given here, subtracting this from the sample size, adding 1, dividing by 2, and ignoring any decimal places (truncating) so as to obtain a whole number.

Notes

Chapter 1: Introduction

[1]Efron and Thisted, 1976.

[2]Mosteller and Wallace, 1972.

[3]Feduccia, 1993.

Chapter 2: The Shape of a Group of Numbers

[1]Data on population and life expectancy are from Appendix 1, Ehrlich, Ehrlich, and Holdren (1977). They cite Population Reference Bureau, 1976. Data on health expenditures are from The world's spending priorities, 1981, p. 4E. They cite *World Military and Social Expenditures,* published by World Priorities.

[2]Percents are computed from vote counts in Hoffman, 1992, p. 90.

[3]Our *categorical* variables are sometimes called *nominal* variables, and our *ordered categorical* variables are sometimes called *ordinal* variables. Our *numerical measured* variables are sometimes split into *interval* variables, for which the distance between values is meaningful, and *ratio* variables, for which there is a natural zero point and the ratio of values is also meaningful.

[4]Data are from Hoffman, 1991, p. 546. Used by permission.

[5]Fisher, 1936, pp. 179–188.

[6]Karl Friedrich Gauss, a German mathematician and scientist, lived from 1777 to 1855. He is commonly regarded as one of the greatest mathematicians in history. It is interesting that the parents of this genius were not themselves in an intellectual line of work; Gauss apparently taught himself reading and mathematics. In addition to his important work in statistics, Gauss is also credited with many important discoveries in mathematics and physics. For further information, see Bell (1955).

[7]Data are from *Military Personnel on Active Duty: 1789 to 1970,* Series Y 915, p. 1141, U.S. Bureau of the Census, 1975.

[8]Data are from Professional golf tournaments in 1991, Hoffman, 1991, p. 881.

[9]Data are from Appendix 2 of Ehrlich et al., 1977. They cite UN World Food Conference (1974), *Assessment of the World Food Situation, Present and Future.*

[10]Data are from Baseball stadiums, Hoffman, 1992, p. 923, and from The National League of Professional Baseball Clubs, 1994.

[11]From Masin, December 1991, p. 105. The author states that ". . . to report on the most exotic of the new 1991 skis, we simply asked the ski factories to send us the skis they were most excited about."

[12]We are grateful to Professor H. V. Daly, Department of Entomology, University of California at Berkeley (personal communication, 1984), for the use of these data.

[13]Data are from Stigler, 1977, Table 6.

[14]Data are from The debt position of the Bell System's local operating companies, 1982, p. D6.

[15]Data are from U.S. Bureau of the Census, 1979, Table 771, *Appalachia—Per Capita and Family Income: 1965 to 1976.*

[16]Data are from 100 largest U.S. commercial banks, Hoffman, 1979, p. 112. Their source is The 300 largest commercial banks in U.S., compiled by the American Banker, New York. Used by permission of *The World Almanac and Book of Facts,* 1980 edition, copyright © Newspaper Enterprise Association, Inc., 1979, New York, NY 10166.

[17]Data are from Banks, in Hoffman, 1992, p. 889.

[18]Hambrecht et al., 1993.

[19]Data from a study of intensive case management for homeless adolescents, provided courtesy of Ana Mari Cauce, Department of Psychology, University of Washington, and Victoria Wagner, Executive Director, YouthCare, Seattle. Data collection was supported by National Institutes of Mental Health/Substance Abuse and Mental Health Services Administration Grant HD5-48085, awarded to Ana Mari Cauce.

[20]Smith, 1993, p. 368, Figure 3.

[21]National Marine Fisheries Service, 1977, Table 23.

[22]Data are modeled after Levy et al., 1993.

Chapter 3: Describing Distributions

[1]U.S. Department of Education, 1990, Table 2:17.

[2]We reuse the data sets throughout the text. This is partly economics on our part, but it is

mostly pedagogical. Analyzing a data set in practice takes time, and you will return to it repeatedly and try new things. We are trying to model this process of successive examinations. We also think that substantial amounts of thought go into the early parts of understanding any data set; once having asked students to invest that amount of thought, we want to build on it.

[3]Masin, November 1991, p. 64. Here is the author's sampling strategy: "We asked manufacturers to send us their do-everything skis—high performance boards that would work well in any snow conditions a skier might encounter."

[4]The outlier thresholds (called "fences") were introduced by J. W. Tukey (1977, p. 43).

[5]Hoaglin, Mosteller, and Tukey, 1983, pp. 63–65.

[6]We are grateful to Professor H. V. Daly, Department of Entomology, University of California at Berkeley (personal communication, 1984), for the use of these data.

[7]Data are from Stigler, 1977, Table 6.

[8]Data are from The debt position of the Bell System's local operating companies, 1982, p. D6.

[9]Data are from U.S. Bureau of the Census, 1979, Table 771, *Appalachia—Per Capita and Family Income: 1965 to 1976.*

[10]Data are from 100 largest U.S. commercial banks, Hoffman, 1979, p. 112. Their source is The 300 largest commercial banks in U.S., compiled by the American Banker, New York. Used by permission of *The World Almanac & Book of Facts,* 1980 edition, copyright © Newspaper Enterprise Association, Inc., 1979, New York, NY 10166.

[11]Data are from Hoffman, 1992, p. 471, for presidents before President Clinton; Clinton's age from contemporary news reports.

[12]Data are from U.S. Department of Education, 1991, Table 11.

Chapter 4: Describing a Normal Distribution and Summarizing Binary Data

[1]Throughout this book, when we speak of an "average," we mean an arithmetic average or mean. There seems to be a trend toward using the term *average* to refer to any measure of central tendency. For example, we see press reports of the "average" prices of homes, when in fact the median (rather than the mean) is being reported. We have not yet gotten onto that bandwagon.

[2]Orange juice data are from Pacific Magazine, *The Seattle Times,* November 7, 1982.

[3]Data are from Morgan, 1979.

[4]Ibid.

[5]Stigler, 1986, p. 201.

[6]In some other books, n is used in place of $n - 1$ in the formula for the variance of yes/no data. We have chosen to present it in this way because $n - 1$ is more consistent with our general approach to computing the sample variance of a group of numbers. In any case, the difference between these two approaches is small and they will yield nearly the same results when n is large.

[7]We are grateful to Professor H. V. Daly, Department of Entomology, University of California at Berkeley (personal communication, 1984), for the use of these data.

[8]Data are from Stigler, 1977, Table 6.

[9]Data are from The debt position of the Bell System's local operating companies, 1982, p. D6.

[10]Data are from Hoffman, 1992, p. 471, for presidents before President Clinton; Clinton's age from contemporary news reports.

[11]From Table 23, National Marine Fisheries Service, 1977.

[12]Data are from U.S. Department of Education, 1991, Table 11.

Chapter 5: Basics of Data Transformation

[1]Data are from Explosions, in Hoffman, 1979, p. 751. Used by permission of *The World Almanac and Book of Facts,* 1980 edition, copyright © Newspaper Enterprise Association, Inc., New York, NY 10166.

[2]Prices are from *New for '92,* October 1991, pp. 54ff.

[3]Data are from Important islands and their areas, Hoffman, 1979, p. 443. Their source is National Geographic Society, Washington, D.C. Used by permission of *The World Almanac and Book of Facts,* 1980 edition, copyright © Newspaper Enterprise Association, Inc., 1979, New York, NY 10166.

[4]See original scale in Display 5.8.

[5]See log scale in Display 5.12.

[6]Data are loosely modeled after Yin, Barnet, and Miller, 1994.

[7]Data courtesy of Maxine Linial, Fred Hutchinson Cancer Research Center; used with permission.

[8]Data modeled after Figure 13 in Berg and Smith, November 1994.

Chapter 6: Choosing a Description

[1]Data are from State finances, in Hoffman, 1992.

[2]Data are from 100 largest U.S. commercial banks, in Hoffman, 1979. Their source is The 300 largest commercial banks in U.S., compiled by the American Banker, New York. Used by permission of *The World Almanac and Book of Facts,* 1980 edition, copyright © Newspaper Enterprise Association, Inc., 1979, New York, NY 10166.

[3]Data are from Baseball stadiums, in Hoffman, 1992, and from National League, 1994.

[4]From Ryan, Joiner, and Ryan, 1985.

[5]Data are from FISH.MTW, distributed with Minitab. Used by permission.

[6]Data are from Table 7.2, U.S. Bureau of the Census, 1991.

[7]From Figure 4.4, Herlihy and Klapisch-Zuber, 1985.

[8]Data are from Table 713, U.S. Bureau of the Census, 1993.

[9]All data are from Tables 327 and 360, U.S. Bureau of the Census, 1979. Their source for the lawyer data was from American Bar Foundation, 1991.

Chapter 7: Probability

[1]Adapted from Table 2:28, Alsalam and Rogers, n.d.

[2]The conditional probability is left undefined if the probability of B is 0, because division by 0 is impossible.

[3]Fortunately, it can be proven that whether we interpret B as providing additonal given

information, as we did here, or we interpret *A* in this way, the answer will be the same. Thus, we need only compare the probability of either event to its conditional probability given the other event.

[4]To show that this definition is equivalent to the previous definition, use the definition of conditional probability: Conditional probability of *A* GIVEN *B* is always equal to the probability of *A* AND *B* divided by the probability of *B*.

[5]vos Savant, November 1991. See also Morgan, J. P. et al., in the same issue, and commentary.

[6]U.S. Department of Education, 1991.

[7]From Copp, 1989.

[8]Olinick, 1991.

[9]Norell et al., 1994.

[10]*The New York Times,* September 5, 1987. Quoted in Truxal, 1989, p. 42.

Chapter 8: Random Variables, Probability Distributions, and the Central Limit Theorem

[1]From Stoppard, 1967, pp. 11–12.

[2]DeMoivre, 1738, p. 235; quoted in Stigler, S. M. 1986.

[3]Data are from Navellier, 1993.

Chapter 9: Toward Statistical Inference

[1]*Webster's New Collegiate Dictionary* (6th ed.). (1961). Springfield, MA: G. and C. Merriam Co., p. 83.

[2]Nunn & Slavin, 1980.

[3]These extra samples are meant to represent further samples from the population identified here and should be thought of in that way. However, for those who are curious, here is a description of how they were obtained. They were found by sampling from the original data set of 12 numbers in the same way as that used by a powerful class of new statistical methods presented by Efron called the "bootstrap" technique. Imagine that each of these 12 numbers is written on a piece of paper. The pieces of paper are mixed up and one is chosen at random, observed, and replaced so that it might be chosen again. This selection process is repeated until 12 values are observed, having been chosen at random from the original 12 values. Note that, strictly speaking, we are sampling from the data and not the population. Nonetheless, Efron and others have shown that the variability from one sample of the data to another sample of the data is a good estimate of the variability from one sample of the population to another sample of the population. The advantage of the bootstrap method is that it may not be necessary to obtain more data (at great expense) from the population to assess variability, even in more general situations. An introduction to the bootstrap technique is provided by Efron and Gong, February 1983, pp. 36–48. A more detailed treatment may be found in Efron, 1982.

[4]In some other books, n is used in place of $n - 1$ in the formula for the standard error of a percentage. We have chosen to present it this way because $n - 1$ is more consistent with our general approach to computing the sample variability measures for a group of numbers. In any case, the difference between these two approaches is small; they will yield nearly the same results when n is large.

[5]Dougherty, 1980, p. D7.

[6]The New York Times/CBS News Poll, March 19, 1982, p. A20.

[7]The tree data set is distributed with Minitab and is used by permission.

[8]Ten largest cable system operators ranked by number of subscribers, September 13, 1983, p. D1. Their source was the *Cable TV Newsletter.*

[9]We are grateful to Professor H. V. Daly, Department of Entomology, University of California at Berkeley (personal communication, 1984), for the use of these data.

[10]See, for example, Aguayo, 1991.

[11]Janus and Janus, 1993.

[12]Laumann et al., 1994.

Chapter 10: Confidence Intervals

[1]We are grateful to Professor H. V. Daly, Department of Entomology, University of California at Berkeley (personal communication, 1984), for the use of the data set from which this example was constructed.

[2]This result is subject to the technical requirements that the population be very large and have a standard deviation that is a finite, nonzero number.

[3]Benjamini, 1983.

[4]Reported in *Time Magazine,* August 16, 1993, p. 23.

[5]Method used in conducting poll, 1982, p. A20.

[6]This data set is adapted from Lee and Krutchkoff, 1980.

[7]The tree data set is distributed with Minitab and is used here by permission.

[8]Data are from Fisher, 1936.

[9]We are grateful to R. D. Stevenson, Department of Zoology, University of Washington at Seattle (personal communication, 1984), for suggesting the circumstances of this problem.

[10]We are grateful to Professor H. V. Daly, Department of Entomology, University of California at Berkeley (personal communication, 1984), for the use of these data.

[11]Wagner, V. A., and Morgan, C. J. (1993). Learning disabilities in runaway youth. Unpublished report.

Chapter 11: Testing a Hypothesis About the Mean

[1]Background information on Gosset and his role in statistics presented in this section is based on J. O. Irwin's contribution to Kruskal and Tanur, 1978, pp. 409–413.

[2]*The Seattle Times,* February 17, 1983, p. G1.

Chapter 12: Comparing Two Groups of Numbers

[1]Data are from Fisher, 1936, pp. 179–188.

[2]Feduccia, 1993.

[3]Data are from Notable tall buildings in North American cities, in Hoffman, 1992, pp. 652–658.

[4]Data are from Voter turnout in presidential elections, in Hoffman, 1979, p. 286. Used by permission of *The World Almanac and Book of Facts,* 1980 edition, copyright © Newspaper Enterprise Association, Inc. New York, NY 10166.

[5]Feduccia, 1993.

[6]Data modeled after Gottman, 1994, pp. 56–61.

[7]Data extracted from or modeled after *Newsweek,* July 26, 1993, p. 48. A technical reference to the work is Mannuzza et al. 1993.

[8]Ibid.

[9]Ibid.

Chapter 13: Analysis of Variance: Several Groups of Numbers

[1]Data are from Census and areas of counties and states, in Hoffman, 1992.

[2]Box, 1976.

[3]Note that the grand mean will not generally be equal to the average of the group means when the groups have different sample sizes. Instead, it is a weighted average of the group means, giving greater importance to the groups with more observations because they have contributed more information to the situation.

[4]Box, 1978, p. 325.

[5]We are grateful to M. J. Fassino, Department of Psychology, Princeton University (personal communication, 1980), for the use of the data set from which this example was constructed.

[6]Miller, 1981, p. 90.

[7]A technical note is in order here. Transforming the five-number summary values is the same as finding the five-number summary for the transformed data only when the summary value is actually a data value rather than an average of two values. However, transforming the five-number summary value still gives us a valuable approximation in all cases.

[8]Data are from Table A1-1, Ehrlich et al., 1977.

[9]Data modeled after Lapp et al., 1994.

[10]Beecher, 1961. Also Beecher, 1966.

[11]This experimental situation and the data are modeled after Wong, 1994.

[12]Devers et al., 1994.

[13]Bowers et al., 1994.

[14]Jones and Rabbitt, 1994.

[15]Claxton, 1994.

Chapter 14: Categorical Data and Chi-Square Analysis

[1]The die might actually be loaded, despite having passed the test.

[2]In the interest of fairness, we must point out that we will not always decide that our fair die is fair, as we have done for this particular data set. Because the Type I error is set at level .05, if we were to do many tests one after another, about 5% of them would wrongly decide that the die is unfair. Because there is such variety (due to randomness) in the way a fair die can come up in repeated tosses, we have to be willing to be wrong *some* of the time to be able to ever decide that a die is unfair.

[3]U.S. Department of Education, 1991.

[4]Data are from Shapiro, R., Strauchen, J. A., Sher, L. S., Tartter, P. I., Siegel, A. F., and Waye, J. D. (1986). Decisions following colonoscopic polypectomy of the malignant adenoma. Unpublished manuscript.

[5]The geographic divisions of the United States used here are from Figure 1, map of the United States showing census divisions and regions, U.S. Bureau of the Census, 1979.

[6]The values in the table are more accurate because they were computed using more precision during the intermediate computations. Instead of 66.390 for "Selected from the North Central region," we will have 66.395 in the table because by using more precision during calculation, we find that

$$\begin{aligned}(63 - 362 \times 216/1{,}099)^2 &= (63 - 71.14831665)^2 \\ &= (-8.148316650)^2 \\ &= 66.39506423\end{aligned}$$

[7]The observant reader may have noticed that the two columns in the previous table of $(\text{Observed} - \text{Expected})^2$ were identical. This will happen any time that we have just two columns. Also, in the case of just two rows, the values will be equal in the rows. This implies that for a two-by-two table (with two rows and two columns), all four numbers will be the same. For three-by-three and larger tables, the numbers will generally all be different. Also, note that when we divide by the expected number, the results are generally different also.

[8]Crimes data are from *San Francisco Chronicle,* October 20, 1982, p. 16. Their source is the Federal Bureau of Investigation.

[9]Recall that we say "approximately" because the chi-square test is only a large-sample approximation.

[10]Data are from O'Reilly, 1986.

[11]Data are from Shapiro et al., 1986.

[12]Data are from Wilford, 1981.

[13]Bronowski, 1973, p. 385.

[14]Lapp et al., 1994.

[15]Chapman and McCauley, 1993.

[16]Beck et al., 1994.

[17]Beecher, 1961. Also, Beecher, 1966.

[18]Rosenthal and Jacobson, 1968.

[19]Ibid.

[20]The problems and the correct percentages are modeled after Berg and Smith, 1994; sample sizes differ.

[21]Ibid.

[22]Ibid.

Chapter 15: Bivariate Data and Regression

[1]U.S. Bureau of the Census, 1975.

[2]We are grateful to J. T. Dwyer, Harvard School of Public Health (personal communication, 1979), for the use of the data from which this graph was constructed. For a detailed statistical analysis of growth data like this, see Dwyer et al., 1983.

[3]Data are from *Bond and Stock Prices: 1900 to 1970,* Series X 495, U.S. Bureau of the Census, 1975.

[4]We are grateful to Professor D. MacLachlan, Department of Marketing and International Business, University of Washington at Seattle (personal communication, 1983), for the use of the data set from which this example was constructed.

[5]Data are from Appendix I of Ehrlich et al., 1977. Their source is *The 1976 World Population Data Sheet.* Washington, D.C.: Population Reference Bureau, Inc.

[6]Data for Figure 15.20 are from The world's spending priorities, 1981, p. 4E. Their source is *World Military and Social Expenditures,* published by World Priorities. Data for Figure 15.21 are from Grobstein et al., 1983.

[7]We are grateful to Professor Erik Trinkaus, Department of Anthropology, Harvard University (personal communication, 1979), for the use of the data set from which this example was constructed.

[8]The assumption here is that there is some variation in *Y*, so that the line has some slope. If all the *Y* values are the same, they will fall perfectly on a straight, but horizontal, line. Their standard deviation will be 0, and the correlation coefficient will be undefined.

[9]Data are from Friedman et al., 1983.

[10]Data are from Proportion of degrees awarded to women, 1981, p. 8. Their source is *Degree Awards to Women,* National Center for Education Statistics.

[11]Data are from Physical growth range for children from 1 to 17 years, Hoffman, 1979. Used by permission of *The World Almanac & Book of Facts,* 1980 edition, copyright © Newspaper Enterprise Association, Inc. New York, NY 10166.

[12]Draper and Smith, 1981.

[13]Fast food for fat watchers. *Consumer Reports,* September 1993, p. 577.

[14]Data are from Monthly normal temperature and precipitation, in Hoffman, 1979, p. 793. Used by permission of *The World Almanac & Book of Facts,* 1980 edition, copyright © Newspaper Enterprise Association, Inc., New York, NY 10166.

[15]Data are from Auto racing, in Hoffman, 1992, p. 915.

[16]Data are from The Bell System operating companies, 1982, p. 35.

[17]We are grateful to S. Hiebert, Department of Zoology, University of Washington at Seattle (personal communication, 1985), for suggesting the circumstances of this problem.

[18]Data are from Motor vehicle traffic deaths by state, in Hoffman, 1979, p. 957. Their source is the National Safety Council. Used by permission of *The World Almanac & Book of Facts.* 1980 edition, copyright © Newspaper Enterprise Association, Inc. New York, NY 10166.

[19]From Levy and Newbourne, 1991.

[20]Martin, 1994, pp. 8–9.

[21]Data are from Tables 505 and 31, U.S. Bureau of the Census, 1993.

[22]Morrison, 1993.

References

Aguayo, R. (1991). *Dr. Deming: The American Who Taught the Japanese About Quality.* New York: Simon & Schuster.

Alsalam, N., and Rogers, G. T. (Eds.) (n.d.). *The Condition of Education 1990: Vol 2. Postsecondary Education.* Washington, D.C.: U.S. Government Printing Office.

American Bar Foundation (1991). *Supplement to the Lawyer Statistical Report: The U.S. Legal Profession in 1988.* Chicago, IL: American Bar Foundation.

Beck, J. G. Stanley, M. A., Baldwin, L. E., Deagle, E. A., and Averill, P. M. (1994, August). Comparison of cognitive therapy and relaxation training for panic disorder. *Journal of Consulting and Clinical Psychology, 62*(4), 818–826.

Beecher, J. K. (1961). Surgery as placebo. *Journal of the American Medical Association, 176,* 1102–1107.

Beecher, J. K. (1966). Pain: One mystery solved. *Science, 151,* 840–841.

Bell, E. T. (1955). Gauss, Karl Friedrich. In *Encyclopedia Britannica,* vol. 10, pp. 75–76. Chicago, IL: Encyclopedia Britannica, Inc.

Bell System operating companies, The. *The New York Times,* January 9, 1982, p. 35.

Benjamini, Y. (1983). Is the *t*-test really conservative when the parent distribution is long-tailed? *Journal of the American Statistical Association, 78,* 645–654.

Berg, C., and Smith, P. (1994, November). Assessing students' abilities to construct and interpret line graphs: Disparities between multiple-choice and free-response instruments. *Science Education, 78*(6), 527–554.

Bowers, R. L., Doran, T. P., Edles, P. A., and May, K. (1994). Paired-associate learning with visual and olfactory cues: Effects of temporal order. *The Psychological Record, 44,* 501–507.

Box, G. E. P. (1976). Science and statistics. *Journal of the American Statistical Association, 71,* 795.

Box, J. F. (1978). *R. A. Fisher, the Life of a Scientist.* New York: John Wiley & Sons.

Bronowski, J. (1973). *The Ascent of Man.* Boston: Little, Brown & Co.

Burton, Joan. (1988). Best hikes with children in Western Washington and the Cascades. Seattle, WA: The Mountaineers.

Chapman, G. B., and McCauley, C. (1993). Early career achievements of National Science Foundation (NSF) graduate applicants: Looking for Pygmalion and Galatea effects on NSF winners. *Journal of Applied Psychology, 78,* 815–820.

Claxton, R. P. (1994). Empirical relationships between birth order and two types of parental feedback. *The Psychological Record, 44,* 475–487.

Copp, N. (1989). *Vaccines: An Introduction to Risk.* Monograph Series of the New Liberal Arts Program, Research Foundation of the State University of New York, Stonybrook, NY.

Debt position of the Bell System's local operating companies, The. *The New York Times,* January 12, 1982, p. D6.

Devers, R., Bradley-Johnson, S., and Johnson, C. M. (1994). The effect of token reinforcement on WISC-R performance for fifth- through ninth-grade American Indians. *The Psychological Record, 44,* 441–449.

Dougherty, P. H. Advertising: Showell chickens in taste test. *The New York Times,* December 29, 1980, p. D7.

Draper, N., and Smith, H. (1981). *Applied Regression Analysis.* New York: John Wiley & Sons.

Dwyer, J. T., Andrew, E. M., Berkey, C., Valadian, I., and Reed, R. B. (1983). Growth in "new" vegetarian preschool children using the Jenss–Bayley curve fitting technique. *The American Journal of Clinical Nutrition, 37,* 815–827.

Efron, B. (1982). *The Jackknife, the Bootstrap, and Other Resampling Plans.* Philadelphia, PA: Society for Industrial and Applied Mathematics.

Efron, B., and Gong, G. (1983, February). A leisurely look at the bootstrap, the jackknife, and cross-validation. *The American Statistician,* 36–48.

Efron, B., and Thisted, R. (1976). Estimating the number of unseen species: How many words did Shakespeare know? *Biometrika, 63,* 435–447.

Ehrlich, P. R., Ehrlich, A. H., and Holdren, J. P. (1977). *Ecoscience: Population, Resources, Environment.* San Francisco, CA: W. H. Freeman and Co.

Fast food for fat watchers. *Consumer Reports,* September 1993, p. 577.

Feduccia, Alan (1993, February 5). Evidence from claw geometry indicating arboreal habits of *Archaeopteryx. Science, 259,* 790–793.

Fisher, R. A. (1936). The use of multiple measurements in taxonomic problems. *Annuals of Eugenics, 7,* 179–188.

Friedman, R. M., Bierbaum, R. M., Catherwood, P. A., Diamond, S. C., Hoberg, G. G., and Lee, V. A. (1983). The acid rain controversy: The limits of confidence. *The American Statistician, 37,* 385–394.

Gottman, J. M., with Silver, N. (1994). *Why Marriages Succeed or Fail: What You Can Learn From the Breakthrough Research to Make Your Marriage Last.* New York: Simon & Schuster.

Grobstein, C., Flower, M., and Mendelhoff (1983, October 14). External human fertilization: An evaluation of policy. *Science, 222*(4620), 127–133.

Hambrecht, R., Neibauer, J., Marburger, C., Grunze, M., Kälberer, B., Hauer, K., Schlierf, G., Kübler, W., and Schuler, G. (1993). Various intensities of leisure time physical activity in patients with coronary artery disease: Effects of cardiorespiratory fitness and progression of coronary atherosclerotic lesions. *Journal of the American College of Cardiology, 22,* 468–477.

Herlihy, D., and Klapisch-Zuber, C. (1985). Tuscans and their families: A study of the Florentine Catasto of 1427. New Haven and London: Yale University Press.

Hoaglin, D. C., Mosteller, F., and Tukey, J. W. (1983). *Understanding Robust and Exploratory Data Analysis.* New York: John Wiley & Sons.

Hoffman, M. S. (Ed.) (1979). *The World Almanac & Book of Facts 1980.* New York: Newspaper Enterprise Association, Inc.

Hoffman, M. S. (Ed.) (1980). *The World Almanac and Book of Facts 1981.* New York: Newspaper Enterprise Association, Inc.

Hoffman, M. S. (Ed.) (1991). *The 1992 World Almanac and Book of Facts.* New York: Pharos Books.

Hoffman, M. S. (Ed.) (1992). *The World Almanac and Book of Facts 1993.* New York: Pharos Books.

Janus, S. S., and Janus, C. L. (1993). *The Janus Report on Sexual Behavior.* New York: John Wiley and Sons.

Jones, S. J., and Rabbitt, P. M. A. (1994). Effects of age on the ability to remember common and rare proper names. *The Quarterly Journal of Experimental Psychology, 47A,* 1001–1014.

Kruskal, W. H., and Tanur, J. M. (Eds.) (1978). *The International Encyclopedia of Statistics.* New York: The Free Press.

Lapp, W. M., Collins, R. L., and Izzo, C. V. (1994). On the enhancement of creativity by alcohol: Pharmacology or expectation? *American Journal of Psychology, 107,* 173–206.

Laumann, E. O., Gagnon, J. H., Michael, R. T., and Michaels, S. (1994). *The Social Organization of Sexuality: Sexual Practices in the United States.* Chicago: University of Chicago Press.

Lee, L., and Krutchkoff, R. G. (1980). Mean and variance of partially truncated distributions. *Biometrics, 36,* 531–536.

Levy, W. C., Cerqueira, M. D., Abrass, I. B., Schwartz, R. S., and Stratton, J. R. (1993). Endurance exercise training augments diastolic filling at rest and during exercise in healthy young and older men. *Circulation, 88,* 116–130.

Levy, D., and Newbourne, M. (1991). *How Computers Play Chess.* New York: W. H. Freeman & Co.

Mannuzza, S., Klein, R., Bessler, A., and Malloy, P. (1993). Adult outcome of hyperactive boys: Educational achievement, occupational rank, and psychiatric status. *Archives of General Psychiatry, 50,* 565–576.

Martin, Molly. Energy bars. *The Seattle Times, Pacific Magazine* supplement, August 14, 1994, pp. 8–9.

Masin, S. (1991, November). The all-terrain gang. *Ski Magazine,* pp. 107ff.

Masin, S. (1991, December). The wild bunch. *Ski Magazine,* pp. 105ff.

Method used in conducting poll. *The New York Times,* March 19, 1982, p. A20.

Miller, R. (1981). *Simultaneous Statistical Inference* (2nd ed.). New York: Springer-Verlag.

Morgan, C. J. (1979). Eskimo hunting groups, social kinship, and the possibility of kin selection in humans. *Ethology and Sociobiology, 1,* 83–86.

Morgan, J. P., Chaganty, N. R., Dahiya, R. C., and Doviak, M. J. (1991, November). Let's make a deal: The player's dilemma. *The American Statistician,* pp. 284–287.

Morrison, E. W. (1993). Longitudinal study of the effects of information seeking on newcomer socialization. *Journal of Applied Psychology, 78,* 173–183.

Mosteller, F., and Wallace, D. L. (1972). Deciding authorship. In J. M. Tanur, F. Mosteller, W. H. Kruskal, R. F. Link, R. S. Pieters, and G. R. Rising (Eds.), *Statistics: A Guide to the Unknown* (pp. 164–175). San Francisco: Holden-Day.

National League of Professional Baseball Clubs. (1994). *National League Green Book.* Los Angeles: MG Book Graphics.

National Marine Fisheries Service. (1977). *Final Environmental Impact Statement, International Whaling Commission's Deletion of Native Exemption for the Subsistence Harvest.* Washington, D.C.: Department of Commerce.

Navellier, Louis. (1993). *MPT Review: Specializing in Modern Portfolio Theory.* Incline Village, NV: Louis Navellier.

New for '92. (1991, October). *Road and Track,* pp. 54ff.

New York Times/CBS News Poll, The: Public views on aid to the poor. *The New York Times,* March 19, 1982, p. A20.

Norell, M. A., Clark, J. M., Dashzeveg, D., Barsbold, R., Chiappe, L. M., Davidson, A. R., McKenna, M. C., Perle, A., and Novacek, M. J. (1994). A theropod dinosaur embryo and the affinities of the Flaming Cliffs dinosaur eggs. *Science, 266,* 779–782.

Nunn, J. F. and Slavin, G. (1980). Posterior intercostal nerve block for pain relief after cholecystectomy. *British Journal of Anaesthesia, 52,* 253–260.

O'Reilly, B. (1986, January 6). Business copes with terrorism. *Fortune, 113*(1), 47–55.

Olinick, Michael. (1991). Evidence of sexism? In John G. Truxal (Ed.), *Quantitative Examples* (p. 105). Monograph Series of the New Liberal Arts Program. Copiaque, NY: The Wexford Press.

Proportion of degrees awarded to women. *The Chronicle of Higher Education,* June 15, 1981, p. 8.

Rosenthal, R., and Jacobson, L. (1968). *Pygmalion in the Classroom: Teacher Expectation and Pupils' Intellectual Development.* New York: Holt, Rinehart & Winston.

Ryan, T., Joiner, B., and Ryan, B. (1985). *Minitab Handbook* (2nd ed.). Distributed with Minitab. Boston: PWS-Kent.

Smith, M. K. (1993). Transition of youth with severe emotional disabilities: Preliminary findings. *The Sixth Annual Research Conference Proceedings, A System of Care for Children's Mental Health: Expanding the Research Base.* Tampa, Florida: Research and Training Center for Children's Mental Health, University of South Florida.

Stigler, S. M. (1977). Do robust estimators work with *real* data? *The Annals of Statistics, 5,* 1055–1098.

Stigler, S. M. (1986). *The History of Statistics.* Cambridge, MA: The Belnap Press of Harvard University Press.

Stoppard, Tom. (1967). *Rosencrantz and Guildenstern Are Dead.* New York: Grove Press.

Ten largest cable system operators ranked by number of subscribers. *The New York Times,* September 13, 1983, p. D1.

Truxal, J. G. (1989). Probability examples. Monograph Series of the New Liberal Arts Program, Research Foundation of the State University of New York, Stonybrook, NY.

Tukey, J. W. (1977). *Exploratory Data Analysis.* Reading, MA: Addison-Wesley.

U.S. Bureau of the Census. (1975). *Historical Statistics of the United States Colonial Times to 1970, Bicentennial Edition, Part 2.* Washington, D.C.: U.S. Government Printing Office.

U.S. Bureau of the Census. (1979). *Statistical Abstract of the United States: 1979* (100th ed.). Washington, D.C.: U.S. Government Printing Office.

U.S. Bureau of the Census. (1991). *Current Population Survey.* Washington, D.C.: U.S. Government Printing Office.

U.S. Bureau of the Census. (1993). *Statistical Abstract of the United States.* Washington, D.C.: Government Printing Office.

U.S. Bureau of the Census. (1994). *Current Population Reports,* P60-180. Washington, D.C.: Government Printing Office.

U.S. Department of Education. (1990). *The Condition of Education, 1990.* Washington, D.C.: National Center for Educational Statistics.

U.S. Department of Education. (1991). *Youth Indicators 1991: Trends in the Well-being of American Youth.* Washington, D.C.: Office of Educational Improvement.

U.S. Department of Energy. (1991). *1992 Gas Mileage Guide: EPA Fuel Economy Estimates.* Washington, D.C.: U.S. Government Printing Office.

vos Savant, M. (1991, November). Letters to the editor. *The American Statistician,* p. 347.

Wilford, J. N. Shuttle weathers its second mission with less damage. *The New York Times,* November 16, 1981, p. 1.

Wong, R. (1994). Response latency of gerbils and hamsters to nuts flavored with bitter-tasting substances. *The Quarterly Journal of Experimental Psychology, 47B,* 173–186.

World's spending priorities, The. *The New York Times,* September 20, 1981, p. 4E.

Yin, H., Barnet, R. C., and Miller, R. R. (1994). Second-order conditioning and Pavlovian conditioned inhibition: Operational similarities and differences. *Journal of Experimental Psychology: Animal Behavior Processes, 20,* 419–428.

Answers to Selected Problems

Chapter 2

2. 55
4. 2
6. 3/24 = 12.5%
8a.

```
0 | 111
0 | 3233222
0 | 444
0 |
0 |
1 | 0
```

b. Yes; Express Gondola
10a. 25 to 30
b. 25 to 30
c. 25 to 30
d. around 260 to 270
e. around 90 to 100 (The graph appears to be close to 100; if actually computed, the number should be 94.)
12a. When doing a stem-and-leaf plot by hand, we usually truncate:

```
3 | 8
4 | 73
5 | 2204658698
6 | 2
```

If you round numbers to two decimals first, rather than truncate, the stem-and-leaf plot looks like this:

```
3 | 8
4 | 74
5 | 2305669698
6 | 2
```

c.

```
3 |
3 | 8
4 | 3
4 | 7
5 | 2204
5 | 658698
6 | 2
6 |
```

d.

```
3 |
3 |
3 |
3 |
3 | 8
4 |
4 | 3
4 |
4 | 7
4 |
5 | 0
5 | 22
5 | 45
5 | 66
5 | 898
6 |
6 | 2
6 |
6 |
6 |
```

e. The first step-and-leaf plot had most of the cases on a single line and lost a lot of detail. The split-stem plot showed more detail, and some skew toward low values. The third plot, which each stem split five times, showed the skew even more clearly, but takes more space. Either of the split-stem plots is acceptable.

14a.

```
4 | 0746527326905
5 | 5595
6 | 056
7 |
8 |
9 | 9
```

b. Skewed, toward high values
c. Yes; the 995
d. A real observation

16a. We simplify the numbers by subtracting 299,000 from each, then using the first two of the remaining three digits. With split stems, we get:

```
8 | 04
8 | 858
9 | 044
9 | 60
```

b. Yes, it appears symmetric

18a.

3	1
3	2
3	
3	7
3	888
4	111010
4	
4	4
4	
4	

b. Skewed, toward low values.
c. If you cut off the low end of a normal distribution, the cut off part would look a lot like this, except for the single observation at 4|4.

20a.

1	9
2	025
3	059
4	
5	6
6	
7	2
8	5

b. Appears to be skewed toward high values.
c. Yes. We often see skewness with counts or amounts that cannot take on negative values. What's more, since we are looking only at the largest banks here, even if the distribution of all banks were nearly normal, the distribution of those at the top end would be skewed.
d. New York

Chapter 3

2a. 4th
b. 23
c. 2.5
d. 16.5, 31
4a. 10(75, 80, 90.5, 109, 126)
10(80, 86, 100.5, 115, 122)
10(90, 93, 106, 114, 153)
8. *I. Virginica* typically has the longest petals; their median length is the largest. *I. Setosa* typically has the shortest petals; they have the smallest median length. *I. Virginica* shows the greatest variation; they have the largest interquartile range.
12b. This data distribution is skewed toward high values. It is centered, as measured by the median, at 1.1. The interquartile range is 1.8 billion, and there is a single high outlier at 5.8 billion.

Chapter 4

2. For reporting, 102.67; for further calculations, 102.6667
4.

data	a. deviations	b. deviations^2
81	−21.6667	469.4459
86	−16.6667	277.7789
96	−6.6667	44.44489
97	−5.6667	32.11149
103	.3333	.111089
111	8.3333	69.44389
114	11.3333	128.4437
115	12.3333	152.1103
121	18.3333	336.1099
sums	−.0003	1,510

c. variance = 188.75
d. SD = 13.73863, rounded to 13.74
6. SD = 17.85, rounded
8. The standard deviation
10a. 100 is zero SDs from the mean. Standard score = 0.
b. 50%
c. 50%
12a. 8.6078
b. .1750
14a. mean = 1.557
b. SD = 1.482
c. mean − SD = .075
d. mean + SD = 3.039
e. 19
f. 82.6%
16a. mean = 3.000
b. SD = 2.287
c. mean = 2.462
d. SD = 1.127
e. The high outlier increases the values of both the mean and the SD.
18a. Mean age = 54.881 years. Rounded to the nearest whole year, 55, it is the same as the median.
b. Mean − SD = 48.622; Mean + SD = 61.140
c. 30 of the 42, or 71.4%, are in this interval. This is close to two-thirds.
20a. ordinal categories
b. Mean before = 4.500; Mean after = 4.833
c. We are calculating a measure of central tendency, the mean, that assumes we have measured numerical data. In fact this works reasonably well with rating scales like this, even though there is an argument about it.

Chapter 5

2. 2, 4, 5, 6
4. −1.69897, .00000, .43457, 1.00000, 2.00000, 4.69897

6. .0004, 1.0, 7.3984, 100, 10,000, 2,500,000,000

8a. 55,000/8 = 6,875, which has four digits, so four orders of magnitude

b. probably

10a. The general answer is yes, this is a good candidate for a transformation. You could support this by a dot plot, stem-and-leaf plot, or histogram; or by noting the number of orders of magnitude spanned by the data.

b. A log transformation does a nice job of making this data set more symmetrical.

12a. Give at least one measure of central tendency and one measure of variability, e.g., "This distribution has a mean of 9.49, and an SD of 11.79." Or "This distribution has a median of 5.15, and an interquartile range of 9.78." Note that on examination, the shape appears to be clearly skewed toward high values, with at least one high outlier.

b. A log transformation works nicely to make this a symmetrical distribution. For log to base e, you could have Mean = 1.797, SD = .934, or Median = 1.639, interquartile range (by computer) = 1.415. For log to base 10 you could have Mean = .780, SD = .406, or Median = .712, interquartile range (by computer) = .614.

14. Give a measure of central tendency and a measure of variability, e.g., "These data have a mean of 3.030 and a standard deviation of 1.918." Or "These data have a median of 3.030 and an interquartile range (by computer) of 2.414." They are not good candidates for transformation. The mean and the median are nearly the same, and the dotplot does not show major outliers or a pronounced skew.

16. No, by eye, these data do not appear to be in need of transformation. They do not span several orders of magnitude, nor do they appear to be highly skewed or to contain outliers. This impression is supported by the dotplot.

18. By eye, these data appear to be reasonable candidates for transformation. The largest value is several hundred times the size of the smallest. The dotplot reinforces this impression. Most values are clustered at the low end, with two pronounced high outliers. A log transformation would be a good candidate.

Chapter 6

2a. skewed toward high values

b. no

c. They appear substantially different.

d. IQR = $Q_3 - Q_1$ = 13.80 − 4.47 = 9.33. SD = 13.42; IQR/SD = 9.33/13.42 = .695. This is not close to the 1.3 expected for a normal distribution.

e. No, calculations based on the assumption of normality might produce misleading results.

4a. By eye, this distribution appears skewed toward high values, but the skew is not extreme. There were somewhat more values in the interval one standard deviation on each side of the mean, and all of the values that were outside of this interval were high. This is perhaps unusual, but not clearly wrong, given the small numbers involved.

b. Possibly, but this is not a strong candidate for transformation.

c. The difference between the mean and the median, due to the skew that is present, is large enough so that the median would be preferred in most cases.

6a. Yes, the normal distribution does a good job here. By eye, the distribution appears symmetric, and the mean and median are close to each other.

b. no

c. The usual normal distribution descriptions—the mean and SD—would be appropriate here.

8a. Yes. They appear to be roughly similar values; at least they are not separated by orders of magnitude.

b. Mean = 53,205; SD = 6,563

c. From the stem-and-leaf plot, the value of the mean is close to the center. Although there is some appearance of a skew toward low values, it is not extreme.

10a. They appear basically symmetric, with perhaps a slight skew toward low values in height and toward high values in diameter and volume. None of the skews are strong enough to require transformation.

b. and **c.**

	N	MEAN	MEDIAN	STDEV	IQR
DIAMETER	31	13.248	12.900	3.138	5.0
HEIGHT	31	76.00	76.00	6.37	8.0
VOLUME	31	30.17	24.20	16.44	19.20

d. For the diameter and height, the means and medians are essentially the same. For volume, the mean appears substantially higher than the median.

e. We have IQR/SD ratios of 1.59, 1.26, and 1.17. These are all reasonably close to the expected value of 1.3.

12a. Males spent an average of 46.8% of their time guarding the nest, with an SD of 6.95%. Or, Males typically spent about 47% of their time guarding the nest (median = 47.25%). By computer, the upper quartile is 51.87% and the lower quartile 41.23%, for an interquartile range of 10.6%. It appears that males spend more time guarding than do females, who average 32.86% (or, who typically spend about a third of their time guarding, with a median of 32.65%).

b. Females spent an average of 51.68% of their time fanning their eggs, with an SD of 7.27%. Or, Females typically spent about 50% of their time fanning their eggs (median = 49.75%). By computer, the upper quartile is 54.77% and the lower quartile 48.02%, for an interquartile range of 6.75%. It appears that females spend more time fanning than do males, who average 7.79% (or, who typically spend about 6% of their time fanning, with a median of 5.95%).

14. Both weight and height appear to be nearly normally distributed, with basically symmetric distributions. By eye, there appears to be a slight skew toward high values for weight, and a slight skew toward low values for height, but these are not extreme. These students average about 44 pounds in weight (mean = 43.908), plus or minus about 8 pounds (standard deviation = 8.333). They average about 44 inches (mean = 43.551), plus or minus about 3 inches (standard deviation = 2.977).

16. The distribution of sailboat prices is skewed toward high values. The median price is about $75,000 (in this data set, it is $74,950), with an interquartile range of $68,000.

18. Ages were calculated by subtracting the year of manufacture from 92, the year of sale. As can be seen in a dotplot, the ages appear to be nearly normally distributed, with perhaps some skew toward high values. The mean (8.917) and the median (8.500) are close, so we choose to describe this distribution as having an average age of about 9 years (mean = 8.917), plus or minus about 5.4 years (standard deviation = 5.385).

20. This distribution is clearly skewed toward high values. There are two problems in showing this from the data as presented: Lower and upper endpoints are not given, and the intervals are uneven. One approach that would be acceptable would be to estimate the quartile and median values from the data given, and use reasonable estimates of low and high values. For instance, the first three categories include, cumulatively, 24.3% of the cases. The 25%tile is then .7/17.4 = .04 of the way into the next interval, so we estimate the lower quartile value at $15,200. Similarly, the median is about halfway into the 25,000 to 34,999 interval. Interpolating, we estimate it at $30,460. Looking at the high end of the income distribution, we see that 25.8% are in the top two intervals, so the 75%tile is just inside the $50,000 to $74,999 interval. This works out to about $51,300.

Chapter 7

2. heads or tails (By convention, we ignore the possibility of bizarre outcomes.)

4. nickel head and dime tail; nickel head and dime head; nickel tail and dime head

6. .55

8. 80%

10. No. From experience, we know that landing on the edge (and staying there) is very, very rare. By convention, we ignore it when discussing coin problems.

12a. The first two outcomes are the only outcomes that produce profits, so together they are the event "making a profit." The probability of making a profit is the sum of their probabilities, or .02 + .47 = .49.

b. The event "lose money" consists of the last two outcomes in the table. Its probability is .21 + .04 = .25.

14a. They are mutually exclusive. One particular monkey cannot be playing in the brook and sleeping in the tree at the same moment.

b. Since these two events are mutually exclusive, we may find the probability of one OR the other by addition: .23 + .37 = .60.

c. Zero. They cannot happen at the same time. (See part **a.**)

18a. $.5^3 = .125$

b. .125

c. P(no heads) = P(three tails) = .125

d. P(no tails) = P(three heads) = .125

e. P(at least one head) = 1 − P(three tails) = 1 − .125 = .875

f. P(at least one tail) = 1 − P(three heads) = 1 − .125 = .875

20a. P(Brown, Brown) = .6 × .6 = .36

b. P(Not brown, Not brown) = .4 × .4 = .16

c. P(at least one brown) = 1 − P(Not brown, Not brown) = 1 − .16 = .84

d. P(exactly one brown) = 1 − (P(two brown) + P(two not brown)) = 1 − (.36 + .16) = 1 − .52 = .48

Chapter 8

2. For the discrete probability distribution, we get mean = 3, variance = 2, and SD = 1.4142.

a. N = 2, Mean = 3, SD of sample averages = 1.4142/sqrt(2) = 1

b. N = 3, Mean = 3, SD of sample averages = 1.4142/sqrt(3) = .8165

c. N = 4, Mean = 3, SD of sample averages = 1.4142/sqrt(4) = .7071

d. N = 9, Mean = 3, SD of sample averages = 1.4142/sqrt(9) = .4714

e. $N = 16$, Mean = 3, SD of sample averages = 1.4142/sqrt(16) = .3536

f. $N = 25$, Mean = 3, SD of sample averages = 1.4142/sqrt(25) = .2828

4a. −3.090

b. −2.326

c. −1.645

d. .000

e. .674

f. 1.282

g. 1.645

h. 2.326

6a. This is the sum of the probabilities of 3, 4, or 5 players: .27 + .43 + .18 = .88

b. This is the sum of the probabilities of 2, 3, or 4 players: .12 + .27 + .43 = .82

c. mean = 3.67

d. variance = .8211

e. standard deviation = .906145

8a.

Winnings	Probability
1	.9
−1	.1

b. expected winnings = ($1)(.9) + (−$1)(.1) = $.80

c. risk = standard deviation = $.60

d. The mean is the expected return, the predicted long-run average. The positive value indicates this is a favorable bet; on average, you will be $.80 ahead with each bet. The standard deviation of $.60 indicates that the amount won can vary considerably from the mean; this is not a sure thing.

10. We use the mean, here equal to $90, as our measure of expected return. We use the standard deviation of the probability distribution, here equal to $103, as our measure of risk.

12. "Should" is a funny word here, since you may just be someone who loves risk, but most people "should" take the gift of $100. This is more than the expected return on the investment, and there is no risk. Of course, there is also no chance of making more than $100, either.

14a. Z = (20,000 − 30,000)/7,500 = −1.333. From the table, the probability of a standard score this low or lower is .091, so the probability of a higher score is 1 − .091 = .909. There is about a 91% probability the company will receive orders for 20,000 or more teddy bears.

b. Z = (15,000 − 30,000)/7,500 = −2. From the table, P = .023. There is little over a 2% probability of a disastrous year.

c. Z = (40,000 − 30,000)/7,500 = 1.33, P = .909. There is about a 91% probability the present plant capacity can handle demand.

d. For 25,000: Z = −.667, P = .251; for 35,000: Z = .667, P = .749. Subtracting, we have .749 − .251 = .498. There is about a 50% probability orders will fall in the desired range.

16a. From the Central Limit Theorem, the mean of a sum equals the mean of a single observation times n. So, Mean total = $58 × 5 = $290.

b. SD of total = SD of single observation × SQRT(n) = 23 × sqrt(5) = 51.43

c. The general justification is the Central Limit Theorem.

d. Z = (250 − 290)/51.43 = −.78. From the table, the probability of this or less is .218, so the probability of at least this is 1 − .218 = .782.

e. Z = (200 − 290)/51.43 = −1.75, P = .040

f. For 200, we have P = .040, from part **e**; for 300, we have Z = (300 − 290)/51.43, P = .572. Subtracting, we have .572 − .040 = .532.

g. The Central Limit Theorem states that the mean of an average is the same as the mean of the individual observations.

h. SD of average = SD of single observations/SQRT(n) = 23/sqrt(5) = 10.286

i. the Central Limit Theorem

j. For $50, Z = −.779, P = .218; for $65, Z = .682, P = .752. Probability of being between 50 and 65 is .752 − .218 = .534. Probability not between 50 and 65 is 1 − .534 = .466

18. The Central Limit Theorem

20. For 12 runs, mean of total = 11 × 12 = 132, SD of total = 3 × SQRT(12) = 10.392. For a total of 100 grams, Z = (100 − 132)/10.392 = −3.079, P = .00124. The probability of not reaching the goal is much less than 1%.

Chapter 9

2. SE = SD/SQRT(n) = 4/SQRT(4) = 2

4a. Number these cartons 1 to 9 in order. First random selection = INTEGER(.9240 × 9) + 1 = INTEGER(8.316) + 1 = 9. Second random selection = INTEGER(.8098 × 9) + 1 = 8. Third = INTEGER(.7841 × 9) + 1 = 8. Already included, do not include again. Next selection = INTEGER(.6155 × 9) + 1 = 6. This completes the sample of 3 with cartons 9, 8, 6. In terms of the original shipping labels, these are G19, F85, D13.

6a. The sample mean diameter, 13.248

b. The standard error of the sample mean, SEMEAN, .564

c. The sample mean height, 76.00

d. The standard error of the sample mean, SEMEAN, 1.14
e. The sample mean volume, 30.17
f. The standard error of the sample mean, SEMEAN, 2.95

12a. 14% = .14 as a proportion, so
Standard error
= SQRT((*X*/*n*) × (1 − *X*/*n*))/(*n* − 1)
= SQRT(.14 × (1 − .14))/(92 − 1)
= SQRT(.0013231)
= .03637, or 3.64%

b. With respect to the 14% of the sample who preferred their toast "very dark," the standard error indicates the plus or minus change we might see in a different sample of the same size from the same population.
c. This also works out to 3.64%.

14a. Average = 8.6078, SD = .1750
b. The sample average, 8.6078
e. .0583

16a. Average = 3,106, SD = 1,676
b. The sample average, 3,106
e. 559

18a. Average = 4,248, SD = 1,476
b. On the basis of a random sample of nine hikes, we estimate the average high point of the hikes in this book to be 4,248 feet, plus or minus 492 feet. (The 492 is the standard error of the sample average.)

20a. No. Looking at the table, we see that it appears to be the first few boats in a larger data set. The brokers appear to be in alphabetical order, and we have only the very start of the alphabet. Within each broker, the boats are arranged in order of length. Clearly, this is not a random sample.

Chapter 10

2. 2.228

4a. For height, 95% CI: 76.00 − 2.042 × 1.14 to 76 + 2.042 × 1.14 or 73.67 to 78.33
b. As above, but now t = 2.750, so the 99% CI is 72.87 to 79.14
c. 24.15 to 36.19
d. 22.06 to 38.28
e. 12.10 to 14.40
f. 11.70 to 14.80

6. 1.398 to 1.525

8. 41.3% to 78.6%

10a. 1.703
b. 2.576
c. 1.746
d. 2.821

12. 90% CI: 34.9% to 39.1%; 99% CI: 33.8% to 40.3%

14. The 13th and 27th ranked values

16a. 8
b. 2.306
c. 8.475 to 8.745

18. 95% CI = 1,976 to 4,836 feet

20a. −1,866 to 16,530
b. Using common logs, we get 2.303 to 3.457, which back transforms to 200.9 to 2,864.2. Or, using natural logs, we get 5.303 to 7.959, which back transforms to 200.9 to 2,861.2. (The small differences in the upper limits of the back-transformed values are due to rounding differences.)
c. The CI calculated on the untransformed data is pretty useless, including the possibility of massive negative deaths, and with an upper bound that excludes only two of the observed earthquakes. The intervals based on the log transformations are much narrower, and include only realistic possibilities.
d. For a sample of 15, the limits are at rank 4 from the top and bottom. These values are 115 and 4,000. This interval is somewhat wider than that produced by the transformations, but is useful, realistic, and directly interpretable.

Chapter 11

2. No, 105 is not significantly different from 100, based on this sample.

4. No, it is not significantly different. The hypothesized value is inside the confidence interval.

6. No, the hypothesized value of 8.6 is not significantly different from the sample average of 8.61.

8. Using the t value for 30 degrees of freedom (since the value for 25 df is not given), we get a 95% CI of 1.37 to 5.59. This interval includes the hypothesized value of 3.48, so it is not significantly different from the sample average.

10. We are 95% sure that the mean weight of a carton of ice cream is between 1.09 and 1.33 lb. The hypothesized value of 1 lb. is below the lower limit of this interval. We therefore reject the hypothesis that the mean weight of the entire day's production is 1 lb.

12. No, we cannot safely conclude that the initiative will pass. The 95% confidence interval is from 49.0% to 59.4%, which includes the value of 50% which separates winning from losing.

14a. average = 2.33, SE(average) = .633167
b. 95% CI: .898 to 3.762
c. The null hypothesis here is that there is no effect produced by the use of medication 2. The hypothesized value of zero does not lie within the confidence interval, so we reject the null hypothesis and accept the hypothesis that some amount of extra sleep is induced.

16a. Nonparametric 95% CI: .1 to 4.6 hours

b. Since the confidence interval does not include the hypothesized value of zero (no extra sleep), we accept the hypothesis that medication 2 produces an increase in sleep.

18a. The 95% CI, 7.97 to 8.63, includes the hypothesized value of 8.0. We conclude that the amount of color in the oil-based paint does not differ significantly from the amount the manufacturer is supposed to insert into each can of paint.

b. For the 10% significance level, the 90% CI is from 8.02 to 8.58. This does not include the hypothesized value of 8.0, so we conclude that the difference is statistically significant at the 10% level.

c. At the 1% significance level the standard value of 8 falls within the confidence interval of 7.86 to 8.74 g/kg of paints. There is no significant difference between the amount of paint the manufacturer was supposed to insert into each can and the sample average.

d. As we compute wider and wider confidence intervals, going from 90% to 95% to 99%, we are more and more likely to include the null hypothesis value and thus to fail to reject the null hypothesis.

20a. 245/5,000 = .049 = 4.9%

b. The 95% upper confidence interval extends from 4.4% to 100%. The hypothesized value of 3% is outside this interval. We conclude that the day's defect rate is significantly above 3%. Adjustments to the machinery are necessary.

Chapter 12

4b. The distributions appear approximately normal.

c. The centers are different, with the voting age percents consistently and substantially lower than the registered voter percents. This is reasonable. Those with little interest in voting probably do not bother to register in the first place.

6a. Average difference = 10.2 − 5.6 = 4.6; SE = 10.79, df = 23. The 95% CI for the difference is from 1.823 to 7.377 sea lions.

b. The hypothesized difference of zero is not included in the confidence interval, so we conclude that there is a statistically significant difference in the number of sea lions at the two locations.

8a. Average difference = .246 − .166 = .08; df = 608, SE = .1590; 95% CI: 1.6% to 14.4%

b. The hypothesized difference of zero is not included in the confidence interval, so we conclude that there is a statistically significant difference in the percent of shoppers buying more than one item.

10a. Yes. Each patient produced two data values.

b. A two-sample test is inappropriate, since the values are not independent. It would also be inefficient, since using the pairing yields a more sensitive test.

c. −1.2, −2.4, −1.3, etc.

d. Average = −1.67, SD = 1.1304, SD = .3575

e. −2.479 to −.861

f. The two medications are significantly different in their effects on sleep.

12. The confidence interval now goes from −1,637 to 18,915. The difference is no longer statistically significant.

14a. Average = 3,085, SD = 780, SE(mean) = 260

b. This sample − sample 1 = 3,085 − 3,106 = −21 feet

c. No. The 95% CI is −1,286 to 1,327, which includes zero.

16a. 4.260

b. 5.552

c. −1.292

d. −1.495 to −1.089

e. yes

f. At the 5% significance level, there is a statistically significant difference in the average length of the petals of *I. versicolor* and *I. virginica.*

18a. 1.462

b. 4.260

c. −2.798

d. −2.939 to −2.657

e. yes

f. At the 5% significance level, there is a statistically significant difference in the average length of the petals of *I. setosa* and *I. versicolor.*

20a. yes

b. yes

Chapter 13

2b. The assumptions of normality and equal variability are clearly violated. The assumption of a random sample is not met in a strict construction—these are censuses of populations. Conceptually, the random sample assumption can be met by arguing that these are samples from larger hypothetical populations.

c. The equal variability assumption is more nearly satisfied.

d. This appears to make things worse, rather than better.

e. The square root transformation, though far from perfect, seems the best in a difficult situation.

4a. Is each group normally distributed? Are the variabilities about the same? Are these random samples?

b. Whether there is a real difference in mean heights between the trees in these three forests, or whether the apparent differences in the sample averages can be explained by chance

c. 73.9

d. between groups mean square = 76,261, with df = 2

e. within groups mean square = 265.45, with df = 1,635

f. $F = 287.28$

g. critical $F = 3.00$

h. 287.28 is larger than 3.00, so we conclude that the groups are significantly different.

i. The forests have been shown to be significantly different in their average tree heights. The apparent differences among the sample averages cannot be explained by chance alone.

6a. Averages: Hyde, 3.5571; Jekyll, 3.9273; Caligari, 3.4556

b. Yes, they are apparently different.

c. $F = 3.12$. The critical F at 5% is 3.49. These averages are not significantly different from each other.

d. no

8. Using the computer and Minitab, you can copy the data on the boats in the 30–35-foot range into new columns (with new names). This lets you restrict the analysis to these cases. You should get $t = 8.71$, $p = .000$, so there are significant differences in prices. On examination, maker 5 is significantly more expensive than any of the others, and maker 1 is significantly more expensive than maker 2.

10. There is a significant difference between the three treatments, with $F = 3.18$, $p = .049$. It appears that the difference is in the effectiveness of cognitive therapy, with a mean of 4.4, compared with means of 9.2 and 9.7 for relaxation therapy and the minimum contact control group.

14. There are no significant differences in the alfalfa yields of different fields. The average yields are all substantially the same, the F value of .05 so small that we would see a value this large or larger, just on the basis of chance, 98% of the time.

16. There are significant differences in the effectiveness of these methods ($F = 5.12$, $p = .004$). Whether or not this is by coincidence, the methods are numbered in order of increasing effectiveness. The averages decrease by 2 to 4 miles an hour, from 64.4 mph to 55.4 mph, as we go from method 1 to method 4.

20. The experimental group does better than the control group in all conditions. It appears that something in the treatment serves to increase motivation and performance. This may be the material reward, or simply the attention of the tester. This data set differs from others we have looked at in that there are only two groups—the treatment and the control—but there are three outcome measures. With the methods we have learned so far, we could do a series of two-sample tests. Note, however, that the "full scale" IQ is not independent of the two other IQs, but is a combination of them.

Chapter 14

2a.

	Male	Female	Total
Salmon	50%	50%	100%
Flounder	33.3%	66.7%	100%
Tuna	20%	80%	100%

b. Different species seem to have different percents of males and females. Species and sex ratio appear to be associated.

4a. Chi-square = 44.882

b. df = 2

c. critical value = 5.991

d. yes

6a. No. The numbers are just codes and have no mathematical meaning.

b.

Code	Count	Percent
1	3	15%
2	5	25%
3	4	20%
4	4	20%
5	4	20%

d. No, you need counts of at least 5 in each cell.

e. Chi-square = 5.956, df = 4, 5% critical value = 9.487. There are no statistically significant differences.

f. These results are consistent with the claim, but do not "prove" it.

8.

	Homicide	Rape	Arson	Total
Berkeley	5.51%	4.89%	2.92%	3.90%
Oakland	34.65%	26.09%	21.69%	24.37%
San Francisco	49.61%	39.95%	21.79%	31.06%
San Jose	10.24%	29.08%	53.60%	40.67%
Total	100%	100%	100%	100%

This table gives us the percent of each type of crime committed in each city: 5.51% of all the homicides were committed in Berkeley, and so on. The answers below assume that you can allocate police resources among the four cities.

a. column percents
b. row percent
c. joint percents

12. It is easy enough to find the value of chi-square and df: chi-square = .606516, df = 1. This value is much smaller than any of the tabled values for df = 1, so we know that the fit is good, but we can't tell how good. We can use the cumulative density function command, CDF, in Minitab to solve for the probability of getting a value this small or smaller. We get p = .5639, that is to say, there is a 56% probability of a fit this good or better.

14. Chi-square = 1.532, df = 1, critical value at 5% is 3.841, not significant

16. Chi-square = 3.003 + 6.050 + 3.860 + 7.776 = 20.688, df = 1, critical value at 5% is 3.841. This is statistically significant.

20. From the information given, we can fill in this table:

	Boys	Girls	Total
fast	38	62	100
medium	52	48	100
slow	69	31	100
Totals	159	141	300

We then calculate
Chi-square = 4.245 + 4.787 + .019 + .021 + 4.830 + 5.447 = 19.350 and df = 2. This is highly statistically significant (p < .001). We accept that there is a relationship between sex and assignment to track.

Chapter 15

4a. Y = 17.33 + .424 × X
b. Y = −144.08 − 17.72 × X

6a. −20
b. 38
c. −111

8a and **b.** You should get two different patterns: The 67–74 data should show a clearly increasing trend; the 86–93 data should show a much more scattered pattern with, by eye, the faint suggestion of a decreasing trend.
c. period one: r = .828; period two: r = −.220; both periods: r = .223
d. For period 1: speed1 = 63.7 + 1.32 year; for period 2: speed2 = 298 − 1.51 year; for both periods: speed = 138 + .268
f. It appears that speed was progressing steadily during the first period. During the second period it did not progress. It became much more variable, and there was no clear relationship with time. It may be that measures were deliberately taken to keep cars from getting faster and faster—new limits may have been put on allowable engines, for instance. Or it may be that there was a series of unusual events—storms or bad accidents, say—that slowed these races. There are other possibilities.

12a. and **b.** The plot should show three distinct clusters.
c. The three clusters are no doubt readily distinguishable by the birds themselves: One group is short and low, one long and low, one of intermediate length and high in pitch. They probably serve different purposes: courtship, warning companions of danger, challenging trespassers, etc.

14a. Y = 4,013.39 + 120.6965 × X
b. $4,013.39
c. $120.70
d. $20,307.42

16a. r = .39
b. no
c. There is not a statistically significant linear relationship, but by eye there is a very clear curvilinear relationship.

18. The plot shows a negative relationship, with one especially important observation high in unit price and low in percent of calories from fat. The correlation of Fat%Cal and UnitPric = −.928. The regression equation is Fat%Cal = 106 − 886UnitPric. The relationship is significant with t-ratio = −4.32, p = .023. The residual plot appears to show some remaining structure.

20. Correlation of Year and Rating = .949. Significant with t-ratio = 7.96, p = .000. The regression equation is Rating = −1,984 + 52.9Year; Significant with t-ratio = 7.96, p = .000.

Index

Continued from front cover

VALUE	PROBABILITY	VALUE	PROBABILITY
1.02	.846	− 1.02	.154
1.03	.848	− 1.03	.152
1.04	.851	− 1.04	.149
1.05	.853	− 1.05	.147
1.06	.855	− 1.06	.145
1.07	.858	− 1.07	.142
1.08	.860	− 1.08	.140
1.09	.862	− 1.09	.138
1.10	.864	− 1.10	.136
1.11	.867	− 1.11	.133
1.12	.869	− 1.12	.131
1.13	.871	− 1.13	.129
1.14	.873	− 1.14	.127
1.15	.875	− 1.15	.125
1.16	.877	− 1.16	.123
1.17	.879	− 1.17	.121
1.18	.881	− 1.18	.119
1.19	.883	− 1.19	.117
1.20	.885	− 1.20	.115
1.21	.887	− 1.21	.113
1.22	.889	− 1.22	.111
1.23	.891	− 1.23	.109
1.24	.893	− 1.24	.107
1.25	.894	− 1.25	.106
1.26	.896	− 1.26	.104
1.27	.898	− 1.27	.102
1.28	.900	− 1.28	.100
1.282	.900**	− 1.282	.100**
1.29	.901	− 1.29	.099
1.30	.903	− 1.30	.097
1.31	.905	− 1.31	.095
1.32	.907	− 1.32	.093
1.33	.908	− 1.33	.092
1.34	.910	− 1.34	.090
1.35	.911	− 1.35	.089
1.36	.913	− 1.36	.087
1.37	.915	− 1.37	.085
1.38	.916	− 1.38	.084
1.39	.918	− 1.39	.082
1.40	.919	− 1.40	.081
1.41	.921	− 1.41	.079
1.42	.922	− 1.42	.078
1.43	.924	− 1.43	.076
1.44	.925	− 1.44	.075
1.45	.926	− 1.45	.074
1.46	.928	− 1.46	.072
1.47	.929	− 1.47	.071
1.48	.931	− 1.48	.069
1.49	.932	− 1.49	.068
1.50	.933	− 1.50	.067
1.51	.934	− 1.51	.066
1.52	.936	− 1.52	.064
1.53	.937	− 1.53	.063
1.54	.938	− 1.54	.062

(continued)